Gary L. Peterson
(817) 483-4079

D0138663

ADVANCED QUANTUM MECHANICS

J. J. SAKURAI
*The Enrico Fermi Institute for Nuclear Studies
and the Department of Physics, University of Chicago*

ADVANCED QUANTUM MECHANICS

ADDISON-WESLEY PUBLISHING COMPANY
Advanced Book Program
Reading, Massachusetts
London · Amsterdam · Don Mills, Ontario · Sydney · Tokyo

This book is in the
ADDISON-WESLEY SERIES IN ADVANCED PHYSICS
MORTON HAMERMESH, *Consulting Editor*

Library of Congress Catalog Card Number 67-19430
International Standard Book Number 0-201-06710-2

First printing, 1967
Second printing, 1969
Third printing, 1972
Fourth printing, with revisions, 1973
Fifth printing, 1976
Sixth printing, 1977
Seventh printing, 1978
Eighth printing, 1980
Ninth printing, 1982

Copyright © 1967 by Addison-Wesley Publishing Company, Inc. Philippines copyright 1973 by Addison-Wesley Publishing Company, Inc. Published simultaneously in Canada.

All rights reserved. No part of this publication may be reproduced, stored in a retrieval system, or transmitted, in any form or by any means, electronic, mechanical, photocopying, recording, or otherwise, without the prior written permission of the publisher.

Printed in the United States of America.

ISBN 0-201-06710-2
IJKLMNOPQR-MA-898765432

PREFACE

The purpose of this book is to present the major advances in the fundamentals of quantum physics from 1927 to the present in a manner that cannot be made any simpler. In selecting the materials covered in this book I have omitted those topics which are discussed in conventional textbooks on nonrelativistic quantum mechanics, group-theoretic methods, atomic and molecular structure, solid-state physics, low-energy nuclear physics, and elementary particle physics. With some regret I have also omitted the formal theory of collision processes; fortunately a careful and detailed treatment of this subject can be found in a companion Addison-Wesley volume, *Advanced Quantum Theory*, by P. Roman. Thus the emphasis is primarily on the quantum theory of radiation, the Dirac theory of leptons, and covariant quantum electrodynamics. No familiarity with relativistic quantum mechanics or quantum field theory is presupposed, but the reader is assumed to be familiar with nonrelativistic quantum mechanics (as covered in Dicke and Wittke or in Merzbacher), classical electrodynamics (as covered in Panofsky and Phillips or in Jackson), and classical mechanics (as covered in Goldstein).

The book has its origin in lecture notes I prepared for the third part of a three-quarter sequence of courses in quantum mechanics required of *all* Ph.D. candidates in physics at the University of Chicago. Twenty years ago such a short course in "advanced quantum mechanics" might have covered the materials discussed in the last three chapters of Schiff. We must realize, however, that forty years have passed since P. A. M. Dirac wrote down the relativistic wave equation for the electron; it was nearly twenty years ago that R. P. Feynman invented the famous graphical techniques that have had profound influences, not only on quantum electrodynamics and high-energy nuclear physics, but also on such remotely related topics as statistical mechanics, superconductivity, and nuclear many-body problems. It is evident that, as the frontier of physics advances, the sort of curriculum adequate for graduate students twenty years ago is no longer satisfactory today.

Chapter 1 of this book is concerned with a very brief introduction to classical field theory needed for the latter parts of the book. The subject matter of Chapter 2 is the quantum theory of radiation. First, the transverse electromagnetic field is quantized in analogy with quantum-mechanical harmonic oscillators. The subsequent parts of the chapter deal with standard topics such as the emission, absorption, and scattering of light by atoms, and thus provide rigorously correct (as opposed to superficial) explanations of a number of atomic phenomena (e. g., spontaneous emission, Planck's radiation law, and the photoelectric effect) with which the students are already familiar from their earlier courses. In addition, we discuss more advanced topics including radiation damping, resonance fluorescence, the Kramers-Kronig (dispersion) relations, the idea of mass renormalization, and Bethe's treatment of the Lamb shift.

It is deplorable that fewer and fewer students nowadays study Heitler's classical treatise on the quantum theory of radiation. As a result, we see a number of sophisticated, yet uneducated, theoreticians who are conversant in the LSZ formalism of the Heisenberg field operators, but do not know why an excited atom radiates, or are ignorant of the quantum-theoretic derivation of Rayleigh's law that accounts for the blueness of the sky. It is hoped that Chapter 2 of this book will fill the missing gap in the education of physicists in the mid-twentieth century.

The wave equation of Dirac is introduced in Chapter 3 by linearizing the relativistic second-order equation involving Pauli matrices, as originally done by B. L. van der Waerden. In addition to presenting standard topics such as the plane-wave solutions, an approximate and the exact treatment of the hydrogen atom, and the physical interpretations of *Zitterbewegung*, we make special attempts to familiarize the reader with the physical meanings of the various gamma matrices. The inadequacy of the single-particle interpretation of the Dirac theory is pointed out, and towards the end of the chapter we quantize the Dirac field using the Jordan-Wigner method. Although a rigorous proof of the spin-statistics connection is not given, we demonstrate that it is difficult to construct a sensible field theory in which the electron does not obey the Pauli exclusion principle. The chapter ends with applications to weak interactions, including short discussions on the two-component neutrino and parity nonconservation in nuclear beta decay, hyperon decay, and pion decay.

Symmetry considerations are emphasized throughout Chapter 3. We not only discuss the formal transformation properties of the Dirac wave function and the quantized Dirac field under Lorentz transformations, parity, and charge conjugation, but also show how the various symmetry operators can actually be used in specific problems (e. g., in constructing momentum and helicity eigenfunctions or in proving that the intrinsic parity of the positron is opposite to that of the electron). In Sections 9 and 10 we attempt to clarify the basic difference between charge conjugation in the unquantized Dirac theory and charge conjugation in the quantized Dirac theory, which is often a source of confusion in the literature.

Covariant perturbation theory is covered in Chapter 4. A distinct feature of this chapter is that we present covariant quantum electrodynamics not as a "new theory" but rather as a natural and almost immediate consequence of relativistic quantum mechanics and *elementary* quantum field theory, whose foundations had been laid down by 1932. In the usual derivation of the Feynman rules from quantum field theory, one first defines five different kinds of invariant functions, three different kinds of ordered products, etc., and during that time the novice has no idea why these concepts are introduced. Instead of deriving the Feynman rules in the most general case from field theory using the Dyson-Wick formalism, we demonstrate how, in a concrete physical example, the vacuum expectation value of the time-ordered product $\langle 0|T\,(\psi(x')\,\bar{\psi}\,(x))|0\rangle$ emerges in a natural manner. It is then pointed out how this vacuum expectation value can be interpreted pictorially in terms of the propagation of an electron going forward or backward in time à la Feynman. The simplicity and elegance of the postwar calculational techniques are explicitly exhibited as we demonstrate how two non-covariant expressions add up to a single covariant expression. The Feynman rules are also discussed from the point of view of the unit source solution (the Green's function) of the wave equation, and Feynman's intuitive space-time approach is compared to the field-theoretic approach. Some electromagnetic processes (e. g., Mott scattering, two-photon annihilation of electron-positron pairs, Møller scattering) are worked out in detail. The last section of Chapter 4 consists of brief discussions of higher-order processes, the mass and charge renormalization, and difficulties with the present field theory. In addition to discussing standard topics such as the electron self-energy and the vertex correction, we demonstrate how the principles of unitarity and causality can be utilized to obtain a sum rule that relates the charge renormalization constant to the probability of pair creation in an external field. The method for evaluating integrals appearing in covariant perturbation theory is discussed in Appendix E; as examples, the self-energy and the anomalous magnetic moment of the electron are calculated in detail.

We present the covariant calculational techniques in such a manner that the reader is least likely to make mistakes with factors of 2π, i, -1, etc. For this reason we employ, throughout the book, the normalization convention according to which there is one particle in a box of volume V; this is more convenient in practice because we know that the various V's must cancel at the very end, whereas the same cannot be said about (2π)'s. A good amount of space is devoted to showing how observable quantities like differential cross sections and decay rates are simply related to the covariant $\mathscr{M}$-matrices, which we can immediately write down just by looking at the "graphs."

Throughout this book the emphasis is on physics with a capital *P*. Complicated mathematical concepts and formalisms that have little relation to physical reality are eliminated as much as possible. For instance, the starting point of the quantization of the Dirac field is the anticommutation relations among the creation and annihilation operators rather than the anticommutation relation between two Dirac fields; this is because the Dirac field itself is not measurable, whereas the anticommutation relation between two creation operators has a simple and direct physical meaning in terms of physically permissible states consistent with the Pauli exclusion principle. In this sense our approach is closer to the "particle" point of view than to the "field" point of view, even though we talk extensively about the quantized Dirac field in the last third of the book.

Whenever there are several alternative methods for deriving the same result, we do not necessarily choose the most elegant, but rather present the one that makes the physics of the problem most transparent at each stage of the derivation.

For example, in discussing the Møller interaction between two electrons we start with the radiation (Coulomb) gauge formalism of E. Fermi and show how this noncovariant but simple method can be used to derive, in an almost miraculous manner, a manifestly covariant matrix element which can be visualized as arising from the exchange of four types of "covariant photons." We prefer this approach to the one based on the Bleuler-Gupta method because the latter introduces artificial concepts, such as the indefinite metric and negative probabilities, which are not very enlightening from the point of view of the *beginner's* physical understanding of quantum electrodynamics.

Wherever possible, we show how the concepts introduced in this book are related to concepts familiar from nonrelativistic quantum mechanics or classical electrodynamics. For example, as we discuss classical electrodynamics in Chapter 1 we review the role of the vector potential in nonrelativistic quantum mechanics and, in particular, consider the Aharonov-Bohm effect and flux quantization. In Chapter 2 the scattering of light by atoms in the quantum theory is compared to its classical analog. In discussing the polarization correlation of the two-photon system resulting from the annihilation of an electron-positron pair, we illustrate some peculiar features of the quantum theory of measurement which have disturbed such great minds as A. Einstein. In Chapter 4 a fair amount of attention is paid to the connection between the calculational methods of the old-fashioned perturbation theory (based on energy denominators) and those of covariant perturbation theory (based on relativistically invariant denominators). In discussing the Møller interaction and the nucleon-nucleon interaction, we try to indicate how the potential concept one learns about in nonrelativistic quantum mechanics is related to the field-theoretic description based on the exchange of quanta.

Although numerous examples from meson theory and nuclear physics are treated throughout the book, it is not our intention to present systematic accounts of nuclear or high-energy phenomena. Nonelectromagnetic processes are discussed solely to illustrate how the ideas and techniques which we acquire in working out electromagnetic problems can readily be applied to other areas of physics.

The forty-seven problems scattered throughout comprise a vital part of the book. The reader who has read the book but cannot work out the problems has learned nothing. Even though some of the problems are more difficult and challenging than others, none are excessively difficult or time-consuming. Nearly every one of them has been worked out by students at the University of Chicago; some, in the final examination of the course on which the book is based.

In recent years several excellent textbooks have appeared on the calculational techniques in relativistic quantum mechanics. The distinct feature of this book is not just to teach the bag of tricks useful only to high-energy physicists or to show how to compute the trace of the product of Dirac matrices, but to make the reader aware of the progress we have made since 1927 in our understanding of fundamental physical processes in the quantum domain. From this point of view we believe it is just as important for the student to know how the quantum description of the radiation field reduces to the familiar classical description in the limit of a large number of quanta, or why the spin-$\frac{1}{2}$ particle "must" obey the exclusion principle, as it is to master the rules that enable us to calculate the magnetic moment of the electron to eight decimals.

To summarize our philosophy: Relativistic quantum mechanics and field theory should be viewed as part of the heroic intellectual endeavor of a large number of twentieth-century theoretical physicists in the finest tradition of M. Planck, A. Einstein, and N. Bohr. It would be catastrophic for the future development of physics if the terminal course in theoretical physics for most Ph.D. level students in physics were nonrelativistic quantum mechanics, the fundamentals of which had essentially been perfected by 1926. For this reason I believe that the topics covered in this book should be studied seriously by every Ph.D. candidate in physics, just as nonrelativistic quantum mechanics has become recognized as a subject matter to be digested by every student of physics and chemistry.

I am grateful to the Alfred P. Sloan Foundation for a fellowship which enabled me to write the last chapter of the book in the congenial atmosphere of CERN (European Organization for Nuclear Research). I wish to thank Drs. J. S. Bell, S. Fenster, and A. Maksymowicz, and Mr. D. F. Greenberg for reading various parts of the book and making many valuable suggestions. Particular thanks are due to Mr. I. Kimel for the painstaking task of filling in the equations.

May 1967
Chicago, Illinois

J. J. S.

CONTENTS

CHAPTER 1

CLASSICAL FIELDS

1–1. PARTICLES AND FIELDS

Nonrelativistic quantum mechanics, developed in the years from 1923 to 1926, provides a unified and logically consistent picture of numerous phenomena in the atomic and molecular domain. Following P.A.M. Dirac, we might be tempted to assert: "The underlying physical laws necessary for the mathematical theory of a large part of physics and the whole of chemistry are completely known."

There are, however, basically two reasons for believing that the description of physical phenomena based on nonrelativistic quantum mechanics is incomplete. First, since nonrelativistic quantum mechanics is formulated in such a way as to yield the nonrelativistic energy-momentum relation in the classical limit, it is incapable of accounting for the fine structure of a hydrogen-like atom. (This problem was treated earlier by A. Sommerfeld, who used a relativistic generalization of N. Bohr's atomic model.) In general, nonrelativistic quantum mechanics makes no prediction about the dynamical behavior of particles moving at relativistic velocities. This defect was amended by the relativistic theory of electrons developed by Dirac in 1928, which will be discussed in Chapter 3. Second, and what is more serious, nonrelativistic quantum mechanics is essentially a single-particle theory in which the probability density for finding a given particle integrated over all space is unity at all times. Thus it is not constructed to describe phenomena such as nuclear beta decay in which an electron and an antineutrino are created as the neutron becomes a proton or to describe even a simpler process in which an excited atom returns to its ground state by "spontaneously" emitting a single photon in the absence of any external field. Indeed, it is no accident that many of the most creative theoretical physicists in the past forty years have spent their main efforts on attempts to understand physical phenomena in which various particles are created or annihilated. The major part of this book is devoted to the progress physicists have made along these lines since the historic 1927 paper of Dirac entitled "The Quantum Theory of the Emission and Absorption of Radiation" opened up a new subject called the *quantum theory of fields*.

The concept of a field was originally introduced in classical physics to account for the interaction between two bodies separated by a finite distance. In classical plysics the electric field $\mathbf{E}(\mathbf{x}, t)$, for instance, is a three-component function defined at each space-time point, and the interaction between two charged bodies, 1 and 2, is to be viewed as the interaction of body 2 with the electric field created by body 1. In the quantum theory, however, the field concept acquires a new dimension. As originally formulated in the late 1920's and the early 1930's, the basic idea of quantum field theory is that we associate *particles* with fields such as the electromagnetic field. To put it more precisely, quantum-mechanical excitations of a field appear as particles of definite mass and spin, a notion we shall illustrate in Section 2–2, where the connection between the transverse electromagnetic field and photons is discussed in detail.

Even before the advent of postwar calculational techniques which enabled us to compute quantities such as the $2s$–$2p_{1/2}$ separation of the hydrogen atom to an accuracy of one part in 10^8, there had been a number of brilliant successes of the quantum theory of fields. First, as we shall discuss in Chapter 2, the quantum theory of radiation developed by Dirac and others provides quantitative understandings of a wide class of phenomena in which real photons are emitted or absorbed. Second, the requirements imposed by quantum field theory, when combined with other general principles such as Lorentz invariance and the probabilistic interpretation of state vectors, severely restrict the class of particles that are permitted to exist in nature. In particular, we may cite the following two rules derivable from *relativistic* quantum field theory:

a) For every charged particle there must exist an antiparticle with opposite charge and with the same mass and lifetime.
b) The particles that occur in nature must obey the spin-statistics theorem (first proved by W. Pauli in 1940) which states that half-integer spin particles (e.g., electron, proton, Λ-hyperon) must obey Fermi-Dirac statistics, whereas integer spin particles (e.g., photon, π-meson, K-meson) must obey Bose-Einstein statistics.

Empirically there is no known exception to these rules. Third, the existence of a nonelectromagnetic interaction between two nucleons at short but finite distances prompts us to infer that a field is responsible for nuclear forces; this, in turn, implies the existence of massive particles associated with the field, a point first emphasized by H. Yukawa in 1935. As is well known, the desired particles, now known as π-mesons or pions, were found experimentally twelve years after the theoretical prediction of their existence.

These considerations appear to indicate that the idea of associating particles with fields and, conversely, fields with particles is not entirely wrong. There are, however, difficulties with the present form of quantum field theory which must be overcome in the future. First, as we shall show in the last section of Chapter 4, despite the striking success of postwar quantum electrodynamics in calculating various observable effects, the "unobservable" modifications in the mass and charge of the electron due to the emission and reabsorption of a virtual photon turn out to diverge logarithmically with the frequency of the virtual photon. Second, the idea of associating a field with each "particle" observed in nature becomes ridiculous and distasteful when we consider the realm of strong interactions where many different kinds of "particles" are known to interact with one another; we know from experiment that nearly 100 "particles" or "resonances" participate in the physics of strong interactions. This difficulty became particularly acute in 1961–1964 when a successful classification scheme of strongly interacting particles was formulated which groups together into a single "family" highly unstable "particles" (lifetimes 10^{-23} sec, often called strong interaction resonances) and moderately metastable particles (lifetimes 10^{-10} sec).‡ Yet, despite these difficulties, it is almost certain that there are many elements in present-day quantum field theory which are likely to survive, say, one hundred years from now.

Before we study quantized fields, we will study classical fields. In part this decision is motivated by the historical fact that prior to the development of quantum electrodynamics there was the classical electrodynamics of Maxwell which, among other things, successfully predicted the existence of Hertzian electromagnetic waves. This chapter is primarily concerned with the elements of *classical* field theory needed for the understanding of *quantized* fields. As a preliminary to the study of quantization we are particularly interested in the dynamical properties of classical fields. For this reason we will follow an approach analogous to Hamilton's formulation of Lagrangian mechanics.

1–2. DISCRETE AND CONTINUOUS MECHANICAL SYSTEMS

The dynamical behavior of a single particle, or more precisely, a mass point in classical mechanics, can be inferred from Lagrange's equation of motion

$$\frac{d}{dt}\left(\frac{\partial L}{\partial \dot{q}_i}\right) - \frac{\partial L}{\partial q_i} = 0, \tag{1.1}$$

which is derivable from Hamilton's variational principle

$$\delta \int_{t_1}^{t_2} L(q_i, \dot{q}_i)\, dt = 0. \tag{1.2}$$

The Lagrangian L (assumed here not to depend explicitly on time) is given by the difference of the kinetic energy T and the potential energy V,

$$L = T - V, \tag{1.3}$$

and the variation in (1.2) is to be taken over an arbitrary path $q_i(t)$ such that δq_i vanishes at t_1 and t_2. The Hamiltonian of the system is

$$H = \sum_i p_i \dot{q}_i - L, \tag{1.4}$$

where the momentum p_i, canonical conjugate to q_i, is given by

$$p_i = \frac{\partial L}{\partial \dot{q}_i}. \tag{1.5}$$

---

‡In fact the one-to-one correspondence between a "field" and a "particle" appears to be lost in a more modern formulation of the field theory of strong interactions as many (if not all) of the so-called "elementary" particles may well be regarded as bound (or resonant) states of each other. The distinction between fundamental particles and composite states, however, is much more clear-cut in the realm of the electromagnetic interactions among electrons, muons, and photons. As an example, in Section 4–4 we shall calculate the lifetime of the ground state of positronium without introducing a field corresponding to the positronium.

These considerations can be generalized to a system with many particles. As a concrete example, let us consider a collection of N particles connected with identical springs of force constant k and aligned in one dimension, as shown in Fig. 1–1.‡ By calling η_i the displacement of the ith particle from its equilibrium position we write the Lagrangian L as follows:

$$
\begin{aligned}
L &= \tfrac{1}{2} \sum_i^N [m\dot{\eta}_i^2 - k(\eta_{i+1} - \eta_i)^2] \\
&= \sum_i^N a \frac{1}{2}\left[\frac{m}{a}\dot{\eta}_i^2 - ka\left(\frac{\eta_{i+1} - \eta_i}{a}\right)^2\right] \\
&= \sum_i^N a\mathscr{L}_i, \qquad (1.6)
\end{aligned}
$$

Fig. 1–1. Particles connected with identical springs.

where a is the separation distance between the equilibrium positions of two neighboring particles and $\mathscr{L}_i$ is the linear Lagrangian density, i.e. the Lagrangian density per unit length.

We can pass from the above discrete mechanical system to a continuous mechanical system as the number of degrees of freedom becomes infinite in such a way that the separation distance becomes infinitesimal:

$$
\begin{aligned}
&a \to dx, \qquad \frac{m}{a} \to \mu = \text{linear mass density}, \\
&\frac{\eta_{i+1} - \eta_i}{a} \to \frac{\partial \eta}{\partial x}, \qquad ka \to Y = \text{Young's modulus}.
\end{aligned}
\tag{1.7}
$$

We now have

$$L = \int \mathscr{L}\, dx, \tag{1.8}$$

where

$$\mathscr{L} = \tfrac{1}{2}\left[\mu\dot{\eta}^2 - Y\left(\frac{\partial\eta}{\partial x}\right)^2\right]. \tag{1.9}$$

We note that η itself has become a function of the continuous parameters x and t. Yet in the Lagrangian formalism η should be treated like a generalized "coordinate" just as q_i in L of Eq. (1.2).

In formulating the variational principle in the continuous case we consider

$$\delta \int_{t_1}^{t_2} L\, dt = \delta \int_{t_1}^{t_2} dt \int dx\, \mathscr{L}\left(\eta, \dot{\eta}, \frac{\partial\eta}{\partial x}\right). \tag{1.10}$$

The variation on η is assumed to vanish at t_1 and t_2 and also at the extremities of the space integration. (In field theory this latter requirement is not stated explicitly since we are usually considering a field which goes to zero sufficiently rapidly at infinity.) Otherwise the nature of the variation is completely arbitrary. The variational integral becomes

$$
\begin{aligned}
\delta \int L\, dt &= \int dt \int dx \left\{\frac{\partial\mathscr{L}}{\partial\eta}\delta\eta + \frac{\partial\mathscr{L}}{\partial(\partial\eta/\partial x)}\delta\left(\frac{\partial\eta}{\partial x}\right) + \frac{\partial\mathscr{L}}{\partial(\partial\eta/\partial t)}\delta\left(\frac{\partial\eta}{\partial t}\right)\right\} \\
&= \int dt \int dx \left\{\frac{\partial\mathscr{L}}{\partial\eta}\delta\eta - \frac{\partial}{\partial x}\left(\frac{\partial\mathscr{L}}{\partial(\partial\eta/\partial x)}\right)\delta\eta - \frac{\partial}{\partial t}\frac{\partial\mathscr{L}}{\partial(\partial\eta/\partial t)}\delta\eta\right\},
\end{aligned}
\tag{1.11}
$$

‡This problem is treated in greater detail in Goldstein (1951), Chapter 11.

where the integrations by parts of the last two terms can be justified since $\delta\eta$ vanishes at the end points of the space and time intervals. If (1.11) is to vanish for any arbitrary variation satisfying the above requirements, we must have

$$\frac{\partial}{\partial x}\frac{\partial \mathscr{L}}{\partial(\partial\eta/\partial x)}+\frac{\partial}{\partial t}\frac{\partial \mathscr{L}}{\partial(\partial\eta/\partial t)}-\frac{\partial \mathscr{L}}{\partial\eta}=0. \tag{1.12}$$

This is called the Euler-Lagrange equation.‡ In our particular example (1.9), Eq. (1.12) becomes

$$Y\frac{\partial^2\eta}{\partial x^2}-\mu\frac{\partial^2\eta}{\partial t^2}=0. \tag{1.13}$$

This is to be identified with the wave equation for the one-dimensional propagation of a disturbance with velocity $\sqrt{Y/\mu}$. We can define the Hamiltonian density $\mathscr{H}$ in analogy with (1.5) as

$$\begin{aligned}\mathscr{H} &= \dot\eta\frac{\partial\mathscr{L}}{\partial\dot\eta}-\mathscr{L}\\ &= \tfrac{1}{2}\mu\dot\eta^2+\tfrac{1}{2}Y\left(\frac{\partial\eta}{\partial x}\right)^2;\end{aligned} \tag{1.14}$$

$\partial\mathscr{L}/\partial\dot\eta$ is called the canonical momentum conjugate to η, and is often denoted by π. The two terms in (1.14) can be identified respectively with the kinetic and potential energy densities.

1–3. CLASSICAL SCALAR FIELDS

Covariant notation. The arguments of the preceding section can readily be generalized to three space dimensions. Consider a field which is assumed to be a real function defined at each space-time point, $\mathbf{x}$, t; $\mathscr{L}$ now depends on ϕ, $\partial\phi/\partial x_k$ ($k = 1, 2, 3$), and $\partial\phi/\partial t$. The Euler-Lagrange equation reads

$$\sum_{k=1}^{3}\frac{\partial}{\partial x_k}\frac{\partial\mathscr{L}}{\partial(\partial\phi/\partial x_k)}+\frac{\partial}{\partial t}\frac{\partial\mathscr{L}}{\partial(\partial\phi/\partial t)}-\frac{\partial\mathscr{L}}{\partial\phi}=0. \tag{1.15}$$

We wish to write (1.15) in a relativistically covariant form, but first let us recall some properties of Lorentz transformations. We introduce a four-vector notation in which the four-vector b_μ with $\mu = 1, 2, 3, 4$ stands for

$$b_\mu = (b_1, b_2, b_3, b_4) = (\mathbf{b}, ib_0), \tag{1.16}$$

where b_1, b_2, and b_3 are real, and $b_4 = ib_0$ is purely imaginary. In general, the Greek indices μ, ν, λ, etc., run from 1 to 4, whereas the italic indices i, j, k, etc.,

---

‡In the literature this equation is sometimes writen in the form

$$\frac{\partial}{\partial t}\frac{\partial\mathscr{L}}{\partial(\partial\eta/\partial t)}-\frac{\delta\mathscr{L}}{\delta\eta}=0,$$

where $\delta\mathscr{L}/\delta\eta$ is called the functional derivative of $\mathscr{L}$ with respect to η. This version is not recommended since (a) it obscures the dependence of $\mathscr{L}$ on the space coordinate, and (b) it singles out time, which is against the spirit of the covariant approach (to be discussed in the next section).

run from 1 to 3. The coordinate vector x_μ is given by

$$\begin{aligned} x_\mu &= (x_1, x_2, x_3, x_4) \\ &= (\mathbf{x}, ict). \end{aligned} \tag{1.17}$$

The symbols x, y, and z may also be used in place of x_1, x_2, and x_3. Under a Lorentz transformation, we have

$$x'_\mu = a_{\mu\nu} x_\nu, \tag{1.18}$$

where the $a_{\mu\nu}$ satisfy

$$a_{\mu\nu} a_{\mu\lambda} = \delta_{\nu\lambda}, \qquad (a^{-1})_{\mu\nu} = a_{\nu\mu}. \tag{1.19}$$

Hence

$$x_\mu = (a^{-1})_{\mu\nu} x'_\nu = a_{\nu\mu} x'_\nu \tag{1.20}$$

when x' and x are related by (1.18). The matrix elements a_{ij}, a_{44} are purely real, whereas a_{j4} and a_{4j} are purely imaginary. A *four-vector*, by definition, transforms in the same way as x_μ under Lorentz transformations. Because of (1.20) we have

$$\frac{\partial}{\partial x'_\mu} = \frac{\partial x_\nu}{\partial x'_\mu} \frac{\partial}{\partial x_\nu} = a_{\mu\nu} \frac{\partial}{\partial x_\nu}; \tag{1.21}$$

so the four-gradient $\partial/\partial x_\mu$ is a four-vector. The scalar product $b \cdot c$ is defined by

$$\begin{aligned} b \cdot c = b_\mu c_\mu &= \sum_{j=1}^{3} b_j c_j + b_4 c_4 \\ &= \mathbf{b} \cdot \mathbf{c} - b_0 c_0. \end{aligned} \tag{1.22}$$

It is unchanged under Lorentz transformations, since

$$\begin{aligned} b' \cdot c' = a_{\mu\nu} b_\nu a_{\mu\lambda} c_\lambda &= \delta_{\nu\lambda} b_\nu c_\lambda \\ &= b \cdot c. \end{aligned} \tag{1.23}$$

A tensor of second rank, $t_{\mu\nu}$, transforms as

$$t'_{\mu\nu} = a_{\mu\lambda} a_{\nu\sigma} t_{\lambda\sigma}. \tag{1.24}$$

Generalizations to tensors of higher rank are straightforward. Note that we make no distinction between a covariant and a contravariant vector, nor do we define the metric tensor $g_{\mu\nu}$. These complications are absolutely unnecessary in the *special* theory of relativity. (It is regrettable that many textbook writers do not emphasize this elementary point.)

Equation (1.15) can now be written as

$$\frac{\partial}{\partial x_\mu}\left[\frac{\partial \mathscr{L}}{\partial(\partial\phi/\partial x_\mu)}\right] - \frac{\partial \mathscr{L}}{\partial \phi} = 0. \tag{1.25}$$

It is seen that the field equation derivable from the Lagrangian density $\mathscr{L}$ is covariant (i.e., the equation "looks the same" in all Lorentz frames) if the Lagrangian density $\mathscr{L}$ is chosen to be a relativistically scalar density. This is an important point because the relativistic invariance of $\mathscr{L}$ is so restrictive that it can be used as a guiding principle for "deriving" a covariant wave equation.

Neutral scalar field. As an illustration let $\phi(x)$ be a scalar field which, by definition, transforms like

$$\phi'(x') = \phi(x), \tag{1.26}$$

under a Lorentz transformation, where ϕ' is the functional form of the field in the primed system. Now the dependence of $\mathscr{L}$ on space-time coordinates is only through the field and its first derivatives, and x_μ cannot appear explicitly in $\mathscr{L}$. This means that $\partial\phi/\partial x_\mu$ is the only four-vector at our disposal; when it appears in $\mathscr{L}$ it must be contracted with itself. Moreover, if we are interested in obtaining a linear wave equation, $\mathscr{L}$ must be a quadratic function of ϕ and $\partial\phi/\partial x_\mu$. A possible candidate for $\mathscr{L}$ consistent with the above requirements is

$$\mathscr{L} = -\frac{1}{2}\left(\frac{\partial\phi}{\partial x_\mu}\frac{\partial\phi}{\partial x_\mu} + \mu^2\phi^2\right). \tag{1.27}$$

From the Euler-Lagrange equation (1.25) we obtain

$$-\frac{1}{2}\frac{\partial}{\partial x_\mu}\left(2\frac{\partial\phi}{\partial x_\mu}\right) + \mu^2\phi = 0, \tag{1.28}$$

or

$$\Box\phi - \mu^2\phi = 0, \tag{1.29}$$

where

$$\Box = \nabla^2 - \frac{1}{c^2}\frac{\partial^2}{\partial t^2}. \tag{1.30}$$

The wave equation (1.29) is called the Klein-Gordon equation. It was considered in the middle 1920's by E. Schrödinger, as well as by O. Klein and W. Gordon, as a candidate for the *relativistic* analog of the *nonrelativistic* Schrödinger wave equation for a free particle. The similarity of (1.29) to the relativistic energy momentum relation for a free particle of mass m,

$$E^2 - |\mathbf{p}|^2c^2 = m^2c^4, \tag{1.31}$$

becomes apparent as we consider heuristic substitutions:

$$E \to i\hbar\frac{\partial}{\partial t}, \qquad p_k \to -i\hbar\frac{\partial}{\partial x_k}. \tag{1.32}$$

The parameter μ in (1.29) has the dimension of inverse length, and, using (1.32), we may make the identification

$$\mu = mc/\hbar. \tag{1.33}$$

Numerically $1/\mu$ is 1.41×10^{-13} cm for a particle of mass 140 MeV/c^2 (corresponding to the mass of the charged pion).

Yukawa potential. So far we have been concerned with a field in the absence of any source. Such a field is often called a *free field*. The interaction of ϕ with a source can easily be incorporated into the Lagrangian formalism by adding

$$\mathscr{L}_{\text{int}} = -\phi\rho, \tag{1.34}$$

to (1.27), where ρ is the source density, which is, in general, a function of space-time coordinates. The field equation now becomes

$$\Box\phi - \mu^2\phi = \rho. \tag{1.35}$$

Let us consider a static (i.e., time-independent) solution to (1.35) where the source is assumed to be a point source at the origin, independent of time. We have

$$(\nabla^2 - \mu^2)\phi = G\delta^{(3)}(\mathbf{x}), \tag{1.36}$$

where G, the numerical constant that characterizes the strength of the coupling of the field to the source, is analogous to the constant e in electrodynamics. Although the solution to (1.36) can be guessed immediately, for pedagogical reasons we solve this equation using the Fourier transform method. First, we define $\tilde{\phi}(\mathbf{k})$ as follows:

$$\begin{aligned}\phi(\mathbf{x}) &= \frac{1}{(2\pi)^{3/2}}\int d^3k\, e^{i\mathbf{k}\cdot\mathbf{x}}\tilde{\phi}(\mathbf{k}),\\ \tilde{\phi}(\mathbf{k}) &= \frac{1}{(2\pi)^{3/2}}\int d^3x\, e^{-i\mathbf{k}\cdot\mathbf{x}}\phi(\mathbf{x}),\end{aligned} \tag{1.37}$$

where d^3k and d^3x, respectively, stand for volume elements in the three-dimensional k-space and the coordinate space. If we multiply both sides of (1.36) by $e^{-i\mathbf{k}\cdot\mathbf{x}}/(2\pi)^{3/2}$ and integrate with d^3x, we obtain, after integrating by parts twice (assuming that ϕ and $\nabla\phi$ go to zero sufficiently rapidly at infinity),

$$(-|\mathbf{k}|^2 - \mu^2)\tilde{\phi}(\mathbf{k}) = \frac{G}{(2\pi)^{3/2}}. \tag{1.38}$$

Thus the differential equation (1.36) has been converted into an algebraic equation which can easily be solved:

$$\tilde{\phi}(\mathbf{k}) = -\frac{G}{(2\pi)^{3/2}}\frac{1}{|\mathbf{k}|^2 + \mu^2}; \tag{1.39}$$

$$\begin{aligned}\phi(\mathbf{x}) &= -\frac{G}{(2\pi)^3}\int d^3k\,\frac{e^{i\mathbf{k}\cdot\mathbf{x}}}{|\mathbf{k}|^2 + \mu^2}\\ &= -\frac{G}{(2\pi)^3}2\pi\int_0^\infty |\mathbf{k}|^2 d|\mathbf{k}|\int_{-1}^{1} d(\cos\theta_k)\,\frac{e^{i|\mathbf{k}|r\cos\theta_k}}{|\mathbf{k}|^2 + \mu^2},\end{aligned} \tag{1.40}$$

where $r = |\mathbf{x}|$ and $\theta_k = \angle(\mathbf{k}, \mathbf{x})$. The integration can be performed to give

$$\phi(\mathbf{x}) = -\frac{G}{4\pi}\frac{e^{-\mu r}}{r}. \tag{1.41}$$

Yukawa proposed that a nucleon is the source of a force field, called the *meson field*, in the same way as an electrically charged object is the source of an electrostatic field. Suppose that the static meson field around a nucleon located at the origin satisfies (1.36). The strength of the meson field at point $\mathbf{x}_2$ due to the presence of a nucleon at point $\mathbf{x}_1$ is given by

$$\phi(\mathbf{x}_2) = -\frac{G}{4\pi}\frac{e^{-\mu|\mathbf{x}_2 - \mathbf{x}_1|}}{|\mathbf{x}_2 - \mathbf{x}_1|}. \tag{1.42}$$

Since the interaction Lagrangian density (1.34) does not involve the time derivative of ϕ, the interaction Hamiltonian density (cf. Eq. 1.14), is given by $\mathscr{H}_{\text{int}} = -\mathscr{L}_{\text{int}}$. Hence the total interaction Hamiltonian is

$$H_{\text{int}} = \int \mathscr{H}_{\text{int}}\, d^3x = \int \phi\rho\, d^3x. \tag{1.43}$$

The interaction energy between two nucleons, one located at point $\mathbf{x}_2$, the other at point $\mathbf{x}_1$, is

$$H_{\text{int}}^{(1,2)} = -\frac{G^2}{4\pi}\frac{e^{-\mu|\mathbf{x}_2-\mathbf{x}_1|}}{|\mathbf{x}_2 - \mathbf{x}_1|}. \tag{1.44}$$

Unlike the Coulomb case, this interaction is attractive‡ and short-ranged; it goes to zero very rapidly for

$$|\mathbf{x}_2 - \mathbf{x}_1| \gg 1/\mu. \tag{1.45}$$

We have seen that by postulating the existence of a field obeying (1.36), we can qualitatively understand the short-ranged force between two nucleons. The mass of a quantum associated with the field was originally estimated by Yukawa to be about 200 times the electron mass. This estimate is not too far from the mass of the *observed* pion (about 270 times the electron mass) discovered by C. F. Powell and his coworkers in 1947. To represent the interaction of the pion field with the nucleon in a more realistic way, we must make a few more modifications. First, we must take into account the spin of the nucleon and the intrinsic odd parity of the pion, both of which will be discussed in Sections 3-10 and 3-11. Second, we must note that the pions observed in nature have three charge states (π^+, π^0, π^-). These considerations naturally lead us to a discussion of a complex field.

Complex scalar field. Suppose we consider two real fields of identical masses. We can always construct complex fields ϕ and ϕ^* by§

$$\phi = \frac{\phi_1 + i\phi_2}{\sqrt{2}}, \qquad \phi^* = \frac{\phi_1 - i\phi_2}{\sqrt{2}}; \tag{1.46}$$

$$\phi_1 = \frac{\phi + \phi^*}{\sqrt{2}}, \qquad \phi_2 = \frac{\phi - \phi^*}{i\sqrt{2}}. \tag{1.47}$$

The free-field Lagrangian density can be written either in terms of the real fields ϕ_1 and ϕ_2 or in terms of the complex fields ϕ and ϕ^*:

$$\begin{aligned}\mathscr{L} &= -\frac{1}{2}\left(\frac{\partial\phi_1}{\partial x_\mu}\frac{\partial\phi_1}{\partial x_\mu} + \mu^2\phi_1^2\right) - \frac{1}{2}\left(\frac{\partial\phi_2}{\partial x_\mu}\frac{\partial\phi_2}{\partial x_\mu} + \mu^2\phi_2^2\right)\\ &= -\left(\frac{\partial\phi^*}{\partial x_\mu}\frac{\partial\phi}{\partial x_\mu} + \mu^2\phi^*\phi\right).\end{aligned} \tag{1.48}$$

The field equations for ϕ and ϕ^* can be obtained from the variational principle by treating ϕ and ϕ^* as two independent fields:

$$\begin{aligned}\frac{\partial}{\partial x_\mu}\frac{\partial\mathscr{L}}{\partial(\partial\phi/\partial x_\mu)} - \frac{\partial\mathscr{L}}{\partial\phi} = 0 &\Rightarrow \Box\phi^* - \mu^2\phi^* = 0,\\ \frac{\partial}{\partial x_\mu}\frac{\partial\mathscr{L}}{\partial(\partial\phi^*/\partial x_\mu)} - \frac{\partial\mathscr{L}}{\partial\phi^*} = 0 &\Rightarrow \Box\phi - \mu^2\phi = 0.\end{aligned} \tag{1.49}$$

‡The reason for the Coulomb repulsion and the Yukawa attraction will be treated in Section 4–6. The difference stems from the fact that the Coulomb field transforms like the fourth component of a vector, whereas our ϕ field is a scalar field.

§Throughout this book the superscript * stands for *complex conjugation*. The superscript † will be used for *Hermitian conjugation*.

What is the physical interpretation of a complex field? It is not difficult to show that if ϕ is a solution to the Klein-Gordon equation in the presence of A_μ with charge e, then ϕ^* is a solution to the Klein-Gordon equation in the presence of the same A_μ but with charge $-e$. This demonstration is left as an exercise (Problem 1–3).

To see further the connection between the complexity of a scalar field and an internal attribute such as electric charge associated with it, we consider the following unitary (actually orthogonal) transformation on ϕ_1 and ϕ_2:

$$\phi_1' = \phi_1 \cos\lambda - \phi_2 \sin\lambda, \qquad \phi_2' = \phi_1 \sin\lambda + \phi_2 \cos\lambda, \tag{1.50}$$

where λ is a real constant independent of space-time points. Since the masses associated with ϕ_1 and ϕ_2 are assumed to be strictly the same in (1.48), the free-field Lagrangian (1.48) is clearly invariant under (1.50). In terms of ϕ and ϕ^*, the transformation (1.50) amounts to

$$\phi' = e^{i\lambda}\phi, \qquad \phi^{*\prime} = e^{-i\lambda}\phi^*. \tag{1.51}$$

Let us consider (1.51) with λ taken to be infinitesimally small. We then have

$$\delta\phi = i\lambda\phi, \qquad \delta\phi^* = -i\lambda\phi^*, \tag{1.52}$$

for the changes in ϕ and ϕ^*. Meanwhile, the variation in $\mathscr{L}$ induced by (1.52) is

$$\begin{aligned}\delta\mathscr{L} &= \left[\frac{\partial\mathscr{L}}{\partial\phi}\delta\phi + \frac{\partial\mathscr{L}}{\partial(\partial\phi/\partial x_\mu)}\delta\left(\frac{\partial\phi}{\partial x_\mu}\right)\right] + \left[\frac{\partial\mathscr{L}}{\partial\phi^*}\delta\phi^* + \frac{\partial\mathscr{L}}{\partial(\partial\phi^*/\partial x_\mu)}\delta\left(\frac{\partial\phi^*}{\partial x_\mu}\right)\right] \\ &= \left[\frac{\partial\mathscr{L}}{\partial\phi} - \frac{\partial}{\partial x_\mu}\left(\frac{\partial\mathscr{L}}{\partial(\partial\phi/\partial x_\mu)}\right)\right]\delta\phi + \left[\frac{\partial\mathscr{L}}{\partial\phi^*} - \frac{\partial}{\partial x_\mu}\left(\frac{\partial\mathscr{L}}{\partial(\partial\phi^*/\partial x_\mu)}\right)\right]\delta\phi^* \\ &\quad + \frac{\partial}{\partial x_\mu}\left[\frac{\partial\mathscr{L}}{\partial(\partial\phi/\partial x_\mu)}\delta\phi + \frac{\partial\mathscr{L}}{\partial(\partial\phi^*/\partial x_\mu)}\delta\phi^*\right] \\ &= -i\lambda\frac{\partial}{\partial x_\mu}\left(\frac{\partial\phi^*}{\partial x_\mu}\phi - \phi^*\frac{\partial\phi}{\partial x_\mu}\right),\end{aligned} \tag{1.53}$$

where we have used the Euler-Lagrange equation. Since the Lagrangian density is known to be unchanged from our earlier argument, $\delta\mathscr{L}$ must be zero. Thus we have the important result

$$\frac{\partial s_\mu}{\partial x_\mu} = 0, \tag{1.54}$$

where

$$s_\mu = i\left(\frac{\partial\phi^*}{\partial x_\mu}\phi - \phi^*\frac{\partial\phi}{\partial x_\mu}\right). \tag{1.55}$$

This means that there exists a conserved four-vector current [i.e., a four-vector density that satisfies the continuity equation (1.54)] associated with a complex field ϕ.

Under the substitution $\phi \rightleftarrows \phi^*$, s_μ changes its sign. This suggests that s_μ is to be interpreted as the charge-current density up to a constant, and that if ϕ is a field corresponding to a particle with charge e, then ϕ^* is a field corresponding to a particle with charge $-e$, in agreement with the interpretation suggested in Problem 1–3. It is a remarkable feature of relativistic field theory that it can readily accommodate a pair of particles with the *same* mass but *opposite* charges. In the formalism, however, there is nothing that compels us to relate s_μ to the charge-current density that appears in electrodynamics. In fact, our formalism can accommodate any conserved internal attribute associated with a complex field.

Let us get back to pions. In order to describe the three charge states observed in nature, we ignore the mass difference between the $\pi^\pm$ and the π^0 (about 5 MeV out of 140 MeV) and start with

$$\mathscr{L} = -\frac{1}{2}\left[\sum_{\alpha=1}^{3}\left(\frac{\partial\phi_\alpha}{\partial x_\mu}\right)\left(\frac{\partial\phi_\alpha}{\partial x_\mu}\right) + \mu^2\phi_\alpha\phi_\alpha\right], \tag{1.56}$$

where the orthogonal linear combinations of ϕ_1 and ϕ_2 given by (1.46) correspond to the charged pions, and ϕ_3 corresponds to the neutral pion. This suggests that we may consider a class of unitary transformations wider than (1.50) in which not only ϕ_1 and ϕ_2 but also all three ϕ_α are mixed with one another. This is essentially the starting point of isospin formalism, a subject which we shall not discuss in this book.

To sum up, the strict mass degeneracy of ϕ_1 and ϕ_2 implies the invariance of $\mathscr{L}$ under (1.50) and (1.51) which in turn gives the conservation law of electric charge, or some similar internal attribute, associated with the complex fields ϕ and ϕ^*. The connection between invariance under a certain transformation and an associated conservation law is well known in both classical and quantum mechanics, e.g., the connection between rotational invariance (isotropy of space) and angular momentum conservation. But here we see that the conservation law of a non-geometrical attribute such as electric charge can also be formulated in terms of invariance under a transformation (1.51) which is called, after W. Pauli, the *gauge transformation of the first kind.*

Perhaps the real significance of what we have accomplished can be appreciated only by considering an example in which the conservation of an internal attribute is approximate. In field theory neutral K mesons created in high-energy collisions must be described by complex fields even though they are electrically neutral. This is because K^0 and its antiparticle $\bar{K}^0$ carry internal attributes called hypercharge, denoted by Y; $Y = +1$ for K^0 with which we may associate a complex field ϕ, and $Y = -1$ for $\bar{K}^0$ with which we may associate ϕ^*. Hypercharge conservation‡ (which is equivalent to the conservation of strangeness, introduced by M. Gell-Mann and K. Nishijima in 1953) is a very useful conservation law, but it is broken by a class of interactions about 10^{12} times weaker than the kind of interactions responsible for the production of K^0 and $\bar{K}^0$ with definite hypercharges. As a result, the particle states known as K_1 and K_2 which essentially correspond to our ϕ_1 and ϕ_2 turn out to have a very small but measurable mass difference ($\sim 10^{-11}$ MeV/c^2). Thanks to the nonconservation of hypercharge we have a realistic example that illustrates the connection between the nonconservation of an internal attribute and a removal of the mass degeneracy.

---

‡For an elementary discussion of hypercharge conservation see, for example, Segrè (1964), Chapter 15. For a more complete discussion consult Nishijima (1964), Chapter 6, and Sakurai (1964), Chapter 10.

1–4. CLASSICAL MAXWELL FIELDS

Basic equations. We shall now discuss electromagnetic fields within the framework of classical electrodynamics. In this chapter and the next we shall use Heaviside-Lorentz (rationalized) units in which the Maxwell equations read:

$$\nabla\cdot\mathbf{E} = \rho, \qquad \nabla\times\mathbf{B} - \frac{1}{c}\frac{\partial\mathbf{E}}{\partial t} = \frac{\mathbf{j}}{c}; \tag{1.57}$$

$$\nabla\cdot\mathbf{B} = 0, \qquad \nabla\times\mathbf{E} + \frac{1}{c}\frac{\partial\mathbf{B}}{\partial t} = 0. \tag{1.58}$$

According to our units the fine-structure constant is given by

$$\frac{e^2}{4\pi\hbar c} \approx \frac{1}{137.04}, \tag{1.59}$$

which is equal to $e^2/\hbar c$ in Gaussian (cgs) units and $e^2/(4\pi\hbar c\epsilon_0)$ in mks rationalized units. The fields and potentials in our units are related to the corresponding fields and potentials in Gaussian units by $1/\sqrt{4\pi}$; for example, $(1/2)(|\mathbf{E}|^2 + |\mathbf{B}|^2)$ in our units should read $(1/8\pi)\,(|\mathbf{E}|^2 + |\mathbf{B}|^2)$ in Gaussian units. Note, however, that expressions such as $\mathbf{p} - e\mathbf{A}/c$ are the same in both units since

$$(\sqrt{4\pi}\,e)\,(\mathbf{A}/\sqrt{4\pi}) = e\mathbf{A}.$$

The Maxwell equations can be written more concisely if we introduce the field tensor $F_{\mu\nu}$, antisymmetric in μ and ν, and the charge-current four-vector j_μ as follows:

$$F_{\mu\nu} = \begin{pmatrix} 0 & B_3 & -B_2 & -iE_1 \\ -B_3 & 0 & B_1 & -iE_2 \\ B_2 & -B_1 & 0 & -iE_3 \\ iE_1 & iE_2 & iE_3 & 0 \end{pmatrix}, \tag{1.60}$$

$$j_\mu = (\mathbf{j},\ ic\rho). \tag{1.61}$$

Equation (1.57) now becomes

$$\frac{\partial F_{\mu\nu}}{\partial x_\nu} = \frac{j_\mu}{c}. \tag{1.62}$$

The simplicity of the covariant form of the Maxwell equations should be noted. In fact, what is now known as Lorentz invariance was first noted by H. Poincaré as he examined the transformation properties of the Maxwell equations.

By virtue of the antisymmetry of $F_{\mu\nu}$, we have the continuity equation for the charge-current density. To show this we just take the four-divergence of both sides of (1.62). We have

$$\frac{\partial}{\partial x_\mu}\frac{\partial F_{\mu\nu}}{\partial x_\nu} = \frac{1}{2}\left(\frac{\partial}{\partial x_\mu}\frac{\partial F_{\mu\nu}}{\partial x_\nu} - \frac{\partial}{\partial x_\mu}\frac{\partial F_{\nu\mu}}{\partial x_\nu}\right) = \frac{1}{2}\left(\frac{\partial}{\partial x_\mu}\frac{\partial F_{\mu\nu}}{\partial x_\nu} - \frac{\partial}{\partial x_\nu}\frac{\partial F_{\mu\nu}}{\partial x_\mu}\right) = 0. \tag{1.63}$$

Hence

$$\frac{\partial j_\mu}{\partial x_\mu} = 0. \tag{1.64}$$

In other words, the Maxwell theory is constructed in such a way that the charge-current conservation is guaranteed automatically once $F_{\mu\nu}$ is introduced. Historically, the conservation of electric charge played a crucial role in the formulation of classical electrodynamics. C. Maxwell introduced the notion of displacement current, the $\partial \mathbf{E}/\partial t$ term in (1.57), so that the charge would be conserved even in nonsteady-state problems.

The vector potential A_μ is introduced by

$$\frac{\partial A_\nu}{\partial x_\mu} - \frac{\partial A_\mu}{\partial x_\nu} = F_{\mu\nu}. \tag{1.65}$$

The second pair of the Maxwell equations (1.58) can be written as

$$t_{\lambda\mu,\nu} + t_{\mu\nu,\lambda} + t_{\nu\lambda,\mu} = 0, \tag{1.66}$$

where a third-rank tensor $t_{\lambda\mu,\nu}$ is defined by

$$t_{\lambda\mu,\nu} = \frac{\partial F_{\lambda\mu}}{\partial x_\nu} = \frac{\partial}{\partial x_\nu}\left(\frac{\partial A_\mu}{\partial x_\lambda} - \frac{\partial A_\lambda}{\partial x_\mu}\right). \tag{1.67}$$

We see that once the vector potential is introduced by (1.65), the second pair of the Maxwell equations are automatically satisfied. Conversely, if there were magnetic monopoles analogous to electric charges so that

$$\nabla\cdot\mathbf{B} = \rho_{\text{mag}} \neq 0, \qquad \nabla\times\mathbf{E} + \frac{1}{c}\frac{\partial \mathbf{B}}{\partial t} = -\mathbf{j}_{\text{mag}} \neq 0, \tag{1.68}$$

then the description of **E** and **B** in terms of A_μ *alone* would be untenable.

Lagrangian and Hamiltonian. The only true scalar density that can be constructed form the field tensor is‡

$$F_{\mu\nu}F_{\mu\nu} = 2(|\mathbf{B}|^2 - |\mathbf{E}|^2). \tag{1.69}$$

We may try the Lagrangian density,

$$\mathscr{L} = -\tfrac{1}{4}F_{\mu\nu}F_{\mu\nu} + (j_\mu A_\mu)/c. \tag{1.70}$$

By regarding each component of A_μ as an independent field, we obtain

$$\begin{aligned}
\frac{\partial}{\partial x_\nu}\frac{\partial\mathscr{L}}{\partial(\partial A_\mu/\partial x_\nu)} &= -\frac{1}{4}\frac{\partial}{\partial x_\nu}\left\{\frac{\partial}{\partial(\partial A_\mu/\partial x_\nu)}\left[\left(\frac{\partial A_\sigma}{\partial x_\lambda} - \frac{\partial A_\lambda}{\partial x_\sigma}\right)\left(\frac{\partial A_\sigma}{\partial x_\lambda} - \frac{\partial A_\lambda}{\partial x_\sigma}\right)\right]\right\} \\
&= -\frac{1}{4}\frac{\partial}{\partial x_\nu}\left[\frac{\partial}{\partial(\partial A_\mu/\partial x_\nu)}\left(2\frac{\partial A_\sigma}{\partial x_\lambda}\frac{\partial A_\sigma}{\partial x_\lambda} - 2\frac{\partial A_\sigma}{\partial x_\lambda}\frac{\partial A_\lambda}{\partial x_\sigma}\right)\right] \\
&= -\frac{1}{4}\frac{\partial}{\partial x_\nu}\left(4\frac{\partial A_\mu}{\partial x_\nu} - 4\frac{\partial A_\nu}{\partial x_\mu}\right) \\
&= -\frac{\partial}{\partial x_\nu}F_{\nu\mu},
\end{aligned} \tag{1.71}$$

‡An alternative form, $(i/8)\epsilon_{\mu\nu\lambda\sigma}F_{\mu\nu}F_{\lambda\sigma} = \mathbf{B}\cdot\mathbf{E}$ (where $\epsilon_{\mu\nu\lambda\sigma}$ is zero unless $\mu, \nu, \lambda, \sigma$ are all different, is 1 for an even permutation of 1, 2, 3, 4, and is -1 for an odd permutation of 1, 2, 3, 4), is not considered here because it is not invariant under space inversion (parity).

and

$$\frac{\partial \mathscr{L}}{\partial A_\mu} = \frac{j_\mu}{c}. \tag{1.72}$$

So the Euler-Lagrange equation for each component of A_μ gives the Maxwell equations (1.62).

The Hamiltonian density $\mathscr{H}_{\text{em}}$ for the free Maxwell field can be evaluated from $\mathscr{L}_{\text{em}} = -\frac{1}{4}F_{\mu\nu}F_{\mu\nu}$ as follows:‡

$$\begin{aligned}\mathscr{H}_{\text{em}} &= \frac{\partial \mathscr{L}_{\text{em}}}{\partial(\partial A_\mu/\partial x_4)}\frac{\partial A_\mu}{\partial x_4} - \mathscr{L}_{\text{em}}\\ &= -F_{4\mu}\left(F_{4\mu} + \frac{\partial A_4}{\partial x_\mu}\right) + \tfrac{1}{2}(|\mathbf{B}|^2 - |\mathbf{E}|^2)\\ &= \tfrac{1}{2}(|\mathbf{B}|^2 + |\mathbf{E}|^2) - i\mathbf{E}\cdot\nabla A_4. \end{aligned}\tag{1.73}$$

In the free-field case the last term of (1.73) has no effect when integrated by parts, since $\nabla\cdot\mathbf{E} = \rho = 0$, and $\mathbf{E}$ as well as A_4 vanish sufficiently rapidly at infinity. In this way we get the familiar expression

$$\begin{aligned}H_{\text{em}} &= \int \mathscr{H}_{\text{em}}\, d^3x\\ &= \tfrac{1}{2}\int (|\mathbf{B}|^2 + |\mathbf{E}|^2)\, d^3x \end{aligned}\tag{1.74}$$

in the free-field case.

Gauge transformations. Let us now go back to the covariant form of the Maxwell equations (1.62) which can be written as

$$\Box A_\mu - \frac{\partial}{\partial x_\mu}\left(\frac{\partial A_\nu}{\partial x_\nu}\right) = -\frac{j_\mu}{c}. \tag{1.75}$$

Suppose that

$$\frac{\partial A_\nu}{\partial x_\nu} \neq 0. \tag{1.76}$$

We may redefine A_μ without changing $F_{\mu\nu}$ as follows:

$$A_\mu^{\text{new}} = A_\mu^{\text{old}} + \frac{\partial \chi}{\partial x_\mu}, \tag{1.77}$$

where

$$\Box\chi = -\frac{\partial A_\mu^{\text{old}}}{\partial x_\mu}. \tag{1.78}$$

Then

$$\frac{\partial A_\mu^{\text{new}}}{\partial x_\mu} = \frac{\partial A_\mu^{\text{old}}}{\partial x_\mu} + \Box\chi = 0. \tag{1.79}$$

---

‡Strictly speaking, the Lagrangian density (1.70) is not suitable for the Hamiltonian formulation of the Maxwell theory. This is because the canonical momentum conjugate to A_4 vanishes identically due to the fact that the Lagrangian density does not contain $\partial A_4/\partial x_4$.

We take the point of view that the $F_{\mu\nu}$ are the only quantities of physical significance; the potential A_μ is introduced merely to simplify computations. So we may as well work with the simpler equation:

$$\Box A_\mu = -j_\mu/c, \tag{1.80}$$

where A_μ satisfies

$$\partial A_\mu/\partial x_\mu = 0. \tag{1.81}$$

Equation (1.81) is known as the Lorentz condition.

Even if we work within the framework of (1.81), the potential A_μ is still not unique. We are free to make a further change:

$$A_\mu \to A'_\mu = A_\mu + \frac{\partial \Lambda}{\partial x_\mu}, \tag{1.82}$$

where Λ now satisfies the homogeneous D'Alembertian equation [in contrast to the inhomogeneous equation (1.78)]:

$$\Box \Lambda = 0. \tag{1.83}$$

The transformation (1.82) is known as a *gauge transformation of the second kind.*

1–5. VECTOR POTENTIALS IN QUANTUM MECHANICS

Charged particles in the Schrödinger theory. We know from classical mechanics that the Hamiltonian of a nonrelativistic mass point of an electronic charge‡ $e = -|e|$ is given by§

$$H = \frac{1}{2m}\left(\mathbf{p} - \frac{e\mathbf{A}}{c}\right)^2 + eA_0, \tag{1.84}$$

$$A_\mu = (\mathbf{A}, iA_0), \tag{1.85}$$

when it is subject to a Lorentz force

$$\mathbf{F} = e[\mathbf{E} + (1/c)(\dot{\mathbf{x}} \times \mathbf{B})]. \tag{1.86}$$

It is important to note that p_k is the momentum conjugate to x_k and is not equal to $m\dot{x}_k$. Rather

$$m\dot{\mathbf{x}} = \mathbf{p} - e\mathbf{A}/c. \tag{1.87}$$

When $\mathbf{A} \neq 0$, the classical velocity one measures is not $\mathbf{p}/m$ but the $\dot{\mathbf{x}}$ which occurs in (1.86). A gauge transformation on $\mathbf{A}$ must be accompanied by a corresponding change in $\mathbf{p}$ so that $m\dot{\mathbf{x}}$ is unchanged. To see explicitly how this comes about, let us recall that the Lorentz force (1.86) can be obtained from

$$L = T - eA_0 + (e/c)\,\mathbf{A}\cdot\dot{\mathbf{x}}, \tag{1.88}$$

and that p_k is equal to $\partial L/\partial \dot{x}_k$.

In nonrelativistic quantum mechanics, if the interaction of the spin magnetic moment is ignored, one starts with a Hamiltonian operator of the same form as (1.84) with $\mathbf{p}$ replaced by the operator p. In the coordinate representation this

---

‡Throughout this book the constant e is taken to be negative.
§See Goldstein (1951), p. 222.

operator is given by $-i\hbar\nabla$. The time-independent Schrödinger equation reads

$$\frac{1}{2m}\left(-i\hbar\nabla - \frac{e\mathbf{A}}{c}\right)^2 \psi + V\psi = E\psi, \tag{1.89}$$

where $V = eA_0$, and both $\mathbf{A}$ and V are assumed to be time independent.

We now show that in a magnetic-field free region ($\mathbf{B} = 0$) the solution ψ when $\mathbf{A}(\mathbf{x}) \neq 0$ can be written in the form

$$\psi(\mathbf{x}) = \psi^{(0)}(\mathbf{x}) \exp\left[\frac{ie}{\hbar c}\int^{\mathbf{s}(\mathbf{x})} \mathbf{A}(\mathbf{x}')\cdot d\mathbf{s}'\right], \tag{1.90}$$

where $\psi^{(0)}(\mathbf{x})$ satisfies the Schrödinger wave equation with the same V but with $\mathbf{A}$ set equal to zero. The line integral can be taken along any path so long as the end point $\mathbf{s}(\mathbf{x})$ is the point $\mathbf{x}$ itself, and the curl of $\mathbf{A}$ vanishes. To prove this we first note that

$$\begin{aligned}\left(-i\hbar\nabla - \frac{e\mathbf{A}}{c}\right)\psi &= \exp\left(\frac{ie}{\hbar c}\int^{\mathbf{s}(\mathbf{x})} \mathbf{A}\cdot d\mathbf{s}'\right)\left[\left(-i\hbar\nabla - \frac{e\mathbf{A}}{c}\right)\psi^{(0)} + \psi^{(0)}(-i\hbar)\left(\frac{ie}{\hbar c}\right)\mathbf{A}(\mathbf{x})\right]\\ &= \exp\left(\frac{ie}{\hbar c}\int^{\mathbf{s}(\mathbf{x})} \mathbf{A}\cdot d\mathbf{s}'\right)(-i\hbar\nabla\psi^{(0)}).\end{aligned} \tag{1.91}$$

Similarly,

$$\left(-i\hbar\nabla - \frac{e\mathbf{A}}{c}\right)^2 \psi = \exp\left[\frac{ie}{\hbar c}\int^{\mathbf{s}(\mathbf{x})} \mathbf{A}\cdot d\mathbf{s}'\right](-\hbar^2\nabla^2\psi^{(0)}). \tag{1.92}$$

Thus $\psi(\mathbf{x})$ satisfies the Schrödinger equation with $\mathbf{A} \neq 0$ so long as $\psi^{(0)}(\mathbf{x})$ satisfies the Schrödinger equation with the same V but with $\mathbf{A} = 0$.

As an application, let us first consider the gauge transformation

$$\mathbf{A} \to \mathbf{A}' = \mathbf{A} + \nabla\Lambda(\mathbf{x}). \tag{1.93}$$

Since for time-independent fields the noncovariant form of (1.65) reads

$$\mathbf{B} = \nabla \times \mathbf{A}, \qquad \mathbf{E} = -\nabla A_0, \tag{1.94}$$

neither $\mathbf{B}$ nor $\mathbf{E}$ is changed. Equation (1.90) tells us that we must change ψ (first emphasized by F. London):

$$\psi' = \psi \exp\left[\frac{ie}{\hbar c}\int^{\mathbf{s}(\mathbf{x})} \nabla'\Lambda(\mathbf{x}')\cdot d\mathbf{s}'\right] = \psi \exp\left[\frac{ie\Lambda(\mathbf{x})}{\hbar c}\right], \tag{1.95}$$

apart from an irrelevant space-time independent phase factor. We thus see that the form of the nonrelativistic wave function depends on the particular gauge we happen to use.

The Aharonov-Bohm effect and flux quantization. We shall now consider a much more startling example (first treated by Y. Aharonov and D. Bohm in 1959) in which a coherent beam of electrons is directed around two sides of a solenoid, as shown in Fig. 1–2. The beam is separated into two parts by a double slit but brought together again in an area denoted in the figure as the "interference region." In the interference region the wave function for the electron can be written as

$$\psi = \psi_1^{(0)} \exp\left[\frac{ie}{\hbar c}\int_{\text{Path 1}}^{\mathbf{s}(\mathbf{x})} \mathbf{A}(\mathbf{x}')\cdot d\mathbf{s}'\right] + \psi_2^{(0)} \exp\left[\frac{ie}{\hbar c}\int_{\text{Path 2}}^{\mathbf{s}(\mathbf{x})} \mathbf{A}(\mathbf{x}')\cdot d\mathbf{s}'\right]. \tag{1.96}$$

It follows that in the interference region there is an observable effect that depends on

$$\begin{Bmatrix}\cos\\ \sin\end{Bmatrix}\left[\frac{e}{\hbar c}\oint \mathbf{A}\cdot d\mathbf{s}\right] = \begin{Bmatrix}\cos\\ \sin\end{Bmatrix}\left[\frac{e}{\hbar c}\int \mathbf{B}\cdot\hat{n}\, dS\right]$$
$$= \begin{Bmatrix}\cos\\ \sin\end{Bmatrix}\frac{e\Phi}{\hbar c}, \tag{1.97}$$

where the closed line integral is along path 1 and then along path 2 in the opposite direction, the surface integral is over an area bounded by paths 1 and 2, and Φ stands for the total magnetic flux enclosed by paths 1 and 2. Note that the amount of interference can be controlled by varying the magnetic flux. This is most remarkable because in this idealized experimental arrangement the electrons never enter the $\mathbf{B} \neq 0$ region. Classically, the dynamical behavior of the electron was thought to depend only on the Lorentz force, which is zero when the electrons go through field-free regions; yet in quantum mechanics, observable effects *do* depend on the strength of the magnetic field in a region inaccessible to the electrons. R. G. Chambers and others actually performed experiments to detect an effect of this type and experimentally established the existence of the interference phenomena indicated by (1.97).

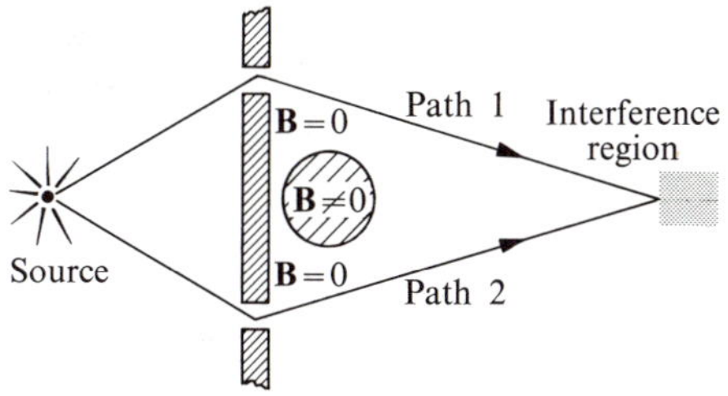

Fig. 1–2. An idealized experimental arrangement to illustrate the Aharonov-Bohm effect.

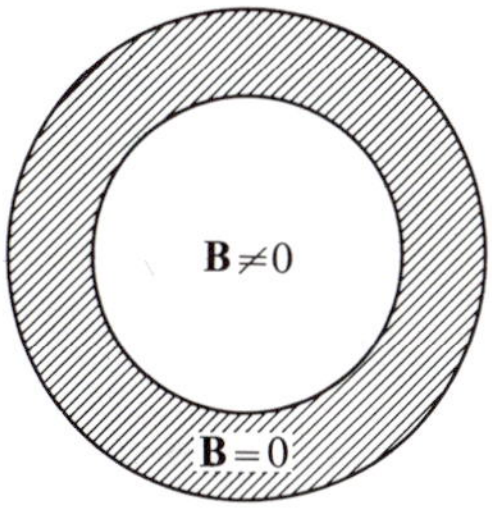

Fig. 1–3. Magnetic flux trapped by a superconductor ring.

Finally, let us consider the quantum-mechanical behavior of a correlated pair of electrons in a superconductor ring. In this example, the ring is assumed to enclose a magnetic flux as shown in Fig. 1–3. There is no magnetic field inside the superconductor material itself. (The absence of the magnetic field in a superconducting material is known as the *Meissner effect.*) This time we write the wave function for an electron pair as a quantum-mechanical "quasi-particle." We have

$$\psi = \psi^{(0)} \exp\left[\frac{2ie}{\hbar c}\int^{\mathbf{s}(\mathbf{x})} \mathbf{A}(\mathbf{x}')\cdot d\mathbf{s}'\right], \tag{1.98}$$

since the electric charge of the quasi-particle is $2e$.‡ Now we must have the same

‡Since the paired electrons are correlated not in coordinate ($\mathbf{x}$) space but in momentum ($\mathbf{p}$) space, some justification is needed to treat them as through they were a single particle of charge $2e$. For a detailed discussion of the pairing effect in a superconductor see, for example, Blatt (1964), Chapter 3.

wave function ψ, whether or not the path of integration encloses the flux; otherwise, the wave function would be multivalued. This imposes a severe restriction on the flux enclosed by the ring. A line integral enclosing the flux once must satisfy

$$(2e/\hbar c)\oint \mathbf{A}\cdot d\mathbf{s} = 2n\pi, \tag{1.99}$$

or

$$\Phi = \pi n\hbar c/e, \qquad n = 0, \pm 1, \pm 2, \ldots \tag{1.100}$$

Thus we arrive at the far-reaching conclusion that the magnetic flux trapped by the superconductor ring must exhibit a step behavior in units of

$$\frac{\pi\hbar c}{e} = \frac{ch}{2e} = 2.07 \times 10^{-7}\ \text{gauss-cm}^2, \tag{1.101}$$

as first discussed by F. London and L. Onsager (apart from the factor 2). The expected behavior was verified experimentally in 1961 by B. S. Deaver and W. M. Fairbank and by R. Doll and M. Näbauer. This experiment also illustrates in a direct and striking manner the existence of pairing effects in superconductors, which forms the basis for currently accepted theories of superconductivity.

We emphasize that throughout this chapter the electromagnetic field has been regarded as *classical*. In particular, the "quantization" of magnetic flux we have just discussed *has nothing to do* with the quantization of the electromagnetic field to be discussed in the next chapter.

PROBLEMS

1–1. (a) Show that the energy-momentum tensor density defined by

$$\mathscr{T}_{\mu\nu} = -\frac{\partial\phi}{\partial x_\nu}\frac{\partial\mathscr{L}}{\partial(\partial\phi/\partial x_\mu)} + \mathscr{L}\delta_{\mu\nu}$$

satisfies the continuity equation

$$\frac{\partial\mathscr{T}_{\mu\nu}}{\partial x_\mu} = 0,$$

when the Euler-Lagrange equation for ϕ is assumed.

(b) Show that each component of the four-vector

$$P_\mu(t) = -i\int \mathscr{T}_{4\mu}d^3x$$

is constant in time if ϕ vanishes sufficiently rapidly at infinity. (The integration is over three-dimensional space at a given instant t.)

(c) Obtain the Hamiltonian density $\mathscr{H} = -\mathscr{T}_{44}$ for the real scalar field.

1–2. Let $\phi(\mathbf{x}, t)$ be a solution to the free-field Klein-Gordon equation. Write

$$\phi(\mathbf{x}, t) = \psi(\mathbf{x}, t)e^{-imc^2t/\hbar}.$$

Under what condition will ψ satisfy the nonrelativistic Schrödinger equation? Interpret your condition physically when ϕ is given by a plane-wave solution.

1–3. Using the prescription

$$-i\hbar\frac{\partial}{\partial x_\mu}\to -i\hbar\frac{\partial}{\partial x_\mu}-\frac{eA_\mu}{c},$$

write the field equation for the charged scalar field interacting with A_μ. Show that if ϕ is a solution with $A_\mu = (0, 0, 0, iA_0)$, then ϕ^* is a solution with A_0 replaced by $-A_0$.

1–4. The Lagrangian density for a massive vector field interacting with a four-vector density j_μ is given by

$$\mathscr{L} = -\left[\frac{1}{4}\left(\frac{\partial\phi_\nu}{\partial x_\mu}-\frac{\partial\phi_\mu}{\partial x_\nu}\right)\left(\frac{\partial\phi_\nu}{\partial x_\mu}-\frac{\partial\phi_\mu}{\partial x_\nu}\right)+\tfrac{1}{2}\mu^2\phi_\mu\phi_\mu\right]+j_\mu\phi_\mu .$$

Obtain the field equation. Show that the continuity equation for j_μ is not guaranteed by the field equation, as it is in the Maxwell case. Show also that the subsidiary condition

$$\partial\phi_\mu/\partial x_\mu = 0$$

necessarily holds if the source is conserved (that is, if j_μ satisfies the continuity equation). *Note:* A massive vector field was first considered by A. Proca.

1–5. Consider a Klein-Gordon particle subject to the four-vector potential A_μ, assumed to be dependent on both **x** and t. Write a relativistic generalization of Eq. (1.91). State explicitly what kind of path in four-dimensional (Minkowski) space is considered.

CHAPTER 2

THE QUANTUM THEORY OF RADIATION

2–1. CLASSICAL RADIATION FIELD

Transversality condition. In order to study radiation phenomena in the quantum domain we shall first discuss in detail the properties of the vector potential satisfying

$$\nabla \cdot \mathbf{A} = 0 \tag{2.1}$$

within the framework of classical electrodynamics. Equation (2.1) is known as the *transversality* condition; it should not be confused with the Lorentz condition (1.81). The electric or magnetic fields derivable from vector potentials satisfying (2.1) are called *transverse fields* or *radiation fields.* Often the term "transverse field" or "radiation field" is used to refer to a vector potential itself satisfying (2.1).

The transversality condition (2.1) is of interest under a variety of circumstances. First, suppose $j_\mu = 0$. We can then consider a gauge transformation that eliminates the fourth component of A_μ and makes $\mathbf{A}$ obey (2.1). Consider

$$\begin{aligned} \mathbf{A} \rightarrow \mathbf{A}' &= \mathbf{A} + \nabla\Lambda, \\ A_0 \rightarrow A_0' &= A_0 + \frac{1}{c}\frac{\partial \Lambda}{\partial t}, \end{aligned} \tag{2.2}$$

such that

$$-\frac{1}{c}\frac{\partial \Lambda}{\partial t} = A_0. \tag{2.3}$$

Since the fourth component of A_μ has been eliminated in the new gauge, the Lorentz condition (1.81) reduces to the transversality condition (2.1).

Let us consider a situation in which $j_\mu \neq 0$, as in the case of mutually interacting electrons. We may first decompose $\mathbf{A}$ so that

$$\begin{aligned} \mathbf{A} = \mathbf{A}_\perp + \mathbf{A}_{||}, \qquad \nabla \cdot \mathbf{A}_\perp &= 0, \\ \nabla \times \mathbf{A}_{||} &= 0. \end{aligned} \tag{2.4}$$

This can always be done.‡ Here $\mathbf{A}_\perp$ and $\mathbf{A}_{||}$ are called respectively the *transverse* and the *longitudinal component* of $\mathbf{A}$. In 1930 E. Fermi was able to show that $\mathbf{A}_{||}$ and A_0 *together* give rise to the instantaneous static Coulomb interactions between

‡See, for example, Morse and Feshbach (1953), pp. 52–54.

the charged particles, whereas $\mathbf{A}_\perp$ accounts for the electromagnetic radiation of moving charged particles. The total Hamiltonian of the charged particles (treated nonrelativistically) and the electromagnetic fields generated by them can be written (apart from the magnetic moment interaction) as:

$$H = \sum_j \frac{1}{2m_j}\left[\mathbf{p}^{(j)} - e_j\frac{\mathbf{A}_\perp(\mathbf{x}^{(j)})}{c}\right]^2 + \sum_{i>j} \frac{e_i e_j}{4\pi\,|\mathbf{x}^{(i)} - \mathbf{x}^{(j)}|} + H_{\text{rad}}, \tag{2.5}$$

where H_{rad} (which we shall discuss later in detail) is the free field Hamiltonian of $\mathbf{A}_\perp$ only. Equation (2.5) is derived in Appendix A. Note that nowhere in (2.5) do A_0 and $\mathbf{A}_{||}$ appear explicitly.

Fermi's formalism based on (2.5) is called the *radiation* (or *Coulomb*) *gauge method.* Since the decomposition (2.5) is not relativistically covariant, nor is the transversality condition itself, the whole formalism appears noncovariant; each time we perform a Lorentz transformation, we must simultaneously make a gauge transformation to obtain a new set of $\mathbf{A}$ and A_0. Yet it is possible to develop manifestly covariant calculational techniques starting with the relativistic analog of the Hamiltonian (2.5), as will be shown in Chapter 4 when we discuss Møller (electron-electron) scattering. It is also possible to construct a formalism which preserves relativistic covariance at every stage.

In any case, it is worth studying a theory of transverse electromagnetic fields before we learn about more sophisticated formalisms. When the theory is quantized, it provides simple and physically transparent descriptions of a variety of processes in which real photons are emitted, absorbed, or scattered. The three basic equations we work with for the free-field case are

$$\mathbf{B} = \nabla \times \mathbf{A}, \qquad \mathbf{E} = -\frac{1}{c}\frac{\partial \mathbf{A}}{\partial t}, \tag{2.6}$$

$$\nabla^2 \mathbf{A} - \frac{1}{c^2}\frac{\partial^2 \mathbf{A}}{\partial t^2} = 0, \tag{2.7}$$

where $\mathbf{A}$ satisfies the transversality condition (2.1).

Fourier decomposition and radiation oscillators. At a given instant, say $t = 0$, we expand $\mathbf{A}$ in Fourier series. We assume the periodic boundary conditions for $\mathbf{A}$ enclosed in a box taken to be a cube of side $L = (V)^{1/3}$. Remembering the reality of $\mathbf{A}$, we have

$$\mathbf{A}(\mathbf{x}, t)\Big|_{t=0} = \frac{1}{\sqrt{V}} \sum_{\mathbf{k}} \sum_{\alpha=1,2} \left(c_{\mathbf{k},\alpha}(0)\mathbf{u}_{\mathbf{k},\alpha}(\mathbf{x}) + c^*_{\mathbf{k},\alpha}(0)\mathbf{u}^*_{\mathbf{k},\alpha}(\mathbf{x})\right), \tag{2.8}$$

where

$$\mathbf{u}_{\mathbf{k},\alpha}(\mathbf{x}) = \boldsymbol{\epsilon}^{(\alpha)} e^{i\mathbf{k}\cdot\mathbf{x}} \tag{2.9}$$

and $\boldsymbol{\epsilon}^{(\alpha)}$, called a (linear) *polarization vector*, is a real unit vector whose direction depends on the propagation direction $\mathbf{k}$ (even though we will not write it in this text as $\boldsymbol{\epsilon}^{(\alpha)}(\mathbf{k})$ as is done in many books). Given $\mathbf{k}$, we choose $\boldsymbol{\epsilon}^{(1)}$ and $\boldsymbol{\epsilon}^{(2)}$ in such a way that $(\boldsymbol{\epsilon}^{(1)}, \boldsymbol{\epsilon}^{(2)}, \mathbf{k}/|\mathbf{k}|)$ form a right-handed set of mutually orthogonal unit vectors. Note that $\boldsymbol{\epsilon}^{(1)}$ and $\boldsymbol{\epsilon}^{(2)}$ are, in general, not along the x- and the y- axes since

$\mathbf{k}$ is, in general, not along the z-axis. Since $\boldsymbol{\epsilon}^{(\alpha)}$ is perpendicular to $\mathbf{k}$, the transversality condition is guaranteed. The Fourier component $\mathbf{u}_{\mathbf{k},\alpha}$ satisfies

$$\frac{1}{V}\int d^3x\, \mathbf{u}_{\mathbf{k},\alpha}\cdot\mathbf{u}^*_{\mathbf{k}',\alpha'} = \delta_{\mathbf{k}\mathbf{k}'}\delta_{\alpha\alpha'},$$
$$\frac{1}{V}\int d^3x \begin{Bmatrix}\mathbf{u}_{\mathbf{k},\alpha}\cdot\mathbf{u}_{\mathbf{k}',\alpha'}\\ \mathbf{u}^*_{\mathbf{k},\alpha}\cdot\mathbf{u}^*_{\mathbf{k}',\alpha'}\end{Bmatrix} = \delta_{\mathbf{k},-\mathbf{k}'}\delta_{\alpha\alpha'}, \tag{2.10}$$

where

$$k_x, k_y, k_z = 2n\pi/L, \qquad n = \pm 1, \pm 2, \ldots \tag{2.11}$$

because of the periodic boundary conditions.

To obtain $\mathbf{A}(\mathbf{x}, t)$ for $t \neq 0$, we simply replace $c_{\mathbf{k},\alpha}(0)$ and $c^*_{\mathbf{k},\alpha}(0)$ by

$$c_{\mathbf{k},\alpha}(t) = c_{\mathbf{k},\alpha}(0)e^{-i\omega t}, \qquad c^*_{\mathbf{k},\alpha}(t) = c^*_{\mathbf{k},\alpha}(0)e^{i\omega t}, \tag{2.12}$$

where

$$\omega = |\mathbf{k}|c. \tag{2.13}$$

With this replacement both the wave equation (2.7) and the reality condition on $\mathbf{A}$ are satisfied. So

$$\mathbf{A}(\mathbf{x}, t) = \frac{1}{\sqrt{V}}\sum_{\mathbf{k}}\sum_{\alpha}(c_{\mathbf{k},\alpha}(t)\boldsymbol{\epsilon}^{(\alpha)}e^{i\mathbf{k}\cdot\mathbf{x}} + c^*_{\mathbf{k},\alpha}(t)\boldsymbol{\epsilon}^{(\alpha)}e^{-i\mathbf{k}\cdot\mathbf{x}})$$
$$= \frac{1}{\sqrt{V}}\sum_{\mathbf{k}}\sum_{\alpha}(c_{\mathbf{k},\alpha}(0)\boldsymbol{\epsilon}^{(\alpha)}e^{ik\cdot x} + c^*_{\mathbf{k},\alpha}(0)\boldsymbol{\epsilon}^{(\alpha)}e^{-ik\cdot x}) \tag{2.14}$$

with

$$k\cdot x = \mathbf{k}\cdot\mathbf{x} - \omega t = \mathbf{k}\cdot\mathbf{x} - |\mathbf{k}|ct. \tag{2.15}$$

The Hamiltonian of the field is

$$H = \tfrac{1}{2}\int (|\mathbf{B}|^2 + |\mathbf{E}|^2)d^3x$$
$$= \tfrac{1}{2}\int [|\nabla\times\mathbf{A}|^2 + |(1/c)(\partial\mathbf{A}/\partial t)|^2]d^3x. \tag{2.16}$$

A typical term we must evaluate for the $|\mathbf{B}|^2$ integration is

$$\int(\nabla\times\mathbf{u}_{\mathbf{k},\alpha})\cdot(\nabla\times\mathbf{u}^*_{\mathbf{k}',\alpha'})d^3x = \int\nabla\cdot[\mathbf{u}_{\mathbf{k},\alpha}\times(\nabla\times\mathbf{u}^*_{\mathbf{k}',\alpha'})]d^3x$$
$$+ \int\mathbf{u}_{\mathbf{k},\alpha}\cdot[\nabla\times(\nabla\times\mathbf{u}^*_{\mathbf{k}',\alpha'})]d^3x$$
$$= -\int\mathbf{u}_{\mathbf{k},\alpha}\cdot\nabla^2\mathbf{u}^*_{\mathbf{k}',\alpha'}d^3x = \left(\frac{\omega}{c}\right)^2 V\delta_{\mathbf{k}\mathbf{k}'}\delta_{\alpha\alpha'}, \tag{2.17}$$

where we have used the periodic boundary conditions and the identity $\nabla\times(\nabla\times\ \ ) = \nabla(\nabla\cdot\ \ ) - \nabla^2$. Similarly, for the $|\mathbf{E}|^2$ integration it is useful to evaluate first

$$\int\left[\frac{1}{c}\frac{\partial}{\partial t}(c_{\mathbf{k},\alpha}\mathbf{u}_{\mathbf{k},\alpha})\right]\cdot\left[\frac{1}{c}\frac{\partial}{\partial t}(c^*_{\mathbf{k}',\alpha'}\mathbf{u}^*_{\mathbf{k}',\alpha'})\right]d^3x = \left(\frac{\omega}{c}\right)^2 V\delta_{\mathbf{k}\mathbf{k}'}\delta_{\alpha\alpha'}c_{\mathbf{k},\alpha}c^*_{\mathbf{k}',\alpha'} \tag{2.18}$$

Using these relations, we obtain

$$H = \sum_{\mathbf{k}} \sum_{\alpha} 2(\omega/c)^2 c^*_{\mathbf{k},\alpha} c_{\mathbf{k},\alpha}, \tag{2.19}$$

where $c_{\mathbf{k},\alpha}$ is a time-dependent Fourier coefficient satisfying

$$\ddot{c}_{\mathbf{k},\alpha} = -\omega^2 c_{\mathbf{k},\alpha} \tag{2.20}$$

(cf. Eq. 2.12). This reminds us of an expression for the energy in a collection of independent and uncoupled harmonic oscillators. To make the analogy more vivid, we define

$$Q_{\mathbf{k},\alpha} = \frac{1}{c}(c_{\mathbf{k},\alpha} + c^*_{\mathbf{k},\alpha}), \qquad P_{\mathbf{k},\alpha} = -\frac{i\omega}{c}(c_{\mathbf{k},\alpha} - c^*_{\mathbf{k},\alpha}). \tag{2.21}$$

Then

$$\begin{aligned} H &= \sum_{\mathbf{k}} \sum_{\alpha} 2\left(\frac{\omega}{c}\right)^2 \left[\frac{c(\omega Q_{\mathbf{k},\alpha} - iP_{\mathbf{k},\alpha})}{2\omega}\right]\left[\frac{c(\omega Q_{\mathbf{k},\alpha} + iP_{\mathbf{k},\alpha})}{2\omega}\right] \\ &= \sum_{\mathbf{k}} \sum_{\alpha} \tfrac{1}{2}(P^2_{\mathbf{k},\alpha} + \omega^2 Q^2_{\mathbf{k},\alpha}). \end{aligned} \tag{2.22}$$

The $P_{\mathbf{k},\alpha}$ and $Q_{\mathbf{k},\alpha}$ are now seen to be canonical variables:

$$\frac{\partial H}{\partial Q_{\mathbf{k},\alpha}} = -\dot{P}_{\mathbf{k},\alpha}, \qquad \frac{\partial H}{\partial P_{\mathbf{k},\alpha}} = +\dot{Q}_{\mathbf{k},\alpha}. \tag{2.23}$$

Thus the radiation field can be regarded as a collection of independent harmonic oscillators each of which is characterized by $\mathbf{k}$, α and whose dynamical variables are orthogonal linear combinations of the Fourier coefficients.

2–2. CREATION, ANNIHILATION, AND NUMBER OPERATORS

Quantization of radiation oscillators. At the end of the nineteenth century it was recognized that the space-time development of the radiation field satisfying the wave equation (2.7) resembles the dynamical behavior of a collection of harmonic oscillators. By assigning an average energy kT to each radiation oscillator, Lord Rayleigh and J. H. Jeans wrote an expressioon for the energy distribution of the radiation field as a function of ω in an ideal situation where the radiation field is enclosed by perfectly absorbing walls. The expression they obtained was in satisfactory agreement with observation for low values of ω at sufficiently high temperatures, but in marked disagreement for high values of ω. This difficulty led M. Planck to take one of the most revolutionary steps ever taken in the history of science. He proposed that:

> The energy of each radiation oscillator is not an arbitrary quantity but must be an integral multiple of $\hbar\omega$, where $\hbar$ is a *new* fundamental constant in nature.

This he did in 1901. Four years later, in order to explain the photoelectric effect, A. Einstein proposed that an electromagnetic wave of wavelength $\lambda = 2\pi c/\omega$ be regarded as a collection of massless particles each of which has energy $\hbar\omega$.

We can now do better than did Planck and Einstein, but only because we know nonrelativistic quantum mechanics. Indeed, no sooner was nonrelativistic quantum mechanics fully developed than P. A. M. Dirac proposed that the canonical dynamical variables of a radiation oscillator be treated as noncommutable operators, just as x and p of a one-dimensional harmonic oscillator are treated in nonrelativistic quantum mechanics.

We postulate that P and Q of the radiation oscillators are no longer mere numbers but are *operators* satisfying

$$[Q_{\mathbf{k},\alpha}, P_{\mathbf{k}',\alpha'}] = i\hbar\,\delta_{\mathbf{kk}'}\delta_{\alpha\alpha'}, \tag{2.24a}$$

$$\begin{aligned}[Q_{\mathbf{k},\alpha}, Q_{\mathbf{k}',\alpha'}] &= 0,\\ [P_{\mathbf{k},\alpha}, P_{\mathbf{k}',\alpha'}] &= 0.\end{aligned} \tag{2.24b}$$

We next consider linear combinations of $P_{\mathbf{k},\alpha}$ and $Q_{\mathbf{k},\alpha}$ given by

$$\begin{aligned} a_{\mathbf{k},\alpha} &= (1/\sqrt{2\hbar\omega})(\omega Q_{\mathbf{k},\alpha} + iP_{\mathbf{k},\alpha}),\\ a^{\dagger}_{\mathbf{k},\alpha} &= (1/\sqrt{2\hbar\omega})(\omega Q_{\mathbf{k},\alpha} - iP_{\mathbf{k},\alpha}). \end{aligned} \tag{2.25}$$

Thus $a_{\mathbf{k},\alpha}$ and $a^{\dagger}_{\mathbf{k},\alpha}$ are seen to be the operator analogs of the Fourier coefficients $c_{\mathbf{k},\alpha}$ and $c^{\dagger}_{\mathbf{k},\alpha}$ when we insert a factor to make $a_{\mathbf{k},\alpha}$ and $a^{\dagger}_{\mathbf{k},\alpha}$ dimensionless:

$$c_{\mathbf{k},\alpha} \to c\sqrt{\hbar/2\omega}\,a_{\mathbf{k},\alpha}.$$

They satisfy the communication relations

$$\begin{aligned}[a_{\mathbf{k},\alpha}, a^{\dagger}_{\mathbf{k}',\alpha'}] &= -\frac{i}{2\hbar}[Q_{\mathbf{k},\alpha}, P_{\mathbf{k}',\alpha'}] + \frac{i}{2\hbar}[P_{\mathbf{k},\alpha}, Q_{\mathbf{k}',\alpha'}]\\ &= \delta_{\mathbf{kk}'}\delta_{\alpha\alpha'},\end{aligned} \tag{2.26a}$$

$$[a_{\mathbf{k},\alpha}, a_{\mathbf{k}',\alpha'}] = [a^{\dagger}_{\mathbf{k},\alpha}, a^{\dagger}_{\mathbf{k}',\alpha'}] = 0. \tag{2.26b}$$

These communication relations are to be evaluated for operators taken at equal times; for example, $[a_{\mathbf{k},\alpha}, a^{\dagger}_{\mathbf{k}',\alpha'}]$ actually stands for $[a_{\mathbf{k},\alpha}(t), a^{\dagger}_{\mathbf{k}',\alpha'}(t)]$.

Before we discuss the physical interpretations of $a_{\mathbf{k},\alpha}$ and $a^{\dagger}_{\mathbf{k},\alpha}$, it is instructive to study the properties of the operator defined by

$$N_{\mathbf{k},\alpha} = a^{\dagger}_{\mathbf{k},\alpha}a_{\mathbf{k},\alpha}. \tag{2.27}$$

We have

$$\begin{aligned}[a_{\mathbf{k},\alpha}, N_{\mathbf{k}',\alpha'}] &= a_{\mathbf{k},\alpha}a^{\dagger}_{\mathbf{k}',\alpha'}a_{\mathbf{k}',\alpha'} - a^{\dagger}_{\mathbf{k}',\alpha'}a_{\mathbf{k}',\alpha'}a_{\mathbf{k},\alpha}\\ &= [a_{\mathbf{k},\alpha}, a^{\dagger}_{\mathbf{k}',\alpha'}]a_{\mathbf{k}',\alpha'} - a^{\dagger}_{\mathbf{k}',\alpha'}[a_{\mathbf{k}',\alpha'}, a_{\mathbf{k},\alpha}]\\ &= \delta_{\mathbf{kk}'}\delta_{\alpha\alpha'}a_{\mathbf{k},\alpha}.\end{aligned} \tag{2.28}$$

Similarly,

$$[a^{\dagger}_{\mathbf{k},\alpha}, N_{\mathbf{k}',\alpha}] = -\delta_{\mathbf{kk}'}\delta_{\alpha\alpha'}a^{\dagger}_{\mathbf{k},\alpha}. \tag{2.29}$$

Unlike a and $a^{\dagger}$, the operator N is Hermitian. (In this paragraph and the next we shall suppress the indices $\mathbf{k}$, α. All the relations are valid for given $\mathbf{k}$, α.) The

Hermiticity of N encourages us to consider a normalized eigenvector of the operator N denoted by $|n\rangle$ such that

$$N|n\rangle = n|n\rangle, \tag{2.30}$$

where n is the eigenvalue of N. Because N is Hermitian, n must be real. Now,

$$\begin{aligned} Na^\dagger|n\rangle &= (a^\dagger N + a^\dagger)|n\rangle \\ &= (n+1)a^\dagger|n\rangle, \end{aligned} \tag{2.31}$$

where we have used (2.29). This can be viewed as a new eigenvalue equation in which the eigenvector $a^\dagger|n\rangle$ is shown to have eigenvalue $n+1$. Similarly,

$$Na|n\rangle = (n-1)a|n\rangle. \tag{2.32}$$

The roles of $a^\dagger$ and a are now clear; $a^\dagger(a)$ acting on $|n\rangle$ gives a new eigenvector with eigenvalue increased (decreased) by one. So

$$\begin{aligned} a^\dagger|n\rangle &= c_+|n+1\rangle, \\ a|n\rangle &= c_-|n-1\rangle, \end{aligned} \tag{2.33}$$

where c_+ and c_- are constants. To determine $c_\pm$ we evaluate

$$\begin{aligned} |c_+|^2 &= |c_+|^2\langle n+1|n+1\rangle = \langle a^\dagger n|a^\dagger n\rangle \\ &= \langle n|aa^\dagger|n\rangle = \langle n|N + [a, a^\dagger]|n\rangle \\ &= n+1; \end{aligned} \tag{2.34}$$

$$|c_-|^2 = \langle an|an\rangle = \langle n|a^\dagger a|n\rangle = n. \tag{2.35}$$

The phases of $c_\pm$ are indeterminate; they may be chosen to be zero at $t = 0$ by convention. We have, at $t = 0$,

$$\begin{aligned} a^\dagger|n\rangle &= \sqrt{n+1}\,|n+1\rangle, \\ a|n\rangle &= \sqrt{n}\,|n-1\rangle. \end{aligned} \tag{2.36}$$

Meanwhile

$$n = \langle n|N|n\rangle = \langle n|a^\dagger a|n\rangle \geq 0, \tag{2.37}$$

since the norm of $a|n\rangle$ must be positive definite. This immediately tells us that n cannot be a noninteger; otherwise, the eigenvalue of $|n\rangle$ could be made to decrease indefinitely as we successively apply a, and n would eventually take negative values, in contradiction to (2.37). On the other hand, if n is a positive integer, successive applications of the operator a proceed as

$$\begin{aligned} a|n\rangle &= \sqrt{n}\,|n-1\rangle \cdots, \qquad a|2\rangle = \sqrt{2}\,|1\rangle, \\ a|1\rangle &= |0\rangle, \qquad a|0\rangle = 0. \end{aligned} \tag{2.38}$$

Note that we obtain a "null vector" (to be distinguished from $|0\rangle$) when a is applied to $|0\rangle$. By applying a to such a null vector, we again obtain a null vector. Hence $n = 0$ is the lowest possible eigenvalue of the operator N.

Explicit matrix representations of a, $a^\dagger$, and N consistent with the commutation relations (2.26), (2.28), and (2.29) can be written as follows:

$$a = \begin{pmatrix} 0 & 1 & 0 & 0 & 0 & \cdots \\ 0 & 0 & \sqrt{2} & 0 & 0 & \cdots \\ 0 & 0 & 0 & \sqrt{3} & 0 & \cdots \\ 0 & 0 & 0 & 0 & \sqrt{4} & \cdots \\ \vdots \end{pmatrix}, \qquad a^\dagger = \begin{pmatrix} 0 & 0 & 0 & 0 & 0 & \cdots \\ 1 & 0 & 0 & 0 & 0 & \cdots \\ 0 & \sqrt{2} & 0 & 0 & 0 & \cdots \\ 0 & 0 & \sqrt{3} & 0 & 0 & \cdots \\ \vdots \end{pmatrix},$$

$$N = \begin{pmatrix} 0 & 0 & 0 & 0 & \cdots \\ 0 & 1 & 0 & 0 & \cdots \\ 0 & 0 & 2 & 0 & \cdots \\ 0 & 0 & 0 & 3 & \cdots \\ \vdots \end{pmatrix}. \tag{2.39}$$

They are assumed to act on a column vector represented by

$$|n\rangle = \begin{pmatrix} 0 \\ 0 \\ \vdots \\ 0 \\ 1 \\ 0 \\ \vdots \end{pmatrix}, \tag{2.40}$$

where only the $(n + 1)$-entry is different from zero.

Photon states. The algebra developed above can be applied to a physical situation in which the number of photons with given momentum and polarization is increased or decreased. The wave vector $\mathbf{k}$ will later be identified with the photon momentum divided by $\hbar$, and α will be shown to represent the polarization state of the photon. We interpret an eigenvector of $N_{\mathbf{k},\alpha}$ as the state vector for a state with a definite number of photons in state $(\mathbf{k}, \alpha)$. To represent a situation in which there are many types of photons with different sets of $(\mathbf{k}, \alpha)$, we consider the direct product of eigenvectors as follows:

$$|n_{\mathbf{k}_1,\alpha_1}, n_{\mathbf{k}_2,\alpha_2}, \ldots, n_{\mathbf{k}_i,\alpha_i}, \ldots\rangle = |n_{\mathbf{k}_1,\alpha_1}\rangle |n_{\mathbf{k}_2,\alpha_2}\rangle \cdots |n_{\mathbf{k}_i,\alpha_i}\rangle \cdots \tag{2.41}$$

This state vector corresponds to the physical situation in which there are $n_{\mathbf{k}_1,\alpha_1}$ photons present in state $(\mathbf{k}_1, \alpha_1)$, $n_{\mathbf{k}_2,\alpha_2}$ photons in state $(\mathbf{k}_2, \alpha_2)$, etc. The number $n_{\mathbf{k},\alpha}$ is called the *occupation number* for state $(\mathbf{k}, \alpha)$.

As an example, the state represented by

$$|0\rangle = |0_{\mathbf{k}_1, \alpha_1}\rangle |0_{\mathbf{k}_2, \alpha_2}\rangle \cdots |0_{\mathbf{k}_i, \alpha_i}\rangle \cdots \tag{2.42}$$

has the property that if $a_{\mathbf{k}, \alpha}$ is applied to it, then we obtain a null vector for any $(\mathbf{k}, \alpha)$. Hence the eigenvalue of $N_{\mathbf{k}, \alpha} = a^{\dagger}_{\mathbf{k}, \alpha} a_{\mathbf{k}, \alpha}$ is zero for any $(\mathbf{k}, \alpha)$. Therefore the state corresponding to (2.42) is called the *vacuum state.* A single-photon state with definite $(\mathbf{k}, \alpha)$ is represented by

$$a^{\dagger}_{\mathbf{k}, \alpha} |0\rangle, \tag{2.43}$$

since the eigenvalue of $N_{\mathbf{k}, \alpha}$ is one (cf. Eq. 2.31). A two-photon state is represented by the normalized eigenvector

$$(1/\sqrt{2}) a^{\dagger}_{\mathbf{k}, \alpha} a^{\dagger}_{\mathbf{k}, \alpha} |0\rangle \tag{2.44}$$

when the two photons are in the *same* state, and

$$a^{\dagger}_{\mathbf{k}, \alpha} a^{\dagger}_{\mathbf{k}', \alpha'} |0\rangle \tag{2.45}$$

when the two photons are in *different* states. More generally,

$$|n_{\mathbf{k}_1, \alpha_1}, n_{\mathbf{k}_2, \alpha_2}, \ldots\rangle = \prod_{\mathbf{k}_i, \alpha_i} \frac{(a^{\dagger}_{\mathbf{k}_i, \alpha_i})^{n_{\mathbf{k}_i, \alpha_i}}}{\sqrt{n_{\mathbf{k}_i, \alpha_i}!}} |0\rangle. \tag{2.46}$$

Note that when $a^{\dagger}_{\mathbf{k}_i, \alpha_i}$ is applied to the most general state vector, we obtain

$$\begin{aligned} a^{\dagger}_{\mathbf{k}_i, \alpha_i} &|n_{\mathbf{k}_1, \alpha_1}, n_{\mathbf{k}_2, \alpha_2}, \ldots, n_{\mathbf{k}_i, \alpha_i}, \ldots\rangle \\ &= \sqrt{n_{\mathbf{k}_i, \alpha_i} + 1}\, |n_{\mathbf{k}_1, \alpha_1}, n_{\mathbf{k}_2, \alpha_2}, \ldots, n_{\mathbf{k}_i, \alpha_i} + 1, \ldots\rangle. \end{aligned} \tag{2.47}$$

Thus $a^{\dagger}_{\mathbf{k}_i, \alpha_i}$ has the property of creating an additional photon in state $(\mathbf{k}_i, \alpha_i)$, leaving the occupation numbers of states other than $(\mathbf{k}_i, \alpha_i)$ unchanged. For this reason $a^{\dagger}_{\mathbf{k}, \alpha}$ is called the creation operator for a photon in state $\mathbf{k}, \alpha$. Similarly, $a_{\mathbf{k}, \alpha}$ can be interpreted as the annihilation or destruction operator for a photon in state $\mathbf{k}, \alpha$. In contrast, $N_{\mathbf{k}, \alpha}$, being diagonal, does not change the occupation number of photons; it simply gives as its eigenvalue the number of photons in state $\mathbf{k}, \alpha$. We might say that the three operators $a^{\dagger}_{\mathbf{k}, \alpha}$, $a_{\mathbf{k}, \alpha}$, and $N_{\mathbf{k}, \alpha}$ correspond respectively to the Creator (Brahma), the Destroyer (Siva), and the Preserver (Vishnu) in Hindu mythology.

Our formalism is capable of describing a physical situation in which the number of photons in a given state is unrestricted. Moreover, any many-photon system we can construct is necessarily symmetric under interchange of any pair of labels. For instance, the two-photon state (2.45) is evidently symmetric under interchange $(\mathbf{k}, \alpha) \longleftrightarrow (\mathbf{k}', \alpha')$ since

$$a^{\dagger}_{\mathbf{k}, \alpha} a^{\dagger}_{\mathbf{k}', \alpha'} |0\rangle = a^{\dagger}_{\mathbf{k}', \alpha'} a^{\dagger}_{\mathbf{k}, \alpha} |0\rangle \tag{2.48}$$

because of the commutation rule. Thus the state vectors we obtain by applying creation operators to $|0\rangle$ are automatically consistent with *Bose-Einstein statistics.* With essentially no modifications our formalism can be applied to physical states made up of indetical particles other than photons as long as they obey Bose-Einstein statistics.

Fermion operators. There exist, in nature, particles that do not obey Bose-Einstein statistics but rather obey *Fermi-Dirac statistics*—electron, muons, protons, etc. For such particles the formalism we have developed is obviously inadequate. We must somehow incorporate the Pauli exclusion principle. This can be done. In 1928 P. Jordan and E. P. Wigner proposed a formalism in which we again consider the operators $b_r^\dagger$ and b_r, but they now satisfy the the "anticommutation relations"

$$\{b_r, b_{r'}^\dagger\} = \delta_{rr'}, \qquad \{b_r, b_{r'}\} = \{b_r^\dagger, b_{r'}^\dagger\} = 0, \tag{2.49}$$

where the term in braces is defined by

$$\{A, B\} = AB + BA. \tag{2.50}$$

The operators $b_r^\dagger$ and b_r are again interpreted as the creation and annihilation operators, and the index r provides a collective description of the momentum state, the spin state, and according to the Dirac hole theory (to be discussed in Sections 3–9 and 3–10), the sign of the energy as well. A single-particle state can be constructed just as before:

$$|1_r\rangle = b_r^\dagger |0\rangle, \tag{2.51}$$

but, since

$$b_r^\dagger b_r^\dagger |0\rangle = \tfrac{1}{2}\{b_r^\dagger, b_r^\dagger\} |0\rangle = 0 \tag{2.52}$$

according to (2.49), we cannot put two particles in the same state. This is just what is needed if the electrons are to satisfy the Pauli exclusion principle. However, if $r \neq r'$, then we can construct a two-particle state

$$b_r^\dagger b_{r'}^\dagger |0\rangle = -b_{r'}^\dagger b_r^\dagger |0\rangle. \tag{2.53}$$

Note that it is necessarily antisymmetric under interchange $r \longleftrightarrow r'$, in conformity with Fermi-Dirac statistics. We define a Hermitian operator N_r by

$$N_r = b_r^\dagger b_r \tag{2.54}$$

just as before. When $b_r^\dagger$ and b_r are interpreted as the creation and annihilation operators, it is natural to regard N_r as the occupation number operator:

$$\begin{aligned} N_r |0\rangle &= b_r^\dagger b_r |0\rangle = 0; \\ N_r b_r^\dagger |0\rangle &= b_r^\dagger (1 - b_r^\dagger b_r) |0\rangle = b_r^\dagger |0\rangle. \end{aligned} \tag{2.55}$$

In general, N_r has the property

$$N_r^2 = b_r^\dagger b_r b_r^\dagger b_r = b_r^\dagger(-b_r^\dagger b_r + 1) b_r = N_r. \tag{2.56}$$

Hence

$$N_r(N_r - 1) = 0, \tag{2.57}$$

which means that the eigenvalue of N_r is either zero or one. Physically speaking, state r is either unoccupied or occupied by just one electron. Explicit matrix representations of b_r, $b_r^\dagger$, and N_r consistent with the anticommunication relations are not difficult to find:

$$b_r = \begin{pmatrix} 0 & 1 \\ 0 & 0 \end{pmatrix}, \qquad b_r^\dagger = \begin{pmatrix} 0 & 0 \\ 1 & 0 \end{pmatrix}, \qquad N_r = \begin{pmatrix} 0 & 0 \\ 0 & 1 \end{pmatrix}. \tag{2.58}$$

They act on the following column vectors:

$$|0\rangle = \begin{pmatrix}1\\0\end{pmatrix} \quad \text{and} \quad |1\rangle = \begin{pmatrix}0\\1\end{pmatrix}. \tag{2.59}$$

Although the algebra of b and $b^\dagger$ is similar to that of a and $a^\dagger$, it is important to note that b and $b^\dagger$ cannot be written as linear combinations of P and Q, satisfying the commutation relations (2.24). This point can be shown to be related to the fact that there is no classically measurable field that corresponds to the quantized fermion field. In this chapter we shall not say anything more about the fermion operators, but we shall come back to this subject in Section 3–10.

2–3. QUANTIZED RADIATION FIELD

Photons as quantum-mechanical excitations of the radiation field. The Fourier coefficients in the expansion of the classical radiation field must be replaced by the corresponding annihilation and creation operators if the canonical variables of the radiation oscillator are to be interpreted as noncommutative quantum-mechanical operators. With the substitutions

$$c_{\mathbf{k},\alpha}(t) \to c\sqrt{\hbar/2\omega}\, a_{\mathbf{k},\alpha}(t) \qquad \text{and} \qquad c^*_{\mathbf{k},\alpha}(t) \to c\sqrt{\hbar/2\omega}\, a^\dagger_{\mathbf{k},\alpha}(t),$$

we have‡

$$\mathbf{A}(\mathbf{x},t) = (1/\sqrt{V})\sum_{\mathbf{k}}\sum_{\alpha} c\sqrt{\hbar/2\omega}\,[a_{\mathbf{k},\alpha}(t)\boldsymbol{\epsilon}^{(\alpha)}e^{i\mathbf{k}\cdot\mathbf{x}} + a^\dagger_{\mathbf{k},\alpha}(t)\boldsymbol{\epsilon}^{(\alpha)}e^{-i\mathbf{k}\cdot\mathbf{x}}]. \tag{2.60}$$

Although this expansion is similar in appearance to (2.14), the meaning of $\mathbf{A}$ is very different. The $\mathbf{A}$ of (2.14) is a classical function with three components defined at each space-time point. In contrast, the $\mathbf{A}$ of (2.60) is an operator that acts on state vectors in occupation number space along the lines discussed in the previous section. Note, however, that it is parametrized by $\mathbf{x}$ and t just as a classical field. Such an operator is called a *field operator* or a *quantized field.*

The *Hamiltonian operator* of the quantized radiation field can be taken as

$$H = \tfrac{1}{2}\int (\mathbf{B}\cdot\mathbf{B} + \mathbf{E}\cdot\mathbf{E})d^3x. \tag{2.61}$$

It can be evaluated using equations such as (2.16) and (2.17) just as before. This time, however, we must be careful with the order of $a_{\mathbf{k},\alpha}$ and $a^\dagger_{\mathbf{k},\alpha}$ since they are no longer mere numbers. We obtain

$$\begin{aligned} H &= \tfrac{1}{2}\sum_{\mathbf{k}}\sum_{\alpha} \hbar\omega(a^\dagger_{\mathbf{k},\alpha}a_{\mathbf{k},\alpha} + a_{\mathbf{k},\alpha}a^\dagger_{\mathbf{k},\alpha}) \\ &= \sum_{\mathbf{k}}\sum_{\alpha}(N_{\mathbf{k},\alpha} + \tfrac{1}{2})\hbar\omega, \end{aligned} \tag{2.62}$$

‡In Gaussian unrationalized units, $c\sqrt{\hbar/2\omega}$ should be replaced by $c\sqrt{2\pi\hbar/\omega}$ because the expression for the total energy has $1/8\pi$ in place of $1/2$ (cf. Eq. 2.16).

with $\omega = |\mathbf{k}|c$. Since the absolute energy scale is arbitrary, we may use the energy scale in which the energy of the vacuum state is zero,

$$H|0\rangle = 0. \tag{2.63}$$

This amounts to subtracting $\sum \sum \hbar\omega/2$ from (2.62), so that (2.62) is replaced by

$$H = \sum_{\mathbf{k}} \sum_{\alpha} \hbar\omega N_{\mathbf{k},\alpha}. \tag{2.64}$$

The Hamiltonian operator H acting on a many-photon state gives

$$H|n_{\mathbf{k}_1,\alpha_1}, n_{\mathbf{k}_2,\alpha_2}, \ldots\rangle = \sum_i n_{\mathbf{k}_i,\alpha_i}\hbar\omega_i|n_{\mathbf{k}_1,\alpha_1}, n_{\mathbf{k}_2,\alpha_2}, \ldots\rangle. \tag{2.65}$$

The total momentum of the radiation field is given in classical electrodynamics as the space integral of the Poynting vector $(\mathbf{E} \times \mathbf{B})/c$.‡ Using an operator expression identical in appearance to the total momentum of the field in classical electrodynamics, we obtain the momentum operator as follows:

$$\begin{aligned}\mathbf{P} = (1/c)\int(\mathbf{E}\times\mathbf{B})d^3x &= \sum_{\mathbf{k}}\sum_{\alpha}\tfrac{1}{2}\hbar\mathbf{k}(a^\dagger_{\mathbf{k},\alpha}a_{\mathbf{k},\alpha} + a_{\mathbf{k},\alpha}a^\dagger_{\mathbf{k},\alpha})\\ &= \sum_{\mathbf{k}}\sum_{\alpha}\hbar\mathbf{k}(N_{\mathbf{k},\alpha} + \tfrac{1}{2}),\end{aligned} \tag{2.66}$$

where we have used relations such as

$$\begin{aligned}&\frac{c^2\hbar}{2\sqrt{\omega\omega'}}\left(\frac{1}{c}\right)a_{\mathbf{k},\alpha}\left(\frac{\omega'}{c}\right)a^\dagger_{\mathbf{k}',\alpha'}\boldsymbol{\epsilon}^{(\alpha)} \times (\mathbf{k}' \times \boldsymbol{\epsilon}^{(\alpha')})\frac{1}{V}\int e^{i(\mathbf{k}-\mathbf{k}')\cdot\mathbf{x}}d^3x\\ &\qquad = \frac{\hbar}{2}a_{\mathbf{k},\alpha}a^\dagger_{\mathbf{k},\alpha}\mathbf{k}\delta_{\mathbf{k}\mathbf{k}'}\delta_{\alpha\alpha'}.\end{aligned} \tag{2.67}$$

(*Note*: $\boldsymbol{\epsilon}^{(1)} \times (\mathbf{k} \times \boldsymbol{\epsilon}^{(2)}) = 0$, $\boldsymbol{\epsilon}^{(1)} \times (\mathbf{k} \times \boldsymbol{\epsilon}^{(1)}) = \mathbf{k}$, etc.) The $\frac{1}{2}$ that appears in (2.66) can be dropped because when we sum over the allowed $\mathbf{k}$, $\hbar\mathbf{k}$ cancels with $-\hbar\mathbf{k}$. Hence

$$\mathbf{P} = \sum_{\mathbf{k}}\sum_{\alpha}\hbar\mathbf{k}N_{\mathbf{k},\alpha}. \tag{2.68}$$

Let us consider the effect of H and $\mathbf{P}$ on a single-photon state:

$$Ha^\dagger_{\mathbf{k},\alpha}|0\rangle = \hbar\omega a^\dagger_{\mathbf{k},\alpha}|0\rangle, \qquad \mathbf{P}a^\dagger_{\mathbf{k},\alpha}|0\rangle = \hbar\mathbf{k}a^\dagger_{\mathbf{k},\alpha}|0\rangle. \tag{2.69}$$

We see that $\hbar\omega = \hbar|\mathbf{k}|c$ and $\hbar\mathbf{k}$ are respectively the energy and the momentum of the photon. The mass of the photon is given by

$$\begin{aligned}(\text{mass})^2 &= (1/c^4)(E^2 - |\mathbf{p}|^2c^2)\\ &= \frac{1}{c^4}[(\hbar\omega)^2 - (\hbar|\mathbf{k}|c)^2]\\ &= 0.\end{aligned} \tag{2.70}$$

The photon state is characterized not just by its momentum but also by the polarization vector $\boldsymbol{\epsilon}^{(\alpha)}$. Since $\boldsymbol{\epsilon}^{(\alpha)}$ transforms like a vector, the general theory of angular momentum encourages us to associate with it one unit of angular momenum. This is what is meant by the statement that the photon has one unit of *spin*

‡This expression can also be obtained by evaluating the 4-k component of the energy momentum tensor of the transverse electromagnetic field (cf. Problem 1–1).

angular momentum.‡ To find spin components we first consider

$$\boldsymbol{\epsilon}^{(\pm)} = \mp(1/\sqrt{2})(\boldsymbol{\epsilon}^{(1)} \pm i\boldsymbol{\epsilon}^{(2)}), \tag{2.71}$$

which are called *circular polarization vectors.* Under an infinitesimal rotation around the propagation direction **k** by an amount $\delta\phi$, the circular polarization vectors are changed by

$$\begin{aligned}\delta\boldsymbol{\epsilon}^{(\pm)} &= \mp(\delta\phi/\sqrt{2})(\boldsymbol{\epsilon}^{(2)} \mp i\boldsymbol{\epsilon}^{(1)}) \\ &= \mp i\delta\phi\boldsymbol{\epsilon}^{(\pm)}.\end{aligned} \tag{2.72}$$

Hence we associate with $\boldsymbol{\epsilon}^{(\pm)}$ the spin component $m = \pm 1$ where the quantization axis has been chosen in the propagation direction. If $\boldsymbol{\epsilon}^{(\alpha)}$ were along **k**, we would associate $m = 0$ (since **k** is unchanged under an infinitesimal rotation about **k**); however, the $m = 0$ state is missing in the expansion of **A** because of the transversality condition $\mathbf{k}\cdot\boldsymbol{\epsilon}^{(\alpha)} = 0$. In other words, the photon spin is either parallel or antiparallel to the propagation direction. We note that the absence of the $m = 0$ state has an invariant meaning only for a particle whose mass is strictly zero. If the photon mass is *different* from zero, we can perform a Lorentz transformation such that in the new frame the photon is at rest; in such a frame the photon spin is "parallel to nothing."

The description of the polarization state with $\boldsymbol{\epsilon}^{\pm}$ as the base vectors is called the *circular polarization representation* in contrast to the linear polarization representation based on $\boldsymbol{\epsilon}^{(1)}$ and $\boldsymbol{\epsilon}^{(2)}$. The orthogonality relations for $\boldsymbol{\epsilon}^{(\pm)}$ are

$$\begin{aligned}\boldsymbol{\epsilon}^{(\pm)}\cdot\boldsymbol{\epsilon}^{(\pm)*} &= -\boldsymbol{\epsilon}^{(\pm)}\cdot\boldsymbol{\epsilon}^{(\mp)} = 1, \\ \boldsymbol{\epsilon}^{(\pm)}\cdot\boldsymbol{\epsilon}^{(\mp)*} &= -\boldsymbol{\epsilon}^{(\pm)}\cdot\boldsymbol{\epsilon}^{(\pm)} = 0,\end{aligned} \tag{2.73a}$$

$$\mathbf{k}\cdot\boldsymbol{\epsilon}^{(\pm)} = 0. \tag{2.73b}$$

We could have started with an expansion of **A** with $\boldsymbol{\epsilon}^{(\pm)}$ in place of $\boldsymbol{\epsilon}^{(1)}$ and $\boldsymbol{\epsilon}^{(2)}$. A single-photon state with definite circular polarization can be constructed by applying the creation operator

$$a^{\dagger}_{\mathbf{k},\pm} = \mp\frac{1}{\sqrt{2}}(a^{\dagger}_{\mathbf{k},1} \mp ia^{\dagger}_{\mathbf{k},2}) \tag{2.74}$$

to the vacuum state. Conversely, $a^{\dagger}_{\mathbf{k},\alpha}|0\rangle$ with $\alpha = 1, 2$ and **k** in the z-direction can be regarded as a 50/50 mixture of the $m = 1$ and $m = -1$ state.

To sum up, the quantum postulate applied to the canonical variables of the radiation oscillator naturally leads to the idea that the quantum-mechanical excitations of the radiation field can be regarded as particles of *mass* zero and *spin* one. It is a general feature of the quantum theory of fields that *with every field we associate a particle of definite mass and spin.* The arguments we have presented can be repeated for the quantization of other fields with essentially no modifications.

---

‡For a further discussion of this point, see Problem 2–2. The spin angular momentum of the radiation field can also be discussed from the point of view of the total angular-momentum operator $(1/c)\int[\mathbf{x} \times (\mathbf{E} \times \mathbf{B})]d^3x$ which can be decomposed into the orbital and the spin part. See, for example, Wentzel (1949), p. 123 and Messiah (1962), pp. 1022–1024, pp. 1032–1034.

To study the time development of the quantized radiation field, let us first note that $a_{\mathbf{k},\alpha}$ and $a^\dagger_{\mathbf{k},\alpha}$ are time-*dependent* operators satisfying the Heisenberg equation of motion:

$$\begin{aligned}\dot{a}_{\mathbf{k},\alpha} &= (i/\hbar)[H, a_{\mathbf{k},\alpha}]\\ &= (i/\hbar)\sum_{\mathbf{k}',\alpha'}[\hbar\omega' N_{\mathbf{k}',\alpha'}, a_{\mathbf{k},\alpha}]\\ &= -\, i\omega a_{\mathbf{k},\alpha}; \end{aligned}\tag{2.75}$$

$$\begin{aligned}\ddot{a}_{\mathbf{k},\alpha} &= (i/\hbar)[H, \dot{a}_{\mathbf{k},\alpha}] = (i/\hbar)[H, -i\omega a_{\mathbf{k},\alpha}]\\ &= -\omega^2 a_{\mathbf{k},\alpha}.\end{aligned}\tag{2.76}$$

These are identical in appearance with the differential equations satisfied by the Fourier coefficients $c_{\mathbf{k},\alpha}(t)$ of the classical radiation field. Similarly,

$$\dot{a}^\dagger_{\mathbf{k},\alpha} = i\omega a^\dagger_{\mathbf{k},\alpha}, \qquad \ddot{a}^\dagger_{\mathbf{k},\alpha} = -\omega^2 a^\dagger_{\mathbf{k},\alpha}. \tag{2.77}$$

From (2.76) and (2.77) it is evident that the quantized field $\mathbf{A}$ satisfies the same wave equation (2.7) as does the classical field. Integrating (2.75) we obtain the explicit time dependence for the annihilation and the creation operator:

$$a_{\mathbf{k},\alpha} = a_{\mathbf{k},\alpha}(0)e^{-i\omega t}, \qquad a^\dagger_{\mathbf{k},\alpha} = a^\dagger_{\mathbf{k},\alpha}(0)e^{i\omega t}. \tag{2.78}$$

Finally, we have

$$\mathbf{A}(\mathbf{x}, t) = \frac{1}{\sqrt{V}}\sum_{\mathbf{k}}\sum_{\alpha} c\sqrt{\frac{\hbar}{2\omega}}[a_{\mathbf{k},\alpha}(0)\boldsymbol{\epsilon}^{(\alpha)}e^{i\mathbf{k}\cdot\mathbf{x}-i\omega t} + a^\dagger_{\mathbf{k},\alpha}(0)\boldsymbol{\epsilon}^{(\alpha)}e^{-i\mathbf{k}\cdot\mathbf{x}+i\omega t}]. \tag{2.79}$$

Note that the quantized field operator $\mathbf{A}$ is Hermitian while the classical field $\mathbf{A}$ is real. It is important to note that the $\mathbf{x}$ and t that appear in the quantized field $\mathbf{A}(\mathbf{x}, t)$ are not quantum-mechanical variables but just parameters on which the field operator depends. In particular, $\mathbf{x}$ and t should not be regarded as the space-time coordinates of the photon.

Fluctuations and the uncertainty relations. We are now in a position to discuss some of the peculiarities arising from the quantum nature of the radiation field. First, let us note that neither the individual occupation number operator $N_{\mathbf{k},\alpha}$ nor the total number operator defined by

$$N = \sum_{\mathbf{k}}\sum_{\alpha} N_{\mathbf{k},\alpha} = \sum_{\mathbf{k}}\sum_{\alpha} a^\dagger_{\mathbf{k},\alpha}a_{\mathbf{k},\alpha} \tag{2.80}$$

commutes with $\mathbf{A}$, $\mathbf{E}$, and $\mathbf{B}$ because of (2.28) and (2.29). As is well known, in quantum mechanics the observables corresponding to noncommutable operators cannot be simultaneously determined to arbitrary degrees of accuracy. In the present case, if the number of photons is approximately fixed, there must necessarily be uncertainties in the field strengths.‡ Such a fluctuation behavior is expected even for the vacuum state. To check this explicitly let us take the

---

‡In general, by "uncertainty" we mean the square root of the deviation of the expectation value of the square of the operator in question from the square of the expectation value of the operator, that is, $\Delta q = \sqrt{\langle q^2\rangle - \langle q\rangle^2}$.

electric field operator $\mathbf{E} = -(1/c)\partial\mathbf{A}/\partial t$. Although the vacuum expectation value of the electric field $\langle 0|\mathbf{E}|0\rangle$ vanishes (because $a_{\mathbf{k},\alpha}|0\rangle = 0$) as expected from symmetry considerations, the mean square fluctuation of the electric field can be readily shown to be infinite:

$$\langle 0|\mathbf{E}\cdot\mathbf{E}|0\rangle - |\langle 0|\mathbf{E}|0\rangle|^2 = \langle 0|\mathbf{E}\cdot\mathbf{E}|0\rangle = \infty. \tag{2.81}$$

This illustrates our assertion that if the occupation number is fixed (zero in this case), then the field strength is completely uncertain. On the other hand, since what we measure by a test body is the field strength averaged over some region in space, it may be more realistic to consider the average field operator about some point, for example the origin, defined by

$$\bar{\mathbf{E}} = (1/\Delta V)\int_{\Delta V} \mathbf{E}\, d^3x, \tag{2.82}$$

where ΔV is a small volume containing the point in question. We can then prove

$$\langle 0|\bar{\mathbf{E}}\cdot\bar{\mathbf{E}}|0\rangle \sim \hbar c/(\Delta l)^4, \tag{2.83}$$

where Δl is the linear dimension of the volume ΔV.‡ Equation (2.83) characterizes the fluctuation in the electric field when no photons are present. In general, if the set of occupation numbers is approximately fixed, then the electric or magnetic field has no precise value but exhibits fluctuations about a certain average value.

Another peculiar feature of the quantized radiation field is that the components of **E** and **B** do not necessarily commute. For instance, we have§

$$[E_x(\mathbf{x}, t), B_y(\mathbf{x}', t)] = ic\hbar\frac{\partial}{\partial z}\delta^3(\mathbf{x} - \mathbf{x}'). \tag{2.84}$$

It follows that we cannot simultaneously determine **E** and **B** at the same point in space to arbitrary degrees of accuracy. However, it is possible to show that

$$[E_x(\mathbf{x}, t), B_y(\mathbf{x}', t')] = 0 \qquad \text{for} \qquad (\mathbf{x} - \mathbf{x}')^2 - (1/c^2)(t - t')^2 \neq 0. \tag{2.85}$$

Thus there is no interference between measurements of field strengths performed at two space-time points separated by a distance that cannot be connected by a light signal. This is in comformity with the principle of causality in the special theory of relativity. The commutation relations (2.84), (2.85), and other similar relations were first obtained by P. Jordan, W. Heisenberg, and W. Pauli.

Using the commutation relation (2.84) we can write an uncertainty relation that involves $\Delta\mathbf{E}$ and $\Delta\mathbf{B}$. However, before we interpret such a relation physically we must first define what is meant in the quantum theory by the measurement of field strength. This would require a rather careful discussion based on a 1933 paper by N. Bohr and L. Rosenfeld. Consequently we shall not discuss this subject any further. An excellent treatment of it can be found in Heitler's book.‖

---

‡We shall not prove (2.83) since in Problem 2–3 the reader is asked to prove a similar relation for the scalar field.

§Again we shall not prove this relation because Problem 2–3 is concerned with a similar relation for the scalar field.

‖Heitler (1954), pp. 76–86.

We know from classical optics that a localized or converging light beam can be constructed by superposing various plane waves with definite phase relations. For this reason we wish to consider the phase operator $\phi_{\mathbf{k},\alpha}$ associated with given $\mathbf{k}$, α such that the observable corresponding to it is the phase of the plane wave in question in the sense of classical optics. This we can do by setting

$$\begin{aligned} a_{\mathbf{k},\alpha} &= e^{i(\phi_{\mathbf{k},\alpha}-\omega t)}\sqrt{N_{\mathbf{k},\alpha}}, \\ a^{\dagger}_{\mathbf{k},\alpha} &= \sqrt{N_{\mathbf{k},\alpha}}\,e^{-i(\phi_{\mathbf{k},\alpha}-\omega t)}, \end{aligned} \tag{2.86}$$

where the operator $\sqrt{N_{\mathbf{k},\alpha}}$ satisfies the property $(\sqrt{N_{\mathbf{k},\alpha}})^2 = N_{\mathbf{k},\alpha}$. The quantized radiation field can be written in such a way that the only operators appearing are $\sqrt{N_{\mathbf{k},\alpha}}$ and $\phi_{\mathbf{k},\alpha}$:

$$\begin{aligned} \mathbf{A}(\mathbf{x}, t) = \frac{1}{\sqrt{V}} \sum_{\mathbf{k}} \sum_{\alpha} c\sqrt{\frac{\hbar}{2\omega}}\Big(&\boldsymbol{\epsilon}^{(\alpha)} \exp\,[i(\mathbf{k}\cdot\mathbf{x} - \omega t + \phi_{\mathbf{k},\alpha})]\sqrt{N_{\mathbf{k},\alpha}} \\ &+ \sqrt{N_{\mathbf{k},\alpha}}\,\boldsymbol{\epsilon}^{(\alpha)} \exp\,[-i(\mathbf{k}\cdot\mathbf{x} - \omega t + \phi_{\mathbf{k},\alpha})]\Big), \end{aligned} \tag{2.87}$$

which is of a form convenient for use in discussing the connection with the classical description. The commutation relation for a and $a^{\dagger}$ (for given $\mathbf{k}$, α) can now be written as

$$\begin{aligned} e^{i\phi} N e^{-i\phi} - N &= 1, \\ e^{i\phi} N - N e^{i\phi} &= e^{i\phi}, \end{aligned} \tag{2.88}$$

where we have suppressed the subscripts $\mathbf{k}$, α. It is easy to prove by expanding each exponential that the commutation relation (2.88) is satisfied whenever $N\phi - \phi N = i$. This leads to a new uncertainty relation for the corresponding observables:

$$\Delta N \Delta\phi \gtrsim 1. \tag{2.89}$$

For example, if the phase difference of two plane-wave components is given precisely, then we cannot tell the occupation number associated with each of the wave components.

The above uncertainty relation can be used to derive a more familiar-looking uncertainty relation for the light beam. Since the momentum operator is written as the sum of the number operators weighted by $\hbar\mathbf{k}$ (cf. Eq. 2.68) the momentum of the light beam is uncertain whenever there are definite phase relations among individual plane-wave components. Meanwhile a localized wave train can be constructed by superposing plane-wave components with definite phase relations. It then follows that we cannot associate a definite momentum to a well localized beam. For simplicity let us consider a one-dimensional wave train confined to a region whose characteristic dimension is $\sim \Delta x$. Although the actual mathematical description of such a localized wave train is somewhat involved, we expect that the phase of a typical plane wave component must be known qualitatively to an accuracy of $k_x \Delta x$ if the localization is to be possible. The uncertainty relation (2.89) then implies

$$\Delta p_x \, \Delta x \gtrsim \hbar, \tag{2.90}$$

since the momentum uncertainty is of the order of $\Delta N \hbar k_x$. This relation is formally identical to Heisenberg's uncertainty relation for a "particle." However, because of (2.89), a single photon cannot be localized in the same sense that a particle in nonrelativistic quantum mechanics can be localized. For this reason, it is better to regard (2.90) as a relation imposed on the light *beam*, as we have done.

It is amusing to note that Heisenberg's uncertainty principle for a material particle, for example, an electron, can be formulated only when the light beam also satisfies the uncertainty relation (2.90). If there were no restriction on Δx and Δp_x for the light beam, we would be able to determine the position of the electron to an infinite degree of accuracy by a very well-localized beam without transferring an appreciable momentum uncertainty to the electron in an uncontrollable way. Two successive measurements of this kind would determine accurately both the momentum and the position of the electron, in contradiction with Heisenberg's relation,

Validity of the classical description. All these peculiarities of the quantized radiation field could be rather disturbing, especially if we recall that many of the conveniences of our civilization rely upon the validity of classical electrodynamics which, as we have seen, must be modified. For instance, having learned about the quantum fluctuations of the radiation field, how can we listen in peace to an FM broadcast of symphonic music? In nonrelativistic quantum mechanics the classical description is trustworthy whenever the noncommutativity of dynamical variables is unimportant. Likewise, in the quantum theory of radiation, if we could ignore the right-hand side of $[a, a^\dagger] = 1$, then we would return to the classical description. Since the nonvanishing matrix elements of a and $a^\dagger$ are of the order of $\sqrt{n}$ (cf. Eq. 2.36), the occupation number must be large if the classical description is to be valid.

To be more specific, let us compare the vacuum fluctuation of a squared-field operator and the square of the field strength for a classical electromagnetic wave of wavelength $2\pi\lambda\!\!\!^{-}$. When there are no photons of any kind, the square of the average electric field operator has the following expectation value (cf. Eq. 2.83):

$$\langle 0|\bar{\mathbf{E}}\cdot\bar{\mathbf{E}}|0\rangle \sim \hbar c/\lambda\!\!\!^{-4}, \tag{2.91}$$

where the average is taken over a volume $\lambda\!\!\!^{-3}$. Meanwhile, according to classical electrodynamics, the time average of $\mathbf{E}^2$ can be equated with the energy density of the electromagnetic wave. So we set

$$(\mathbf{E}^2)_{\text{average}} = \bar{n}\hbar(c/\lambda\!\!\!^{-}), \tag{2.92}$$

where $\bar{n}$ stands for the number of photons per unit volume. For the validity of the classical description, purely quantum effects such as (2.91) must be completely negligible in comparison to (2.92). For this we must have

$$\bar{n} \gg 1/\lambda\!\!\!^{-3}. \tag{2.93}$$

In other words, the description of physical phenomena based on classical electrodynamics is reliable when the number of photons per volume $\lambda\!\!\!^{-3}$ is *much greater* than one. As an example, for a Chicago FM station (WFMT), which broadcasts at

98.7 Mc ($\lambda\!\!\!^{-} \approx 48$ cm) with a power of 135,000 watts, the number of photons per volume $\lambda\!\!\!^{-3}$ at a distance five miles from the antenna is about 10^{17}. Thus the classical approximation is an extremely good one.

The historical development of quantum mechanics was guided by an analogy between the electron and the photon both of which were recognized to exhibit the famous wave-particle duality. As Heitler correctly emphasized in his treatise, this similarity can be somewhat misleading. The classical limit of the quantum theory of radiation is achieved when the number of photons becomes so large that the occupation number may as well be regarded as a continuous variable. The space-time development of the classical electromagnetic wave approximates the dynamical behavior of trillions of photons. In contrast, the classical limit of Schrödinger's wave mechanics is the mechanics of a *single* mass point obeying Newton's equation of motion. Thus it was no coincidence that in the very beginning only the *wave nature* of light and the *particle nature* of the electron were apparent.

2-4. EMISSION AND ABSORPTION OF PHOTONS BY ATOMS

Basic matrix elements for emission and absorption. We now have the necessary machinery to deal with the emission and absorption of photons by nonrelativistic atomic electrons. The interaction Hamiltonian between the atomic electrons and the radiation field is assumed to be obtainable from the standard prescription $\mathbf{p} \to \mathbf{p} - e\mathbf{A}/c$, where $\mathbf{A}$ now stands for the quantized radiation field. We have

$$H_{\text{int}} = \sum_i \left[-\frac{e}{2mc}(\mathbf{p}_i \cdot \mathbf{A}(\mathbf{x}_i, t) + \mathbf{A}(\mathbf{x}_i, t) \cdot \mathbf{p}_i) + \frac{e^2}{2mc^2}\mathbf{A}(\mathbf{x}_i, t) \cdot \mathbf{A}(\mathbf{x}_i, t) \right], \tag{2.94}$$

where the summation is over the various atomic electrons that participate in the interaction. The expression $\mathbf{A}(\mathbf{x}_i, t)$ is now a field operator assumed to act on a photon state or a many-photon state at $\mathbf{x}_i$, where $\mathbf{x}_i$ refers to the coordinate of the ith electron. The operator $\mathbf{p}_i$ in $\mathbf{p}_i \cdot \mathbf{A}$ is a differential operator that acts on everything that stands to the right; however, because of the transversality condition $\nabla \cdot \mathbf{A} = 0$, it is legitimate to replace $\mathbf{p}_i \cdot \mathbf{A}$ by $\mathbf{A} \cdot \mathbf{p}_i$. Since the spin magnetic moment is known to interact with the magnetic field, there is an additional interaction of the form

$$H_{\text{int}}^{(\text{spin})} = -\sum_i \frac{e\hbar}{2mc}\boldsymbol{\sigma}_i \cdot [\nabla \times \mathbf{A}(\mathbf{x}, t)]_{\mathbf{x}=\mathbf{x}_i}. \tag{2.95}$$

As we saw in the previous section, $\mathbf{A}(\mathbf{x}, t)$ is a linear combination of the creation and annihilation operators for photons. The question of which particular creation or annihilation operator gives rise to a nonvanishing matrix element depends entirely on the nature of the initial and final states in question.

From the time-dependent perturbation theory (which we will review shortly), it is well known that the transition matrix element for $A \longrightarrow B$ (where A and B may, for example, be atomic states) can be computed to first order in the interaction Hamiltonian H_I by simply taking the matrix element of H_I between A and

B. The H_{int} that appears in (2.94) and (2.95), however, acts not only on atomic states but also on photon states. In a typical process appearing in the quantum theory of radiation the state vector for the initial (or final) state is the *direct product* of the state vector for an atomic state (denoted by A, B, etc.) and the state vector for a single- or a multi-photon state (characterized by $n_{\mathbf{k},\alpha}$). With this point in mind we can still evaluate the transition matrix element to lowest order by taking the matrix element of (2.94) or (2.95) between the initial and final states.

Let us first consider the absorption of a light quantum characterized by $\mathbf{k}, \alpha$. An atom which is initially in state A makes a radiative transition to state B. For simplicity we shall assume that there are only photons of the kind $(\mathbf{k}, \alpha)$ present. If there are $n_{\mathbf{k},\alpha}$ photons in the initial state, then there are $n_{\mathbf{k},\alpha} - 1$ photons in the final state. Although $\mathbf{A}$ contains both $a_{\mathbf{k},\alpha}$ and $a^{\dagger}_{\mathbf{k},\alpha}$, only $a_{\mathbf{k},\alpha}$ gives rise to a non-vanishing matrix element so long as we are computing the absorption process to lowest order. The quadratic term $\mathbf{A}\cdot\mathbf{A}$ makes no contribution to this process in lowest order since it changes the total number of photons by either 0 or ± 2. So, ignoring the spin magnetic moment interaction, we have

$$
\begin{aligned}
&\langle B; n_{\mathbf{k},\alpha} - 1 \,|\, H_{\text{int}} \,|\, A; n_{\mathbf{k},\alpha} \rangle \\
&\qquad = -\frac{e}{mc} \langle B; n_{\mathbf{k},\alpha} - 1 \,|\, \sum_i c\sqrt{\frac{\hbar}{2\omega V}} a_{\mathbf{k},\alpha}(0) e^{i\mathbf{k}\cdot\mathbf{x}_i - i\omega t} \mathbf{p}_i \cdot \boldsymbol{\epsilon}^{(\alpha)} \,|\, A; n_{\mathbf{k},\alpha} \rangle \\
&\qquad = -\frac{e}{m}\sqrt{\frac{n_{\mathbf{k},\alpha}\hbar}{2\omega V}} \sum_i \langle B \,|\, e^{i\mathbf{k}\cdot\mathbf{x}_i} \mathbf{p}_i \cdot \boldsymbol{\epsilon}^{(\alpha)} \,|\, A \rangle e^{-i\omega t},
\end{aligned}
\tag{2.96}
$$

where we have used (2.36) and (2.79). Note that the annihilation operator for a photon with momentum and polarization different from $\hbar\mathbf{k}$ and $\boldsymbol{\epsilon}^{(\alpha)}$ gives a zero matrix element.

It is instructive to compare this expression with the expression we obtain using the semiclassical theory of radiation in which the vector potential $\mathbf{A}$ is treated classically. Within the framework of the semiclassical theory we may define an equivalent classical vector potential denoted by $\mathbf{A}^{(\text{abs})}$ such that when it is used for an absorption process we obtain a matrix element identical to (2.96). Evidently

$$
\mathbf{A}^{(\text{abs})} = \mathbf{A}_0^{(\text{abs})} e^{i\mathbf{k}\cdot\mathbf{x} - i\omega t} \tag{2.97}
$$

will do, where

$$
\mathbf{A}_0^{(\text{abs})} = c\sqrt{\frac{n_{\mathbf{k},\alpha}\hbar}{2\omega V}}\,\boldsymbol{\epsilon}^{(\alpha)}. \tag{2.98}
$$

According to the semiclassical theory the absorption probability is proportional to the intensity $|\mathbf{A}_0|^2$; in the quantum theory it is proportional to $n_{\mathbf{k},\alpha}$. We expect on general grounds that the semiclassical theory approximates reality for large values of $n_{\mathbf{k},\alpha}$, but since the matrix element (2.96) is linear in $\sqrt{n_{\mathbf{k},\alpha}}$, the semiclassical theory gives the correct answer even for small values of $n_{\mathbf{k},\alpha}$ or for weak radiation.

The above happy accident no longer takes place for emission processes which we will now discuss. This time it is the creation operator $a^{\dagger}_{\mathbf{k},\alpha}$ that gives rise to a nonvanishing contribution since the photon occupation number is increased by one. We have (cf. Eqs. 2.36 and 2.79):

$$\langle B; n_{\mathbf{k},\alpha}+1 \,|\, H_{\text{int}} \,|\, A; n_{\mathbf{k},\alpha}\rangle = -\frac{e}{m}\sqrt{\frac{(n_{\mathbf{k},\alpha}+1)\hbar}{2\omega V}}\sum_i \langle B \,|\, e^{-i\mathbf{k}\cdot\mathbf{x}_i}\mathbf{p}_i\cdot\boldsymbol{\epsilon}^{(\alpha)} \,|\, A\rangle e^{i\omega t}. \tag{2.99}$$

If $n_{\mathbf{k},\alpha}$ is very large, the semiclassical treatment based on the complex conjugate of the equivalent potential, (2.97) and (2.98), is adequate since, in practice, there is no difference between $\sqrt{n_{\mathbf{k},\alpha}+1}$ and $\sqrt{n_{\mathbf{k},\alpha}}$.‡ Therefore, for very intense radiation, the problem can again be treated using the semiclassical theory. On the other hand, for small values of $n_{\mathbf{k},\alpha}$ the correspondence with the semiclassical description fails completely. In particular, the transition matrix element for emission need not be zero when there are no photons at all present initially ($n_{\mathbf{k},\alpha}=0$). This accounts for the emission of a photon by an isolated excited atom when there are no electromagnetic waves incident on it, a phenomenon known as "spontaneous emission." In contrast, the emission of a photon in the case where $n_{\mathbf{k},\alpha}\neq 0$ is known as "induced (or stimulated) emission." Note that in the quantum field-theoretic treatment *spontaneous* emission and *induced* emission are discussed on the same footing. The single expression (2.99) suffices for both.

The fact that we can account for spontaneous emission in such a natural way is one of the triumphs of the quantum theory of radiation as opposed to the semiclassical theory of radiation. So long as the electromagnetic field is described by a *classical* potential, the transition matrix element for emission is zero when there is no electromagnetic wave incident on the atom; $\mathbf{A}\cdot\mathbf{p}=0$ when $\mathbf{A}=0$, and that is that. In the *quantum* theory of radiation this is not the case because the matrix element of the field operator $\mathbf{A}(\mathbf{x}, t)$ taken between the vacuum state and the one-photon state does not vanish.

In general, in the classical theory, $\mathbf{A}$ is an externally applied potential that influences the charged particles but is not influenced by them; $\mathbf{A}$ itself does not change as an atom makes a radiative transition. This description is satisfactory, even within the framework of quantum theory, whenever the occupation number is so large that the radiation field can be regarded as an inexhaustible source and sink of photons. It should be clear that, for intense radiation, taking one photon away from the radiation field or adding one photon to it does not make much practical difference. On the other hand, for weak or no incident radiation the classical description runs into difficulty because the change in the radiation field brought about by the emission and absorption of a photon by an atom is quite noticeable.

With the above limitations in mind, we may still define an equivalent classical potential appropriate for emission processes as follows:

$$\mathbf{A}^{(\text{emis})} = \mathbf{A}_0^{(\text{emis})} e^{-i\mathbf{k}\cdot\mathbf{x}+i\omega t}, \tag{2.100}$$

‡We recall that the total time-dependent potential that appears in nonrelativistic quantum mechanics must be Hermitian. In our particular case, the classical vector potential $\mathbf{A}$ must be real anyway.

with

$$\mathbf{A}_0^{(\text{emis})} = c\sqrt{(n_{\mathbf{k},\alpha}+1)\hbar/2\omega V}\,\boldsymbol{\epsilon}^{(\alpha)}. \tag{2.101}$$

By treating (2.100) as the time-dependent vector potential that appears in the Schrödinger equation we can obtain the correct matrix elements (2.99) for emission processes (including spontaneous emission). The fact that $\mathbf{A}^{(\text{emis})}$ is no longer linear in $\sqrt{n_{\mathbf{k},\alpha}}$ and is not quite the complex conjugate of $\mathbf{A}^{(\text{abs})}$ given by (2.97) and (2.98), reflects the failure of the classical concepts.

To sum up, starting with the quantum theory of radiation, we have rigorously derived the following very useful rule.

> The emission or absorption of a light quantum by a charged particle is completely equivalent to an interaction of the charged particle with the equivalent unquantized vector potential given below:

$$\begin{aligned} \text{absorption:}\quad & c\sqrt{\frac{n_{\mathbf{k},\alpha}\hbar}{2\omega V}}\,\boldsymbol{\epsilon}^{(\alpha)}\,e^{i\mathbf{k}\cdot\mathbf{x}-i\omega t}, \\ \text{emission:}\quad & c\sqrt{\frac{(n_{\mathbf{k},\alpha}+1)\hbar}{2\omega V}}\,\boldsymbol{\epsilon}^{(\alpha)}\,e^{-i\mathbf{k}\cdot\mathbf{x}+i\omega t}, \end{aligned} \tag{2.102}$$

> where $n_{\mathbf{k},\alpha}$ is the photon occupation number in the initial state.

This simple rule is one of the most important results of this book.‡

Time-dependent perturbation theory. Before we start computing the transition probabilities for various processes let us briefly review the time-dependent perturbation theory developed by Dirac. An atomic wave function ψ can be expanded as

$$\psi = \sum_k c_k(t)u_k(\mathbf{x})e^{-iE_kt/\hbar}, \tag{2.103}$$

where $u_k(\mathbf{x})$ is the energy eigenfunction with energy E_k satisfying

$$H_0u_k(\mathbf{x}) = E_ku_k(\mathbf{x}) \tag{2.104}$$

in the absence of a time-dependent perturbation. The time-dependent Schrödinger equation in the presence of a time-dependent potential $H_I(t)$ is

$$\begin{aligned} (H_0+H_I)\psi &= i\hbar(\partial\psi/\partial t) \\ &= i\hbar\sum_k(\dot{c}_ku_ke^{-iE_kt/\hbar} - i(E_k/\hbar)c_ku_ke^{-iE_kt/\hbar}). \end{aligned} \tag{2.105}$$

In our case, in addition to the kinetic energies of the electrons the unperturbed Hamiltonian H_0 contains the Coulomb interactions between the electrons and the nucleus, whereas $H_I(t)$ accounts for the interaction of the atomic electrons with the equivalent vector potential (2.102). Using (2.104), we have

$$\sum_k H_Ic_ku_ke^{-iE_kt/\hbar} = i\hbar\sum_l \dot{c}_lu_le^{-iE_lt/\hbar}. \tag{2.106}$$

‡We again emphasize that in Gaussian unrationalized units $c\sqrt{\hbar/2\omega}$ is to be replaced by $c\sqrt{2\pi\hbar/\omega}$.

Multiplying $u_m^* e^{iE_m t/\hbar}$ and integrating over the space coordinates, we obtain the differential equation

$$\dot{c}_m = \sum_k (1/i\hbar)\langle m | H_I(t) | k \rangle e^{i(E_m - E_k)t/\hbar} c_k(t). \tag{2.107}$$

Suppose only state l is populated when the time-dependent perturbation is turned on at $t = 0$; then

$$c_k(0) = \delta_{kl}. \tag{2.108}$$

We may then approximate c_m by

$$c_m^{(1)}(t) = \frac{1}{i\hbar} \int_0^t dt' \langle m | H_I(t') | l \rangle e^{i(E_m - E_l)t'/\hbar}. \tag{2.109}$$

If $c_m^{(1)}$ vanishes for some reason or if a better approximation is called for, we use

$$c_m \simeq c_m^{(1)} + c_m^{(2)}, \tag{2.110}$$

where

$$c_m^{(2)}(t) = \frac{1}{i\hbar} \sum_n \int_0^t dt'' \langle m | H_I(t'') | n \rangle e^{i(E_m - E_n)t''/\hbar} c_n^{(1)}(t'')$$

$$= \frac{1}{(i\hbar)^2} \sum_n \int_0^t dt'' \int_0^{t''} dt' \langle m | H_I(t'') | n \rangle e^{i(E_m - E_n)t''/\hbar} \langle n | H_I(t') | l \rangle e^{i(E_n - E_l)t'/\hbar}, \tag{2.111}$$

and so on.

When we are dealing with the emission and absorption of a photon by an atom, we may work with just $c_m^{(1)}$. The time-dependent potential H_I can be written as

$$H_I(t) = H_I' e^{\mp i\omega t} \qquad \text{for} \quad \begin{Bmatrix} \text{absorption} \\ \text{emission} \end{Bmatrix}, \tag{2.112}$$

where H_I' is a time-independent operator. Hence

$$c_m^{(1)} = \frac{1}{i\hbar} \langle m | H_I' | l \rangle \int_0^t dt' e^{i(E_m - E_l \mp \hbar\omega)t'/\hbar}. \tag{2.113}$$

We can readily perform the time integration and obtain

$$|c_m^{(1)}(t)|^2 = (2\pi/\hbar) |\langle m | H_I' | l \rangle|^2 t \delta(E_m - E_l \mp \hbar\omega), \tag{2.114}$$

where we have used

$$\lim_{\alpha \to \infty} \frac{1}{\pi} \frac{\sin^2 \alpha x}{\alpha x^2} = \delta(x). \tag{2.115}$$

Note that the transition probability per unit time is $|c_m^{(1)}(t)|^2/t$, which is independent of t.

Let us now apply our formalism to emission processes. The photon states allowed by the periodic boundary conditions form a continuous energy spectrum as the normalization volume becomes infinite (cf. Eq. 2.11). For a photon emitted into a solid angle element $d\Omega$, the number of allowed states in an energy interval $[\hbar\omega, \hbar(\omega + d\omega)]$ can be written as $\rho_{\hbar\omega, d\Omega} d(\hbar\omega)$, where

$$\rho_{\hbar\omega, d\Omega} = \frac{V|\mathbf{k}|^2 d|\mathbf{k}| d\Omega}{(2\pi)^3 d(\hbar\omega)} = \frac{V\omega^2 d\Omega}{(2\pi)^3 \hbar c^3}. \tag{2.116}$$

So for the transition probability per unit time into a solid angle element $d\Omega$ we obtain the famous *Golden Rule*

$$w_{d\Omega} = \int (|c_m^{(1)}|^2/t)\rho_{\hbar\omega,\, d\Omega}\, d(\hbar\omega)$$
$$= (2\pi/\hbar)\,|\langle m|\, H_I' \,|l\rangle|^2 \rho_{\hbar\omega,\, d\Omega}, \tag{2.117}$$

where $\hbar\omega$ must satisfy

$$E_m - E_l + \hbar\omega = 0. \tag{2.118}$$

Spontaneous emission in the dipole approximation. In the case of spontaneous emission, an atomic state A makes a radiative transition to a state B in the absence of any incident electromagnetic wave. The matrix element $\langle B\,|H_I'\,|\,A\rangle$ in this case is just (2.99) with $e^{i\omega t}$ omitted and $n_{\mathbf{k},\alpha}$ set equal to zero. Hence for $w_{d\Omega}$ we obtain

$$w_{d\Omega} = \frac{2\pi}{\hbar}\frac{e^2\hbar}{2m^2\omega V}\Big|\sum_i \langle B|\, e^{-i\mathbf{k}\cdot\mathbf{x}_i}\boldsymbol{\epsilon}^{(\alpha)}\cdot\mathbf{p}_i\,|A\rangle\Big|^2 \frac{V\omega^2\, d\Omega}{(2\pi)^3\,\hbar c^3}, \tag{2.119}$$

where ω satisfies the energy conservation $E_A = E_B + \hbar\omega$. The normalization volume V cancels out as it should.

In a typical atomic transition in the optical region the wavelength of the emitted photon is much greater than the linear dimension of the atom:

$$\lambda\!\!\!^{-}_{\text{photon}} = 1/|\mathbf{k}| \gg r_{\text{atom}}, \tag{2.120}$$

since $\lambda\!\!\!^{-}_{\text{photon}}$ is typically of the order of several thousand angstrom units whereas the atomic radius is of the order of one angstrom unit. This means that we can replace

$$e^{-i\mathbf{k}\cdot\mathbf{x}_i} = 1 - i\mathbf{k}\cdot\mathbf{x}_i - (\mathbf{k}\cdot\mathbf{x}_i)^2/2 + \cdots \tag{2.121}$$

by its leading term 1. It turns out that the spin-magnetic-moment interaction is also negligible. To see this, just note that the matrix element of $(e/mc)\boldsymbol{\epsilon}^{(\alpha)}\cdot\mathbf{p}_i$ is larger than that of $(e\hbar/2mc)\boldsymbol{\sigma}_i\cdot(\mathbf{k}\times\boldsymbol{\epsilon}^{(\alpha)})$, again by $\lambda\!\!\!^{-}_{\text{photon}}/r_{\text{atom}}$ since the matrix element of $\mathbf{p}_i$ is of the order of $\hbar/r_{\text{atom}}$. An approximation in which only the $\boldsymbol{\epsilon}^{(\alpha)}\cdot\mathbf{p}_i$ term is kept is called the electric dipole ($E1$) approximation.

To further simplify the problem let us assume that only one of the atomic electrons participates in spontaneous emission, as in the case of a hydrogen-like atom (an atom in which there is only one valence electron).‡ Omitting the sum over i we have

$$w_{d\Omega} = \frac{e^2\omega}{8\pi^2 m^2\hbar c^3}|\langle B|\,\mathbf{p}\,|A\rangle\cdot\boldsymbol{\epsilon}^{(\alpha)}|^2 d\Omega. \tag{2.122}$$

Meanwhile, using the commutation relation between $\mathbf{p}^2$ and $\mathbf{x}$,

$$[\mathbf{p}^2, \mathbf{x}] = -2i\hbar\mathbf{p}, \tag{2.123}$$

‡Many of the results we derive for one-electron atoms can readily be generalized to many-electron atoms.

we can rewrite the matrix element $\langle B|\mathbf{p}|A\rangle$ as follows:

$$\begin{aligned}\langle B|\mathbf{p}|A\rangle &= \left\langle B\left|\frac{im}{\hbar}[H_0, \mathbf{x}]\right|A\right\rangle \\ &= -\frac{im(E_B - E_A)}{\hbar}\langle B|\mathbf{x}|A\rangle \\ &= im\omega\mathbf{x}_{BA}.\end{aligned} \tag{2.124}$$

This matrix element is precisely what we would directly compute if the interaction Hamiltonian were replaced as follows:

$$-\frac{e\mathbf{A}\cdot\mathbf{p}}{mc} \longrightarrow \frac{e}{c}\mathbf{x}\cdot\frac{\partial\mathbf{A}}{\partial t} = -e\mathbf{x}\cdot\mathbf{E}, \tag{2.125}$$

provided that we again kept only the leading term in the plane-wave expansion of **E**. The origin of the term "electric dipole transition" is now evident.

The angular momentum selection rule for $E1$ is

$$|J_B - J_A| = 1, 0; \qquad \text{no} \quad 0 \longrightarrow 0. \tag{2.126}$$

To derive this we first note that the components of **x** can be rearranged as follows:

$$\begin{aligned}V^{\pm 1} &= \mp\frac{1}{\sqrt{2}}(x \pm iy) = r\sqrt{\frac{4\pi}{3}}\,Y_1^{\pm 1}, \\ V^0 &= z = r\sqrt{\frac{4\pi}{3}}\,Y_1^0.\end{aligned} \tag{2.127}$$

Then the Wigner-Eckart theorem gives

$$\langle J_B, m_B|V^q|J_A, m_A\rangle = \frac{1}{\sqrt{2J_A + 1}}\langle J_A 1 m_A q|J_B m_B\rangle\langle B||V^q||A\rangle, \tag{2.128}$$

where $\langle J_A 1 m_A q|J_B m_B\rangle$ is a Clebsch-Gordan coefficient.‡ The condition that the Clebsch-Gordan coefficient shall not vanish leads immediately to the angular momentum selection rule (2.126). Physically this selection rule is a consequence of the fact that one unit of angular momentum is carried away by the spin of the emitted photon in an $E1$ transition.

We also must consider the parity selection rule: parity change "yes." To prove this merely note that

$$\begin{aligned}\langle B|\mathbf{x}|A\rangle &= -\langle B|\Pi^{-1}\mathbf{x}\Pi|A\rangle \\ &= -\Pi_B\Pi_A\langle B|\mathbf{x}|A\rangle,\end{aligned} \tag{2.129}$$

where Π and $\Pi_{A,B}$ stand for the parity operator and its eigenvalue.

Equation (2.122) can also be written as

$$w_{d\Omega} = \frac{e^2\omega^3}{8\pi^2\hbar c^3}|\mathbf{x}_{BA}|^2\cos^2\Theta^{(\alpha)}\,d\Omega, \tag{2.130}$$

where we have defined the angle $\Theta^{(\alpha)}$ by

$$\cos\Theta^{(\alpha)} = |\mathbf{x}_{BA}\cdot\boldsymbol{\epsilon}^{(\alpha)}|/|\mathbf{x}_{BA}|. \tag{2.131}$$

‡Those who are unfamiliar with the Wigner-Eckart theorem and Clebsch-Gordan coefficients may consult Merzbacher (1961), Chapter 22; Messiah (1962), Chapter 13.

Note that

$$|\mathbf{x}_{BA}|^2 = |x_{BA}|^2 + |y_{BA}|^2 + |z_{BA}|^2$$
$$= \left| -\frac{(x_{BA} + iy_{BA})}{\sqrt{2}} \right|^2 + \left| \frac{x_{BA} - iy_{BA}}{\sqrt{2}} \right|^2 + |z_{BA}|^2. \tag{2.132}$$

So far we have been concerned with a radiative transition in which a photon with definite $\mathbf{k}$, α is emitted. We must now sum over two independent polarization states for given $\mathbf{k}$ and integrate over all propagation directions. From Fig. 2–1 it is evident that

$$\cos \Theta^{(1)} = \sin \theta \cos \phi, \qquad \cos \Theta^{(2)} = \sin \theta \sin \phi. \tag{2.133}$$

The sum over the two polarization states just gives $\sin^2 \theta$. Integrating over all possible propagation directions with the direction of $\mathbf{x}_{BA}$ fixed in space, we obtain

$$2\pi \int_{-1}^{1} \sin^2 \theta \, d(\cos \theta) = 8\pi/3. \tag{2.134}$$

Finally, for the integrated transition probability for spontaneous emission we have

$$w = \frac{e^2 \omega^3}{3\pi \hbar c^3} |\mathbf{x}_{BA}|^2$$
$$= \left(\frac{e^2}{4\pi \hbar c}\right) \frac{4}{3} \frac{\omega^3}{c^2} |\mathbf{x}_{BA}|^2, \tag{2.135}$$

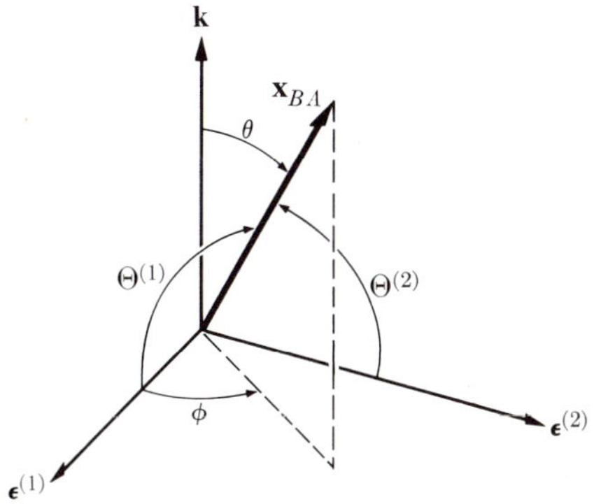

Fig. 2–1. The orientation of $\mathbf{x}_{BA}$.

with $(e^2/4\pi\hbar c) \simeq 1/137$. Note that the dimension of w is $(\text{time})^{-1}$ as expected. This expression was obtained prior to the advent of quantum field theory by W. Heisenberg, who used the correspondence principle.

In order to compute the mean lifetime τ of state A we must sum the transition probabilities into all possible final states allowed by the selection rule and energy conservation:

$$1/\tau_A = \sum_i w_{A \to B_i}. \tag{2.136}$$

In particular, it is important to sum over the magnetic quantum number of the final state.

As a concrete example, we may mention that the mean lifetime of a radiative $E1$ transition from the state having the quantum numbers (n, l, m) to the state having (n', l', m') where $m' = m, m \pm 1$, is given by the reciprocal‡ of

$$\sum_{m'} w[(nlm) \longrightarrow (n'l'm')]$$
$$= \frac{e^2 \omega^3}{3\pi \hbar c^3} \begin{Bmatrix} (l+1)/(2l+1) \\ l/(2l+1) \end{Bmatrix} \left| \int_0^\infty R_{n'l'}(r) R_{nl}(r) r^3 \, dr \right|^2 \quad \text{for} \quad l' = \begin{cases} l+1 \\ l-1 \end{cases} \tag{2.137}$$

‡See Merzbacher (1961), p. 481, for a detailed derivation of this formula.

where $R_{nl}(r)$ is the normalized radial wave function of a hydrogen-like atom characterized by n and l. Note that the lifetime is independent of m. This result, which can be proved using the orthogonality properties of the Clebsch-Gordan coefficients (cf. Eq. 2.128), is reasonable in view of rotational invariance; the lifetime of an excited state in the absence of an external field should not depend on its orientation in space. In general, because the lifetime of an isolated state is independent of its m value, in computing the lifetime we can either sum over *just* the *final* magnetic quantum numbers for a fixed initial magnetic quantum number or sum over *both* the *initial and* the *final* magnetic quantum numbers and divide by the multiplicity of the initial state. For instance, in (2.137) we are free to make the replacement

$$\sum_{m'} \longrightarrow \frac{1}{2J+1}\sum_{m'}\sum_{m}, \tag{2.138}$$

where J (in this case, just l) is the angular momentum of the decaying state.

As an application of (2.137) we may compute the lifetime of the $2p$ state of the hydrogen atom which decays into the ground ($1s$) state. The radial wave functions are

$$R_{nl} = \frac{1}{\sqrt{24a_0^3}}\frac{r}{a_0}e^{-r/2a_0} \qquad \text{and} \qquad R_{n'l'} = \frac{2}{\sqrt{a_0^3}}e^{-r/a_0}, \tag{2.139}$$

where a_0 is the Bohr radius. We obtain

$$\tau(2p \longrightarrow 1s) = 1.6 \times 10^{-9}\ \text{sec}. \tag{2.140}$$

The mean lifetimes of other excited states of the hydrogen atom are tabulated in the literature.‡ In general, the lifetime of an excited hydrogen-like atom goes up with increasing n, roughly as n^3 for a fixed value of l, and as

$$\tau_n \propto \left(\sum_l \frac{(2l+1)}{n^5}\right)^{-1} \sim n^{4.5} \tag{2.141}$$

for the average over l.

Occasionally the symmetry of atomic states may be such that the electric dipole emission of a photon is forbidden. This occurs when $\mathbf{x}_{BA} = 0$ for every state B with an energy lower than A. It is then necessary to go back to the plane-wave expansion (2.121) and take seriously the term $\mathbf{k}\cdot\mathbf{x}$ which we previously ignored. The matrix element we must evaluate can be decomposed as follows:

$$\begin{aligned}\langle B|(\mathbf{k}\cdot\mathbf{x})(\boldsymbol{\epsilon}^{(\alpha)}\cdot\mathbf{p})|A\rangle &= \tfrac{1}{2}\langle B|(\mathbf{k}\cdot\mathbf{x})(\boldsymbol{\epsilon}^{(\alpha)}\cdot\mathbf{p}) + (\mathbf{k}\cdot\mathbf{p})(\boldsymbol{\epsilon}^{(\alpha)}\cdot\mathbf{x})|A\rangle \\ &\quad + \tfrac{1}{2}\langle B|(\mathbf{k}\cdot\mathbf{x})(\boldsymbol{\epsilon}^{(\alpha)}\cdot\mathbf{p}) - (\mathbf{k}\cdot\mathbf{p})(\boldsymbol{\epsilon}^{(\alpha)}\cdot\mathbf{x})|A\rangle.\end{aligned} \tag{2.142}$$

The first term can be rewritten using

$$\tfrac{1}{2}[(\mathbf{k}\cdot\mathbf{x})(\boldsymbol{\epsilon}^{(\alpha)}\cdot\mathbf{p}) + (\mathbf{k}\cdot\mathbf{p})(\boldsymbol{\epsilon}^{(\alpha)}\cdot\mathbf{x})] = \tfrac{1}{2}\mathbf{k}\cdot(\mathbf{xp}+\mathbf{px})\cdot\boldsymbol{\epsilon}^{(\alpha)}, \tag{2.143}$$

where $\mathbf{xp}+\mathbf{px}$ is a symmetric dyadic. The radiative transition due to this term is known as an *electric quadrupole* ($E2$) transition since

$$\mathbf{xp}+\mathbf{px} = (im/\hbar)[H_0, \mathbf{xx}], \tag{2.144}$$

‡Bethe and Salpeter (1957), p. 266.

and
$$\frac{\mathbf{k}}{2}\cdot\langle B|\mathbf{xp}+\mathbf{px}|A\rangle\cdot\boldsymbol{\epsilon}^{(\alpha)} = -\frac{im\omega}{2}\mathbf{k}\cdot\langle B|\mathbf{xx}|A\rangle\cdot\boldsymbol{\epsilon}^{(\alpha)}. \tag{2.145}$$

Because of the transversality condition $\mathbf{k}\cdot\boldsymbol{\epsilon}^{(\alpha)} = 0$, it is legitimate to replace $\mathbf{xx}$ by its traceless part whose individual components are

$$T_{ij} = x_i x_j - (\delta_{ij}/3)|\mathbf{x}|^2. \tag{2.146}$$

Note that it has five independent components which can be written as linear combinations of Y_2^m. It then follows from the Wigner-Eckhart theorem that the angular momentum selection rule for an $E2$ transition is

$$|J_B - J_A| \leq 2 \leq J_B + J_A. \tag{2.147}$$

The second term in (2.142) can be written as follows:

$$(\mathbf{k}\cdot\mathbf{x})(\boldsymbol{\epsilon}^{(\alpha)}\cdot\mathbf{p}) - (\mathbf{k}\cdot\mathbf{p})(\boldsymbol{\epsilon}^{(\alpha)}\cdot\mathbf{x}) = (\mathbf{k}\times\boldsymbol{\epsilon}^{(\alpha)})\cdot(\mathbf{x}\times\mathbf{p}). \tag{2.148}$$

Now $\mathbf{k}\times\boldsymbol{\epsilon}^{(\alpha)}$ is the leading term in the plane-wave expansion of the magnetic field $\mathbf{B}$ while $\mathbf{x}\times\mathbf{p}$ is just the orbital angular momentum operator of the atomic electron which gives the orbital magnetic moment operator when combined with $e/(2mc)$. Hence the radiative transition due to this term is called a *magnetic dipole* (*M*1) *transition*. Together with this term we should consider the leading term from the spin magnetic moment interaction (2.95) $(e\hbar/2mc)\,\boldsymbol{\sigma}\cdot(\mathbf{k}\times\boldsymbol{\epsilon}^{(\alpha)})$, which is of the same order. The angular momentum selection rule for an $M1$ transition is $|J_B - J_A| \leq 1$, no $0\rightarrow 0$, just as in the $E1$ case. In contrast to an $E1$ transition, the parity of the atomic states changes neither in an $M1$ nor in an $E2$ transition.

We can treat higher multipole transitions by considering the higher powers of $\mathbf{k}\cdot\mathbf{x}$ in (2.121). However, it is better to use a formalism that employs *vector spherical harmonics*, a more powerful technique based on an expansion of $\boldsymbol{\epsilon}^{(\alpha)}e^{i\mathbf{k}\cdot\mathbf{x}}$ in terms of the eigenfunctions of the angular momentum operator of the radiation field. We shall not discuss this method; it is treated in standard textbooks on nuclear physics.‡

The atomic states for which electric dipole transitions are forbidden have long lifetimes. In contrast to the lifetime of order 10^{-8} sec characteristic of a typical $E1$ transition, the lifetimes of typical $M1$ or $E2$ transitions are of the order of 10^{-3} sec, roughly $(\lambda\!\!\!^{-}/r_{\text{atom}})^2$ times the typical $E1$ lifetime, as expected. As an example in which $E1$, $M1$, $E2$, etc., are all forbidden for practical purposes, we may mention the radiative transition between the metastable $2s$ state and the ground ($1s$) state of the hydrogen atom. For this transition $E1$ is forbidden by parity, the $M1$ matrix element vanishes when the nonrelativistic wave functions are used, and $E2$ and all other higher multipole transitions are forbidden by angular momentum conservation. Its decay mode turns out to be the simultaneous emission of two photons which can be calculated to be $\frac{1}{7}$ sec (cf. Problem 2–6). Note that this lifetime is extremely long compared to the lifetime of the $2p$ state, 1.6×10^{-9} sec.

‡Consult, for example, Blatt and Weisskopf (1952), Chapter 12.

Planck's radiation law. We wish to conclude this section with a derivation of Planck's radiation law from the point of view of quantum field theory. Suppose we have atoms and a radiation field which can freely exchange energy by the reversible process

$$\mathrm{A} \rightleftarrows \gamma + \mathrm{B} \tag{2.149}$$

in such a way that thermal equilibrium is established. If the populations of the upper and lower atomic levels are denoted by $N(A)$ and $N(B)$ respectively, we have the equilibrium condition

$$N(B)w_{\text{abs}} = N(A)w_{\text{emis}}, \tag{2.150}$$

$$\frac{N(B)}{N(A)} = \frac{e^{-E_B/kT}}{e^{-E_A/kT}} = e^{\hbar\omega/kT}, \tag{2.151}$$

where w_{abs} and w_{emis} are respectively the transition probabilities for $B + \gamma \to A$ and $A \to B + \gamma$. According to (2.96), (2.99), and (2.114) we have

$$\frac{w_{\text{emis}}}{w_{\text{abs}}} = \frac{(n_{\mathbf{k},\alpha} + 1)\,|\sum_i \langle B|\, e^{-i\mathbf{k}\cdot\mathbf{x}_i}\boldsymbol{\epsilon}^{(\alpha)}\cdot\mathbf{p}_i\,|A\rangle|^2}{n_{\mathbf{k},\alpha}\,|\sum_i \langle A|\, e^{i\mathbf{k}\cdot\mathbf{x}_i}\boldsymbol{\epsilon}^{(\alpha)}\cdot\mathbf{p}_i\,|B\rangle|^2}. \tag{2.152}$$

But

$$\langle B|\,e^{-i\mathbf{k}\cdot\mathbf{x}_i}\boldsymbol{\epsilon}^{(\alpha)}\cdot\mathbf{p}_i\,|A\rangle = \langle A|\,\mathbf{p}_i\cdot\boldsymbol{\epsilon}^{(\alpha)}e^{i\mathbf{k}\cdot\mathbf{x}_i}\,|B\rangle^* = \langle A|\,e^{i\mathbf{k}\cdot\mathbf{x}_i}\boldsymbol{\epsilon}^{(\alpha)}\cdot\mathbf{p}_i\,|B\rangle^*. \tag{2.153}$$

Hence

$$\frac{N(B)}{N(A)} = \frac{w_{\text{emis}}}{w_{\text{abs}}} = \frac{n_{\mathbf{k},\alpha} + 1}{n_{\mathbf{k},\alpha}}. \tag{2.154}$$

Using (2.151) and (2.154) we obtain

$$n_{\mathbf{k},\alpha} = \frac{1}{e^{\hbar\omega/kT} - 1}. \tag{2.155}$$

All this is for photon states which satisfy $\hbar\omega = E_A - E_B$. Suppose the radiation field is enclosed by "black" walls which are made up of various kinds of atoms and are capable of absorbing and re-emitting photons of any energy. The energy of the radiation field, per volume, in the angular frequency interval $(\omega, \omega + d\omega)$ is given by (cf. Eq. 2.116)

$$\begin{aligned} U(\omega)\,d\omega &= \frac{1}{L^3}\left(\frac{\hbar\omega}{e^{\hbar\omega/kT} - 1}\right)2\left(\frac{L}{2\pi}\right)^3 4\pi k^2\,dk \\ &= \frac{8\pi\hbar}{c^3}\left(\frac{\omega}{2\pi}\right)^3\left(\frac{1}{e^{\hbar\omega/kT} - 1}\right)d\omega. \end{aligned} \tag{2.156}$$

The energy distribution per frequency per volume is

$$U(\nu) = U(\omega)\frac{d\omega}{d\nu} = \frac{8\pi h\nu^3}{c^3}\frac{1}{e^{h\nu/kT} - 1}. \tag{2.157}$$

This is Planck's famous law which opened up the twentieth-century physics of the quantum domain.

It is instructive to compare our derivation of Planck's law with Einstein's 1917 derivation.‡ They are both based on thermal equilibrium between the atoms and the radiation field. In Einstein's derivation the principle of detailed balance is *explicitly* invoked; by contrast, in our derivation the physics of detailed balance is contained in (2.153) which is an automatic consequence of the hermiticity of the Hamiltonian used in the quantum theory of radiation. Note also that in our derivation we do not distinguish between the contributions from spontaneous emission and induced emission.

Although our attention has been focused in this section on the radiative transitions between two atomic states, the techniques we have acquired can readily be applied to a host of other phenomena. For instance, the reader may calculate the cross section for the photoelectric effect (Problem 2–4) or the lifetime of the Σ^0 hyperon (Problem 2–5):

$$\Sigma^0 \xrightarrow{M1} \Lambda + \gamma.$$

2–5. RAYLEIGH SCATTERING, THOMSON SCATTERING, AND THE RAMAN EFFECT

Kramers-Heisenberg formula. Let us now examine the field-theoretic treatment of the scattering of photons by atomic electrons. Before the scattering, the atom is in state A, and the incident photon is characterized by $(\mathbf{k}, \boldsymbol{\epsilon}^{(\alpha)})$. After the scattering, the atom is left in state B, and the outgoing photon is characterized by $(\mathbf{k}', \boldsymbol{\epsilon}^{(\alpha')})$. For simplicity let us again consider a one-electron atom and neglect the spin-magnetic-moment interaction.

The interaction Hamiltonian (2.94) is made up of a linear $(\mathbf{A}\cdot\mathbf{p})$ term and a quadratic $(\mathbf{A}\cdot\mathbf{A})$ term. Since $\mathbf{A}$ changes the number of photons by one, $\mathbf{A}\cdot\mathbf{p}$ makes no contribution in first order to a scattering process in which there is no *net* change in the number of photons. On the other hand, the $\mathbf{A}\cdot\mathbf{A}$ term contains $aa^\dagger$, $a^\dagger a$, $a\,a$, and $a^\dagger a^\dagger$, the first two of which *do* give nonvanishing contributions provided that $a^\dagger$ and a, respectively, represent the creation operator for $(\mathbf{k}', \alpha')$ and the annihilation operator for $(\mathbf{k}, \alpha)$, e.g., $\langle \mathbf{k}', \alpha' \,|\, a_{\mathbf{k},\alpha} a^\dagger_{\mathbf{k}',\alpha'} \,|\, \mathbf{k}, \alpha\rangle = 1$. Hence

$$
\begin{aligned}
&\langle B; \mathbf{k}', \boldsymbol{\epsilon}^{(\alpha')} \,|\, H_{\text{int}} \,|\, A; \mathbf{k}, \boldsymbol{\epsilon}^{(\alpha)}\rangle \\
&\quad = \left\langle B; \mathbf{k}', \boldsymbol{\epsilon}^{(\alpha')} \left| \frac{e^2}{2mc^2} \mathbf{A}(\mathbf{x}, t)\cdot\mathbf{A}(\mathbf{x}, t) \right| A; \mathbf{k}, \boldsymbol{\epsilon}^{(\alpha)} \right\rangle \\
&\quad = \left\langle B; \mathbf{k}', \boldsymbol{\epsilon}^{(\alpha')} \left| \frac{e^2}{2mc^2} (a_{\mathbf{k},\alpha} a^\dagger_{\mathbf{k}',\alpha'} + a^\dagger_{\mathbf{k}',\alpha'} a_{\mathbf{k},\alpha}) \frac{c^2\hbar}{2V\sqrt{\omega\omega'}} \boldsymbol{\epsilon}^{(\alpha)}\cdot\boldsymbol{\epsilon}^{(\alpha')} \right.\right. \\
&\qquad\qquad \times \exp\left[i(\mathbf{k}-\mathbf{k}')\cdot\mathbf{x} - i(\omega-\omega')t\right] \Big| \, A; \mathbf{k}, \boldsymbol{\epsilon}^{(\alpha)} \Big\rangle \\
&\quad = \frac{e^2}{2mc^2} \frac{c^2\hbar}{2V\sqrt{\omega\omega'}} 2\boldsymbol{\epsilon}^{(\alpha)}\cdot\boldsymbol{\epsilon}^{(\alpha')} \exp\left[-i(\omega-\omega')t\right] \langle B \,|\, A\rangle,
\end{aligned}
\tag{2.158}
$$

‡See, for example, Kittel (1958), pp. 175–176.

where we have replaced $e^{i\mathbf{k}\cdot\mathbf{x}}$ and $e^{-i\mathbf{k}'\cdot\mathbf{x}}$ by 1, since in the long-wave approximation the atomic electron may be assumed to be situated at the origin. For the first-order transition amplitude $c^{(1)}(t)$ we have

$$c^{(1)}(t) = \frac{1}{i\hbar}\frac{e^2}{2mc^2}\frac{c^2\hbar}{2V\sqrt{\omega\omega'}}2\delta_{AB}\boldsymbol{\epsilon}^{(\alpha)}\cdot\boldsymbol{\epsilon}^{(\alpha')}\int_0^t \exp\left[i(\hbar\omega' + E_B - \hbar\omega - E_A)t_1/\hbar\right]dt_1, \tag{2.159}$$

with $\omega = |\mathbf{k}|\,c$ and $\omega' = |\mathbf{k}'|\,c$ as usual.

Although the $\mathbf{A}\cdot\mathbf{p}$ term makes no contribution in first order, the $\mathbf{A}\cdot\mathbf{p}$ term taken *twice* is of the same order as the $\mathbf{A}\cdot\mathbf{A}$ term, so far as powers of e are concerned. Therefore we must treat a double $\mathbf{A}\cdot\mathbf{p}$ interaction and a single $\mathbf{A}\cdot\mathbf{A}$ interaction simultaneously. The $\mathbf{A}\cdot\mathbf{p}$ interaction acting at t_1 can either annihilate the incident photon $(\mathbf{k}, \alpha)$ or create the outgoing photon $(\mathbf{k}', \alpha')$. When the $\mathbf{A}\cdot\mathbf{p}$ interaction acts again at a time t_2 which is later than t_1 it must necessarily create the outgoing photon $(\mathbf{k}', \alpha')$ if the outgoing photon has not yet been created. Otherwise we would end up with a zero matrix element. On the other hand, if the outgoing photon has already been created but the incoming photon has not yet been annihilated, the $\mathbf{A}\cdot\mathbf{p}$ interaction acting at $t_2 > t_1$ must annihilate the incoming photon $(\mathbf{k}, \alpha)$. Between t_1 and t_2 the atom is in state I which is, in general, different from A and B. To summarize, two types of intermediate states are possible. In the first type the atom is in state I and no photons are present. In the second type the atom is in state I and both the incident and the outgoing photon are present.‡

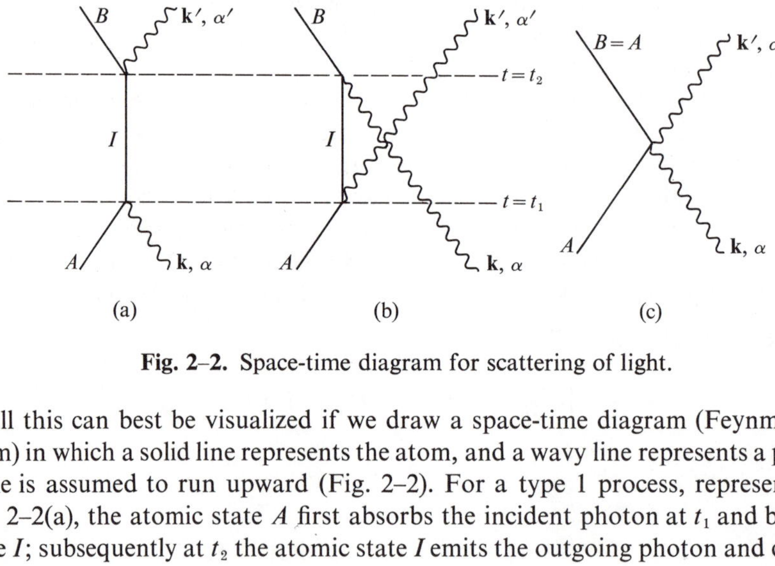

Fig. 2–2. Space-time diagram for scattering of light.

All this can best be visualized if we draw a space-time diagram (Feynman diagram) in which a solid line represents the atom, and a wavy line represents a photon. Time is assumed to run upward (Fig. 2–2). For a type 1 process, represented by Fig. 2–2(a), the atomic state A first absorbs the incident photon at t_1 and becomes state I; subsequently at t_2 the atomic state I emits the outgoing photon and changes

‡Strictly speaking, we should also consider the case where I stands for a continuum state. The relevant matrix element then corresponds to a photo-effect matrix element (cf. Problem 2–4). In practice such "distant" intermediate states are not important because the energy denominators become large (cf. Eq. 2.160 below).

into state B. For a type 2 process, represented by Fig. 2–2(b), state A first emits the outgoing photon at t_1 and changes into state I; subsequently at t_2 state I absorbs the incident photon (which has not yet been annihilated) and becomes state B. In contrast, the lowest-order $\mathbf{A}\cdot\mathbf{A}$ interaction, discussed earlier, is represented by Fig. 2–2(c) ("seagull graph").

As emphasized in the previous section, the emission and absorption of a photon by an atomic electron are equivalent to interactions of the atomic electron with the time-dependent potentials (2.102). Using this rule, we can readily write down the second-order transition amplitude $c^{(2)}(t)$ as follows:

$$
\begin{aligned}
c^{(2)}(t) &= \frac{1}{(i\hbar)^2}\frac{c^2\hbar}{2V\sqrt{\omega\omega'}}\left(-\frac{e}{mc}\right)^2 \int_0^t dt_2 \int_0^{t_2} dt_1 \\
&\quad \times \Big[\sum_I \langle B|\mathbf{p}\cdot\boldsymbol{\epsilon}^{(\alpha')}|I\rangle \exp[i(E_B - E_I + \hbar\omega')t_2/\hbar] \\
&\quad \times \langle I|\mathbf{p}\cdot\boldsymbol{\epsilon}^{(\alpha)}|A\rangle \exp[i(E_I - E_A - \hbar\omega)t_1/\hbar] \\
&\quad + \sum_I \langle B|\mathbf{p}\cdot\boldsymbol{\epsilon}^{(\alpha)}|I\rangle \exp[i(E_B - E_I - \hbar\omega)t_2/\hbar] \\
&\quad \times \langle I|\mathbf{p}\cdot\boldsymbol{\epsilon}^{(\alpha')}|A\rangle \exp[i(E_I - E_A + \hbar\omega')t_1/\hbar] \\
&= -\frac{c^2\hbar}{i\hbar 2V\sqrt{\omega\omega'}}\left(\frac{e}{mc}\right)^2 \sum_I \left(\frac{(\mathbf{p}\cdot\boldsymbol{\epsilon}^{(\alpha')})_{BI}(\mathbf{p}\cdot\boldsymbol{\epsilon}^{(\alpha)})_{IA}}{E_I - E_A - \hbar\omega} + \frac{(\mathbf{p}\cdot\boldsymbol{\epsilon}^{(\alpha)})_{BI}(\mathbf{p}\cdot\boldsymbol{\epsilon}^{(\alpha')})_{IA}}{E_I - E_A + \hbar\omega'}\right) \\
&\quad \times \int_0^t dt_2 \exp[i(E_B - E_A + \hbar\omega' - \hbar\omega)t_2/\hbar],
\end{aligned} \tag{2.160}
$$

where we have made the dipole approximation and ignored a term that depends on the artificial sudden turning of the perturbation (which is negligible if the energy conservation, $E_B - E_A + \hbar\omega' - \hbar\omega = 0$, is nearly satisfied). Combining $c^{(1)}(t)$ and $c^{(2)}(t)$, we have the transition probability

$$
\begin{aligned}
w_{d\Omega} &= \int (|c^{(1)} + c^{(2)}|^2/t)\rho_{E,d\Omega}\, dE \\
&= \frac{2\pi}{\hbar}\left(\frac{c^2\hbar}{2V\sqrt{\omega\omega'}}\right)^2 \left(\frac{e^2}{mc^2}\right)^2 \frac{V}{(2\pi)^3}\frac{\omega'^2}{\hbar c^3}\, d\Omega \\
&\times \left|\delta_{AB}\boldsymbol{\epsilon}^{(\alpha)}\cdot\boldsymbol{\epsilon}^{(\alpha')} - \frac{1}{m}\sum_I \left(\frac{(\mathbf{p}\cdot\boldsymbol{\epsilon}^{(\alpha')})_{BI}(\mathbf{p}\cdot\boldsymbol{\epsilon}^{(\alpha)})_{IA}}{E_I - E_A - \hbar\omega} + \frac{(\mathbf{p}\cdot\boldsymbol{\epsilon}^{(\alpha)})_{BI}(\mathbf{p}\cdot\boldsymbol{\epsilon}^{(\alpha')})_{IA}}{E_I - E_A + \hbar\omega'}\right)\right|^2 .
\end{aligned} \tag{2.161}
$$

To obtain the differential cross section we must divide this transition probability by the flux density which is just c/V, since initially there is one photon in the normalization box of volume V. Finally, we have for the differential cross section

$$
\begin{aligned}
\frac{d\sigma}{d\Omega} = r_0^2\left(\frac{\omega'}{\omega}\right)\Bigg|\delta_{AB}\boldsymbol{\epsilon}^{(\alpha)}\cdot\boldsymbol{\epsilon}^{(\alpha')} &- \frac{1}{m}\sum_I \Bigg(\frac{(\mathbf{p}\cdot\boldsymbol{\epsilon}^{(\alpha')})_{BI}(\mathbf{p}\cdot\boldsymbol{\epsilon}^{(\alpha)})_{IA}}{E_I - E_A - \hbar\omega} \\
&+ \frac{(\mathbf{p}\cdot\boldsymbol{\epsilon}^{(\alpha)})_{BI}(\mathbf{p}\cdot\boldsymbol{\epsilon}^{(\alpha')})_{IA}}{E_I - E_A + \hbar\omega'}\Bigg)\Bigg|^2,
\end{aligned} \tag{2.162}
$$

where r_0 stands for the classical radius of the electron, and

$$
r_0 = \frac{e^2}{4\pi mc^2} \simeq \frac{1}{137}\frac{\hbar}{mc} \simeq 2.82 \times 10^{-13}\ \text{cm}. \tag{2.163}
$$

A formula equivalent to (2.162) was first obtained by H. A. Kramers and W. Heisenberg using the correspondence principle in 1925; hence it is called the Kramers-Heisenberg formula.

Rayleigh scattering. There are certain special cases of (2.162) worth examining in detail. Let us first discuss the case in which $A = B$, $\hbar\omega = \hbar\omega'$. This situation corresponds to elastic scattering of light. It is also called Rayleigh scattering because this problem was treated classically by Lord Rayleigh. To simplify (2.162) we rewrite $\boldsymbol{\epsilon}^{(\alpha)}\cdot\boldsymbol{\epsilon}^{(\alpha')}$, using the commutation relation between $\mathbf{x}$ and $\mathbf{p}$, the completeness of the intermediate states I, and (2.124):‡

$$\begin{aligned}\boldsymbol{\epsilon}^{(\alpha)}\cdot\boldsymbol{\epsilon}^{(\alpha')} &= \frac{1}{i\hbar}\sum_I [(\mathbf{x}\cdot\boldsymbol{\epsilon}^{(\alpha)})_{AI}(\mathbf{p}\cdot\boldsymbol{\epsilon}^{(\alpha')})_{IA} - (\mathbf{p}\cdot\boldsymbol{\epsilon}^{(\alpha)})_{AI}(\mathbf{x}\cdot\boldsymbol{\epsilon}^{(\alpha')})_{IA}] \\ &= \frac{1}{m\hbar}\sum_I \frac{1}{\omega_{IA}}[(\mathbf{p}\cdot\boldsymbol{\epsilon}^{(\alpha)})_{AI}(\mathbf{p}\cdot\boldsymbol{\epsilon}^{(\alpha')})_{IA} + (\mathbf{p}\cdot\boldsymbol{\epsilon}^{(\alpha')})_{AI}(\mathbf{p}\cdot\boldsymbol{\epsilon}^{(\alpha)})_{IA}] \qquad (2.164)\end{aligned}$$

where $\omega_{IA} = (E_I - E_A)/\hbar$. We now see that the three terms in (2.162) combine so that

$$\begin{aligned}&\delta_{AA}\boldsymbol{\epsilon}^{(\alpha)}\cdot\boldsymbol{\epsilon}^{(\alpha')} - \frac{1}{m\hbar}\sum_I\left[\frac{(\mathbf{p}\cdot\boldsymbol{\epsilon}^{(\alpha')})_{AI}(\mathbf{p}\cdot\boldsymbol{\epsilon}^{(\alpha)})_{IA}}{\omega_{IA}-\omega} + \frac{(\mathbf{p}\cdot\boldsymbol{\epsilon}^{(\alpha)})_{AI}(\mathbf{p}\cdot\boldsymbol{\epsilon}^{(\alpha')})_{IA}}{\omega_{IA}+\omega}\right] \\ &\quad = -\frac{1}{m\hbar}\sum_I\left[\frac{\omega(\mathbf{p}\cdot\boldsymbol{\epsilon}^{(\alpha')})_{AI}(\mathbf{p}\cdot\boldsymbol{\epsilon}^{(\alpha)})_{IA}}{\omega_{IA}(\omega_{IA}-\omega)} - \frac{\omega(\mathbf{p}\cdot\boldsymbol{\epsilon}^{(\alpha)})_{AI}(\mathbf{p}\cdot\boldsymbol{\epsilon}^{(\alpha')})_{IA}}{\omega_{IA}(\omega_{IA}+\omega)}\right]. \qquad (2.165)\end{aligned}$$

Using the expansion $1/(\omega_{IA} \mp \omega) \approx [1 \pm (\omega/\omega_{IA})]/\omega_{IA}$, valid for small values of ω, and

$$\begin{aligned}&\sum_I \frac{1}{\omega_{IA}^2}[(\mathbf{p}\cdot\boldsymbol{\epsilon}^{(\alpha')})_{AI}(\mathbf{p}\cdot\boldsymbol{\epsilon}^{(\alpha)})_{IA} - (\mathbf{p}\cdot\boldsymbol{\epsilon}^{(\alpha)})_{AI}(\mathbf{p}\cdot\boldsymbol{\epsilon}^{(\alpha')})_{IA}] \\ &\quad = m^2\sum_I [(\mathbf{x}\cdot\boldsymbol{\epsilon}^{(\alpha')})_{AI}(\mathbf{x}\cdot\boldsymbol{\epsilon}^{(\alpha)})_{IA} - (\mathbf{x}\cdot\boldsymbol{\epsilon}^{(\alpha)})_{AI}(\mathbf{x}\cdot\boldsymbol{\epsilon}^{(\alpha')})_{IA}] \\ &\quad = m^2([\mathbf{x}\cdot\boldsymbol{\epsilon}^{(\alpha')}, \mathbf{x}\cdot\boldsymbol{\epsilon}^{(\alpha)}])_{AA} \\ &\quad = 0, \qquad (2.166)\end{aligned}$$

we obtain the Rayleigh cross section for $\omega \ll \omega_{IA}$:

$$\begin{aligned}\frac{d\sigma}{d\Omega} &= \left(\frac{r_0}{m\hbar}\right)^2\omega^4\left|\sum_I\left(\frac{1}{\omega_{IA}}\right)^3[(\mathbf{p}\cdot\boldsymbol{\epsilon}^{(\alpha')})_{AI}(\mathbf{p}\cdot\boldsymbol{\epsilon}^{(\alpha)})_{IA} + (\mathbf{p}\cdot\boldsymbol{\epsilon}^{(\alpha)})_{AI}(\mathbf{p}\cdot\boldsymbol{\epsilon}^{(\alpha')})]_{IA}\right|^2 \\ &= \left(\frac{r_0 m}{\hbar}\right)^2\omega^4\left|\sum_I\left(\frac{1}{\omega_{IA}}\right)[(\mathbf{x}\cdot\boldsymbol{\epsilon}^{(\alpha')})_{AI}(\mathbf{x}\cdot\boldsymbol{\epsilon}^{(\alpha)})_{IA} + (\mathbf{x}\cdot\boldsymbol{\epsilon}^{(\alpha)})_{AI}(\mathbf{x}\cdot\boldsymbol{\epsilon}^{(\alpha')})_{IA}]\right|^2. \qquad (2.167)\end{aligned}$$

Thus we see that *the scattering cross section at long wavelengths varies as the inverse fourth power of the wavelength* (Rayleigh's law). For atoms in ordinary colorless gases the light wave corresponding to a typical ω_{IA} is in the ultraviolet region.

‡The intermediate states I form a complete set only when we include the continuum states as well as the discrete (bound) states.

Hence the approximation $\omega \ll \omega_{IA}$ is good for ω in the visible optical region. This theory explains why the sky is blue and the sunset is red.

Thomson scattering. Let us now consider the opposite case in which the incident photon energy is much larger than the atomic binding energy. It is then legitimate to ignore the second and third term of (2.162), since $\hbar\omega(=\hbar\omega')$ is much larger than $(\mathbf{p}\cdot\boldsymbol{\epsilon}^{(\alpha')})_{AI}(\mathbf{p}\cdot\boldsymbol{\epsilon}^{(\alpha)})_{IA}/m$, so the scattering is due solely to the matrix element corresponding to the "seagull graph" (Fig. 2–2c). Now the $\delta_{AB}\boldsymbol{\epsilon}^{(\alpha)}\cdot\boldsymbol{\epsilon}^{(\alpha')}$ term is insensitive to the nature of the binding of the atomic electron. The cross section we compute in this case coincides with the cross section for the scattering of light by a free (unbound) electron, first obtained classically by J. J. Thomson:

$$\frac{d\sigma}{d\Omega} = r_0^2\,|\boldsymbol{\epsilon}^{(\alpha)}\cdot\boldsymbol{\epsilon}^{(\alpha')}|^2. \tag{2.168}$$

Note that this expression is independent of ω.

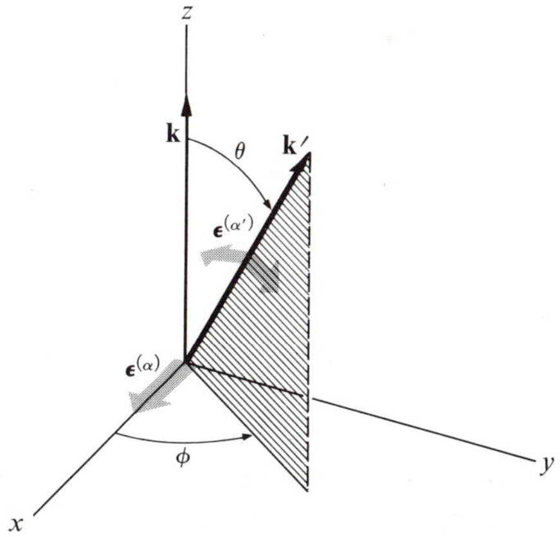

Fig. 2–3. Polarization in Thomson scattering.

To study the polarization dependence of Thomson scattering we consider a coordinate system in which $\boldsymbol{\epsilon}^{(\alpha)}$ and $\mathbf{k}$ are taken along the x-and the z-axes respectively, as shown in Fig. 2–3. The orientation of $\mathbf{k}'$ is characterized by the spherical coordinate angles θ and ϕ. The final polarization vector $\boldsymbol{\epsilon}^{(\alpha')}$ may be taken to be *normal* to the shaded plane (the plane determined by $\mathbf{k}$ and $\mathbf{k}'$) for $\alpha' = 1$; $\boldsymbol{\epsilon}^{(\alpha')}$ with $\alpha' = 2$ must then lie *in* the shaded plane. The Cartesian components of $\boldsymbol{\epsilon}^{(\alpha')}$ are given by

$$\boldsymbol{\epsilon}^{(\alpha')} = \begin{cases} (\sin\phi, -\cos\phi, 0) & \text{for} \quad \alpha' = 1, \\ (\cos\theta\cos\phi, \cos\theta\sin\phi, -\sin\theta) & \text{for} \quad \alpha' = 2. \end{cases} \tag{2.169}$$

Hence

$$\frac{d\sigma}{d\Omega} = r_0^2 \begin{cases} \sin^2\phi & \text{for} \quad \alpha' = 1, \\ \cos^2\theta\cos^2\phi & \text{for} \quad \alpha' = 2. \end{cases} \tag{2.170}$$

For initially unpolarized photons we may either integrate (2.170) over the angle ϕ and divide by 2π or evaluate

$$\left(\frac{d\sigma}{d\Omega}\right)_{\text{unpolarized}} = \frac{1}{2}\left[\frac{d\sigma}{d\Omega}(\phi = 0) + \frac{d\sigma}{d\Omega}\left(\phi = \frac{\pi}{2}\right)\right]. \tag{2.171}$$

The two procedures are completely equivalent. Note that even if the initial polarization vector is randomly oriented, the final photon emitted with $\cos\theta \neq \pm 1$ is polarized, since the differential cross section is $r_0^2/2$ for $\boldsymbol{\epsilon}^{(\alpha')}$ normal to the plane determined by $\mathbf{k}$ and $\mathbf{k}'$ and $(r_0^2/2)\cos^2\theta$ for $\boldsymbol{\epsilon}^{(\alpha')}$ lying in the plane. It is remarkable that the polarization of the scattered photon is complete for $\theta = \pi/2$. We find, then, that a completely unpolarized light beam, when scattered through 90°, results in a 100 % linearly polarized beam whose polarization vector is normal to the plane determined by $\mathbf{k}$ and $\mathbf{k}'$.

If the initial photon is polarized but the final photon polarization is not observed, we must sum over the two possible states of polarization. We have

$$\left.\frac{d\sigma}{d\Omega}\right|_{\substack{\text{final}\\ \text{polarization}\\ \text{summed}}} = r_0^2(\sin^2\phi + \cos^2\theta\cos^2\phi). \tag{2.172}$$

If the initial photon is not polarized and the final photon polarization is not measured, we have

$$\left.\frac{d\sigma}{d\Omega}\right|_{\substack{\text{unpolarized;}\\ \text{final polarization}\\ \text{summed}}} = \frac{r_0^2}{2}(1 + \cos^2\theta). \tag{2.173}$$

The total cross section for Thomson scattering is

$$\sigma_{\text{tot}} = \frac{8\pi r_0^2}{3} = 6.65 \times 10^{-25}\ \text{cm}^2. \tag{2.174}$$

As we emphasized earlier, this expression for the cross section is valid at photon energies much greater than the atomic binding energy. However, the foregoing derivation breaks down if the photon energy is so high that it actually becomes comparable to the rest energy of the electron. We must then take into account the relativistic nature of the electron, as we shall do in Section 4–4, discussing Compton scattering.

The quantum-theoretic treatment of Rayleigh and Thomson scattering can be compared to the classical counterpart. The scattering of an electromagnetic wave can be visualized in classical mechanics by the following two-step process:

a) A bound electron oscillates when it is exposed to a time-dependent electric field.
b) The oscillating charge in turn radiates an electromagnetic wave.

For a model of the electron bound by a force obeying Hooke's law, the displacement $\mathbf{x}$ of the electron in the presence of an applied electric field $\mathbf{E}_0 e^{-i\omega t}$ satisfies the differential equation

$$\ddot{\mathbf{x}} + \omega_0^2\mathbf{x} = (e/m)\mathbf{E}_0\, e^{-i\omega t}, \tag{2.175}$$

where ω_0 is the characteristic angular frequency of the oscillator. Knowing that the acceleration of the electron is given by

$$\ddot{\mathbf{x}} = -\left(\frac{e}{m}\right)\left(\frac{\omega^2}{\omega_0^2 - \omega^2}\right)\mathbf{E}_0 e^{-i\omega t}, \tag{2.176}$$

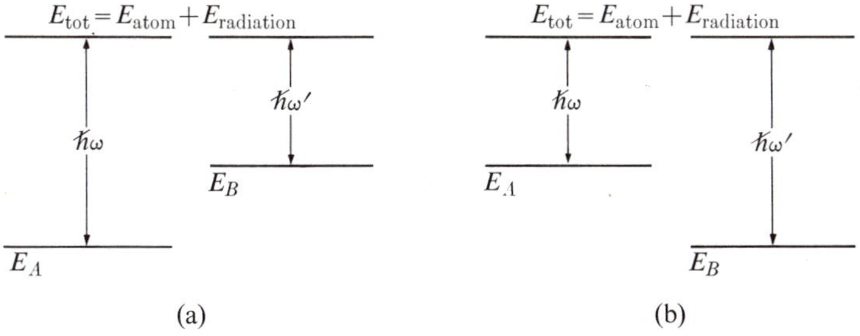

Fig. 2-4. (a) Stokes' line, (b) anti-Stokes' line.

we can readily compute the total scattering cross section in a straightforward manner.‡ We obtain

$$\sigma_{tot} = \frac{8\pi r_0^2}{3} \frac{\omega^4}{(\omega_0^2 - \omega^2)^2}. \tag{2.177}$$

For $\omega \ll \omega_0$ we have the ω^4 dependence of (2.167), whereas for $\omega \gg \omega_0$ we recover the frequency independent cross section (2.174).

The Raman effect. The Kramers-Heisenberg formula (2.162) can also be applied to inelastic scattering of light in which $\omega \neq \omega'$ and $A \neq B$. In atomic physics this phenomenon is called the *Raman effect* after C. V. Raman who observed a shift in the frequency of radiation scattered in liquid solutions, an effect predicted earlier by A. Smekal. If the initial atomic state A is the ground state, then the energy of the final photon $\hbar\omega'$ cannot be greater than the incident photon energy $\hbar\omega$ because $\hbar\omega + E_A = \hbar\omega' + E_B$ (Fig. 2-4a). This accounts for the presence of a *Stokes' line* in atomic spectra, a spectral line more reddish than that of incident radiation. On the other hand, if the atom is in an excited state, ω' can be larger than ω (Fig. 2-4b). This leads to an *anti-Stokes' line* which is more violet than the spectral line of the incident radiation.

2-6. RADIATION DAMPING AND RESONANCE FLUORESCENCE

The Kramers-Heisenberg formula we derived in the previous section is clearly inadequate if $\hbar\omega$ becomes equal to $E_I - E_A$ for some state I. The cross section according to (2.162) is then infinite, a phenomenon not observable in nature, of course. It is nevertheless true that the scattering cross section becomes very large and goes through a very sharp maximum in the neighborhood of $E_I - E_A = \hbar\omega$. This is a phenomenon known as resonance scattering of light or *resonance fluorescence.*

Where did our theory go wrong? When we use the second-order time-dependent perturbation theory, we assumed that the intermediate state I is a stationary state with an infinitely long lifetime. In other words, we did not take into account the

‡Panofsky and Phillips (1955), p. 326; Jackson (1962), pp. 602–604.

instability of state I due to spontaneous emission. In this connection, it is amusing to note that the classical expression for the Rayleigh scattering cross section (2.177) also blows up for $\omega \approx \omega_0$. This catastrophe can be avoided by introducing a damping force in the differential equation (2.175) as follows:

$$\ddot{\mathbf{x}} + \gamma\dot{\mathbf{x}} + \omega_0^2\mathbf{x} = (e/m)\mathbf{E}_0 e^{-i\omega t}. \tag{2.178}$$

The scattering cross section is then given by

$$\sigma_{\text{tot}} = \frac{8\pi r_0^2}{3}\frac{\omega^4}{(\omega_0^2 - \omega^2)^2 + \gamma^2/4}, \tag{2.179}$$

which is large but finite at $\omega = \omega_0$. As we shall see in a moment, in the quantum-theoretic treatment the depletion of state I due to spontaneous emission plays a role analogous to the damping force in the classical treatment; hence it is known as *radiation damping*. The quantum theory of unstable states was first developed by V. F. Weisskopf and E. P. Wigner precisely in this connection.

To simplify, let us assume that the initial atomic state A is a stable (ground) state. As the light beam hits the atom, the coefficient $c_I(t)$ builds up because state A makes a transition to state I as it absorbs the incident photon. Mathematically the differential equation for $c_I(t)$ due to this absorption process is (cf. Eq. 2.107)

$$\dot{c}_I = \frac{1}{i\hbar} H_{IA}^{(\text{abs})}(t) c_A e^{i(E_I - E_A)/\hbar}, \tag{2.180}$$

where $H_{IA}^{\text{abs}}(t)$ is the time-dependent matrix element characteristic of the photon absorption. But this is not the whole story. The amplitude $c_I(t)$ changes with time because of spontaneous emission even if state I is left alone in the absence of any incident radiation. The probability of finding the excited state I decreases with time as $e^{-\Gamma_I t/\hbar}$, where $\Gamma_I = \hbar/\tau_I$, τ_I being the mean life of state I given by (2.135) and (2.136). Since the amplitude for state I must vary as $e^{-\Gamma_I t/2\hbar}$ in the absence of any incident radiation, we can take into account the depletion of state I by adding to the right-hand side of the differential equation (2.180) a new term as follows:

$$\dot{c}_I = \frac{1}{i\hbar} H_{IA}^{(\text{abs})}(t) c_A e^{i(E_I - E_A)t/\hbar} - \frac{\Gamma_I}{2\hbar} c_I. \tag{2.181}$$

The first term of (2.181) characterizes the growth of state I due to the absorption of light, while the second term describes the depletion of state I due to spontaneous emission. Admittedly our approach is somewhat phenomenological; a more rigorous justification of the second term will be given in Section 2–8 when we discuss the second-order level shift due to the emission and absorption of a virtual photon.

We solve the differential equation (2.181) subject to the initial condition $c_I(0) = 0$, $c_A(0) = 1$. By direct substitution we find that the solution to this differential equation satisfying the above initial condition is

$$c_I^{(1)}(t) = \frac{H'_{IA}(\exp[-\Gamma_I t/2\hbar] - \exp[i(E_I - E_A - \hbar\omega)t/\hbar])}{E_I - E_A - \hbar\omega - i\Gamma_I/2}, \tag{2.182}$$

where H'_{IA} is the time-independent matrix element given by

$$H_{IA}^{(\text{abs})}(t) = H'_{AI} e^{-i\omega t} = -(e/mc)\langle I | c\sqrt{\hbar/2\omega V}\,\mathbf{p}\cdot\boldsymbol{\epsilon}^{(\alpha)} | A\rangle e^{-i\omega t}. \tag{2.183}$$

The lowest-order probability for finding state I is just $|c_I^{(1)}(t)|^2$. Note that, unlike (2.114) of Section 2–4, it is *not* proportional to t times $\delta(E_I - E_A - \hbar\omega)$ as $t \to \infty$. Whether (2.182) or (2.113) is the more realistic expression for describing a given absorption process depends on whether the lifetime of state I is short or long compared to the observation time. From (2.182) we obtain

$$|c_I^{(1)}(\infty)|^2 = \frac{|H'_{IA}|^2}{(E_I - E_A - \hbar\omega)^2 + \Gamma_I^2/4}. \tag{2.184}$$

State I is found copiously in abundance when the resonance condition $E_I - E_A \approx \hbar\omega$ is satisfied. If (2.184) is regarded as a function of the incident photon energy $\hbar\omega$, it has the familiar Lorentz shape whose full width at half maximum is given by $\Gamma_I = \hbar/\tau_I$. When the atom is irradiated by a continuous spectrum of radiation, we expect an absorption line whose width corresponds in the energy scale to Γ_I, often called the *natural width.* Using a similar argument, we can show that the frequency dependence of an emission line for an excited state I returning to the ground state A is also given by the same form, as we would expect from considerations based on equilibrium (Kirchhoff's law).‡

Let us now return to the scattering problem. To compute the scattering cross section for $\gamma + A \to \gamma + B$, where $\hbar\omega \simeq E_I - E_A$ for some I, we must re-evaluate $c_B^{(2)}$ using our modified $c_I^{(1)}$ in the general formula (cf. Eq. 2.111).

$$c_B^{(2)}(t) = \frac{1}{i\hbar}\sum_I \int_0^t dt'' \langle B|H_I(t'')|I\rangle\, e^{i(E_B - E_I)t''/\hbar} c_I^{(1)}(t''). \tag{2.185}$$

For a sufficiently large value of t we can omit the transient term $e^{-\Gamma_I t/2\hbar}$ in (2.182). Denoting by R those intermediate states I for which the resonance condition $\hbar\omega \simeq E_I - E_A$ is satisfied, we obtain the second-order amplitude $c_B^{(2)}$ as follows:

$$c_B^{(2)}(t) = -\frac{1}{i\hbar}\left(-\frac{e}{mc}\right)^2 \sum_R \frac{\langle B|c\sqrt{\hbar/2\omega' V}\,\mathbf{p}\cdot\boldsymbol{\epsilon}^{(\alpha')}|R\rangle\langle R|c\sqrt{\hbar/2\omega V}\,\mathbf{p}\cdot\boldsymbol{\epsilon}^{(\alpha)}|A\rangle}{E_R - E_A - \hbar\omega - i\Gamma_R/2}$$

$$\times \int_0^t dt'' \exp\left[i(E_B - E_A + \hbar\omega' - \hbar\omega)t''/\hbar\right] + \text{nonresonant terms}, \tag{2.186}$$

where the "nonresonant terms" stand for (2.160) except that in the sum over I for the type 1 intermediate states those states (R) which satisfy the resonance condition are missing. We see that the only change necessary is the substitution

$$E_I \longrightarrow E_I - i(\Gamma_I/2) \tag{2.187}$$

‡The foregoing discussion of line width ignores the broadenings of spectral lines due to the Doppler effect and atomic collisions which, in many instances, are more important than the broadening due to spontaneous emission. However, our phenomenological treatment can easily be generalized to the case where the depletion of state I is due to inelastic collisions. Compare our treatment with the discussion on absorption of radiation found in Dicke and Wittke (1960), pp. 273–275, where the main depletion is assumed to be due to atomic collisions rather than spontaneous emission.

for the type 1 intermediate states satisfying the resonance condition. Hence for any arbitrary ω we have the modified Kramers-Heisenberg formula

$$\frac{d\sigma}{d\Omega} = r_0^2\left(\frac{\omega'}{\omega}\right)\Bigg|\delta_{AB}\boldsymbol{\epsilon}^{(\alpha)}\boldsymbol{\epsilon}^{(\alpha')} - \frac{1}{m}\sum_I\left[\frac{(\mathbf{p}\cdot\boldsymbol{\epsilon}^{(\alpha')})_{BI}(\mathbf{p}\cdot\boldsymbol{\epsilon}^{(\alpha)})_{IA}}{E_I - E_A - \hbar\omega - i\Gamma_I/2} + \frac{(\mathbf{p}\cdot\boldsymbol{\epsilon}^{(\alpha)})_{BI}(\mathbf{p}\cdot\boldsymbol{\epsilon}^{(\alpha')})_{IA}}{E_I - E_A + \hbar\omega'}\right]\Bigg|^2. \tag{2.188}$$

In practice Γ_I can be ignored except when $E_I - E_A \simeq \hbar\omega$.

In general, the resonant amplitude is much larger than the sum of the nonresonant amplitudes. This is because the magnitude of the resonant amplitude is of the order of $\lambda\!\!\!{}^-=c/\omega$ (cf. Problem 2–8) while the magnitude of the nonresonant amplitude is of the order of r_0. Ignoring the nonresonant amplitudes, we can obtain a single-level resonance formula applicable to the scattering of light in the vicinity of a nondegenerate resonance state:

$$\frac{d\sigma}{d\Omega} = r_0^2\left(\frac{\omega'}{\omega}\right)\left(\frac{1}{m^2}\right)\frac{|(\mathbf{p}\cdot\boldsymbol{\epsilon}^{(\alpha')})_{BR}|^2\,|(\mathbf{p}\cdot\boldsymbol{\epsilon}^{(\alpha)})_{RA}|^2}{(E_R - E_A - \hbar\omega)^2 + \Gamma_R^2/4}. \tag{2.189}$$

It is amusing to note that this expression is identical with the probability of finding state R formed by the absorption of photon $(\mathbf{k}, \alpha)$, multiplied by the spontaneous emission probability per solid angle for $R \to B$ with the emission of photon $(\mathbf{k}', \alpha)$, and divided by the flux density. To prove this statement we merely note that from (2.184) and (2.119) we obtain

$$\frac{(\text{absorption probability}) \times (\text{emission probability per solid angle})}{\text{flux density}}$$

$$= \frac{\dfrac{e^2\hbar}{2\omega m^2 V}\left[\dfrac{|(\mathbf{p}\cdot\boldsymbol{\epsilon}^{(\alpha)})_{RA}|^2}{(E_R - E_A - \hbar\omega^2) + \Gamma_R^2/4}\right]\left[\dfrac{2\pi}{\hbar}\dfrac{e^2\hbar}{2m^2\omega' V}|(\mathbf{p}\cdot\boldsymbol{\epsilon}^{(\alpha')})_{BR}|^2\dfrac{V\omega'^2}{(2\pi)^3\hbar c^3}\right]}{c/V}, \tag{2.190}$$

which is identical to (2.189). Thus resonance scattering can be visualized as the formation of the resonance state R due to the absorption of the incident photon followed by the spontaneous emission of the outgoing photon.

The above simple interpretation of resonance fluorescence as the formation of resonance followed by its decay is subject to the usual peculiarities of quantum mechanics. First of all, if the nonresonant amplitudes are appreciable, (2.188) tells us that they can interfere with the resonant amplitude. Interference between the nonresonant amplitudes and the resonant amplitude is frequently observed in nuclear resonance fluorescence. Further, if the resonant state has spin or, more generally, if there are several resonant states of the same or approximately the same energy, it is important to sum over the amplitudes corresponding to the various resonant states as indicated in (2.186) before squaring the amplitude. This point will be illustrated in Problem 2–8.

These considerations naturally lead to the following question: How can we decide whether a particular physical phenomenon is best described as a single quantum-mechanical process of resonance scattering or as two independent quantum-mechanical processes of absorption followed by emission? The answer to this question depends critically on how the lifetime of the metastable state compares with the collision time, that is, the time interval during which the atom is exposed to the incident beam. To illustrate this point, let us first consider an experimental arrangement in which the temporal duration of the incident photon beam is short compared to the lifetime of the resonance. The energy resolution of such a photon beam is necessarily ill-defined, $\Delta(\hbar\omega) \gg \Gamma_R$, because of the uncertainty principle. It is then possible to select the excited resonant state by observing the exponential decay of the resonance long after the irradiation of the atom has stopped. In such a case, it is legitimate to regard the formation of the metastable state and its subsequent decay as *two independent* quantum-mechanical processes.

On the other hand, the situation is very different with scattering of a photon beam of a very well-defined energy. If the energy resolution of the incident beam is so good (or if the lifetime of the resonant state is so short) that the relation $\Delta(\hbar\omega) \ll \Gamma_R$ is satisfied, it is not possible to select the excited resonant state by a delayed lifetime measurement or any other measurement without disturbing the system. This is because the uncertainty principle demands that such a monochromatic photon beam must last for a time interval *much longer* than the lifetime of the resonance. We can still determine the lifetime of the resonance by measuring the decay width Γ_R through the variation of the scattering cross section as a function of ω, using *monochromatic* beams of *variable* energies. Under such a circumstance the interference effect between the nonresonant amplitudes and the resonant amplitude discussed earlier can become important. In such a case, it is meaningless to ask whether or not a particular photon has come from the resonance state. If we are considering an everlasting, monochromatic, incident beam, resonance fluorescence must be regarded as a single quantum-mechanical process so long as the atom is undisturbed, or so long as no attempt is made to determine the nature of the intermediate states.

In nuclear physics it is sometimes possible to obtain a sufficiently monochromatic photon beam so that we can study the Lorentzian energy dependence of the cross section in the vicinity of a resonance whose lifetime is $\lesssim 10^{-10}$ sec. For a resonance with lifetime $\gtrsim 10^{-10}$ sec, the mean life may be more readily determined by observing the exponential decay. For a nuclear level with lifetime $\sim 10^{-10}$ sec (corresponding to a decay width of about 10^{-7} eV) we can determine the lifetime both by a delayed lifetime measurement and by a study of the energy dependence of the cross section. Whenever both procedures are available, the agreement between the lifetimes determined by the two entirely different methods is quite satisfactory.

2–7. DISPERSION RELATIONS AND CAUSALITY

Real and imaginary parts of the forward scattering amplitude. In this section we shall discuss the analytic properties of the amplitude for the elastic scattering of photons by atoms in the forward direction, with no change in polarization. We defined the coherent forward scattering amplitude $f(\omega)$ by

$$(d\sigma/d\Omega)_{\theta=0,\,\epsilon^{(\alpha)}=\epsilon^{(\alpha')}} = |f(\omega)|^2. \tag{2.191}$$

The sign of $f(\omega)$ is taken to be opposite to that of the amplitude inside the absolute value sign of (2.188). This is because the amplitude computed in Born approximation is positive if the interaction is attractive and negative if it is repulsive.‡

The explicit form of $f(\omega)$ can be obtained directly from (2.188) if the scatterer is assumed to be the ground state of a single-electron atom:

$$f(\omega) = -r_0\left[1 - \frac{1}{m}\sum_I\left(\frac{|(\mathbf{p}\cdot\boldsymbol{\epsilon}^{(\alpha)})_{IA}|^2}{E_I - E_A - \hbar\omega - i\Gamma_I/2} + \frac{|(\mathbf{p}\cdot\boldsymbol{\epsilon}^{(\alpha)})_{IA}|^2}{E_I - E_A + \hbar\omega}\right)\right]. \tag{2.192}$$

We now evaluate the real and imaginary parts of $f(\omega)$. In atomic physics the level width Γ_I is of the order of 10^{-7} eV, while the level spacing is of the order of 1 eV. Hence in computing the real part we may as well ignore the presence of Γ_I. This approximation breaks down only in very narrow intervals near $\hbar\omega = E_I - E_A$. Using the trick already employed in obtaining (2.165), we have

$$\text{Re}[f(\omega)] = \sum_I \frac{2r_0\omega^2}{m\hbar\omega_{IA}} \frac{|(\mathbf{p}\cdot\boldsymbol{\epsilon}^{(\alpha)})_{IA}|^2}{(\omega_{IA}^2 - \omega^2)}, \tag{2.193}$$

where $\hbar\omega_{IA} = E_I - E_A$. In computing the imaginary part we can make a narrow-width approximation in which the very sharp peak at $\hbar\omega \simeq E_I - E_A$ is replaced by a δ-function. Thus

$$\begin{aligned}\text{Im}[f(\omega)] &= \frac{r_0}{m}\sum_I \frac{|(\mathbf{p}\cdot\boldsymbol{\epsilon}^{(\alpha)})_{IA}|^2(\Gamma_I/2)}{(E_I - E_A - \hbar\omega)^2 + \Gamma_I^2/4} \\ &= \sum_I \frac{\pi r_0}{m\hbar}|(\mathbf{p}\cdot\boldsymbol{\epsilon}^{(\alpha)})_{IA}|^2\delta(\omega_{IA} - \omega),\end{aligned} \tag{2.194}$$

where we have used

$$\lim_{\epsilon\to 0}\frac{\epsilon}{x^2+\epsilon^2} = \pi\delta(x). \tag{2.195}$$

We note that $f(\omega)$ acquires an imaginary part only when there is an intermediate state into which the initial system $\gamma + A$ can make a transition *without violating energy conservation.*

Because of (2.193) and (2.194) the real and imaginary parts of $f(\omega)$ satisfy the relation

$$\text{Re}[(f(\omega)] = \frac{2\omega^2}{\pi}\int_0^\infty \frac{\text{Im}[f(\omega')]\,d\omega'}{\omega'(\omega'^2 - \omega^2)}. \tag{2.196}$$

This relation is known as the *dispersion relation for scattering of light.* It is equivalent to a similar relation for the complex index of refraction (to be discussed later) written by H. A. Kramers and R. Kronig in 1926-1927; hence it is sometimes referred to as the *Kramers-Kronig relation.*§ The relation (2.196) can be shown to be a direct consequence of a very general principle, loosely referred to as the *causality principle* which we shall discuss presently. Its validity is independent of the various approximations we have made, such as the narrow width approximation for states I, the dipole approximation for the emission and absorption matrix elements, and the perturbation-theoretic expansion.

---

‡Merzbacher (1961), pp. 491–493.

§The history of the Kramers-Kronig relation was traced back by M. L. Goldberger to an 1871 paper by Sellmeier.

Causality and analyticity. What do we mean by the causality principle? In the classical wave theory it simply means that *no signal can travel with a velocity faster than that of light*. When applied to a scattering problem, it amounts to the requirement that no outgoing disturbance shall start until the incoming disturbance hits the scatterer. In quantum field theory "causality" is often equated with the vanishing of the commutator of two field operators taken at two space-time points separated by a spacelike distance (cf. Eq. 2.85). As mentioned in Section 2–3, this requirement, technically referred to as local commutativity, is directly related to the notion that two measurements cannot influence each other if they are made at a spacelike separation. Historically, the connection between the classical causality principle and the Kramers-Kronig relation was discussed by R. Kronig, N. G. van Kampen, and J. S. Toll. In 1953 M. Gell-Mann, M. L. Goldberger, and W. Thirring proved the dispersion relation for the scattering of light within the framework of quantum field theory.

The derivation of the dispersion relation from the requirement of local commutativity in quantum field theory is beyond the scope of this book. Instead, we shall now demonstrate how the dispersion relation can be obtained from the causality principle in the classical wave theory.

In studying the properties of the coherent forward amplitude for scattering of light, we find that the photon spin is an inessential complication. So we shall consider a scattering problem within the framework of the classical field theory of a scalar zero-mass field ϕ. Let us suppose that there is an incident monochromatic wave whose propagation vector $\mathbf{k}$ is oriented along the z-axis. The asymptotic form of the wave at large r is

$$\phi \sim e^{i\mathbf{k}\cdot\mathbf{x}-i\omega t} + f(\omega;\theta)e^{i|\mathbf{k}|r-i\omega t}/r = e^{i\omega[(z/c)-t]} + f(\omega;\theta)e^{i\omega[(r/c)-t]}/r, \quad (2.197)$$

where $f(\omega;\theta)$ is the scattering amplitude. We also use the notation $f(\omega)$ for the forward scattering amplitude as in (2.191),

$$f(\omega) = f(\omega;\theta)\,|_{\theta=0}. \quad (2.198)$$

If we are to take advantage of the causal connection between the outgoing wave and the incident wave, it is not so convenient to use a plane wave for the incident disturbance. This is because the plane wave is everlasting both in space and time. It is much better to use an incident wave train with a very sharp front. For this reason we consider a δ-function pulse moving in the positive z-direction with the property that the disturbance vanishes everywhere except at $z = ct$. Because of the well-known relation

$$\delta\left(\frac{z}{c} - t\right) = \frac{1}{2\pi}\int_{-\infty}^{\infty} e^{i\omega[(z/c)-t]}\,d\omega, \quad (2.199)$$

the δ-function pulse contains negative as well as positive frequency components. To obtain the scattering amplitude for negative values of ω we exploit the reality of the wave equation which tells us that, if ϕ is a solution to the scattering problem, then ϕ^* is also a solution with the same scatterer. Clearly, the form (2.197) is preserved if we set

$$f(-\omega;\theta) = f^*(\omega;\theta). \quad (2.200)$$

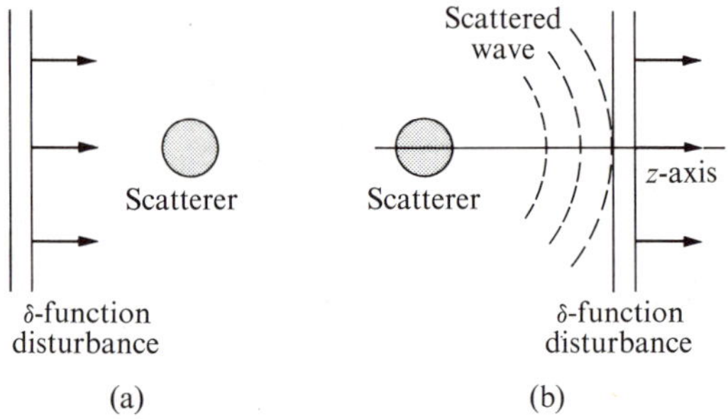

Fig. 2–5. Causal relation between the scattered wave and the incident pulse: (a) before scattering, (b) after scattering.

Using the superposition principle, we can write an asymptotic solution to the scattering problem when the incident wave is given by a δ-function pulse instead of a plane wave. All we have to do is integrate (2. 197) with respect to ω. Denoting by $F(\mathbf{x}, t)$ the outgoing wave at $\mathbf{x}$ (characterized by r, θ, and ϕ), we have the asymptotic form

$$\delta\left(\frac{z}{c} - t\right) + F(\mathbf{x}, t) \sim \frac{1}{2\pi}\int_{-\infty}^{\infty} e^{i\omega[(z/c)-t]}\, d\omega + \frac{1}{r}\int_{-\infty}^{\infty} f(\omega; \theta) e^{i\omega[(r/c)-t]}\, d\omega. \tag{2.201}$$

Initially, we have only the first term, the incident δ-function pulse whose "trajectory" is given by $z = ct$. The scattered wave or the outgoing disturbance should not start until the incident pulse hits the scatterer; otherwise the causality principle would be violated (Fig. 2–5). This requirement takes a particularly simple form in the strictly forward direction $r = z$, $\cos\theta = 1$. We have

$$F(\mathbf{x}, t)|_{\mathbf{x} \text{ on the } z\text{-axis}} = 0 \qquad \text{for} \quad z > ct, \tag{2.202}$$

which because of (2.201) implies the following very stringent requirement for the Fourier transform of the forward scattering amplitude:

$$\tilde{f}(\tau) = [1/(2\pi)^{1/2}]\int_{-\infty}^{\infty} f(\omega) e^{-i\omega\tau}\, d\omega = 0 \qquad \text{for} \quad \tau < 0. \tag{2.203}$$

Assuming that $\tilde{f}(\tau)$ is square integrable, we can invert the Fourier transform as follows:

$$f(\omega) = [1/(2\pi)^{1/2}]\int_{\infty}^{\infty} \tilde{f}(\tau) e^{i\omega\tau}\, d\tau = [1/(2\pi)^{1/2}]\int_{0}^{\infty} \tilde{f}(\tau) e^{i\omega\tau}\, d\tau. \tag{2.204}$$

In other words, $f(\omega)$ is the Fourier transform of a function which vanishes for $\tau < 0$.

So far $f(\omega)$ is defined only for real values of ω. To study its analytic properties we now *define* $f(\omega)$ in the entire upper half of a complex ω-plane by (2.204), that is,

$$f(\omega_r + i\omega_i) = [1/(2\pi)^{1/2}]\int_{0}^{\infty} \tilde{f}(\tau) e^{i(\omega_r + i\omega_i)\tau}\, d\tau \qquad \text{for} \quad \begin{cases} -\infty < \omega_r < +\infty, \\ 0 \leq \omega_i < +\infty, \end{cases} \tag{2.205}$$

where ω_r and ω_i are the real and imaginary parts, respectively, of the complex angular frequency ω. Clearly, $f(\omega)$ with $\omega_i > 0$ is better behaved than $f(\omega)$ with ω purely real because of the exponential damping factor $e^{-\omega_i \tau}$ with $\tau > 0$. Thus $f(\omega)$ is a function analytic in the entire upper half of the ω-plane.

We now consider a line integral

$$\oint \frac{f(\omega')\,d\omega'}{\omega' - \omega_r}, \tag{2.206}$$

whose contour is shown in Fig. 2–6. The point ω_r is an arbitrary point on the real axis, and the line integral along the real axis is to be understood as the principal value integral

$$\int_{-\infty}^{\infty} = \lim_{\epsilon \to 0} \left[\int_{-\infty}^{\omega_r - \epsilon} + \int_{\omega_r + \epsilon}^{\infty} \right]. \tag{2.207}$$

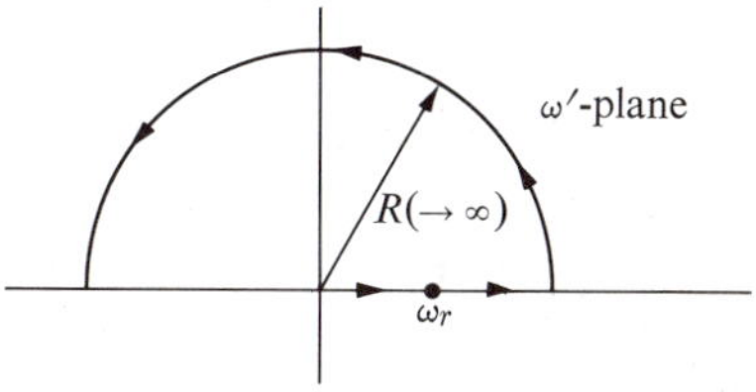

Fig. 2–6. Contour of the line integral (2.206).

There is no contribution from the semicircular path if $|f(\omega)| \to 0$ at infinity faster than $1/|\omega|$. Using the Cauchy relation we obtain‡

$$f(\omega_r) = \frac{1}{\pi i} \int_{-\infty}^{\infty} \frac{f(\omega_r')\,d\omega_r'}{\omega_r' - \omega_r}. \tag{2.208}$$

This leads to a pair of relations

$$\begin{aligned} \text{Re}[f(\omega)] &= \frac{1}{\pi} \int_{-\infty}^{\infty} \frac{\text{Im}[f(\omega')]\,d\omega'}{\omega' - \omega}, \\ \text{Im}[f(\omega)] &= -\frac{1}{\pi} \int_{-\infty}^{\infty} \frac{\text{Re}[f(\omega')]\,d\omega'}{\omega' - \omega}, \end{aligned} \tag{2.209}$$

where, although we have dropped subscripts r, the frequencies ω and ω' are to be understood as real. Whenever a function $f(\omega)$ satisfies (2.209), the real and imaginary parts of $f(\omega)$ are said to be the *Hilbert transforms* of each other.§

In the realistic case of the scattering of light by atoms the coherent forward scattering amplitude $f(\omega)$ does not vanish at infinity. But we can repeat the foregoing analyticity argument for the function $f(\omega)/\omega$. Provided that $f(0) = 0$ [which is satisfied for the scattering of light by a *bound* electron; cf. Rayleigh's law (2.167)] we have

$$\text{Re}[f(\omega)] = \frac{\omega}{\pi} \int_{-\infty}^{\infty} \frac{\text{Im}[f(\omega')]\,d\omega'}{\omega'(\omega' - \omega)}. \tag{2.210}$$

---

‡In (2.208), $1/\pi i$ rather than $1/2\pi i$ appears because the point ω_r lies right on the contour; see Morse and Feshbach (1953), p. 368.

§Titchmarsh (1937), pp. 119–128, gives a mathematically rigorous discussion of the connection among (a) the vanishing of the Fourier transform for $\tau < 0$, (b) the analytical extension of the function of a real variable into the entire upper half plane, and (c) the Hilbert transform relation.

Because of the reality condition (2.200) this can be rewritten as

$$\mathrm{Re}[f(\omega)] = \frac{\omega}{\pi}\int_0^\infty \frac{\mathrm{Im}[f(\omega')]}{\omega'}\left[\frac{1}{\omega'-\omega} - \frac{1}{\omega'+\omega}\right]d\omega'$$

$$= \frac{2\omega^2}{\pi}\int_0^\infty \frac{\mathrm{Im}f(\omega')\,d\omega'}{\omega'(\omega'^2-\omega^2)}, \tag{2.211}$$

which is identical to (2.197).‡

Index of refraction; the optical theorem. The above relation for the scattering amplitude can be converted into an equivalent relation for the *complex index of refraction.* To demonstrate this, we first derive a relation between the forward scattering amplitude and the index of refraction. Let us consider a plane wave $e^{i\omega z/c}$ normally incident on a slab of an infinitesimal thickness δ. (We are omitting the time dependence $e^{-i\omega t}$ throughout.) The slab, assumed to be in the xy-plane, is made up of N scatterers per unit volume. It is not difficult to show that the transmitted wave at $z = l$ is given by

$$e^{i\omega l/c}\left[1 + \frac{2\pi ic}{\omega}N\delta f(\omega)\right], \tag{2.212}$$

for sufficiently large $l \gg c/\omega$. The first term, of course, is due to the original incident wave. To prove that the waves scattered in the various parts of the slab actually add up to the second term, we simply note that the expression

$$N\delta\int_0^\infty \frac{e^{i(\omega/c)\sqrt{\rho^2+l^2}}}{\sqrt{\rho^2+l^2}}f(\omega,\theta)2\pi\rho\,d\rho \tag{2.213}$$

[where $\rho^2 = x^2 + y^2$, $\theta = \tan^{-1}(\rho/l)$] can be integrated by parts to give the second term of (2.212) plus a term that varies as $(c/\omega l)$, which is negligible for large l.§ Consider now a slab of finite thickness D. The phase change of the transmitted wave can be computed by compounding the infinitesimal phase change given by (2.212). We have

$$\lim_{n\to\infty}\left[1 + \frac{2\pi icNDf(\omega)}{\omega n}\right]^n = e^{(2\pi icN/\omega)f(\omega)D}. \tag{2.214}$$

This should be compared to $e^{i\omega n(\omega)D/c}$, where $n(\omega)$ is recognized to be the complex index of refraction. In this manner, we obtain an equation that relates the microscopic quantity, the forward scattering amplitude, to the macroscopically measurable quantity, the complex index of refraction:

$$n(\omega) = 1 + 2\pi(c/\omega)^2Nf(\omega) \tag{2.215}$$

(first derived by H. A. Lorentz).

---

‡If $f(\omega)$ does not vanish at $\omega = 0$, simply replace $\mathrm{Re}[f(\omega)]$ by $\mathrm{Re}[f(\omega) - f(0)]$. In this connection we mention that $f(0)$ for the scattering of light by any *free* particle of mass M and electric charge q can rigorously be shown to be given by the Thomson amplitude $-(q^2/4\pi Mc^2)$.

§Strictly speaking, the integral (2.213) is ill defined because the integrand oscillates with a finite amplitude even for large values of ρ. Such an integral can be made convergent by setting $\omega \to \omega + i\epsilon$, where ϵ is an infinitesimal real positive parameter.

It is now evident that we can write an integral representation for $n(\omega)$ analogous to (2.211):

$$\operatorname{Re}[n(\omega)] = 1 + \frac{2}{\pi}\int_0^\infty \frac{\omega' \operatorname{Im}[n(\omega')]\, d\omega'}{\omega'^2 - \omega^2} = 1 + \frac{c}{\pi}\int_0^\infty \frac{\alpha(\omega')d\omega'}{\omega'^2 - \omega^2}, \tag{2.216}$$

where $\alpha(\omega)$ is the absorption coefficient defined by

$$\alpha(\omega) = (2\omega/c) \operatorname{Im}[n(\omega)]. \tag{2.217}$$

The relation (2.216) is precisely the original dispersion relation derived by Kramers and Kronig.

As another useful application of (2.215) we can show that the imaginary part of $f(\omega)$ is related to the total cross section σ_{tot}. First let us note that the intensity of the transmitted wave attenuates as

$$e^{-\sigma_{\text{tot}} N z} = |e^{i(\omega n z/c)}|^2 = e^{-(2\omega/c) z \operatorname{Im}(n)}. \tag{2.218}$$

It then follows from (2.215) that

$$\operatorname{Im}[f(\omega)] = (1/4\pi)(\omega/c)\sigma_{\text{tot}}(\omega). \tag{2.219}$$

If we write it as

$$\sigma_{\text{tot}}(\omega) = \frac{4\pi \operatorname{Im}[f(\omega)]}{|\mathbf{k}|}, \tag{2.220}$$

we see that it is identical in form to the *optical theorem* (the Bohr-Peierls-Placzek relation) in wave mechanics which can be derived from the requirement of the conservation law of probability.‡ Because of this theorem we can rewrite the dispersion relation as

$$\operatorname{Re}[f(\omega)] = \frac{\omega^2}{2\pi^2 c}\int_0^\infty \frac{\sigma_{\text{tot}}(\omega')\, d\omega'}{\omega'^2 - \omega^2}. \tag{2.221}$$

To avoid any possible misunderstanding we emphasize that the optical theorem (2.219) is exact, whereas the expression for the imaginary part given by (2.194), being based on perturbation theory, is approximate. According to (2.194) the imaginary part of $f(\omega)$ vanishes when $\omega \neq \omega_{IA}$; yet (2.188) tells us that there must be a finite amount of scattering even when $\omega \neq \omega_{IA}$. There is no inconsistency here. The nonresonant cross section obtained by squaring the perturbation amplitude is of the order of e^4; hence within the framework of an approximation in which only terms of order e^2 are kept, σ_{tot}, or equivalently, $\operatorname{Im}[f(\omega)]$, does vanish except at $\omega \approx \omega_{IA}$. In this connection we may recall that when we originally inferred the validity of the dispersion relation using (2.193) and (2.194), we were merely equating terms of order e^2, ignoring the e^4 contributions. In any event, the approximation implied in (2.194) is a very good one, since the cross section at resonance is greater, by many orders of magnitude, than the nonresonant cross section.

The utility of relations such as (2.209) and (2.211) is not limited to the scattering of light. Similar relations can be written for a wide class of problems in physics whenever there is a causal connection between the input and the output, or, more

‡Merzbacher (1961), p. 499.

precisely, whenever (a) the response (the output) of a system to a disturbance (the input) vanishes before the disturbance is applied, and (b) the output is a linear functional of the input. For example, in ac network theory where the input and the output are an impressed current and the resulting voltage, respectively, the real and imaginary parts of the complex impedance are the Hilbert transforms of each other. In ferromagnetism it is the magnetic susceptibility (which relates an applied magnetic field to the resulting magnetization) that satisfies a Kramers-Kronig relation. Similar relations are valid for the dielectric constant, the electrical conductivity, and so forth.

Finally, we should mention that the real and imaginary parts of the amplitudes for the forward scattering of $\pi^{\pm}$ mesons on protons satisfy relations somewhat similar to (2.21), as proved by M. L. Goldberger and others on the basis of very general principles of quantum field theory.‡ These relations are particularly useful since they supply nontrivial connections among directly observable quantities (the real part of the forward scattering amplitude and the total cross section) in a domain where not much is known about the detailed form of the dynamics.

2–8. THE SELF-ENERGY OF A BOUND ELECTRON; THE LAMB SHIFT

Self-energy problem. Suppose we consider an electron (bound or free). The very presence of the electron implies that there is an electromagnetic field accompanying the electron. Classically the electromagnetic field due to the presence of the electron can react on the electron itself, a point discussed extensively by H. Poincaré, M. Abraham, and H. A. Lorentz.§ In the quantum theory of radiation there is an analogous effect which can be visualized as the following two-step process:

a) the electron emits a (virtual) photon, and

b) the photon is subsequently absorbed by the same electron.

The interaction energy between the electron and the electromatnetic field (classical or quantized) due to the presence of the electron itself is called the *self-energy* of the electron. For all practical purposes the self-energy forms an inseparable part of the observable rest energy of the electron since there is no way of turning off the electromagnetic interaction. We intend to show in this section that the dynamical mechanisms that give rise to the (unobservable) self-energy have profound consequences on the (observable) energy levels of atomic states.

The problem of self-energy can be discussed at several levels. To start with, this problem arises even in classical electrostatics. Suppose we consider a model of the electron which has a charge distribution ρ. Let ϕ be the electrostatic potential due to the presence of the electron. The interaction energy is then given by

$$E_{\text{int}} = \frac{1}{2} \int \rho\phi \, d^3x. \tag{2.222}$$

‡For the proof of the pion-nucleon dispersion relation see, for example, Källén (1964), Chapter 5.

§Jackson (1962), pp. 578–597.

If the electron were a point particle of charge e located at the origin, we would have $\rho = e\delta^{(3)}(\mathbf{x})$ and $\phi = e/4\pi r$; hence E_{int} would be infinite. Instead, we may consider a spherical charge distribution of uniform density. In such a model, if a substantial part of the electron rest energy is to be attributed to the self-energy, then the "radius" of the electron must be of the order of the classical radius of the electron $r_0 = (e^2/4\pi mc^2)$. If the self-energy of the electron is computed quantum-mechanically using the relativistic electron theory of Dirac, it is still infinite, but the divergence involved can be shown to be much weaker. We shall come back to this point in Section 4–7.

Atomic level shifts. We shall now discuss the self-energy of a bound, atomic electron within the framework of the quantum theory of radiation we have developed in which the transverse electromagnetic field is quantized but the electron is treated nonrelativistically. Consider an atomic state A (which is not necessarily assumed to be a stable ground state). The self-energy of the bound electron manifests itself as the energy shift of state A due to its interaction with the quantized radiation field. As we emphasized earlier, according to the quantum theory of radiation, even if there is no incident radiation present, the atomic state A can emit a photon. Let us now suppose that the atom subsequently absorbs the emitted photon and returns to state A. To order e^2 there are two classes of processes, as represented by Fig. 2–7. Of the two diagrams, the result shown in Fig. 2–7(a) due to the $\mathbf{A}\cdot\mathbf{A}$ interaction term in the interaction Hamiltonian can be shown to be uninteresting from the point of view of atomic level shifts (cf. Problem 2–11). The basic perturbation matrix elements in Fig. 2–7(b) are the *emission* matrix element

$$H_{IA}^{(\text{emis})} = H'_{IA}e^{i\omega t} \tag{2.223}$$

for $A \to I + \gamma$ and the *absorption* matrix element

$$H_{IA}^{(\text{abs})} = H'_{AI}e^{-i\omega t} \tag{2.224}$$

for $I + \gamma \to A$, where the time-independent matrix elements H'_{IA} and H'_{AI} are given by

$$H'_{IA} = (H'_{AI})^* = -c\sqrt{\hbar/2\omega V}(e/mc)\langle I|\mathbf{p}\cdot\boldsymbol{\epsilon}^{(\alpha)}|A\rangle \tag{2.225}$$

just as before if the dipole approximation is assumed.

To obtain the time-dependent amplitudes c_I and c_A we must solve the following coupled differential equations:

$$i\hbar\dot{c}_I = \sum_{\text{photon}} H_{IA}^{(\text{emis})} c_A e^{i(E_I - E_A)t/\hbar}, \tag{2.226}$$

$$i\hbar\dot{c}_A = \sum_{\text{photon}} \sum_I H_{AI}^{(\text{abs})} c_I e^{i(E_A - E_I)t/\hbar}. \tag{2.227}$$

Since the quantized field operator $\mathbf{A}$ has a nonvanishing matrix element between the vacuum state and a single photon state of *any* momentum and polarization, the sum implied by the symbol

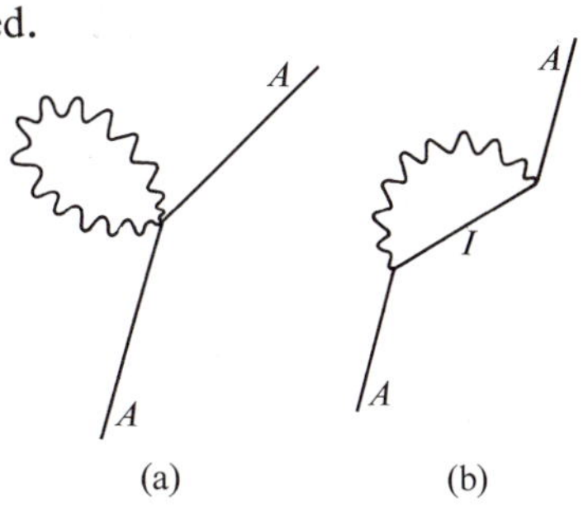

Fig. 2–7. The self-energy of a bound electron.

$\sum_{\text{photon}}$ is over all possible momenta and polarizations, even those for which the photon energy is *not* equal to $E_I - E_A$.

We are interested in obtaining an expression for the energy shift of state A. For this reason we propose to solve (2.226) and (2.227) by making the *ansatz*

$$c_A = \exp[-i\Delta E_A t/\hbar]. \tag{2.228}$$

With this ansatz the complete wave function varies as

$$\psi \sim u_A(\mathbf{x}) \exp[-i(E_A + \Delta E_A)t/\hbar], \tag{2.229}$$

so that the time dependence of the wave function is precisely what we expect from a quantum-mechanical state of energy $E_A + \Delta E_A$. Inserting (2.228) into (2.226) and integrating, we have

$$\begin{aligned} c_I &= \sum_{\text{photon}} \frac{H'_{IA}}{i\hbar} \int_0^t \exp[i\omega t'] \exp\left[-\frac{i\Delta E_A t'}{\hbar}\right] \exp\left[\frac{i(E_I - E_A)t'}{\hbar}\right] dt' \\ &= \sum_{\text{photon}} \frac{H'_{IA}(\exp[i(E_I - E_A - \Delta E_A + \hbar\omega)t/\hbar] - 1)}{E_A + \Delta E_A - E_I - \hbar\omega}. \end{aligned} \tag{2.230}$$

To obtain the energy shift for state A we substitute (2.228) and (2.230) into (2.227) and divide by $\exp[-i\Delta E_A t/\hbar]$:

$$\Delta E_A = \sum_{\text{photon}} \sum_I \frac{|H'_{IA}|^2 (1 - \exp[i(E_A + \Delta E_A - E_I - \hbar\omega)t/\hbar])}{E_A + \Delta E_A - E_I - \hbar\omega}. \tag{2.231}$$

Since we are computing ΔE_A to second order in e, it is legitimate to drop ΔE_A on the right-hand side of (2.231). Hence

$$\Delta E_A = \sum_{\text{photon}} \sum_I \frac{|H'_{IA}|^2 (1 - \exp[i(E_A - E_I - \hbar\omega)t/\hbar])}{E_A - E_I - \hbar\omega}. \tag{2.232}$$

The time interval during which the perturbation acts may now be made to go to infinity. The expression as it stands, however, indicates oscillation as $t \to \infty$. This is not surprising since we obtained (2.232) from an integral of the form (cf. Eq. 2.230)

$$\int_0^t e^{ixt'} dt', \tag{2.233}$$

which is obviously ill defined as $t \to \infty$. However, such an integral can be made convergent by adding a small positive imaginary part to x. In this way we obtain a very useful set of relations:

$$\begin{aligned} \lim_{t\to\infty} \frac{1 - e^{ixt}}{x} &= -\lim_{\epsilon\to 0+} i \int_0^\infty e^{i(x+i\epsilon)t'} dt' \\ &= \lim_{\epsilon\to 0+} \frac{1}{x + i\epsilon} \\ &= \lim_{\epsilon\to 0+} \left[\frac{x}{x^2 + \epsilon^2} - \frac{i\epsilon}{x^2 + \epsilon^2}\right] \\ &= \frac{1}{x} - i\pi\delta(x). \end{aligned} \tag{2.234}$$

These are directly applicable to (2.233) with $t \to \infty$.

The energy shift ΔE_A is now seen to have both a real and an imaginary part:

$$\mathrm{Re}\,(\Delta E_A) = \sum_{\text{photon}} \sum_I \frac{|H'_{IA}|^2}{E_A - E_I - \hbar\omega}, \tag{2.235}$$

$$\mathrm{Im}\,(\Delta E_A) = -\pi \sum_{\text{photon}} \sum_I |H'_{IA}|^2 \delta(E_A - E_I - \hbar\omega). \tag{2.236}$$

The photon sum in (2.235) is over photons of all possible momenta and polarizations, and the atomic level sum $\sum_I$ in (2.235) is also unrestricted in the sense that $E_I < E_A$ need not be satisfied (in contrast with a "real" emission process). This means that the emission and absorption of the photons in (2.235) do not in general satisfy energy conservation, hence they are not like the emission and absorption processes observed for actual photons (discussed in Section 2–4). Such photons are said to be "virtual." We may imagine that the physical atom is part of the time in a dissociated state "atom + virtual photon." It is the interaction energy associated with the energy-nonconserving emission and absorption of virtual photons of all possible momenta and polarizations that gives rise to the real part of the energy shift.

In sharp contrast with the photon sum in (2.235), the photon sum in (2.236) is over only those photons which satisfy energy conservation $E_A = E_I + \hbar\omega$. In other words the photons that appear in (2.236) are "real" as opposed to "virtual." The energy shift ΔE_A acquires an imaginary part only when it is possible for state A to decay into state I via spontaneous emission without violating energy conservation. More quantitatively, we note that

$$-\frac{2}{\hbar}\,\mathrm{Im}\,(\Delta E_A) = \sum_{\text{photon}} \sum_I \frac{2\pi}{\hbar} |H'_{IA}|^2 \delta(E_A - E_I - \hbar\omega). \tag{2.237}$$

But the right-hand side of this expression is precisely the expression for the transition probability for spontaneous emission computed according to the Golden Rule with all the energetically allowed final states summed over (cf. Eqs. 2.114 through 2.117). Hence it is equal to the reciprocal of the mean lifetime of state A. Thus we have the very important result

$$-\frac{2}{\hbar}\,\mathrm{Im}\,[\Delta E_A] = \frac{1}{\tau_A} = \frac{\Gamma_A}{\hbar}. \tag{2.238}$$

The physical significance of $\mathrm{Im}\,[\Delta E_A]$ is now clear. Going back to (2.229) we see that the complete wave function is given by

$$\psi \sim u_A(\mathbf{x}) \exp\left[-\frac{i(E_A + \mathrm{Re}\,[\Delta E_A])t}{\hbar} - \frac{\Gamma_A t}{2\hbar}\right]. \tag{2.239}$$

From this follows the familiar conclusion: the probability of finding the unstable state A decreases as

$$|\psi|^2 \sim e^{-\Gamma_A t/\hbar}. \tag{2.240}$$

To sum up, the real part of the energy shift ΔE_A (arising from the emission and absorption of virtual photons) is what we normally call the level shift, whereas the imaginary part of ΔE_A (arising from the emission and absorption of real photons) characterizes the decay width, or the reciprocal lifetime, of the unstable state in question. We also note that our earlier phenomenological treatment of radiation damping, given in Section 2–6, is now justified since the second term in the differential equation (2.181) can be viewed as arising from the imaginary part of ΔE_A.

From now on we shall concentrate on only the real part of ΔE_A. (In the following the expression ΔE_A shall mean its real part.) Inserting the explicit form (2.225) into (2.235), we have

$$\Delta E_A = \frac{c^2\hbar}{V}\left(\frac{e}{mc}\right)^2 \sum_I \int \frac{d^3k V}{(2\pi)^3}\frac{1}{2\omega} \sum_\alpha \frac{|(\mathbf{p}\cdot\boldsymbol{\epsilon}^{(\alpha)})_{IA}|^2}{E_A - E_I - \hbar\omega}. \tag{2.241}$$

The polarization sum and the angular integration are identical with those encountered in our earlier discussion of spontaneous emission (cf. Eqs. 2.130 through 2.132),

$$\int d\Omega \sum_\alpha |(\mathbf{p}\cdot\boldsymbol{\epsilon}^{(\alpha)})_{IA}|^2 = (8\pi/3)|(\mathbf{p})_{IA}|^2. \tag{2.242}$$

We are thus left with just the energy integration

$$\Delta E_A = \frac{2}{3\pi}\left(\frac{e^2}{4\pi\hbar c}\right)\frac{1}{(mc)^2} \sum_I \int \frac{E_\gamma |(\mathbf{p})_{IA}|^2 \, dE_\gamma}{E_A - E_I - E_\gamma}, \tag{2.243}$$

where $E_\gamma = \hbar\omega = \hbar|\mathbf{k}|c$. This energy integration is over all possible values of E_γ, from zero to infinity. But the integral clearly diverges linearly. We may argue that we cannot take too seriously the contributions from virtual photons of very high energies since the nonrelativistic approximation for the electron must break down for the emission of a photon of $E_\gamma \gtrsim mc^2$. So we replace

$$\int_0^\infty dE_\gamma \longrightarrow \int_0^{E_\gamma^{(\max)}} dE_\gamma, \tag{2.244}$$

where $E_\gamma^{(\max)}$ is known as the *cut-off energy*. Unfortunately, the energy shift computed in this way is very sensitive to the arbitrary parameter $E_\gamma^{(\max)}$ which we introduced.

Mass renormalization. At this stage it will be advantageous to discuss the self-energy of a free electron of a fixed momentum $\mathbf{p}$ due to the process shown in Fig. 2–7(b). This time the wave functions for states A and I are represented by $e^{i\mathbf{p}\cdot\mathbf{x}/\hbar}/\sqrt{V}$ and $e^{i\mathbf{p}'\cdot\mathbf{x}/\hbar}/\sqrt{V}$, respectively. The time-independent perturbation matrix element H'_{IA} in (2.235) is given by

$$\begin{aligned} H'_{IA} &= -c\sqrt{\frac{\hbar}{2\omega}}\frac{e}{mc}\frac{1}{V^{3/2}} \int e^{-i\mathbf{p}'\cdot\mathbf{x}/\hbar} e^{-i\mathbf{k}\cdot\mathbf{x}} (\mathbf{p}\cdot\boldsymbol{\epsilon}^{(\alpha)}) e^{i\mathbf{p}\cdot\mathbf{x}/\hbar}\, d^3x \\ &= -\sqrt{\frac{\hbar}{2\omega}}\frac{e}{m}\frac{1}{V^{1/2}}\mathbf{p}\cdot\boldsymbol{\epsilon}^{(\alpha)}\,\delta_{\mathbf{p}',\mathbf{p}-\hbar\mathbf{k}}. \end{aligned} \tag{2.245}$$

The energy denominator is

$$E_A - E_I - \hbar\omega = \mathbf{p}^2/2m - (\mathbf{p} - \hbar\mathbf{k})^2/2m - \hbar\omega \approx -\hbar\omega \tag{2.246}$$

in the energy regions where the electron can be treated nonrelativistically. The remaining steps are the same as before. We obtain for the self-energy of the free electron

$$\Delta E_{\text{free}} = -\left(\frac{e^2}{4\pi\hbar c}\right)\frac{2\mathbf{p}^2}{3\pi(mc)^2}\int_0^{E_\gamma^{(\max)}} dE_\gamma = C\mathbf{p}^2, \tag{2.247}$$

where

$$C = -\left(\frac{e^2}{4\pi\hbar c}\right)\frac{2}{3\pi}\frac{E_\gamma^{(\max)}}{(mc)^2}\,. \tag{2.248}$$

Note that (2.247) is also linearly divergent as $E_\gamma^{(\max)} \to \infty$.

We see that Fig. 2–7(b) for the free electron simply gives rise to an energy proportional to $\mathbf{p}^2$. To the extent that we can never switch off the electromagnetic interaction, this additional energy cannot be separated from the usual kinetic energy term $\mathbf{p}^2/2m$. In fact, the observable kinetic energy is the sum of the kinetic energy in a hypothetical world in which there is no interaction of the type represented by Fig. 2–7(b) and the $C\mathbf{p}^2$ we computed. Meanwhile, if the electron is regarded as a nonrelativistic particle, the electron mass is given by the energy momentum relation

$$\partial E/\partial \mathbf{p}^2 = 1/2m, \tag{2.249}$$

where E is the energy of a free particle. The observed mass m_{obs} of the electron is then related to C by

$$\mathbf{p}^2/2m_{\text{obs}} = \mathbf{p}^2/2m_{\text{bare}} + C\mathbf{p}^2 \approx \mathbf{p}^2/[2m_{\text{bare}}(1 - 2m_{\text{bare}}C)], \tag{2.250}$$

where m_{bare} stands for the electron mass we would measure in the absence of the interaction represented by Fig. 2–7(b). To summarize, the net effect of Fig. 2–7(b) is only to change the definition of the electron mass:

$$m_{\text{bare}} \longrightarrow m_{\text{obs}} \approx \left[1 + \left(\frac{1}{137}\right)\frac{4}{3\pi}\frac{E_\gamma^{(\max)}}{mc^2}\right]m_{\text{bare}}. \tag{2.251}$$

The mass we observe in our laboratory is not m_{bare} but m_{obs}, sometimes known as the *renormalized mass*.‡ Note that $\Delta E^{(\text{free})}$ in (2.247) is negative while $\Delta m = m_{\text{obs}} - m_{\text{bare}}$ is positive. From (2.251) we see that for a cut-off $E_\gamma^{(\max)}$ of order mc^2, about 0.3% of the observed electron mass is attributed to the self-energy. We cannot, however, attach much significance to this kind of calculation since we obtain a very different expression for the electron self-energy if Dirac's relativistic electron theory is used.

Now let us return to the level shifts of atomic states. When we solve the Schrödinger equation to compute the energy levels of an atom, the kinetic energy we use is $\mathbf{p}^2/2m_{\text{obs}}$, which already includes the correction term $C\mathbf{p}^2$. In estimating the observable energy shifts, we must make sure to subtract that part of the energy which is already accounted for, since we used $\mathbf{p}^2/2m_{\text{obs}}$ instead of $\mathbf{p}^2/2m_{\text{bare}}$ in computing the unperturbed energy levels. This is the idea of *mass renormalization*

‡The reader who is familiar with solid-state physics may recognize that the notion of the observed or renormalized mass in the quantum theory of radiation is somewhat like that of the effective mass of an electron in interaction with lattice vibrations of solids.

proposed by H. A. Kramers. Quantitatively, the *observable* energy shift $\Delta E_A^{(\text{obs})}$ is given by

$$\begin{aligned}\Delta E_A^{(\text{obs})} &= \Delta E_A - \langle A\,|(\mathbf{p}^2/2m_{\text{obs}} - \mathbf{p}^2/2m_{\text{bare}})|\,A\rangle \\ &= \Delta E_A - C\langle A\,|\,\mathbf{p}^2\,|\,A\rangle. \end{aligned} \tag{2.252}$$

In other words, what is observable is the difference of the bound-electron self-energy and the free-electron self-energy.

Bethe's treatment of the Lamb shift. Stimulated by precise level-shift measurements made possible by the wartime development in microwave techniques, H. A. Bethe applied Kramer's idea of mass renormalization in computing the energy shifts of the hydrogen atom due to the interaction shown in Fig. 2–7(b). From (2.243) and (2.247) we obtain

$$\begin{aligned}\Delta E_A^{(\text{obs})} &= \left(\frac{e^2}{4\pi\hbar c}\right)\frac{2}{3\pi(mc)^2}\int_0^{E_\gamma^{(\max)}}\left(\sum_I \frac{E_\gamma\,|(\mathbf{p})_{IA}|^2}{E_A - E_I - E_\gamma} + (\mathbf{p}^2)_{AA}\right)dE_\gamma \\ &= \left(\frac{e^2}{4\pi\hbar c}\right)\frac{2}{3\pi(mc)^2}\int_0^{E_\gamma^{(\max)}}\sum_I \frac{|(\mathbf{p})_{IA}|^2(E_A - E_I)}{E_A - E_I - E_\gamma}\,dE_\gamma \\ &= \left(\frac{e^2}{4\pi\hbar c}\right)\frac{2}{3\pi(mc)^2}\sum_I |(\mathbf{p})_{IA}|^2(E_I - E_A)\log\left(\frac{E_\gamma^{(\max)}}{|E_I - E_A|}\right) \\ &= \left(\frac{e^2}{4\pi\hbar c}\right)\frac{2}{3\pi(mc)^2}\log\left(\frac{E_\gamma^{(\max)}}{\langle E_I - E_A\rangle_{\text{average}}}\right)\sum_I |(\mathbf{p})_{IA}|^2(E_I - E_A), \end{aligned} \tag{2.253}$$

where we have used

$$(\mathbf{p}^2)_{AA} = \sum_I |(\mathbf{p})_{IA}|^2 \tag{2.254}$$

and assumed that $E_\gamma^{(\max)}$ is much greater than $E_I - E_A$. Note that $\Delta E_A^{(\text{obs})}$ now depends much less sensitively on $E_\gamma^{(\max)}$. Our expression still diverges as $E_\gamma^{(\max)} \to \infty$, but the divergence is logarithmic rather than linear.

Summing over I is facilitated by the observation that

$$\sum_I |(\mathbf{p})_{IA}|^2(E_A - E_I) = -\tfrac{1}{2}\hbar^2 \int |\psi_A|^2 \nabla^2 V\, d^3x, \tag{2.255}$$

which can be proved by means of the following three steps. First, note that for an unperturbed Hamiltonian of the form $(\mathbf{p}^2/2m) + V(\mathbf{x})$, we have

$$\mathbf{p}H_0 - H_0\mathbf{p} = -i\hbar\nabla V. \tag{2.256}$$

Second, take the matrix element of (2.256) between I and A:

$$\mathbf{p}_{IA}E_A - E_I\mathbf{p}_{IA} = -i\hbar(\nabla V)_{IA} \tag{2.257}$$

Third, multiply $\mathbf{p}_{AI}$ from the left, sum over I, and note that the resulting expression must be real:

$$\begin{aligned}\sum_I |(\mathbf{p})_{IA}|^2(E_A - E_I) &= -\sum_I i\hbar\mathbf{p}_{AI}\cdot(\nabla V)_{IA} \\ &= \sum_I i\hbar(\nabla V)_{AI}\cdot\mathbf{p}_{IA} \\ &= -(i\hbar/2)[\mathbf{p}, \nabla V]_{AA} \\ &= -(\hbar^2/2)(\nabla^2 V)_{AA}, \end{aligned} \tag{2.258}$$

which yields (2.255). For the hydrogen atom we have

$$\nabla^2 V = e^2 \delta^{(3)}(\mathbf{x}). \tag{2.259}$$

Let us also recall that the atomic wave function vanishes at the origin except for s states. The integral that appears in (2.255) is

$$\int |\psi_A|^2 \nabla^2 V d^3x = e^2 |\psi_A(0)|^2 = \begin{cases} (e^2/\pi n^3 a_0^3) & \text{for } s \text{ states,} \\ 0 & \text{otherwise,} \end{cases} \tag{2.260}$$

where

$$a_0 = \frac{(\hbar/mc)}{(e^2/4\pi\hbar c)} \tag{2.261}$$

is the Bohr radius of the hydrogen atom. The observable energy level shift now becomes

$$\Delta E_A^{(\text{obs})} = \frac{8}{3\pi}\left(\frac{e^2}{4\pi\hbar c}\right)^3 \left(\frac{e^2}{8\pi a_0}\right)\frac{1}{n^3} \log\left(\frac{E_\gamma^{(\max)}}{\langle E_I - E_A\rangle_{\text{ave}}}\right) \tag{2.262}$$

for s states, where we may recall that $(e^2/8\pi a_0)$ is the ionization energy for the ground state of the hydrogen atom (sometimes denoted by Ry, the Rydberg energy). For states with $l \neq 0$, no observable level shifts are expected.

It is well known that in the Schrödinger theory the energy levels of the hydrogen atom depend only on the principal quantum number n. As we shall see in Chapter 3, if we solve the hydrogen atom problem using the Dirac equation, the above degeneracy is partially removed by the spin-orbit force, but states with the same n and j are still degenerate. In particular, the $2s\frac{1}{2}$ and $2p\frac{1}{2}$ states are exactly degenerate according to the Dirac theory. Although the possible existence of an energy difference between the $2s\frac{1}{2}$ and $2p\frac{1}{2}$ states was speculated upon by S. Pasternack and others even before the 1940's, no precise measurement was made of this energy difference until the development of microwave techniques during the war. In 1947 W. E. Lamb and R. C. Retherford showed that the radio frequency corresponding to the $2s$–$2p\frac{1}{2}$ separation is about 1060 Mc, the $2s$ state being the higher of the two.‡ Bethe derived the formula (2.262) to account for the observation of Lamb and Retherford, usually referred to as the Lamb shift. The average excitation energy $\langle E_I - E_A\rangle_{\text{ave}}$ was estimated numerically for the $2s$ state to be 17.8 times the Rydberg energy, and when $E_\gamma^{(\max)}$ was set to be mc^2 (to allow for the breakdown of the whole approach in the relativistic domain), the calculated upward shift for the $2s$ state turned out to be 1040 Mc, in remarkable agreement with observation.

Subsequent relativistic calculations, which will be discussed in Section 4–7, have led to a completely convergent result for this separation.§ The *theoretical*

---

‡To appreciate the order of magnitude of the Lamb shift note that the reciprocal wavelength corresponding to 1060 Mc is 0.035 cm^{-1}, compared with 27,000 cm^{-1} corresponding to the ionization energy of the ground state.

§This is because the self-energy of the free or bound electron is only logarithmically divergent in the relativistic treatment, as first shown by V. F. Weisskopf. Crudely speaking, the difference between the two logarithmically divergent expressions that appear in the relativistic approach turns out to be finite just as the difference between the two linearly divergent expressions in (2.253) is only logarithmically divergent.

value 1057.7 $\pm$ 0.2 Mc is to be compared with the *experimental* value 1057.8 $\pm$ 0.1 Mc.‡ In any case, the semiquantitative success of Bethe's treatment indicates that the Lamb shift is basically a nonrelativistic low-frequency effect. What is more important, this treatment illustrates how we can extract sensible numbers that can be compared with observation despite the divergence difficulties inherent in the present form of field theory. The idea of mass renormalization which is so basic to the success of Bethe's calculation turns out to play a role just as essential in the more sophisticated relativistic treatment of the Lamb shift.

PROBLEMS

2–1. The annihilation and creation operators of the electron denoted by $b_{\mathbf{k}s}$ and $b^{\dagger}_{\mathbf{k}s}$ are expected to satisfy the anticommutation relation

$$\{b_{\mathbf{k}s}, b_{\mathbf{k}'s'}\}=0, \qquad \{b_{\mathbf{k}s}, b^{\dagger}_{\mathbf{k}'s'}\} = \delta_{\mathbf{k}\mathbf{k}'}\delta_{ss'},$$

where s characterizes the spin state up ($\uparrow$) or down ($\downarrow$). In the Bardeen-Cooper-Schrieffer theory of superconductivity, the annihilation and creation operators for a correlated pair of electrons are defined by

$$c_{\mathbf{k}} = b_{-\mathbf{k}\downarrow} b_{\mathbf{k}\uparrow}, \qquad c^{\dagger}_{\mathbf{k}} = b^{\dagger}_{\mathbf{k}\uparrow} b^{\dagger}_{-\mathbf{k}\downarrow}.$$

Show that $[c_{\mathbf{k}}, c_{\mathbf{k}'}] = 0$, $[c^{\dagger}_{\mathbf{k}}, c^{\dagger}_{\mathbf{k}'}] = 0$. Evaluate $[c_{\mathbf{k}}, c^{\dagger}_{\mathbf{k}'}]$.

2–2. Show that the three independent nonvanishing components of an antisymmetric dyadic

$$\mathbf{S}_{k} = -i\hbar(\boldsymbol{\epsilon}^{(i)}\boldsymbol{\epsilon}^{(j)} - \boldsymbol{\epsilon}^{(j)}\boldsymbol{\epsilon}^{(i)})$$
$$= i\hbar(\boldsymbol{\epsilon}^{(k)} \times \quad)$$

satisfy the angular momentum commutation relation

$$\mathbf{S}_i\cdot\mathbf{S}_j - \mathbf{S}_j\cdot\mathbf{S}_i = i\hbar\mathbf{S}_k,$$

where (ijk) are cyclic permutations of (1, 2, 3), and $\boldsymbol{\epsilon}^{(3)}$ is defined by $\mathbf{k}/|\mathbf{k}|$. Show also that

$$\mathbf{S}_3\cdot\mathbf{u}_{\mathbf{k}\pm} = \pm\hbar\mathbf{u}_{\mathbf{k}\pm}, \qquad \sum_i^3 (\mathbf{S}_i\cdot\mathbf{S}_i)\mathbf{u}_{\mathbf{k}\pm} = 2\hbar^2\mathbf{u}_{\mathbf{k}\pm},$$

where

$$\mathbf{u}_{\mathbf{k}\pm} = \mp\frac{\boldsymbol{\epsilon}^{(1)} \pm i\boldsymbol{\epsilon}^{(2)}}{\sqrt{2}}e^{i\mathbf{k}\cdot\mathbf{x}}.$$

2–3. When quantized, the neutral scalar field can be expanded as follows:

$$\phi(\mathbf{x}, t) = \sum_{\mathbf{k}} c\sqrt{\frac{\hbar}{2\omega V}}(a_{\mathbf{k}}(t)e^{i\mathbf{k}\cdot\mathbf{x}} + a^{\dagger}_{\mathbf{k}}(t)e^{-i\mathbf{k}\cdot\mathbf{x}});$$

$$\omega/c = \sqrt{\mathbf{k}^2 + (mc/\hbar)^2}, \qquad a_{\mathbf{k}}(t) = a_{\mathbf{k}}(0)e^{-i\omega t},$$

$$[a_{\mathbf{k}}, a_{\mathbf{k}'}] = [a^{\dagger}_{\mathbf{k}}, a^{\dagger}_{\mathbf{k}'}] = 0, \qquad [a_{\mathbf{k}}, a^{\dagger}_{\mathbf{k}'}] = \delta_{\mathbf{k}\mathbf{k}'}.$$

---

‡See, however, the first footnote on p. 294.

a) Prove the equal-time commutation relation

$$[\phi(\mathbf{x}, t), \pi(\mathbf{x}', t)] = i\hbar\delta^{(3)}(\mathbf{x} - \mathbf{x}'),$$

where

$$\pi(\mathbf{x}, t) = (1/c^2)(\partial\phi/\partial t).$$

b) Suppose we define the *average field operator* $\bar{\phi}$ in the neighborhood of the origin by

$$\bar{\phi} = [1/(2\pi b^2)^{3/2}] \int d^3x e^{-r^2/2b^2} \phi(\mathbf{x}, t).$$

Show that apart from a numerical factor the vacuum expectation value of $(\bar{\phi})^2$ is given by $c\hbar/b^2$, provided $b \ll \hbar/mc$.

2–4. Consider the photoelectric effect of the ground state of the hydrogen atom.

a) Using the quantum theory of radiation (as opposed to the semiclassical arguments found in many texts), write the transition matrix element in lowest order.

b) Show that the differential cross section is

$$\frac{d\sigma}{d\Omega} = 32\left(\frac{e^2}{4\pi\hbar c}\right)\left(\frac{\hbar}{mc}\right)\left(\frac{c}{\omega}\right)\frac{1}{(|\mathbf{k}_f|a_0)^5}\frac{\sin^2\theta\cos^2\phi}{[1-(v/c)\cos\theta]^4},$$

if the energy of the incident photon is so large that the final-state wave function of the ejected electron can be approximated by a plane wave. The spherical coordinate variables θ and ϕ are defined in such a way that the incident photon momentum and polarization are along the z- and x-axes respectively, and a_0 stands for the Bohr radius.

2–5. The phenomenological interaction Hamiltonian responsible for the decay of the Σ^0 hyperon ($\Sigma^0 \to \Lambda + \gamma$) located at $\mathbf{x} = 0$ can be taken as

$$\frac{\kappa e\hbar}{(m_\Lambda + m_\Sigma)c}\tau_{\Lambda\Sigma}\boldsymbol{\sigma}\cdot(\nabla \times \mathbf{A})\Big|_{\mathbf{x}=0},$$

where $\tau_{\Lambda\Sigma}$ is an operator that converts Σ^0 into Λ, leaving the spin state unchanged, and κ is a dimensionless constant assumed to be of the order of unity.

a) Show that the angular distribution of the decay is isotropic even when the parent Σ^0 is polarized.

b) Find the mean lifetime (in seconds) for $\kappa = 1$ ($m_\Lambda = 1115$ MeV/c^2, $m_\Sigma = 1192$ MeV/c^2).

2–6. The metastable $2s$ state of the hydrogen atom turns into the ground state by emitting two photons. The Golden Rule in this case takes the form

$$dw = \frac{2\pi}{\hbar}|T_{fi}|^2\frac{Vd^3k_1}{(2\pi)^3}\frac{Vd^3k_2}{(2\pi)^3}\delta(E_{2s} - E_{1s} - \hbar\omega_1 - \hbar\omega_2)$$

for the emission of photons characterized by $(\mathbf{k}_1, \alpha_1)$ and $(\mathbf{k}_2, \alpha_2)$. Write out the expression for T_{fi}. State explicitly what kind of intermediate states characterized by (n, l, m) give rise to nonvanishing contributions. Simplify your expression as far as you can.

2–7. Prove that the scattering amplitude of a neutral scalar meson on an infinitely heavy nucleon which has no discrete excited states is zero up to order g^2 when the interaction density is given by

$$\mathscr{H} = g\phi(\mathbf{x}, t)\delta^{(3)}(\mathbf{x} - \mathbf{x}'), \qquad \mathbf{x}' = \text{position of the nucleon.}$$

2–8. A stable, spinless nucleus A of even parity has a spin-one, odd-parity excited state R whose only significant decay mode is $R \to A + \gamma$.

a) Assuming that excitations to intermediate states other than R are unimportant and ignoring the nuclear Thomson term (due to $\mathbf{A}\cdot\mathbf{A}$), show that the differential and total scattering cross sections of a γ-ray by the ground state A are given by

$$\frac{d\sigma}{d\Omega} = \frac{9}{16}\left(\frac{c}{\omega}\right)^2 (\boldsymbol{\epsilon}^{(\alpha)}\cdot\boldsymbol{\epsilon}^{(\alpha')})^2 \frac{\Gamma_R^2}{(E_R - E_A - \hbar\omega)^2 + (\Gamma_R^2/4)},$$

$$\sigma_{\text{tot}} = \frac{3}{2}(4\pi\lambda\!\!\!^{-2}) \frac{(\Gamma_R^2/4)}{(E_R - E_A - \hbar\omega)^2 + (\Gamma_R^2/4)},$$

where $\lambda\!\!\!^{-} = c/\omega$. (The second formula can be generalized to resonance scattering involving any multipole transition if we replace 3/2 by $(2J_R + 1)/[2(2J_A + 1)]$.)

b) Verify that the above expression for the total cross section is equal to $4\pi(c/\omega)\,\mathrm{Im}f(\omega)$.

c) The nucleus C^{14} whose ground state is known to be a 0^+ state has an excited 1^- state (denoted by C^{14*}) 6.1 MeV above the ground state. The only decay mode of C^{14*} is known to be $C^{14} + \gamma$. Compute the total cross section at exact resonance and compare it with the cross section for nuclear Thomson scattering due to the C^{14} nucleus as a whole.

2–9. Assuming that $f(\omega)$ for the scattering of a high-energy photon by the hydrogen atom is given by the Thomson amplitude, derive the sum rule

$$2\pi^2 c r_0 = \int_0^\infty \sigma_{\text{tot}}(\omega)\, d\omega.$$

Show that within the framework of the approximations made in this chapter, the above sum rule is equivalent to the well-known Thomas-Reiche-Kuhn sum rule:‡

$$\sum_I \frac{2m\omega_{IA}}{\hbar} |\mathbf{x}_{IA}|^2 = 3.$$

2–10. Assuming the validity of perturbation theory, use the fixed-source neutral scalar theory of Problem 2–7 to obtain an expression for the probability of finding one virtual meson of energy $<\hbar\omega^{(\max)}$ around the nucleon.

2–11. Why is it legitimate to ignore Fig. 2–7(a) in estimating the Lamb shift?

‡Merzbacher (1961) p. 446.

CHAPTER 3

RELATIVISTIC QUANTUM MECHANICS OF SPIN-$\frac{1}{2}$ PARTICLES

3-1. PROBABILITY CONSERVATION IN RELATIVISTIC QUANTUM MECHANICS

It is almost traditional to start an exposition of the Dirac theory of spin-$\frac{1}{2}$ particles by discussing some of the "difficulties" of the Klein-Gordon theory. As we shall see later, there is actually nothing wrong in the Klein-Gordon equation if it is properly interpreted. We shall sketch, however, the usual arguments against the Klein-Gordon equation since they played an important historic role in the formulation of relativistic quantum mechanics.

In Schrödinger's wave mechanics we associate a complex-valued wave function ψ with a single particle, such that $|\psi|^2 d^3x$ gives the probability of finding the particle in a volume element d^3x. This interpretation is possible because the probability density P and the flux density $\mathbf{S}$ given by

$$P = |\psi|^2 > 0 \tag{3.1}$$

and

$$\mathbf{S} = -(i\hbar/2m)(\psi^* \nabla\psi - \psi\nabla\psi^*) \tag{3.2}$$

satisfy the continuity equation

$$\partial P/\partial t + \nabla \cdot \mathbf{S} = 0 \tag{3.3}$$

by virtue of the Schrödinger equation. Using Gauss' theorem, we also see that the integral over all space $\int P d^3x$ is a constant of the motion which can be set to unity by appropriately normalizing ψ.

If one is to construct a relativistic quantum mechanics in analogy with non-relativistic quantum mechanics, it may appear natural to impose the following requirements on the theory. First, with the relativistic wave function we must be able to construct bilinear forms which can be interpreted as the probability density and the flux density satisfying a continuity equation of type (3.3). The probability density we form must, of course, be positive definite. In addition, the special theory of relativity requires that P must be the fourth component of a four-vector density. To see this last point, we recall that $d^3x \longrightarrow d^3x\sqrt{1-(v/c)^2}$ under a Lorentz transformation because of the well-known Lorentz contraction of the volume element; if Pd^3x is to remain invariant, it is essential that P transforms like the fourth component of a four-vector $P \longrightarrow P/\sqrt{1-(v/c)^2}$. The continuity equation (3.3) takes the following covariant form,

$$(\partial/\partial x_\mu)s_\mu = 0, \tag{3.4}$$

where

$$s_\mu = (\mathbf{S}, icP). \tag{3.5}$$

Let us now see whether the relativistic quantum mechanics based on the Klein-Gordon equation satisfies the above requirements. Consider a four-vector density given by

$$s_\mu = A\left(\phi^* \frac{\partial \phi}{\partial x_\mu} - \frac{\partial \phi^*}{\partial x_\mu}\phi\right), \tag{3.6}$$

where ϕ is a solution to the free particle Klein-Gordon equation, and A is a multiplicative constant. The four-divergence of (3.6) vanishes,

$$\frac{\partial s_\mu}{\partial x_\mu} = A\left[\frac{\partial \phi^*}{\partial x_\mu}\frac{\partial \phi}{\partial x_\mu} - (\Box\phi^*)\phi - \phi^*\Box\phi - \frac{\partial \phi^*}{\partial x_\mu}\frac{\partial \phi}{\partial x_\mu}\right] = 0, \tag{3.7}$$

by virtue of the Klein-Gordon equation. For a Klein-Gordon particle moving at nonrelativistic velocities ($E \approx mc^2$),

$$\phi \sim \psi e^{-imc^2t/\hbar}, \tag{3.8}$$

where ψ is the corresponding Schrödinger solution (cf. Problem 1–2). The components of s_μ are then given by

$$s_0 = -is_4 \approx (2imc/\hbar)A\,|\psi|^2, \qquad \mathbf{s} = A[\psi^*\nabla\psi - (\nabla\psi^*)\psi]. \tag{3.9}$$

If we set $A = -i\hbar/2m$, then $\mathbf{s}$ and s_0 are precisely the flux density and c times the probability density in the Schrödinger theory. Thus we obtain a four-vector current density from a solution of the Klein-Gordon equation with the following properties: (i) the current density satisfies the continuity equation, and (ii) the components of the current density coincide with the flux density and c times the probability density in the nonrelativistic limit.

So far everything appears satisfactory. There is, however, a difficulty in interpreting

$$P = \frac{i\hbar}{2mc^2}\left(\phi^*\frac{\partial \phi}{\partial t} - \frac{\partial \phi^*}{\partial t}\phi\right) \tag{3.10}$$

as the probability density. In the Schrödinger theory, in which the time derivative appears only linearly in the wave equation, the sign of the frequency is determined by the eigenvalue of the Hamiltonian operator. In contrast, because the Klein-Gordon equation is of second order in the time derivative, both $u(\mathbf{x})e^{-iEt/\hbar}$ and $u^*(\mathbf{x})e^{+iEt/\hbar}$ are equally good solutions for a given physical situation (cf. Problem 1–3). This means that P given by (3.10) can be positive or negative. We may arbitrarily omit all solutions of the form $u(\mathbf{x})e^{-iEt/\hbar}$ with $E < 0$. But this would be unjustified because solutions of the form $u(\mathbf{x})e^{-iEt/\hbar}$ with $E > 0$ *alone* do not form a complete set. It appears that we must either abandon the interpretation of (3.10) as the probability density or abandon the Klein-Gordon equation altogether.

Let us analyze the origin of this difficulty a little more closely. From the derivation of the continuity equation (3.7) we may infer that the appearance of the linear time derivative in s_0 is unavoidable so long as the wave function satisfies a partial differential equation *quadratic* in the time derivative. Perhaps we could avoid this difficulty if we wrote a relativistic wave equation *linear* in the time derivative.

In 1928, in what is undoubtedly one of the most significant papers in the physics of the twentieth century, P.A.M. Dirac succeeded in devising a relativistic wave equation starting with the requirement that the wave equation be linear in $\partial/\partial t$. Using his equation, known to us as the Dirac equation, he was able to construct a conserved four-vector density whose zeroth component is positive-definite. For this reason, from 1928 until 1934 the Dirac equation was considered to be the only correct wave equation in relativistic quantum mechanics.

In 1934 the Klein-Gordon equation was revived by W. Pauli and V. F. Weisskopf. Their proposal was that, up to a proportionality factor, s_μ given by (3.6) be interpreted as the charge-current density rather than as the probability-current density. As we saw in Section 1–3, an interpretation of this kind is reasonable in the classical field theory of a complex scalar field. The fact that the sign of s_0 changes when $u^*(\mathbf{x})e^{-iEt/\hbar}$ is substituted for $u(\mathbf{x})e^{iEt/\hbar}$ makes good sense if the *negative-energy* solution is interpreted as the wave function for a particle with *opposite electric charge* (cf. Problem 1–3).

The interpretation of s_μ as the charge-current density is even more satisfactory for a theory in which a solution to the Klein-Gordon equation is to be interpreted as a quantized field operator. In analogy with what we did for the complex scalar field in classical field theory we form a non-Hermitian field operator $\phi(\neq \phi^\dagger)$ such that

$$\phi = \frac{\phi_1 + i\phi_2}{\sqrt{2}}, \qquad \phi^\dagger = \frac{\phi_1 - i\phi_2}{\sqrt{2}}, \tag{3.11}$$

where ϕ_1 and ϕ_2 are Hermitian operators whose properties are given in Problem 2–3. Consider now a four-current *operator*

$$j_\mu = e\left(\phi^\dagger \frac{\partial \phi}{\partial x_\mu} - \frac{\partial \phi^\dagger}{\partial x_\mu}\phi\right). \tag{3.12}$$

It is easy to show (Problem 3–1) that the fourth component of j_μ has the property

$$\int (j_4/ic)\, d^3x = e \sum_{\mathbf{k}} (N_{\mathbf{k}}^{(+)} - N_{\mathbf{k}}^{(-)}), \tag{3.13}$$

where

$$N^{(\pm)} = a_{\mathbf{k}\pm}^\dagger a_{\mathbf{k}\pm}, \tag{3.14}$$

with

$$a_{\mathbf{k}\pm} = \frac{1}{\sqrt{2}}(a_{\mathbf{k}}^{(1)} \pm i a_{\mathbf{k}}^{(2)}), \qquad a_{\mathbf{k}\pm}^\dagger = \frac{1}{\sqrt{2}}(a_{\mathbf{k}}^{(1)} \mp i a_{\mathbf{k}}^{(2)}). \tag{3.15}$$

Physically, $N^{(+)}$ and $N^{(-)}$ are the number operators for the Klein-Gordon particle of charge e and for its antiparticle with charge $-e$. So the eigenvalue of the operator expression (3.13) is the total charge of the field. Usually the four-vector j_μ is regarded as the charge-current density operator.

As emphasized in Chapter 2, quantum field theory accommodates physical situations in which particles are created or annihilated. When there are processes like $\gamma \longrightarrow \pi^+ + \pi^-$ which take place in the Coulomb field of a nucleus, what is conserved is not the probability of finding a given particle integrated over all space but rather the total charge of the field given by the eigenvalue of (3.13).

Coming back to the original argument against the Klein-Gordon equation, we see that if we are to reject the Klein-Gordon equation on the ground that we cannot form a positive-definite probability density, we might as well give up the Maxwell theory which, as the reader may verify, cannot accommodate any conserved four-vector density bilinear in the electromagnetic field.

3–2. THE DIRAC EQUATION

Derivation of the Dirac equation. Even though the Klein-Gordon equation is quite satisfactory when properly interpreted, there is reason for rejecting it for the description of an electron. The Klein-Gordon equation cannot accommodate the spin-$\frac{1}{2}$ nature of the electron as naturally as the Dirac equation can. In this connection, let us first study how to incorporate the electron spin in nonrelativistic quantum mechanics.

In nonrelativistic quantum mechanics, in order to account for the interaction of the electron spin magnetic moment with the magnetic field, it is customary to add a term

$$H^{(\mathrm{spin})} = -(e\hbar/2mc)\boldsymbol{\sigma}\cdot\mathbf{B} \tag{3.16}$$

to the usual Hamiltonian, as done originally by W. Pauli. This procedure appears somewhat artificial, especially if we subscribe to the philosophy that the only "fundamental" electromagnetic interactions are those which can be generated by the substitution $p_\mu \rightarrow p_\mu - eA_\mu/c$. There is, however, a slightly less *ad hoc* way of introducing the spin magnetic moment interaction. In the usual wave-mechanical treatment of the electron, the kinetic energy operator in the absence of the vector potential is taken to be

$$H^{(\mathrm{KE})} = \mathbf{p}^2/2m. \tag{3.17}$$

However, for a spin-$\frac{1}{2}$ particle we may just as well start with the expression

$$H^{(\mathrm{KE})} = (\boldsymbol{\sigma}\cdot\mathbf{p})(\boldsymbol{\sigma}\cdot\mathbf{p})/2m. \tag{3.18}$$

This alternative form is indistinguishable from (3.17) for all practical purposes when there is no vector potential.‡ There is, however, a difference when we make the substitution $\mathbf{p} \rightarrow \mathbf{p} - e\mathbf{A}/c$. The expression (3.18) then becomes

$$\begin{aligned}
\frac{1}{2m}\boldsymbol{\sigma}\cdot\left(\mathbf{p}-\frac{e\mathbf{A}}{c}\right)\boldsymbol{\sigma}\cdot\left(\mathbf{p}-\frac{e\mathbf{A}}{c}\right)
&= \frac{1}{2m}\left(\mathbf{p}-\frac{e\mathbf{A}}{c}\right)^2 + \frac{i}{2m}\boldsymbol{\sigma}\cdot\left[\left(\mathbf{p}-\frac{e\mathbf{A}}{c}\right)\times\left(\mathbf{p}-\frac{e\mathbf{A}}{c}\right)\right] \\
&= \frac{1}{2m}\left(\mathbf{p}-\frac{e\mathbf{A}}{c}\right)^2 - \frac{e\hbar}{2mc}\boldsymbol{\sigma}\cdot\mathbf{B},
\end{aligned} \tag{3.19}$$

where we have used

$$\mathbf{p}\times\mathbf{A} = -i\hbar(\boldsymbol{\nabla}\times\mathbf{A}) - \mathbf{A}\times\mathbf{p}. \tag{3.20}$$

(The operator $\mathbf{p}$ is assumed to act on everything that stands to the right; in contrast,

‡We recall that the formula $(\boldsymbol{\sigma}\cdot\mathbf{A})(\boldsymbol{\sigma}\cdot\mathbf{B}) = \mathbf{A}\cdot\mathbf{B} + i\boldsymbol{\sigma}\cdot(\mathbf{A}\times\mathbf{B})$ holds even if $\mathbf{A}$ and $\mathbf{B}$ are operators.

the ∇ operator in (3.20) acts only on $\mathbf{A}$.) Note that the spin magnetic moment generated in this way has the correct gyromagnetic ratio $g = 2$.‡

Our object is to derive a relativistic wave equation for a spin-$\frac{1}{2}$ particle. Just as we incorporated the electron spin into the nonrelativistic theory by using the kinetic energy operator (3.18), we can incorporate the electron spin into the general framework of relativistic quantum mechanics by taking the operator analog of the classical expression

$$(E^2/c^2) - \mathbf{p}^2 = (mc)^2 \tag{3.21}$$

to be

$$\left(\frac{E^{(\text{op})}}{c} - \boldsymbol{\sigma}\cdot\mathbf{p}\right)\left(\frac{E^{(\text{op})}}{c} + \boldsymbol{\sigma}\cdot\mathbf{p}\right) = (mc)^2, \tag{3.22}$$

where

$$E^{(\text{op})} = i\hbar\frac{\partial}{\partial t} = i\hbar c\frac{\partial}{\partial x_0}, \tag{3.23}$$

and $\mathbf{p} = -i\hbar\nabla$ as before. This enables us to write a second-order equation (due to B. L. van der Waerden),

$$\left(i\hbar\frac{\partial}{\partial x_0} + \boldsymbol{\sigma}\cdot i\hbar\nabla\right)\left(i\hbar\frac{\partial}{\partial x_0} - \boldsymbol{\sigma}\cdot i\hbar\nabla\right)\phi = (mc)^2\phi, \tag{3.24}$$

for a free electron, where ϕ is now a *two-component* wave function.

We are interested in obtaining a wave equation of first order in the time derivative. Relativistic covariance suggests that the wave equation linear in $\partial/\partial t$ must be linear in ∇ also. An analogy with the Maxwell theory may now be helpful. The free-field D'Alembertian equation $\Box A_\mu = 0$ is a second-order equation, while the free-field Maxwell equation $(\partial/\partial x_\mu)\, F_{\mu\nu} = 0$ is a first-order equation. Note that $F_{\mu\nu}$ obtained by differentiating A_μ has more components than A_μ. This increase in the number of components is the price we have to pay when we work with the first-order equation.

Motivated by this analogy, we can define two *two-component* wave functions $\phi^{(R)}$ and $\phi^{(L)}$:

$$\phi^{(R)} = \frac{1}{mc}\left(i\hbar\frac{\partial}{\partial x_0} - i\hbar\boldsymbol{\sigma}\cdot\nabla\right)\phi, \qquad \phi^{(L)} = \phi. \tag{3.25}$$

The total number of components has now been increased to four. The superscripts R and L come from the fact that as $m \longrightarrow 0$, $\phi^{(R)}$ and $\phi^{(L)}$, respectively, describe a right-handed (spin *parallel* to the momentum direction) and a left-handed (spin *antiparallel* to the momentum direction) state of the spin-$\frac{1}{2}$ particle, as we shall see later. The second-order equation (3.24) is now equivalent to two first-order equations

$$\begin{aligned} [i\hbar\boldsymbol{\sigma}\cdot\nabla - i\hbar(\partial/\partial x_0)]\phi^{(L)} &= -mc\phi^{(R)}, \\ [-i\hbar\boldsymbol{\sigma}\cdot\nabla - i\hbar(\partial/\partial x_0)]\phi^{(R)} &= -mc\phi^{(L)}. \end{aligned} \tag{3.26}$$

---

‡Historically, all this was first obtained by working out the nonrelativistic limit of the Dirac theory, as we shall show in the next section. For this reason, most textbooks state that the $g = 2$ relation is a consequence of the Dirac theory. We have seen, however, that the $g = 2$ relation follows just as naturally from the nonrelativistic Schrödinger-Pauli theory if we start with the kinetic energy operator (3.18). This point was emphasized particularly by R. P. Feynman.

Note that unless the particle is massless, these first-order equations couple $\phi^{(R)}$ and $\phi^{(L)}$ just as the Maxwell equations, also first-order equations, couple **E** and **B**.

Equation (3.26) is equivalent to the celebrated wave equation of Dirac (cf. Problem 3–5). To bring it to the form originally written by Dirac, we take the sum and the difference of (3.26). We then have

$$\begin{aligned} -i\hbar(\boldsymbol{\sigma}\cdot\boldsymbol{\nabla})(\phi^{(R)} - \phi^{(L)}) - i\hbar(\partial/\partial x_0)(\phi^{(L)} + \phi^{(R)}) &= -mc(\phi^{(L)} + \phi^{(R)}), \\ i\hbar(\boldsymbol{\sigma}\cdot\boldsymbol{\nabla})(\phi^{(L)} + \phi^{(R)}) + i\hbar(\partial/\partial x_0)(\phi^{(R)} - \phi^{(L)}) &= -mc(\phi^{(R)} - \phi^{(L)}), \end{aligned} \tag{3.27}$$

or, denoting the sum and the difference of $\phi^{(R)}$ and $\phi^{(L)}$ by ψ_A and ψ_B, we have

$$\begin{pmatrix} -i\hbar(\partial/\partial x_0) & -i\hbar\boldsymbol{\sigma}\cdot\boldsymbol{\nabla} \\ i\hbar\boldsymbol{\sigma}\cdot\boldsymbol{\nabla} & i\hbar(\partial/\partial x_0) \end{pmatrix} \begin{pmatrix} \psi_A \\ \psi_B \end{pmatrix} = -mc \begin{pmatrix} \psi_A \\ \psi_B \end{pmatrix}. \tag{3.28}$$

Defining a *four*-component wave function ψ by

$$\psi = \begin{pmatrix} \psi_A \\ \psi_B \end{pmatrix} = \begin{pmatrix} \phi^{(R)} + \phi^{(L)} \\ \phi^{(R)} - \phi^{(L)} \end{pmatrix}, \tag{3.29}$$

we can rewrite (3.28) more concisely as

$$\left(\boldsymbol{\gamma}\cdot\boldsymbol{\nabla} + \gamma_4 \frac{\partial}{\partial(ix_0)}\right)\psi + \frac{mc}{\hbar}\psi = 0, \tag{3.30}$$

or

$$\left(\gamma_\mu \frac{\partial}{\partial x_\mu} + \frac{mc}{\hbar}\right)\psi = 0, \tag{3.31}$$

where γ_μ with $\mu = 1, 2, 3, 4$ are 4×4 matrices given by

$$\gamma_k = \begin{pmatrix} 0 & -i\sigma_k \\ i\sigma_k & 0 \end{pmatrix}, \qquad \gamma_4 = \begin{pmatrix} I & 0 \\ 0 & -I \end{pmatrix}, \tag{3.32}$$

which really mean

$$\gamma_3 = \begin{pmatrix} 0 & 0 & -i & 0 \\ 0 & 0 & 0 & i \\ i & 0 & 0 & 0 \\ 0 & -i & 0 & 0 \end{pmatrix}, \qquad \gamma_4 = \begin{pmatrix} 1 & 0 & 0 & 0 \\ 0 & 1 & 0 & 0 \\ 0 & 0 & -1 & 0 \\ 0 & 0 & 0 & -1 \end{pmatrix}, \quad \text{etc.}‡ \tag{3.33}$$

Equation (3.31) is the famous Dirac equation.§

---

‡We use the standard form of the 2×2 Pauli matrices

$$\sigma_1 = \begin{pmatrix} 0 & 1 \\ 1 & 0 \end{pmatrix}, \qquad \sigma_2 = \begin{pmatrix} 0 & -i \\ i & 0 \end{pmatrix}, \qquad \sigma_3 = \begin{pmatrix} 1 & 0 \\ 0 & -1 \end{pmatrix}.$$

The symbol I stands for the 2×2 identity matrix

$$I = \begin{pmatrix} 1 & 0 \\ 0 & 1 \end{pmatrix}.$$

§In deriving the Dirac equation we have not followed the path originally taken by Dirac. The first published account of the derivation given here is found in B. L. van der Waerden's 1932 book on group theory. The original approach of Dirac can be found in many books, for example, Rose (1961), pp. 39–44; Bjorken and Drell (1964), pp. 6–8.

We wish to emphasize that (3.31) is actually four differential equations that couple the four components of ψ represented by a single-column matrix

$$\psi = \begin{pmatrix} \psi_1 \\ \psi_2 \\ \psi_3 \\ \psi_4 \end{pmatrix}. \tag{3.34}$$

A four-component object of this kind is known as a bispinor or, more commonly, as a *Dirac spinor*. If there is any confusion as to the real meaning of (3.31), the reader may write the matrix indices explicitly as follows:

$$\sum_{\mu=1}^{4} \sum_{\beta=1}^{4} \left[(\gamma_\mu)_{\alpha\beta} \frac{\partial}{\partial x_\mu} + \left(\frac{mc}{\hbar}\right) \delta_{\alpha\beta} \right] \psi_\beta = 0. \tag{3.35}$$

The fact that ψ has four components has nothing to do with the four-dimensional nature of space-time; ψ_β does *not* transform like a four-vector under a Lorentz transformation, as we shall see in Section 3–4.

The 4×4 matrices γ_μ we introduced are called the *gamma matrices* or the *Dirac matrices*. They satisfy the following anticommutation relations as we may readily verify:

$$\{\gamma_\mu, \gamma_\nu\} = \gamma_\mu \gamma_\nu + \gamma_\nu \gamma_\mu = 2\delta_{\mu\nu}. \tag{3.36a}$$

For instance,

$$\gamma_4^2 = \begin{pmatrix} I & 0 \\ 0 & -I \end{pmatrix}^2 = \begin{pmatrix} I & 0 \\ 0 & I \end{pmatrix},$$

$$\begin{aligned} \gamma_1\gamma_2 + \gamma_2\gamma_1 &= \begin{pmatrix} 0 & -i\sigma_1 \\ i\sigma_1 & 0 \end{pmatrix}\begin{pmatrix} 0 & -i\sigma_2 \\ i\sigma_2 & 0 \end{pmatrix} + \begin{pmatrix} 0 & -i\sigma_2 \\ i\sigma_2 & 0 \end{pmatrix}\begin{pmatrix} 0 & -i\sigma_1 \\ i\sigma_1 & 0 \end{pmatrix} \\ &= \begin{pmatrix} \sigma_1\sigma_2 + \sigma_2\sigma_1 & 0 \\ 0 & \sigma_1\sigma_2 + \sigma_2\sigma_1 \end{pmatrix} = 0. \end{aligned} \tag{3.36b}$$

Moreover, each γ_μ is seen to be Hermitian,

$$\gamma_\mu^\dagger = \gamma_\mu, \tag{3.37}$$

and traceless.‡

Multiplying (3.30) by γ_4, we see that the Dirac equation can also be written in the Hamiltonian form

$$H\psi = i\hbar(\partial\psi/\partial t), \tag{3.38}$$

where

$$H = -ic\hbar\boldsymbol{\alpha}\cdot\boldsymbol{\nabla} + \beta mc^2, \tag{3.39}$$

with

$$\beta = \gamma_4 = \begin{pmatrix} I & 0 \\ 0 & -I \end{pmatrix}, \qquad \alpha_k = i\gamma_4\gamma_k = \begin{pmatrix} 0 & \sigma_k \\ \sigma_k & 0 \end{pmatrix}. \tag{3.40}$$

---

‡In the literature some people define gamma matrices that do not quite satisfy (3.35) and (3.37). This is deplorable. Our notation agrees with that of Dirac's original paper and Pauli's *Handbuch* article. The alternative notation sometimes found in the literature is summarized in Appendix B.

The matrices $\boldsymbol{\alpha}$ and β satisfy‡

$$\{\alpha_k, \beta\} = 0, \qquad \beta^2 = 1, \qquad \{\alpha_k, \alpha_l\} = 2\delta_{kl}. \tag{3.41}$$

In Section 3–6, we shall make extensive use of the Hamiltonian formalism based on (3.39).

Conserved current. We shall now derive the differential law of current conservation. First let us define

$$\bar{\psi} = \psi^\dagger \gamma_4, \tag{3.42}$$

where $\bar{\psi}$ is called the adjoint spinor, as distinguished from the Hermitian conjugate spinor $\psi^\dagger$. Explicitly $\bar{\psi}$ as well as $\psi^\dagger$ can be represented by single-row matrices, that is, if

$$\psi = \begin{pmatrix} \psi_1 \\ \psi_2 \\ \psi_3 \\ \psi_4 \end{pmatrix}, \tag{3.43}$$

then

$$\begin{aligned} \psi^\dagger &= (\psi_1^*, \psi_2^*, \psi_3^*, \psi_4^*), \\ \bar{\psi} &= (\psi_1^*, \psi_2^*, -\psi_3^*, -\psi_4^*). \end{aligned} \tag{3.44}$$

To obtain the wave equation for $\bar{\psi}$ we start with the Hermitian conjugate of the Dirac equation,

$$\frac{\partial}{\partial x_k} \psi^\dagger \gamma_k + \frac{\partial}{\partial x_4^*} \psi^\dagger \gamma_4 + \frac{mc}{\hbar} \psi^\dagger = 0. \tag{3.45}$$

Multiplying (3.45) by γ_4 from the right, we obtain the *adjoint equation*

$$-\frac{\partial}{\partial x_\mu} \bar{\psi} \gamma_\mu + \frac{mc}{\hbar} \bar{\psi} = 0, \tag{3.46}$$

where we have used

$$\frac{\partial}{\partial x_4^*} = \frac{\partial}{\partial (ict)^*} = -\frac{\partial}{\partial x_4}, \tag{3.47}$$

and $\gamma_k \gamma_4 = -\gamma_4 \gamma_k$. We now multiply the original Dirac equation (3.31) from the left by $\bar{\psi}$, the adjoint equation (3.46) from the right by ψ, and subtract. Then

$$(\partial / \partial x_\mu)(\bar{\psi} \gamma_\mu \psi) = 0. \tag{3.48}$$

Thus we see that

$$s_\mu = ic\bar{\psi}\gamma_\mu \psi = (c\psi^\dagger \boldsymbol{\alpha} \psi, ic\psi^\dagger \psi) \tag{3.49}$$

---

‡In addition to γ_μ, α_k, and β, one also finds in the literature

$$\rho_1 = \begin{pmatrix} 0 & I \\ I & 0 \end{pmatrix}, \qquad \rho_2 = \begin{pmatrix} 0 & -iI \\ iI & 0 \end{pmatrix}, \qquad \text{and} \qquad \rho_3 = \begin{pmatrix} I & 0 \\ 0 & -I \end{pmatrix}.$$

We shall, however, not use the rho matrices in this book.

satisfies the continuity equation.‡ Using Green's theorem we see that

$$\int \bar{\psi}\gamma_4\psi d^3x = \int \psi^\dagger\psi d^3x = \text{const}, \tag{3.50}$$

which may be set to unity by appropriately normalizing ψ. Now, unlike (3.9),

$$\bar{\psi}\gamma_4\psi = \psi^\dagger\psi = \sum_{\beta=1}^{4} \psi_\beta^*\psi_\beta \tag{3.51}$$

is necessarily positive-definite. So we may be tempted to identify $\bar{\psi}\gamma_4\psi = \psi^\dagger\psi$ with the probability density as we did in the Schrödinger theory. With this interpretation,

$$s_k = ic\bar{\psi}\gamma_k\psi = c\psi^\dagger\alpha_k\psi \tag{3.52}$$

is identified with the flux density. For the covariance of the continuity equation, we must, of course, prove rigorously that s_μ indeed transforms like a four-vector; this will be done in Section 3–4.

For the time being, we assume that a solution ψ to the Dirac equation is a single-particle wave function subject to the above interpretation of $\psi^\dagger\psi$ as the probability density. Toward the end of this chapter, however, we shall point out some of the difficulties associated with this interpretation. For a more satisfactory interpretation of the Dirac theory, it is necessary to quantize the Dirac field. This we shall do in Section 3–10.

Representation independence. In concluding this section we shall mention briefly the notion of the representation independence of the Dirac equation. Suppose somebody writes

$$\left(\gamma'_\mu \frac{\partial}{\partial x_\mu} + \frac{mc}{\hbar}\right)\psi' = 0, \tag{3.53}$$

where the only defining property of γ'_μ is that γ'_μ with $\mu = 1 \ldots 4$ forms a set of 4×4 matrices satisfying

$$\{\gamma'_\mu, \gamma'_\nu\} = 2\delta_{\mu\nu}. \tag{3.54}$$

Our assertion is that this equation is equivalent to the Dirac equation (3.31) where the matrices γ_μ are explicitly given by (3.32). Note that we do not assert that for a given physical situation ψ and ψ' are the same. Different sets of 4×4 matrices satisfying (3.54) are referred to as sets of the gamma matrices in different representations.

We can prove the representation independence of the Dirac equation by appealing to what is known as Pauli's fundamental theorem: Given two sets of 4×4 matrices satisfying

$$\{\gamma_\mu, \gamma_\nu\} = 2\delta_{\mu\nu} \qquad \text{and} \qquad \{\gamma'_\mu, \gamma'_\nu\} = 2\delta_{\mu\nu}$$

‡All this can be readily proved even in the presence of the electromagnetic interaction generated by the substitution

$$-i\hbar\frac{\partial}{\partial x_\mu} \longrightarrow -i\hbar\frac{\partial}{\partial x_\mu} - \frac{eA_\mu}{c}$$

in (3.31).

with $\mu, \nu = 1 \ldots 4$, there exists a nonsingular 4×4 matrix S such that

$$S\gamma_\mu S^{-1} = \gamma'_\mu. \tag{3.55}$$

Moreover, S is unique up to a multiplicative constant. The proof of the theorem can be found in Appendix C. Assuming the validity of Pauli's theorem, we rewrite (3.53) as

$$\left(S\gamma_\mu S^{-1}\frac{\partial}{\partial x_\mu} + \frac{mc}{\hbar}\right) SS^{-1}\psi' = 0, \tag{3.56}$$

where S is the matrix that relates the set $\{\gamma'_\mu\}$ and the set $\{\gamma_\mu\}$ via a relation of the form (3.55). Multiplying by S^{-1} from the left, we obtain

$$\left(\gamma_\mu\frac{\partial}{\partial x_\mu} + \frac{mc}{\hbar}\right) S^{-1}\psi' = 0. \tag{3.57}$$

This is the same as the original Dirac equation with $S^{-1}\psi'$ as solution. In other words, (3.53) is equivalent to the Dirac equation, (3.31), and the wave functions ψ' and ψ are related by

$$\psi' = S\psi. \tag{3.58}$$

Let us consider the case where the γ'_μ are also Hermitian. By taking the Hermitian conjugate of (3.55) we see that S can be chosen to be unitary $S^\dagger = S^{-1}$. With a unitary S we see that expressions like the probability density and the flux density are the same:

$$\begin{aligned} \bar{\psi}'\gamma'_\mu\psi' &= \psi'^\dagger\gamma'_4\gamma'_\mu\psi' \\ &= \psi^\dagger S^\dagger S\gamma_4 S^{-1}S\gamma_\mu S^{-1}S\psi \\ &= \bar{\psi}\gamma_\mu\psi. \end{aligned} \tag{3.59}$$

Evidently all the physical consequences are the same regardless of whether we use (3.31) or (3.53). Note, however, that the wave functions for the same physical situation *look* different when different representations are used.‡

In practice three representations of the gamma matrices are found in the literature:

a) The standard (Dirac-Pauli) representation given explicitly by (3.32).

b) The Weyl representation in which not only γ_k but also γ_4 are off-diagonal (cf. Problem 3–5).

‡Actually we are already familiar with an analogous situation in the Pauli two-component theory. In nonrelativistic quantum mechanics it is customary to use the standard representation of the σ matrices in which σ_3 is diagonal. From the point of view of the commutation relations, however, one may as well work with what we may call "a nonconformist's representation,"

$$\sigma_1 = \begin{pmatrix} 0 & -i \\ i & 0 \end{pmatrix}, \quad \sigma_2 = \begin{pmatrix} 1 & 0 \\ 0 & -1 \end{pmatrix}, \quad \sigma_3 = \begin{pmatrix} 0 & 1 \\ 1 & 0 \end{pmatrix}.$$

The spin-up spinor (the spin in the positive z-direction) is then given by

$$\frac{1}{\sqrt{2}}\begin{pmatrix} 1 \\ 1 \end{pmatrix} \quad \text{rather than by} \quad \begin{pmatrix} 1 \\ 0 \end{pmatrix}.$$

c) The Majorana representation in which the γ_k are purely real and γ_4 is purely imaginary, hence $\gamma_\mu(\partial/\partial x_\mu)$ is purely real.

In this book, whenever explicit forms of γ_μ or ψ are called for, only the standard (Dirac-Pauli) representation is used.

3–3. SIMPLE SOLUTIONS; NONRELATIVISTIC APPROXIMATIONS; PLANE WAVES

Large and small components. Before we study the behavior of Dirac's wave function ψ under Lorentz transformations, let us examine the kind of physics buried in the harmless-looking equation (3.31).

In the presence of electromagnetic couplings, the Dirac equation reads

$$\left(\frac{\partial}{\partial x_\mu} - \frac{ie}{\hbar c}A_\mu\right)\gamma_\mu\psi + \frac{mc}{\hbar}\psi = 0, \tag{3.60}$$

where the usual replacement $-i\hbar(\partial/\partial x_\mu) \to -i\hbar(\partial/\partial x_\mu) - eA_\mu/c$ is assumed to be valid. Assuming that A_μ is time independent, we let the time dependence of ψ be given by

$$\psi = \psi(\mathbf{x}, t)\,|_{t=0}\,e^{-iEt/\hbar} \tag{3.61}$$

(which, of course, means that ψ is an eigenfunction of $i\hbar\,\partial/\partial t$ with eigenvalue E). We can then write the coupled equations for the upper and lower two components, ψ_A and ψ_B, as follows (Eq. 3.28):

$$\begin{aligned}\left[\boldsymbol{\sigma}\cdot\left(\mathbf{p} - \frac{e\mathbf{A}}{c}\right)\right]\psi_B &= \frac{1}{c}(E - eA_0 - mc^2)\psi_A,\\ -\left[\boldsymbol{\sigma}\cdot\left(\mathbf{p} - \frac{e\mathbf{A}}{c}\right)\right]\psi_A &= -\frac{1}{c}(E - eA_0 + mc^2)\psi_B,\end{aligned} \tag{3.62}$$

where $A_\mu = (\mathbf{A}, iA_0)$ as before. Using the second equation, we can readily eliminate ψ_B in the first equation to obtain

$$\left[\boldsymbol{\sigma}\cdot\left(\mathbf{p} - \frac{e\mathbf{A}}{c}\right)\right]\left[\frac{c^2}{E - eA_0 + mc^2}\right]\left[\boldsymbol{\sigma}\cdot\left(\mathbf{p} - \frac{e\mathbf{A}}{c}\right)\right]\psi_A = (E - eA_0 - mc^2)\psi_A. \tag{3.63}$$

Up to now we have made no approximations. We now assume that

$$E \approx mc^2, \qquad |eA_0| \ll mc^2. \tag{3.64}$$

Defining the energy measured from mc^2 by

$$E^{(\mathrm{NR})} = E - mc^2, \tag{3.65}$$

we can make the following expansion:

$$\frac{c^2}{E - eA_0 + mc^2} = \frac{1}{2m}\left[\frac{2mc^2}{2mc^2 + E^{(\mathrm{NR})} - eA_0}\right] = \frac{1}{2m}\left[1 - \frac{E^{(\mathrm{NR})} - eA_0}{2mc^2} + \cdots\right]. \tag{3.66}$$

This can be regarded as an expansion in powers of $(v/c)^2$ since $E^{(\mathrm{NR})} - eA_0$ is roughly $[\mathbf{p} - (e\mathbf{A}/c)]^2/2m \approx mv^2/2$. Keeping only the leading term in (3.66), we obtain

$$\frac{1}{2m}\boldsymbol{\sigma}\cdot\left(\mathbf{p}-\frac{e\mathbf{A}}{c}\right)\boldsymbol{\sigma}\cdot\left(\mathbf{p}-\frac{e\mathbf{A}}{c}\right)\psi_A=(E^{(\mathrm{NR})}-eA_0)\psi_A, \tag{3.67}$$

which, as we have already seen (cf. Eq. 3.19), becomes

$$\left[\frac{1}{2m}\left(\mathbf{p}-\frac{e\mathbf{A}}{c}\right)^2-\frac{e\hbar}{2mc}\boldsymbol{\sigma}\cdot\mathbf{B}+eA_0\right]\psi_A=E^{(\mathrm{NR})}\psi_A. \tag{3.68}$$

Thus to zeroth order in $(v/c)^2$, ψ_A is nothing more than the Schrödinger-Pauli two-component wave function in nonrelativistic quantum mechanics multiplied by $e^{-imc^2t/\hbar}$. Using the second expression of (3.62), we see that ψ_B is "smaller" than ψ_A by a factor of roughly $|\mathbf{p}-e(\mathbf{A}/c)|/2mc \approx v/2c$, provided that (3.64) is valid. For this reason with $E \sim mc^2$, ψ_A and ψ_B are respectively known as the *large* and *small components* of the Dirac wave function ψ.

Approximate Hamiltonian for an electrostatic problem. We shall now study the consequences of *keeping* the second term in (3.66). For simplicity, let us treat the case $\mathbf{A}=0$. The equation we must work with is

$$H_A^{(\mathrm{NR})}\psi_A=E^{(\mathrm{NR})}\psi_A, \tag{3.69}$$

where

$$H_A^{(\mathrm{NR})}=(\boldsymbol{\sigma}\cdot\mathbf{p})\frac{1}{2m}\left(1-\frac{E^{(\mathrm{NR})}-eA_0}{2mc^2}\right)(\boldsymbol{\sigma}\cdot\mathbf{p})+eA_0. \tag{3.70}$$

At first sight it might appear that we can regard (3.69) as the time-independent Schrödinger equation for ψ_A. There are, however, three difficulties with this interpretation. First, if we are working to order $(v/c)^2$, ψ_A no longer satisfies the normalization requirement because the probabilistic interpretation of the Dirac theory requires that

$$\int(\psi_A^\dagger\psi_A+\psi_B^\dagger\psi_B)\,d^3x=1, \tag{3.71}$$

where ψ_B is already of the order of v/c. Second, by expanding (3.70) it is easy to see that $H_A^{(\mathrm{NR})}$ contains a non-Hermitian term $i\hbar\mathbf{E}\cdot\mathbf{p}$. Third, (3.69) is not an eigenvalue equation for the energy since $H_A^{(\mathrm{NR})}$ contains $E^{(\mathrm{NR})}$ itself.

To overcome these difficulties, let us first note that the normalization requirement (3.71) can be written as

$$\int\psi_A^\dagger\left(1+\frac{\mathbf{p}^2}{4m^2c^2}\right)\psi_A\,d^3x\approx 1 \tag{3.72}$$

to order $(v/c)^2$ since, according to the second expression of (3.62),

$$\psi_B\approx\frac{\boldsymbol{\sigma}\cdot\mathbf{p}}{2mc}\psi_A. \tag{3.73}$$

This suggests that we should work with a new *two-component* wave function Ψ defined by

$$\Psi=\Omega\psi_A, \tag{3.74}$$

where

$$\Omega=1+(\mathbf{p}^2/8m^2c^2). \tag{3.75}$$

With this choice, Ψ is normalized to order $(v/c)^2$ since

$$\int \Psi^\dagger \Psi d^3x \approx \int \psi_A^\dagger [1 + (\mathbf{p}^2/4m^2c^2)]\psi_A d^3x, \tag{3.76}$$

where we have used (3.71). Multiplying (3.69) from the left by $\Omega^{-1} = 1 - (\mathbf{p}^2/8m^2c^2)$, we obtain

$$\Omega^{-1}H_A\Omega^{-1}\Psi = E^{(\mathrm{NR})}\Omega^{-2}\Psi. \tag{3.77}$$

Explicitly, to order $(v/c)^2$ we have

$$\left[\frac{\mathbf{p}^2}{2m} + eA_0 - \left\{\frac{\mathbf{p}^2}{8m^2c^2}, \left(\frac{\mathbf{p}^2}{2m} + eA_0\right)\right\} - \frac{(\boldsymbol{\sigma}\cdot\mathbf{p})}{2m}\left(\frac{E^{(\mathrm{NR})} - eA_0}{2mc^2}\right)(\boldsymbol{\sigma}\cdot\mathbf{p})\right]\Psi$$
$$= E^{(\mathrm{NR})}\left(1 - \frac{\mathbf{p}^2}{4m^2c^2}\right)\Psi; \tag{3.78}$$

or, writing $E^{(\mathrm{NR})}\,\mathbf{p}^2$ as $\frac{1}{2}\{E^{(\mathrm{NR})}, \mathbf{p}^2\}$, we have

$$\left[\frac{\mathbf{p}^2}{2m} + eA_0 - \frac{\mathbf{p}^4}{8m^3c^2} + \frac{1}{8m^2c^2}(\{\mathbf{p}^2, (E^{(\mathrm{NR})} - eA_0)\} - 2(\boldsymbol{\sigma}\cdot\mathbf{p})(E^{(\mathrm{NR})} - eA_0)(\boldsymbol{\sigma}\cdot\mathbf{p}))\right]\Psi$$
$$= E^{(\mathrm{NR})}\Psi. \tag{3.79}$$

In general, for any pair of operators A and B, we have

$$\{A^2, B\} - 2ABA = [A, [A, B]]. \tag{3.80}$$

This very useful formula can be employed to simplify (3.79) where we set $\boldsymbol{\sigma}\cdot\mathbf{p} = A$ and $E^{(\mathrm{NR})} - eA_0 = B$. Using

$$[\boldsymbol{\sigma}\cdot\mathbf{p}, (E^{(\mathrm{NR})} - eA_0)] = -ie\hbar\boldsymbol{\sigma}\cdot\mathbf{E}, \tag{3.81}$$

and

$$[\boldsymbol{\sigma}\cdot\mathbf{p}, -ie\hbar\boldsymbol{\sigma}\cdot\mathbf{E}] = -e\hbar^2\nabla\cdot\mathbf{E} - 2e\hbar\boldsymbol{\sigma}\cdot(\mathbf{E}\times\mathbf{p}), \tag{3.82}$$

both of which are valid since $\nabla A_0 = -\mathbf{E}$ and $\nabla \times \mathbf{E} = 0$, we finally obtain

$$\left[\frac{\mathbf{p}^2}{2m} + eA_0 - \frac{\mathbf{p}^4}{8m^3c^2} - \frac{e\hbar\boldsymbol{\sigma}\cdot(\mathbf{E}\times\mathbf{p})}{4m^2c^2} - \frac{e\hbar^2}{8m^2c^2}\nabla\cdot\mathbf{E}\right]\Psi = E^{(\mathrm{NR})}\Psi. \tag{3.83}$$

This equation is free of the difficulties mentioned earlier and can therefore be regarded as the Schrödinger equation for the two-component wave function.

The physical significance of each term in (3.83) will now be given. The first two need no explanation. The third term is due to the relativistic correction to the kinetic energy as seen from the expansion

$$\sqrt{(mc^2)^2 + |\mathbf{p}|^2c^2} - mc^2 = |\mathbf{p}|^2/2m - (|\mathbf{p}|^4/8m^3c^2) + \cdots \tag{3.84}$$

The fourth term represents the spin interaction of the moving electron with the electric field. Crudely speaking, we say this arises because the moving electron "sees" an apparent magnetic field given by $\mathbf{E} \times (\mathbf{v}/c)$. Naively, we expect in this way $-(e\hbar/2mc)\,\boldsymbol{\sigma}\cdot[\mathbf{E}\times(\mathbf{v}/c)]$, which is just twice the fourth term of (3.83). That this argument is incomplete was shown within the framework of classical electrodynamics two years before the advent of the Dirac theory by L. H. Thomas, who argued that a more careful treatment which would take into account the energy associated with the precession of the electron spin would result in reduction by a factor of two, in agreement with the fourth term of (3.83).‡ Hence the fourth term of (3.83) is called the Thomas term. For a central potential

$$eA_0 = V(r), \tag{3.85}$$

we obtain

$$-\frac{e\hbar}{4m^2c^2}\boldsymbol{\sigma}\cdot(\mathbf{E}\times\mathbf{p}) = -\frac{\hbar}{4m^2c^2}\left(-\frac{1}{r}\frac{dV}{dr}\right)\boldsymbol{\sigma}\cdot(\mathbf{x}\times\mathbf{p}) = \frac{1}{2m^2c^2}\frac{1}{r}\frac{dV}{dr}\mathbf{S}\cdot\mathbf{L} \tag{3.86}$$

with $\mathbf{S} = \hbar\boldsymbol{\sigma}/2$. Thus the well-known spin-orbit force in atomic physics, represented by (3.86), is an automatic consequence of the Dirac theory.

As for the last term of (3.83) we note that $\boldsymbol{\nabla}\cdot\mathbf{E}$ is just the charge density. For the hydrogen atom where $\boldsymbol{\nabla}\cdot\mathbf{E} = -e\delta^{(3)}(\mathbf{x})$, it gives rise to an energy shift

$$\int \frac{e^2\hbar^2}{8m^2c^2}\delta^{(3)}(\mathbf{x})\,|\psi^{(\mathrm{Schrö})}|^2\,d^3x = \frac{e^2\hbar^2}{8m^2c^2}\,|\psi^{(\mathrm{Schrö})}_{(\mathbf{x})}|^2\Big|_{\mathbf{x}=0}, \tag{3.87}$$

which is nonvanishing only for the s states [in contrast to (3.86) which affects all but the s states]. The last term of (3.86) was first studied in detail by C. G. Darwin; hence it is called the Darwin term. We shall postpone the physical interpretations of the Darwin term until Section 3–7.

Using the third, fourth, and fifth terms of (3.83) as the perturbation Hamiltonian and the wave functions for the hydrogen atom in nonrelativistic quantum mechanics as the unperturbed wave functions, we can compute the lowest-order relativistic correction to the energy levels of the hydrogen atom. Since this calculation based on first-order time-independent perturbation theory is straightforward, we present only the results:§

$$\Delta E = -\left(\frac{e^2}{4\pi\hbar c}\right)^2\left(\frac{e^2}{8\pi a_0}\right)\frac{1}{n^3}\left(\frac{1}{j+\frac{1}{2}} - \frac{3}{4n}\right), \tag{3.88}$$

which is to be added to the unperturbed energy levels

$$E^{(0)} = (e^2/8\pi a_0 n^2). \tag{3.89}$$

The sum $E^{(0)} + \Delta E$ correctly describes the fine structure of the hydrogen atom to order $(\frac{1}{137})^2$ times the Rydberg energy $(e^2/8\pi a_0)$. Note that states with the same n but different j (for example, $2p_{1/2} - 2p_{3/2}$) are now split. On the other hand, states with the same n and the same j (for example, $2s_{1/2} - 2p_{1/2}$) are still degenerate. This degeneracy persists even in the exact treatment of the Coulomb potential problem, as we shall see in Section 3–8.||

‡For the derivation of the Thomas factor in classical electrodynamics consult Jackson (1962), pp. 364–368.

§See, for example, Bethe and Salpeter (1957), pp. 59–61.

||In practice it is of little interest to work out an expansion more accurate than (3.88) using only the Dirac equation with $V = -e^2/(4\pi r)$, because other effects such as the Lamb shift (discussed in the last chapter) and the hyperfine structure (to be discussed in Section 3–8) become more significant.

Free particles at rest. Let us now turn our attention to problems where exact solutions to the Dirac equation are possible. The simplest solvable problem deals with a free particle. We will first demonstrate that each component of the four-component wave function satisfies the Klein-Gordon equation if the particle is free. Although the validity of this statement is rather obvious if we go back to $\phi^{(L)}$ and $\phi^{(R)}$, we prove it by starting with the Dirac equation (3.31). Multiplying (3.31) from the left by $\gamma_\nu\,(\partial/\partial x_\nu)$, we have

$$\frac{\partial}{\partial x_\nu}\frac{\partial}{\partial x_\mu}\gamma_\nu\gamma_\mu\psi - \left(\frac{mc}{\hbar}\right)^2\psi = 0. \tag{3.90}$$

Adding to (3.90) the same equation written in a form in which the summation indices μ and ν are interchanged, we obtain

$$\frac{\partial}{\partial x_\nu}\frac{\partial}{\partial x_\mu}(\gamma_\nu\gamma_\mu + \gamma_\mu\gamma_\nu)\psi - 2\left(\frac{mc}{\hbar}\right)^2\psi = 0, \tag{3.91}$$

which reduces to

$$\Box\psi - (mc/\hbar)^2\psi = 0 \tag{3.92}$$

by virtue of the anticommutation relation of the gamma matrices (3.35). Note that (3.92) is to be understood as four separate *uncoupled* equations for each component of ψ. Because of (3.92), the Dirac equation admits a free-particle solution of the type

$$\psi \sim u(\mathbf{p})\exp\left[i(\mathbf{p}\cdot\mathbf{x}/h) - i(Et/\hbar)\right] \tag{3.93}$$

with

$$E = \pm\sqrt{|\mathbf{p}|^2c^2 + m^2c^4}, \tag{3.94}$$

where $u(\mathbf{p})$ is a *four*-component spinor independent of $\mathbf{x}$ and t. Note that (3.93) is a simultaneous eigenfunction of $-i\hbar\nabla$ and $i\hbar(\partial/\partial t)$ with eigenvalues $\mathbf{p}$ and E, respectively.

For a particle at rest ($\mathbf{p} = 0$), the equation we must solve is

$$\gamma_4\frac{\partial}{\partial(ict)}\psi = -\frac{mc}{\hbar}\psi. \tag{3.95}$$

In accordance with (3.93) and (3.94), we first try the time-dependence $e^{-imc^2t/\hbar}$. Equation (3.95) then becomes

$$-\frac{imc^2}{ic\hbar}\begin{pmatrix} I & 0 \\ 0 & -I \end{pmatrix}\begin{pmatrix} u_A(0) \\ u_B(0) \end{pmatrix} = -\frac{mc}{\hbar}\begin{pmatrix} u_A(0) \\ u_B(0) \end{pmatrix}, \tag{3.96}$$

which is satisfied only if the lower two-component spinor $u_B(0)$ vanishes. But, using a similar argument, we see that (3.95) can be satisfied equally well by the time-dependence $e^{+imc^2t/\hbar}$ provided that the *upper* two-component spinor $u_A(0)$ vanishes. As in the Pauli theory, the nonvanishing two-component spinors can be taken as

$$\begin{pmatrix} 1 \\ 0 \end{pmatrix} \quad \text{and} \quad \begin{pmatrix} 0 \\ 1 \end{pmatrix}.$$

So there are four independent solutions to (3.95):

$$\begin{pmatrix}1\\0\\0\\0\end{pmatrix} e^{-imc^2t/\hbar}, \qquad \begin{pmatrix}0\\1\\0\\0\end{pmatrix} e^{-imc^2t/\hbar},$$
$$\begin{pmatrix}0\\0\\1\\0\end{pmatrix} e^{+imc^2t/\hbar}, \qquad \begin{pmatrix}0\\0\\0\\1\end{pmatrix} e^{+imc^2t/\hbar}. \tag{3.97}$$

If we insist on the interpretation that $i\hbar(\partial/\partial t)$ is the Hamiltonian operator, the first two are "positive-energy" solutions while the last two are "negative-energy" solutions. Note that the eigenvalues of the Hamiltonian operator are $\pm mc^2$, depending on whether the eigenvalues of $\gamma_4 = \beta$ are ± 1; this also follows directly from the expression for the Hamiltonian (3.39). In Section 3–9 we shall show that the existence of negative-energy solutions is intimately related to the fact that the Dirac theory can accommodate a positron.

Earlier in this section we showed that in the nonrelativistic limit $E \approx mc^2$, the upper two-component spinor ψ_A coincides with the Schrödinger wave function apart from $e^{-imc^2t/\hbar}$. Therefore the first solution in (3.97) is the Dirac wave function for a particle at rest with spin *up*, since the eigenvalue of σ_3 is $+1$ when applied to $\begin{pmatrix}1\\0\end{pmatrix}$. This leads us to define a 4×4 matrix

$$\Sigma_3 = \frac{\gamma_1\gamma_2 - \gamma_2\gamma_1}{2i} = \begin{pmatrix}\sigma_3 & 0\\0 & \sigma_3\end{pmatrix}, \tag{3.98}$$

whose eigenvalue is to be interpreted as the spin component in the positive z-direction in units of $\hbar/2$. This interpretation will be justified in later parts of this chapter where we shall show that

a) the operator $\Sigma_3/2$ is the infinitesimal generator of a rotation about the z-axis acting on the space-time independent part of the Dirac wave function; and that
b) for a central force problem the sum of $\mathbf{x} \times \mathbf{p}$ and $\hbar\Sigma/2$ is indeed a constant of the motion, to be identified with the total angular momentum, where Σ_k is defined as it was in (3.98), that is,

$$\Sigma_k = \frac{\gamma_i\gamma_j - \gamma_j\gamma_i}{2i} = \begin{pmatrix}\sigma_k & 0\\0 & \sigma_k\end{pmatrix} \qquad (ijk) \text{ cyclic.‡} \tag{3.99}$$

Evidently the solutions (3.97) are eigenfunctions of Σ_3 with eigenvalues $+1$ for the first and third and -1 for the second and fourth. Thus for each sign of the energy there are two independent solutions corresponding to two spin states, as expected from the spin-$\frac{1}{2}$ nature of the particle described by the Dirac equation.

‡The notation "(ijk) cyclic" stands for $(i, j, k) = (1, 2, 3)$, $(2, 3, 1)$, or $(3, 1, 2)$.

It has sometimes been stated that we need a $2 \times 2 = 4$ component wave function in the relativistic electron theory to take into account the two energy states and the two spin states of the electron for given $\mathbf{p}$. This argument is incomplete (if not incorrect). To see this we recall that although the wave function in the Klein-Gordon theory is a single-component wave function, it can accommodate the two energy states of a spin-zero particle. Furthermore, the two-component second-order equation (3.24) of Waerden, which we started with, is a perfectly valid equation for the electron; it can accommodate the two energy and the two spin states of the electron just as the Dirac equation can.‡ In this connection we note an important difference between a differential equation linear in $\partial/\partial t$, such as the diffusion (heat) equation or the Schrödinger equation, and a differential equation quadratic in the time derivative, such as the wave equation (for a vibrating string) or the Klein-Gordon equation. When we solve the problem of the temperature distribution of a potato placed in boiling water, it is sufficient to specify initially only the temperature of the inside region of the potato; we do *not* need to know *how fast* the potato is warming up. On the other hand, if we want to predict the time development of a vibrating string for $t > 0$, it is *not* sufficient to know only the displacement of the string at $t = 0$; we must also know how fast the various parts of the string are moving at $t = 0$. In general, then, when solving a partial differential equation quadratic in the time derivative, we must specify as initial conditions both the function and its time derivative. However, when solving a partial differential equation linear in the time derivative, we need specify *only* the function itself. Thus in the Dirac theory if we know ψ at $t = 0$, we can predict the time development of ψ for $t > 0$. By contrast, the two-component equation (3.24) of Waerden is second order in the time derivative; so we must know both ϕ and $\partial\phi/\partial t$ at $t = 0$ to predict the future development of ϕ. For an energy eigenstate, specifying $\partial\phi/\partial t$ (in addition to ϕ) amounts to specifying the sign of the energy. The number of independent components we must specify is *four* whether we use the four-component Dirac equation or the two-component Waerden equation.

Plane-wave solutions. Now let us return to the free-particle problem, this time with $\mathbf{p} \neq 0$. Substituting

$$\psi = \begin{pmatrix} \psi_A \\ \psi_B \end{pmatrix} = \begin{pmatrix} u_A(\mathbf{p}) \\ u_B(\mathbf{p}) \end{pmatrix} \exp\left(i\mathbf{p}\cdot\frac{\mathbf{x}}{\hbar} - i\frac{Et}{\hbar}\right) \tag{3.100}$$

into (3.62) with $A_0 = \mathbf{A} = 0$, we obtain

$$u_A(\mathbf{p}) = \frac{c}{E - mc^2}(\boldsymbol{\sigma}\cdot\mathbf{p})u_B(\mathbf{p}), \qquad u_B(\mathbf{p}) = \frac{c}{E + mc^2}(\boldsymbol{\sigma}\cdot\mathbf{p})u_A(\mathbf{p}). \tag{3.101}$$

‡One of the reasons that the two-component formalism based on (3.24) is not used so extensively as the Dirac theory is that a solution to (3.24) with the electromagnetic interaction added has a rather complicated transformation property under parity (cf. Problem 3–5). Nevertheless, the two-component equation can be used for solving practical problems in quantum electrodynamics just as the Dirac equation can, as shown by R. P. Feynman and L. M. Brown.

For $E = \sqrt{|\mathbf{p}|^2 c^2 + m^2c^4} > 0$ we may try, apart from a normalization constant,

$$\begin{pmatrix} 1 \\ 0 \end{pmatrix} \quad \text{and} \quad \begin{pmatrix} 0 \\ 1 \end{pmatrix}$$

for $u_A(\mathbf{p})$. We can easily find the lower two-component spinor $u_B(\mathbf{p})$ by using the second equation of (3.101) if we recall that

$$\boldsymbol{\sigma}\cdot\mathbf{p} = \begin{pmatrix} p_3 & p_1 - ip_2 \\ p_1 + ip_2 & -p_3 \end{pmatrix}. \tag{3.102}$$

In this way we get two independent solutions for $E > 0$,

$$u^{(1)}(\mathbf{p}) = N\begin{pmatrix} 1 \\ 0 \\ p_3c/(E+mc^2) \\ (p_1+ip_2)c/(E+mc^2) \end{pmatrix} \quad \text{and} \quad u^{(2)}(\mathbf{p}) = N\begin{pmatrix} 0 \\ 1 \\ (p_1-ip_2)c/(E+mc^2) \\ -p_3c/(E+mc^2) \end{pmatrix}, \tag{3.103}$$

where the normalization constant N is to be determined later. For $E = -\sqrt{|\mathbf{p}|^2 c^2 + m^2c^4} < 0$, we may start with the lower two-component spinor u_B set to

$$\begin{pmatrix} 1 \\ 0 \end{pmatrix} \quad \text{and} \quad \begin{pmatrix} 0 \\ 1 \end{pmatrix}$$

so that in the case where $\mathbf{p} = 0$, ψ reduces to the third and fourth solutions of (3.97). Using the first of (3.101) to obtain the upper spinor u_A, we have

$$u^{(3)}(\mathbf{p}) = N\begin{pmatrix} -p_3c/(|E|+mc^2) \\ -(p_1+ip_2)c/(|E|+mc^2) \\ 1 \\ 0 \end{pmatrix} \quad \text{and} \quad u^{(4)}(\mathbf{p}) = N\begin{pmatrix} -(p_1-ip_2)c/(|E|+mc^2) \\ p_3c/(|E|+mc^2) \\ 0 \\ 1 \end{pmatrix}. \tag{3.104}$$

Since

$$u^{(r)}(\mathbf{p}) \exp\left[i\frac{\mathbf{p}\cdot\mathbf{x}}{\hbar} - \frac{iEt}{\hbar}\right]$$

satisfies the free-field Dirac equation, it is evident that each free-particle spinor $u^{(r)}$ with $r = 1, \ldots, 4$ satisfies

$$(i\gamma\cdot p + mc)u^{(r)}(\mathbf{p}) = 0 \tag{3.105}$$

with $\gamma\cdot p = \gamma_\mu p_\mu$, $p = (\mathbf{p}, iE/c)$ regardless of whether $E > 0$ or $E < 0$. This can, of course, be checked by direct substitution.

As shown earlier, the four independent free-particle solutions with $\mathbf{p} = 0$ written in the form (3.97) are eigenspinors of the 4×4 matrix Σ_3. This is not true for the free-particle solutions we have written for the case where $\mathbf{p} \neq 0$, as we can directly verify by applying Σ_3 to (3.103) and (3.104). But suppose we choose the z-axis in the direction of momentum $\mathbf{p}$ so that $p_1 = p_2 = 0$. We then see that $u^{(r)}$ with $r = 1, \ldots, 4$ *are* eigenspinors of Σ_3 with eigenvalues of $+1, -1, +1, -1$, respectively. In general, although free-particle plane-wave solutions can always be chosen so that they are eigenfunctions of $\Sigma \cdot \hat{\mathbf{p}}$ (where $\hat{\mathbf{p}} = \mathbf{p}/|\mathbf{p}|$), it is not possible to choose solutions in such a way that they are eigenfunctions of $\Sigma \cdot \hat{\mathbf{n}}$ with an *arbitrary* unit vector $\hat{\mathbf{n}}$.‡ As we shall see in Section 3–5, this peculiarity of the plane-wave solution in the Dirac theory stems from the fact that the operator $\Sigma \cdot \hat{\mathbf{n}}$ does not commute with the free-particle Hamiltonian unless $\hat{\mathbf{n}} = \pm\hat{\mathbf{p}}$ or $\mathbf{p} = 0$. The operator $\Sigma \cdot \hat{\mathbf{p}}$ which can be diagonalized simultaneously with the free-particle Hamiltonian is called the *helicity operator*. The eigenstates of helicity with eigenvalues $+1$ and -1 are referred to, respectively, as the *right-handed state* (spin parallel to motion) and the *left-handed state* (spin opposite to motion).

It is easy to see that for a *given fixed* $\mathbf{p}$, the free-particle spinors $u^{(r)}(\mathbf{p})$ given by (3.103) and (3.104) with $r = 1, \ldots, 4$ are orthogonal to each other;

$$u^{(r)\dagger}(\mathbf{p})u^{(r')}(\mathbf{p}) = 0 \quad \text{for} \quad r \neq r'. \tag{3.106}$$

For the normalization of $u^{(r)}$, two conventions are found in the literature:

(a)
$$u^{(r)}(\mathbf{p})^\dagger u^{(r)}(\mathbf{p}) = 1, \tag{3.107}$$

which implies

$$[1 + (c^2|\mathbf{p}|^2)/(|E| + mc^2)^2]N^2 = 1, \tag{3.108}$$

hence

$$N = \sqrt{(|E| + mc^2)/2|E|}; \tag{3.109}$$

(b)
$$u^{(r)\dagger}(\mathbf{p})u^{(r)}(\mathbf{p}) = |E|/mc^2, \tag{3.110}$$

which means

$$N = \sqrt{(|E| + mc^2)/2mc^2}. \tag{3.111}$$

The second normalization convention which says that $u^\dagger u$ transforms like the zeroth component of a four-vector appears somewhat artificial at this stage. However, we shall see in the next section that this convention is quite natural from the relativistic point of view. Throughout this book we shall use the normalization condition given by the second form (3.110).

To summarize, the normalized plane-wave solutions for given $\mathbf{p}$ are:

$$\psi = \sqrt{\frac{mc^2}{EV}}\, u^{(1 \text{ or } 2)}(\mathbf{p}) \exp\left[i\mathbf{p}\cdot\frac{\mathbf{x}}{\hbar} - i\frac{Et}{\hbar}\right], \tag{3.112}$$

for $E = \sqrt{|\mathbf{p}|^2c^2 + m^2c^4} > 0$, and

$$\psi = \sqrt{\frac{mc^2}{|E|V}}\, u^{(3 \text{ or } 4)}(\mathbf{p}) \exp\left[i\frac{\mathbf{p}\cdot\mathbf{x}}{\hbar} + \frac{i|E|t}{\hbar}\right], \tag{3.113}$$

---

‡This situation is in sharp contrast to the nonrelativistic Pauli theory in which *any* space-time independent two-component spinor can be regarded as an eigenspinor of $\boldsymbol{\sigma}\cdot\hat{\mathbf{n}}$, where $\hat{\mathbf{n}}$ is a unit vector in *some* direction, that is,

$$(\boldsymbol{\sigma}\cdot\hat{\mathbf{n}})\begin{pmatrix} a \\ b \end{pmatrix} = \begin{pmatrix} a \\ b \end{pmatrix}.$$

Assuming that the spinor is normalized, all we have to do is set $a = \cos(\theta_0/2)e^{-i\phi_0/2}$ and $b = \sin(\theta_0/2)e^{+i\phi_0/2}$, where θ_0 and ϕ_0 characterize the orientation of the unit vector along which the spin component is sharp.

for $E = -\sqrt{|\mathbf{p}|^2 c^2 + m^2 c^4} < 0$, where

$$u^{(1 \text{ or } 2)}(\mathbf{p}) = \sqrt{(E+mc^2)/2mc^2}\left[\begin{pmatrix} 1 \\ 0 \\ p_3 c/(E+mc^2) \\ (p_1+ip_2)c/(E+mc^2) \end{pmatrix} \text{ or } \begin{pmatrix} 0 \\ 1 \\ (p_1-ip_2)c/(E+mc^2) \\ -p_3 c/(E+mc^2) \end{pmatrix}\right], \tag{3.114}$$

for $E > 0$, and

$$u^{(3 \text{ or } 4)}(\mathbf{p}) = \sqrt{(|E|+mc^2)/2mc^2}\left[\begin{pmatrix} -p_3 c/(|E|+mc^2) \\ -(p_1+ip_2)c/(|E|+mc^2) \\ 1 \\ 0 \end{pmatrix} \text{ or } \begin{pmatrix} -(p_1-ip_2)c/(|E|+mc^2) \\ p_3 c/(|E|+mc^2) \\ 0 \\ 1 \end{pmatrix}\right], \tag{3.115}$$

for $E < 0$. The square root factors in (3.112) and (3.113) merely compensate for $|E|^2/mc^2$ in (3.110), so that

$$\int_V \psi^\dagger \psi \, d^3x = 1. \tag{3.116}$$

As $V \longrightarrow \infty$, the allowed values of E form a continuous spectrum. For positive-energy free-particle solutions, $mc^2 \le E < \infty$, whereas for negative-energy solutions, $-\infty < E \le -mc^2$, as shown in Fig. 3–1.

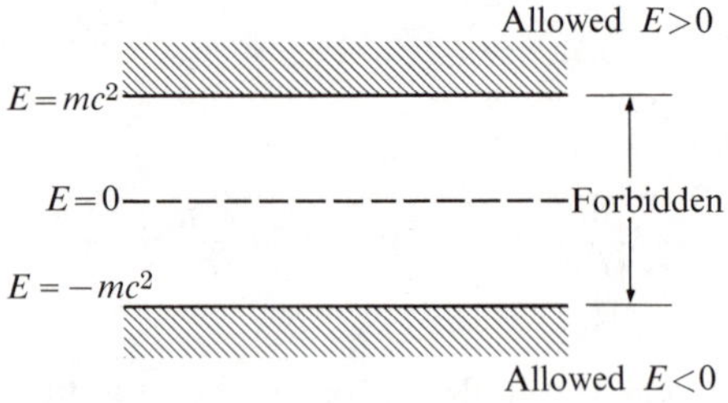

Fig. 3–1. The allowed values of E for free particles.

When the particles are not free, problems for which the Dirac equation can be solved exactly are rather scarce. In Section 3–8 we shall treat the classical problem of the electron in the Coulomb potential for which exact solutions can be found. Another problem that can be solved exactly is that of an electron in a uniform magnetic field, which is left as an exercise (Problem 3–2).

3–4. RELATIVISTIC COVARIANCE

Lorentz transformations and rotations. Before we establish the relativistic covariance of the Dirac equation, we shall review briefly the properties of Lorentz transformations. Since a Lorentz transformation is essentially a rotation in Minkowski space, let us first examine ordinary rotations in the usual three-dimensional space.

Consider a rotation through an angle ω, of the coordinate system in the 1-2 plane about the third axis. The sense of rotation is that of a right-handed screw advancing in the positive x_3-direction if $\omega > 0$. A point described by (x_1, x_2, x_3) in the old coordinate system is described in the new (primed) coordinate system by (x_1', x_2', x_3') where

$$x_1' = x_1 \cos\omega + x_2 \sin\omega, \qquad x_2' = -x_1 \sin\omega + x_2 \cos\omega. \tag{3.117}$$

Meanwhile, when we perform a Lorentz transformation along the x_1-direction such that the primed system moves with velocity $v = \beta c$, $(\mathbf{x}, ix_0)$ and $(\mathbf{x}', ix_0')$ are related by

$$x_1' = \frac{x_1}{\sqrt{1-\beta^2}} - \frac{\beta x_0}{\sqrt{1-\beta^2}}, \qquad x_0' = -\frac{\beta x_1}{\sqrt{1-\beta^2}} + \frac{x_0}{\sqrt{1-\beta^2}}, \tag{3.118}$$

or

$$x_1' = x_1 \cosh\chi + ix_4 \sinh\chi, \qquad x_4' = -ix_1 \sinh\chi + x_4 \cosh\chi, \tag{3.119}$$

where we have set

$$\tanh\chi = \beta. \tag{3.120}$$

Since the cosine and sine of an *imaginary* angle $i\chi$ are $\cosh\chi$ and $i \sinh\chi$, the above Lorentz transformation may be viewed as a rotation in Minkowski space by an angle $i\chi$ "in" the 1-4 plane, just as the transformation (3.117) represents a rotation in the 1-2 plane by a *real* angle ω. For both three-dimensional rotations and Lorentz transformations we have

$$a_{\mu\nu} a_{\lambda\nu} = \delta_{\mu\lambda}, \tag{3.121}$$

where $a_{\mu\nu}$ is defined by

$$x_\mu' = a_{\mu\nu} x_\nu. \tag{3.122}$$

From now on, the term Lorentz transformation will be used both for a three-dimensional rotation of the form (3.117) and for a "pure" Lorentz transformation of the form (3.119). At this stage we shall consider only those coordinate transformations which can be obtained by successive applications of a transformation that differs infinitesimally from the identity transformation. For example, (3.117) can be obtained by successively compounding an infinitesimal rotation

$$x_1' = x_1 + \delta\omega x_2, \qquad x_2' = -\delta\omega x_1 + x_2, \tag{3.123}$$

where $\delta\omega$ is the infinitesimal angle of rotation. This means that we restrict our considerations to the cases where

$$\det(a_{\mu\nu}) = 1, \qquad a_{44} > 0. \tag{3.124}$$

Covariance of the Dirac equation. What is meant by the covariance of the Dirac equation under Lorentz transformations? First of all, if someone, working with the primed system, were to formulate a relativistic electron theory, his first-order wave equation would *look* like the Dirac equation. Second, there exists an explicit prescription that relates $\psi(x)$ and $\psi'(x')$, where $\psi(x)$ and $\psi'(x')$ are the wave functions corresponding to a given physical situation viewed in the unprimed and the primed system, respectively. Third, using the foregoing prescription, we must be able to show that the "Dirac-like" equation in the primed system not only looks like the Dirac equation in the unprimed system but actually is equivalent to it.

We take the point of view that the gamma matrices are introduced merely as useful short-hand devices that enable us to keep track of how the various components of ψ are coupled to each other. Hence the explicit forms of the gamma matrices are assumed to be unchanged under Lorentz transformations.‡ Note, in particular, that the gamma matrices *themselves* are not to be regarded as components of a four-vector even though, as we shall show in a moment, $\bar{\psi}\gamma_\mu\psi$ does transform like a four-vector. On the other hand, we expect that for a given physical situation the wave functions in different Lorentz frames are no longer the same.

With this point of view, if the Dirac equation is to look the same in the primed system, we must have

$$\gamma_\mu \frac{\partial}{\partial x'_\mu} \psi'(x') + \frac{mc}{\hbar} \psi'(x') = 0. \tag{3.125}$$

Note that the gamma matrices themselves are *not* primed. The question we must now ask is: How are ψ and ψ' related? An analogy with electrodynamics may be helpful here. When we perform a Lorentz transformation without a simultaneous change in the gauge, the components of $A_\mu(x)$ and $A'_\mu(x')$ are related by

$$A'_\mu(x') = a_{\mu\nu} A_\nu(x), \tag{3.126}$$

where $a_{\mu\nu}$ given by (3.122) depends only on the nature of the transformation and is independent of the space-time coordinates. Similarly, we may assume that the prescription that relates $\psi(x)$ and $\psi'(x')$ is a linear one that can be written as

$$\psi'(x') = S\psi(x), \tag{3.127}$$

where S is a 4×4 matrix which depends only on the nature of the Lorentz transformation and is completely independent of $\mathbf{x}$ and t. We rewrite (3.125), using $\partial/\partial x'_\mu = a_{\mu\nu}(\partial/\partial x_\nu)$ [which was proved in Chapter 1 (cf. Eq. 1.21)] as follows:

$$\gamma_\mu a_{\mu\nu} \frac{\partial}{\partial x_\nu} S\psi + \frac{mc}{\hbar} S\psi = 0 \tag{3.128}$$

or, multiplying S^{-1} from the left, we have

$$S^{-1}\gamma_\mu S a_{\mu\nu} \frac{\partial}{\partial x_\nu} \psi + \frac{mc}{\hbar} \psi = 0. \tag{3.129}$$

We now ask: Is (3.129) equivalent to the Dirac equation? The answer to this question is affirmative provided that we can find an S that satisfies

$$S^{-1}\gamma_\mu S\, a_{\mu\nu} = \gamma_\nu \tag{3.130}$$

‡This means that we are considering a Lorentz transformation not accompanied by a simultaneous change in the representation of the gamma matrices.

or, multiplying by $a_{\lambda\nu}$ and summing over ν, we obtain

$$S^{-1}\gamma_\lambda S = \gamma_\nu a_{\lambda\nu}, \tag{3.131}$$

where we have used (3.121).‡ The problem of demonstrating the relativistic covariance of the Dirac equation is now reduced to that of finding an S that satisfies (3.131).

We shall first treat the case of an ordinary rotation in three dimensions which we refer to as a "pure rotation." In looking for an S appropriate for the transformation (3.117), we recall that for a particle at rest the upper two-component spinor u_A is identical with the corresponding two-component spinor in the nonrelativistic Pauli theory. So in this particular case we know from nonrelativistic quantum mechanics how u_A transforms under three-dimensional rotations: if u_A is the Pauli spinor in the unprimed system, then

$$\left(\cos\frac{\omega}{2} + i\sigma_3 \sin\frac{\omega}{2}\right) \tag{3.132}$$

times u_A is the Pauli spinor corresponding to the same physical situation seen in the primed system defined by (3.117).§ Since the form of S is independent of whether it acts on the "at-rest" spinor or a more general wave function, it is natural to try for S the 4×4 analog of (3.132) given by

$$\begin{aligned} S_{\text{rot}} &= \cos\frac{\omega}{2} + i\,\Sigma_3 \sin\frac{\omega}{2} \\ &= \cos\frac{\omega}{2} + \gamma_1\gamma_2 \sin\frac{\omega}{2}, \end{aligned} \tag{3.133}$$

which is now assumed to act on a four-component wave function. Since $(\gamma_1\gamma_2)^2 = \gamma_1\gamma_2\gamma_1\gamma_2 = -\gamma_1^2\gamma_2^2 = -1$, S_{rot}^{-1} is given by

$$S_{\text{rot}}^{-1} = \cos\frac{\omega}{2} - \gamma_1\gamma_2 \sin\frac{\omega}{2}. \tag{3.134}$$

---

‡Note that the matrices $a_{\lambda\nu}$ rearrange the coordinates $\mathbf{x}$ and x_0, whereas the gamma matrices and the 4×4 matrix S rearrange the components of ψ. Although $a_{\mu\nu}$ and γ_μ are both 4×4 matrices, they are in entirely different spaces. The relation (3.131), for instance, really means

$$\sum_{\beta,\gamma} (S^{-1})_{\alpha\beta}(\gamma_\lambda)_{\beta\gamma}(S)_{\gamma\delta} = \sum_\nu a_{\lambda\nu}(\gamma_\nu)_{\alpha\delta}.$$

§See, for example, Dicke and Wittke (1960), p. 255. As an example, let us consider an electron whose spin is in the positive x_1-direction. Its wave function is represented by

$$\frac{1}{\sqrt{2}}\begin{pmatrix}1\\1\end{pmatrix},$$

which is evidently an eigenspinor of σ_1 with eigenvalue $+1$. In a primed system obtained by a rotation of 90° about the x_3-axis, the electron spin is in the negative x_2'-direction. Accordingly, the wave function in the primed system is

$$\left(\cos\frac{\pi}{4} + i\sigma_3 \sin\frac{\pi}{4}\right)\frac{1}{\sqrt{2}}\begin{pmatrix}1\\1\end{pmatrix} = \frac{1}{2}\begin{pmatrix}1+i\\1-i\end{pmatrix}.$$

This is indeed seen to be an eigenspinor of σ_2 with eigenvalue -1.

So the relation to be checked is (cf. Eq. 3.131)

$$\left(\cos\frac{\omega}{2} - \gamma_1\gamma_2 \sin\frac{\omega}{2}\right)\gamma_\lambda\left(\cos\frac{\omega}{2} + \gamma_1\gamma_2\sin\frac{\omega}{2}\right) = \gamma_\nu a_{\lambda\nu}. \tag{3.135}$$

For $\lambda = 3, 4$ we have $a_{\lambda\nu} = \delta_{\lambda\nu}$; hence (3.135) is trivially satisfied by virtue of $\gamma_1\gamma_2\gamma_{3,4} = \gamma_{3,4}\gamma_1\gamma_2$ and $\gamma_1\gamma_2\gamma_{3,4}\gamma_1\gamma_2 = -\gamma_{3,4}$. For $\lambda = 1$,

$$\begin{aligned}&\left(\cos\frac{\omega}{2} - \gamma_1\gamma_2\sin\frac{\omega}{2}\right)\gamma_1\left(\cos\frac{\omega}{2} + \gamma_1\gamma_2\sin\frac{\omega}{2}\right)\\ &= \gamma_1\cos^2\frac{\omega}{2} + 2\gamma_2\sin\frac{\omega}{2}\cos\frac{\omega}{2} - \gamma_1\sin^2\frac{\omega}{2}\\ &= \gamma_1\cos\omega + \gamma_2\sin\omega.\end{aligned} \tag{3.136}$$

But according to (3.117) $\cos\omega = a_{11}$, $\sin\omega = a_{12}$. So (3.135) is indeed satisfied for $\lambda = 1$; similarly for $\lambda = 2$. Thus we have proved that the S_{rot} that satisfies (3.131) is given correctly by (3.133).

According to (3.119) a "pure" Lorentz transformation is nothing more than a rotation in the 1-4 plane by an imaginary angle $i\chi$. Hence for S corresponding to (3.119) we try a matrix analogous to (3.133) with $\omega \to i\chi$, $\gamma_2 \to \gamma_4$:

$$S_{\text{Lor}} = \cosh(\chi/2) + i\gamma_1\gamma_4\sinh(\chi/2). \tag{3.137}$$

The relation we must check this time is

$$\left(\cosh\frac{\chi}{2} - i\gamma_1\gamma_4\sinh\frac{\chi}{2}\right)\gamma_\lambda\left(\cosh\frac{\chi}{2} + i\gamma_1\gamma_4\sinh\frac{\chi}{2}\right) = \gamma_\nu a_{\lambda\nu}. \tag{3.138}$$

Using similar techniques, we can verify this. For instance,

$$\begin{aligned}&\left(\cosh\frac{\chi}{2} - i\gamma_1\gamma_4\sinh\frac{\chi}{2}\right)\gamma_4\left(\cosh\frac{\chi}{2} + i\gamma_1\gamma_4\sinh\frac{\chi}{2}\right)\\ &= \gamma_4\cosh^2\frac{\chi}{2} - 2i\gamma_1\cosh\frac{\chi}{2}\sinh\frac{\chi}{2} + \gamma_4\sinh^2\frac{\chi}{2}\\ &= \gamma_4\cosh\chi + \gamma_1(-i\sinh\chi)\\ &= \gamma_4 a_{44} + \gamma_1 a_{41}.\end{aligned} \tag{3.139}$$

Since we have found S that satisfy (3.131) for the transformation (3.117) and (3.119), we have established the covariance of the Dirac equation under (3.117) and (3.119). In general, the S that relates ψ and ψ' via (3.127) is given by

$$S_{\text{rot}} = \cos\frac{\omega}{2} + i\sigma_{ij}\sin\frac{\omega}{2} \tag{3.140}$$

for a rotation about the kth-axis (ijk cyclic) by an angle ω, and

$$S_{\text{Lor}} = \cosh\frac{\chi}{2} - \sigma_{k4}\sinh\frac{\chi}{2} \tag{3.141}$$

for a Lorentz transformation along the kth-axis with $\beta = \tanh\chi$, where we have introduced new 4×4 matrices

$$\sigma_{\mu\nu} = (1/2i)[\gamma_\mu, \gamma_\nu] = -i\gamma_\mu\gamma_\nu, \qquad \mu \neq \nu. \tag{3.142}$$

More explicitly,

$$\sigma_{ij} = -\sigma_{ji} = \Sigma_k = \begin{pmatrix} \sigma_k & 0 \\ 0 & \sigma_k \end{pmatrix}, \qquad (ijk) \text{ cyclic},$$
$$\sigma_{k4} = -\sigma_{4k} = \alpha_k = \begin{pmatrix} 0 & \sigma_k \\ \sigma_k & 0 \end{pmatrix}. \tag{3.143}$$

It is easily verified that

$$S^\dagger_{\text{rot}} = \cos\frac{\omega}{2} - i\sigma_{ij}\sin\frac{\omega}{2} = S^{-1}_{\text{rot}}, \tag{3.144}$$

but

$$S^\dagger_{\text{Lor}} = \cosh\frac{\chi}{2} - \sigma_{k4}\sinh\frac{\chi}{2} = S_{\text{Lor}} \neq S^{-1}_{\text{Lor}}. \tag{3.145}$$

Thus, unlike S_{rot}, S_{Lor} is not unitary. This is not catastrophic, however; S_{Lor} should not be unitary if $\psi^\dagger\psi$ is to transform like the fourth component of a four-vector under Lorentz transformations. It is very important to note that for both pure rotations and pure Lorentz transformations we have

$$S^{-1} = \gamma_4 S^\dagger \gamma_4, \qquad S^\dagger = \gamma_4 S^{-1}\gamma_4, \tag{3.146}$$

since γ_4 commutes with σ_{ij} but anticommutes with σ_{k4}. We shall make extensive use of (3.146) in the next section.

Space inversion. So far we have considered only those transformations which can be obtained by compounding transformations that differ infinitesimally from the identity transformation. Under such transformations the right-handed coordinate system remains right-handed. In contrast, space inversion (often called the *parity operation*) represented by

$$\mathbf{x}' = -\mathbf{x}, \qquad t' = t \tag{3.147}$$

changes a right-handed coordinate system into a left-handed one; hence it is outside the class of transformations considered so far. We shall now demonstrate that the Dirac theory is covariant under space inversion even in the presence of A_μ.

According to the Maxwell theory the four-vector potential transforms under space inversion as‡

$$\mathbf{A}'(\mathbf{x}', t') = -\mathbf{A}(\mathbf{x}, t), \qquad A_4'(\mathbf{x}', t') = A_4(\mathbf{x}, t). \tag{3.148}$$

---

‡To prove this we first note that if the equation for the Lorentz force

$$m\frac{d}{dt}\left(\frac{\mathbf{v}}{\sqrt{1-(v/c)^2}}\right) = e[\mathbf{E} + (\mathbf{v}/c)\times\mathbf{B}]$$

is to take the same form in the space-inverted system, the electromagnetic fields must transform as $\mathbf{E}' = -\mathbf{E}$, $\mathbf{B}' = \mathbf{B}$ since $\mathbf{v}' = -\mathbf{v}$. Meanwhile, $\mathbf{E}$ and $\mathbf{B}$ are related to $\mathbf{A}$ and $A_0 = iA_4$ by

$$\mathbf{E} = -\nabla A_0 - \frac{1}{c}\frac{\partial A}{\partial t}, \qquad \mathbf{B} = \nabla \times \mathbf{A}.$$

Hence we must have (3.148).

If the Dirac equation is to look similar in the space-inverted system, we have

$$\left(\frac{\partial}{\partial x'_\mu} - \frac{ie}{\hbar c} A'_\mu\right)\gamma_\mu \psi' + \frac{mc}{\hbar}\psi' = 0, \tag{3.149}$$

or, equivalently,

$$\left[-\left(\frac{\partial}{\partial x_k} - \frac{ie}{\hbar c} A_k\right)\gamma_k + \left(\frac{\partial}{\partial x_4} - \frac{ie}{\hbar c} A_4\right)\gamma_4\right]\psi' + \frac{mc}{\hbar}\psi' = 0. \tag{3.150}$$

We try as before

$$\psi'(\mathbf{x}', t') = S_P \psi(\mathbf{x}, t), \tag{3.151}$$

where S_P is a 4×4 matrix independent of the space-time coordinates. Multiplying (3.150) by S_P^{-1} from the left, we see that (3.149) is equivalent to the Dirac equation if there exists S_P with the property

$$S_P^{-1}\gamma_k S_P = -\gamma_k, \qquad S_P^{-1}\gamma_4 S_P = \gamma_4. \tag{3.152}$$

Since γ_4 commutes with γ_4 but anticommutes with γ_k,

$$S_P = \eta\gamma_4, \qquad S_P^{-1} = \gamma_4/\eta \tag{3.153}$$

will do the job, where η is some multiplicative constant. The invariance of the probability density $\psi^\dagger\psi$ further requires $|\eta|^2 = 1$, or $\eta = e^{i\phi}$ with ϕ real. It is customary to set this phase factor η equal to 1 even though no experiment in the world can uniquely determine what this phase factor is.‡ We shall take

$$S_P = \gamma_4 \tag{3.154}$$

for an electron wave function.

Closely related to the parity operation is an operation known as *mirror reflection*, for example,

$$(x'_1, x'_2, x'_3) = (x_1, x_2, -x_3), \qquad x'_0 = x'_0. \tag{3.155}$$

Since (3.155) is nothing more than the parity operation followed by a 180° rotation about the third axis, the covariance of the Dirac equation under mirror reflection has already been demonstrated implicitly. Similar statements hold for more general transformations with

$$\det(a_{\mu\nu}) = -1, \qquad a_{44} > 0, \tag{3.156}$$

which are called *improper orthochronous Lorentz transformations* in contrast to proper orthochronous Lorentz transformations that satisfy

$$\det(a_{\mu\nu}) = 1, \qquad a_{44} > 0. \tag{3.157}$$

Simple examples. To appreciate the real physical significance of S_{rot}, S_{Lor}, S_P, it is instructive to work out some examples at this stage. As a first example, let us consider an infinitesimal form of (3.117) in which $\cos\omega$ and $\sin\omega$ are set, respectively, to 1 and $\delta\omega$. The wave functions in the two systems are related by

$$\psi'(x') = [1 + i\,\Sigma_3\,(\delta\omega/2)]\psi(x), \tag{3.158}$$

‡Some authors argue that one can narrow down the choice for η to ± 1, $\pm i$ by requiring that four successive inversion operations return the wave function to itself. However, there does not appear to be any deep physical significance attached to such a requirement.

where $\mathbf{x}' = \mathbf{x} + \delta\mathbf{x}$ with

$$\delta\mathbf{x} = (x_2\delta\omega, \; -x_1\delta\omega, 0). \tag{3.159}$$

But

$$\psi'(x') = \psi'(x) + \delta x_1 \frac{\partial\psi'}{\partial x_1} + \delta x_2 \frac{\partial\psi'}{\partial x_2}. \tag{3.160}$$

Consequently,

$$\begin{aligned}\psi'(x) &= \left[1 + i\,\Sigma_3\frac{\delta\omega}{2} - \left(x_2\delta\omega\frac{\partial}{\partial x_1} - x_1\delta\omega\frac{\partial}{\partial x_2}\right)\right]\psi(x)\\ &= \psi(x) + i\delta\omega\left[\frac{\Sigma_3}{2} + \frac{1}{\hbar}\left(-i\hbar x_1\frac{\partial}{\partial x_2} + i\hbar x_2\frac{\partial}{\partial x_1}\right)\right]\psi(x).\end{aligned} \tag{3.161}$$

We see that the change in the *functional form of* ψ induced by the infinitesimal rotation consists of two parts: the space-time independent operator $i\Sigma_3\delta\omega/2$ acting on the "internal" part of $\psi(x)$ and the familiar $iL_3\delta\omega/\hbar$ operator affecting just the spatial part of the wave function. The sum

$$(1/\hbar)[(\hbar\,\Sigma_3/2) + L_3] \tag{3.162}$$

is to be identified with the third component of the total angular-momentum operator in units of $\hbar$ since it generates an infinitesimal rotation around the third axis.‡

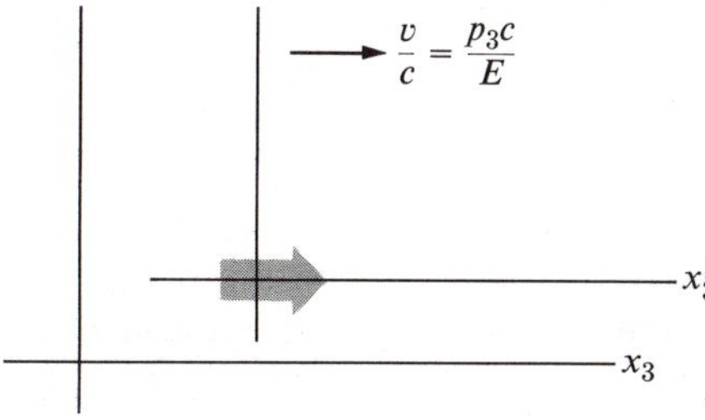

Fig. 3–2. A positive-helicity electron moving with momentum **p** along the x_3-axis. The electron is at rest in the primed system. The gray arrow indicates the spin direction.

As a second example, let us consider a free positive-energy electron of helicity +1 and momentum **p** along the positive x_3-direction. We choose a primed system in such a way that it will coincide with the rest system of the electron (Fig. 3–2). In the primed system the electron wave function can be written

$$\psi'(x') = \frac{1}{\sqrt{V'}}\begin{pmatrix}1\\0\\0\\0\end{pmatrix} e^{-imc^2t/\hbar}. \tag{3.163}$$

‡Usually an operator that rotates the *physical system* around the third axis by an angle $\delta\omega$ is $1 - i\delta\omega(J_3/h)$. But in our case we are rotating the coordinate system rather than the physical system. This explains why we have $1 + i\delta\omega J_3/h$ instead of the above operator.

The question is: What is the wave function for the same physical situation in the unprimed system? According to (3.141) and (3.137)

$$\psi(x) = S_{\text{Lor}}^{-1}\psi'(x'), \tag{3.164}$$

where

$$S_{\text{Lor}}^{-1} = \cosh(\chi/2) - i\gamma_3\gamma_4 \sinh(\chi/2), \tag{3.165}$$

with χ given by

$$\cosh\chi = E/mc^2, \qquad \sinh\chi = p_3/mc. \tag{3.166}$$

Since

$$\begin{aligned}\cosh(\chi/2) &= \sqrt{(1+\cosh\chi)/2} = \sqrt{(E+mc^2)/2mc^2},\\ \sinh(\chi/2) &= \sqrt{\cosh^2(\chi/2)-1} = p_3c/\sqrt{2mc^2(E+mc^2)},\end{aligned} \tag{3.167}$$

we obtain

$$S_{\text{Lor}}^{-1}\begin{pmatrix}1\\0\\0\\0\end{pmatrix} = \left(\begin{array}{c|c}\sqrt{\dfrac{E+mc^2}{2mc^2}} & \dfrac{p_3c\sigma_3}{\sqrt{2mc^2(E+mc^2)}}\\ \hline \dfrac{p_3c\sigma_3}{\sqrt{2mc^2(E+mc^2)}} & \sqrt{\dfrac{E+mc^2}{2mc^2}}\end{array}\right)\begin{pmatrix}1\\0\\0\\0\end{pmatrix}$$

$$= \sqrt{\frac{E+mc^2}{2mc^2}}\begin{pmatrix}1\\0\\ \dfrac{p_3c}{E+mc^2}\\0\end{pmatrix}. \tag{3.168}$$

This result is in complete agreement with $u^{(1)}(\mathbf{p})$, with $p_1 = p_2 = 0$ obtained earlier by solving directly the Dirac equation (cf. Eq. 3.114). It is amusing that the normalization constant which we pick up automatically is precisely the one that appears when $u(\mathbf{p})$ is normalized according to (3.110), which says that $u^\dagger u$ is the fourth component of a four-vector. As for the space-time dependence of the wave function, we merely note that

$$\begin{aligned}t' &= t\cosh\chi - (x_3/c)\sinh\chi\\ &= (E/mc^2)t - (p_3/mc^2)x_3.\end{aligned} \tag{3.169}$$

So we find that

$$\begin{aligned}\psi(x) &= S_{\text{Lor}}^{-1}\psi'(x')\\ &= \frac{1}{\sqrt{V'}}u^{(1)}(\mathbf{p})\exp\left[i\frac{p_3x_3}{\hbar} - \frac{iEt}{\hbar}\right]\\ &= \sqrt{\frac{mc^2}{EV}}u^{(1)}(\mathbf{p})\exp\left[i\frac{\mathbf{p}\cdot\mathbf{x}}{\hbar} - \frac{iEt}{\hbar}\right],\end{aligned} \tag{3.170}$$

where we have used $V = (mc^2/E)V'$ that follows from the Lorentz contraction of the normalization volume along the direction of motion. Thus we see that once we know the form of the wave function for a particle at rest, the correct wave function for a moving particle of definite momentum can be constructed just by applying S_{Lor}^{-1}. This operation is sometimes known as the *Lorentz boost*.

To work out an example that involves S_P, let us look at a Dirac wave function with a definite parity:

$$\Pi\,\psi(\mathbf{x}, t) = \pm\psi(\mathbf{x}, t). \tag{3.171}$$

According to (3.151) and (3.154) the functional form of the wave function in the space-inverted system is

$$\psi'(\mathbf{x}, t) = \gamma_4\psi(-\mathbf{x}, t), \tag{3.172}$$

which is to be identified with $\Pi\psi(x, t)$. Thus

$$\begin{pmatrix} I & 0 \\ 0 & -I \end{pmatrix}\begin{pmatrix} \psi_A(-\mathbf{x}, t) \\ \psi_B(-\mathbf{x}, t) \end{pmatrix} = \pm\begin{pmatrix} \psi_A(\mathbf{x}, t) \\ \psi_B(\mathbf{x}, t) \end{pmatrix}. \tag{3.173}$$

If, in addition, ψ_A and ψ_B can be assumed to be eigenstates of orbital angular momentum, then

$$\begin{aligned} \psi_A(-\mathbf{x}, t) &= (-1)^{l_A}\psi_A(\mathbf{x}, t) = \pm\psi_A(\mathbf{x}, t), \\ -\psi_B(-\mathbf{x}, t) &= -(-1)^{l_B}\psi_B(\mathbf{x}, t) = \pm\psi_B(\mathbf{x}, t), \end{aligned} \tag{3.174}$$

where l_A and l_B are the orbital angular momenta of the two-component wave functions. Thus

$$(-1)^{l_A} = -(-1)^{l_B}. \tag{3.175}$$

At first sight this appears to be a peculiar result, since it implies that if ψ_A is a two-component wave function with an even (odd) orbital angular momentum, then ψ_B is a two-component wave function with an odd (even) orbital angular momentum. Actually, this is not too surprising in view of the second part of (3.62), which, for a central force problem with $\mathbf{A} = 0$, takes the form

$$\psi_B = \frac{c}{E - V(r) + mc^2}(\sigma\cdot\mathbf{p})\psi_A. \tag{3.176}$$

Let us suppose that ψ_A is an $s_{1/2}$ state wave function with spin up so that

$$\psi_A = R(r)\begin{pmatrix} 1 \\ 0 \end{pmatrix} e^{-i(Et/\hbar)}. \tag{3.177}$$

Then we find that

$$\begin{aligned} \psi_B &= -\frac{i\hbar c}{E - V + mc^2}\begin{pmatrix} \dfrac{\partial}{\partial x_3} & \dfrac{\partial}{\partial x_1} - i\dfrac{\partial}{\partial x_2} \\ \dfrac{\partial}{\partial x_1} + i\dfrac{\partial}{\partial x_2} & -\dfrac{\partial}{\partial x_3} \end{pmatrix}\begin{pmatrix} R(r) \\ 0 \end{pmatrix} e^{-i(Et/\hbar)} \\ &= -\frac{i\hbar c}{E - V + mc^2}\frac{1}{r}\frac{dR}{dr}\begin{pmatrix} x_3 & x_1 - ix_2 \\ x_1 + ix_2 & -x_3 \end{pmatrix}\begin{pmatrix} 1 \\ 0 \end{pmatrix} e^{-i(Et/\hbar)} \\ &= \frac{i\hbar c}{E - V + mc^2}\frac{dR}{dr}\left[-\sqrt{\frac{4\pi}{3}}\,Y_1^0\begin{pmatrix} 1 \\ 0 \end{pmatrix} + \sqrt{\frac{8\pi}{3}}\,Y_1^1\begin{pmatrix} 0 \\ 1 \end{pmatrix}\right] e^{-i(Et/\hbar)}, \end{aligned} \tag{3.178}$$

which is recognized as a wave function whose angular part is that of a $p_{1/2}$ wave function with $j_3 = \frac{1}{2}$. Thus ψ_A and ψ_B have opposite parities in agreement with (3.175). We could have guessed that this would be so, since the operator that multiplies ψ_A in (3.176) is a pseudoscalar operator which does not change j and

j_3 but does change the parity. We shall make use of this property in Section 3–8 when we discuss central-force problems in detail.

An even more striking consequence of $S_P = \gamma_4$ may now be discussed. Consider a positive-energy free-particle wave function and a negative-energy free-particle wave function both with $\mathbf{p} = 0$:

$$\begin{pmatrix} \chi^{(s)} \\ 0 \end{pmatrix} e^{-imc^2t/\hbar} \quad \text{and} \quad \begin{pmatrix} 0 \\ \chi^{(s)} \end{pmatrix} e^{imc^2t/\hbar}, \tag{3.179}$$

where $\chi^{(s)}$ may stand for

$$\begin{pmatrix} 1 \\ 0 \end{pmatrix} \text{ or } \begin{pmatrix} 0 \\ 1 \end{pmatrix}.$$

Since they are eigenstates of γ_4 with eigenvalues $+1$ and -1, respectively, we have the following far-reaching result: *a positive energy electron at rest and a negative energy electron at rest have opposite parities.* This will be shown to imply that an electron and a positron have opposite "intrinsic" parities when a negative energy state is properly interpreted in Section 3–9. For instance, the parity of an e^+e^--system in a relative s state must be *odd* despite its *even* orbital parity. This remarkable prediction of the Dirac theory has been checked experimentally in the decay of a positronium and will be discussed in Chapter 4.

3–5. BILINEAR COVARIANTS

Transformation properties of bilinear densities. We are in a position to discuss bilinear densities of the form $\bar{\psi}\Gamma\psi$, where Γ is a product of gamma matrices. Such densities are called bilinear covariants since they have definite transformation properties under Lorentz transformations, as will be shown in a moment. Let us first note that because of (3.146) the relation $\psi'(x') = S\,\psi(x)$ implies that

$$\bar{\psi}'(x') = \psi^\dagger(x)S^\dagger\gamma_4 = \psi^\dagger(x)\gamma_4\gamma_4S^\dagger\gamma_4 = \bar{\psi}(x)S^{-1}, \tag{3.180}$$

whether S stands for S_{rot} or S_{Lor}. Clearly this relation holds also for S_P. Using (3.180) we immediately see that $\bar{\psi}\psi$ is invariant:

$$\bar{\psi}'(x')\,\psi'(x') = \bar{\psi}(x)\,\psi(x) \tag{3.181}$$

under pure rotations, pure Lorentz transformations, and space inversion; hence $\bar{\psi}\psi$ (not $\psi^\dagger\psi$) is a scalar density. To investigate the transformation properties of $\bar{\psi}\gamma_\mu\psi$, it is sufficient to recall (3.131). We have

$$\bar{\psi}'(x')\gamma_\mu\psi'(x') = \bar{\psi}(x)S^{-1}\gamma_\mu S\psi(x) = a_{\mu\nu}\bar{\psi}(x)\gamma_\nu\psi(x) \tag{3.182}$$

under pure rotations and pure Lorentz transformations. For the behavior under space inversion, we obtain

$$\bar{\psi}'\begin{Bmatrix}\gamma_k \\ \gamma_4\end{Bmatrix}\psi' = \bar{\psi}S_P^{-1}\begin{Bmatrix}\gamma_k \\ \gamma_4\end{Bmatrix}S_P\psi = \begin{Bmatrix}-\bar{\psi}\gamma_k\psi \\ \bar{\psi}\gamma_4\psi\end{Bmatrix}. \tag{3.183}$$

Hence $\bar{\psi}\gamma_\mu\psi$ is a four-vector density whose space components change under parity. Consequently the flux density and the probability density defined earlier (cf. Eq. 3.49) do indeed form a four-vector. Using similar techniques we find

Table 3–1

BEHAVIOR OF BILINEAR COVARIANTS UNDER LORENTZ TRANSFORMATIONS

		Proper orthochronous Lorentz transformations	Space inversion
Scalar	$\bar{\psi}\psi$	$\bar{\psi}\psi$	$\bar{\psi}\psi$
Vector	$\bar{\psi}\gamma_\mu\psi$	$a_{\mu\nu}\bar{\psi}\gamma_\nu\psi$	$\begin{Bmatrix} -\bar{\psi}\gamma_k\psi \\ \bar{\psi}\gamma_4\psi \end{Bmatrix}$
Tensor (antisymmetric, second rank)	$\bar{\psi}\sigma_{\mu\nu}\psi$	$a_{\mu\lambda}a_{\nu\sigma}\bar{\psi}\sigma_{\lambda\sigma}\psi$	$\begin{Bmatrix} \bar{\psi}\sigma_{kl}\psi \\ -\bar{\psi}\sigma_{k4}\psi \end{Bmatrix}$
Axial vector (pseudovector)	$i\bar{\psi}\gamma_5\gamma_\mu\psi$	$a_{\mu\nu}i\bar{\psi}\gamma_5\gamma_\nu\psi$	$\begin{Bmatrix} i\bar{\psi}\gamma_5\gamma_k\psi \\ -i\bar{\psi}\gamma_5\gamma_4\psi \end{Bmatrix}$
Pseudoscalar	$\bar{\psi}\gamma_5\psi$	$\bar{\psi}\gamma_5\psi$	$-\bar{\psi}\gamma_5\psi$

that $\bar{\psi}\sigma_{\mu\nu}\psi = -i\bar{\psi}\gamma_\mu\gamma_\nu\psi$ with $\mu \neq \nu$ (which is necessarily antisymmetric in μ and ν) is a second-rank *tensor* density. At this point it is advantageous to define a Hermitian 4×4 matrix

$$\gamma_5 = \gamma_1\gamma_2\gamma_3\gamma_4. \tag{3.184}$$

This γ_5 matrix has the remarkable property that it anticommutes with every one of γ_μ with $\mu = 1, \ldots, 4$,

$$\{\gamma_\mu, \gamma_5\} = 0, \qquad \mu \neq 5, \tag{3.185}$$

as seen, for instance, from $\gamma_2\gamma_1\gamma_2\gamma_3\gamma_4 = (-1)^3\gamma_1\gamma_2\gamma_3\gamma_4\gamma_2$, where we have used the fact that γ_2 commutes with γ_2 but anticommutes with γ_1, γ_3, γ_4. Note also that

$$\gamma_5^2 = 1 \tag{3.186}$$

just as γ_μ with $\mu = 1, \ldots, 4$. As for its explicit form it is easy to show that

$$\gamma_5 = \begin{pmatrix} 0 & -I \\ -I & 0 \end{pmatrix} \tag{3.187}$$

in the standard (Dirac-Pauli) representation. Using (3.185), we see that

$$S_{\text{Lor}}^{-1}\gamma_5 S_{\text{Lor}} = \gamma_5, \qquad S_{\text{rot}}^{-1}\gamma_5 S_{\text{rot}} = \gamma_5, \tag{3.188}$$

since γ_5 commutes with $\sigma_{\mu\nu}$, but

$$S_P^{-1}\gamma_5 S_P = -\gamma_5, \tag{3.189}$$

since γ_5 anticommutes with γ_4. Hence $\bar{\psi}\gamma_5\psi$ transforms exactly like the scalar density $\bar{\psi}\psi$ under proper orthochronous Lorentz transformations but changes its sign under space inversion. This transformation is characteristic of a *pseudoscalar* density. Finally, using similar arguments, we can easily see that $i\bar{\psi}\gamma_5\gamma_\mu\psi$ transforms in the same way as $\bar{\psi}\gamma_\mu\psi$ under proper orthochronous Lorentz transformations but in exactly the opposite way under space inversion. This is expected of an *axial vector* (pseudovector) density. Table 3–1 summarizes the results.

The question naturally arises: Have we listed all possible bilinear covariants of the form $\bar{\psi}\Gamma\psi$? To answer this question let us start multiplying the γ_μ. If we multiply any pair of gamma matrices, we get either $\gamma_\mu^2 = 1$ when the two matrices are the same or $\gamma_\mu \gamma_\nu = -\gamma_\nu \gamma_\mu = i\sigma_{\mu\nu}$ when the two matrices are different. When we multiply three gamma matrices, we get back only one of the γ_μ up to sign unless all three are different (for example, $\gamma_1\gamma_2\gamma_1 = -\gamma_1\gamma_1\gamma_2 = -\gamma_2$); when the three matrices are different, we do get a new matrix $\gamma_\mu\gamma_\nu\gamma_\lambda$. But $\gamma_\mu\gamma_\nu\gamma_\lambda$ with $\mu \neq \nu \neq \lambda$ can always be written in the form $\gamma_5\gamma_\sigma$ up to sign, where $\sigma \neq \mu, \nu, \lambda$ (for example, $\gamma_1\gamma_2\gamma_3\gamma_4\gamma_4 = \gamma_5\gamma_4$). Finally, when we multiply four gamma matrices, we get only one new matrix, $\gamma_5 = \gamma_1\gamma_2\gamma_3\gamma_4$ (which is, of course, equal to $-\gamma_2\gamma_3\gamma_4\gamma_1$, $\gamma_3\gamma_4\gamma_1\gamma_2$, etc.). Needless to say, when we multiply five or more gamma matrices, we obtain nothing new. So $= -\frac{i}{2}(\gamma_\mu\gamma_\nu - \gamma_\nu\gamma_\mu)$

$$\Gamma_A = 1, \quad \gamma_\mu, \quad \sigma_{\mu\nu} = -i\gamma_\mu\gamma_\nu \quad (\mu \neq \nu), \quad i\gamma_5\gamma_\mu, \quad \text{and} \quad \gamma_5 \tag{3.190}$$

represents all we can get. This means that there are in all sixteen independent 4×4 matrices (as we might have guessed): the identity matrix, the four γ_μ matrices, the six $\sigma_{\mu\nu}$ matrices (antisymmetric in μ and ν), the four $i\gamma_5\gamma_\mu$ matrices, and the γ_5 matrix. The factors $\pm i$ in (3.190) are inserted so that

$$\Gamma_A^2 = 1, \tag{3.191}$$

for $A = 1, \ldots, 16$. The Γ_A are all traceless with the obvious exception of the identity matrix, as the reader may easily verify by using the explicit forms of Γ_A in the standard (Dirac-Pauli) representation‡ (cf. Appendix B). Moreover, they are all linearly independent. Consequently any 4×4 matrix can be written as a unique linear combination of the sixteen Γ_A. We can find the coefficient λ_A in the expansion of an arbitrary 4×4 matrix Λ

$$\Lambda = \sum_A^{16} \lambda_A \Gamma_A, \tag{3.192}$$

by simply evaluating

$$Tr(\Lambda\Gamma_A) = Tr(\sum_B \lambda_B \Gamma_B \Gamma_A) = 4\lambda_A, \tag{3.193}$$

where we have used the fact that $\Gamma_B\Gamma_A$ is traceless when $B \neq A$ and is equal to the identity matrix when $A = B$. The algebra generated by Γ_A is called *Clifford algebra* after W. K. Clifford, who studied generalized quaternions half a century before the advent of the Dirac theory.

Let us return now to our discussion of the bilinear covariants. It is worth keeping in mind that Table 3–1 exhausts all possible bilinear covariants of the form $\bar{\psi}\Gamma\psi$, as first shown by J. von Neumann in 1928. For instance, note that we have no way to write a symmetric second-rank tensor of the form $\bar{\psi}\Gamma\psi$. This does not mean, however, that we cannot form a symmetric second-rank tensor in the Dirac

‡To prove this without appealing to any particular representation, first verify that for every Γ_A there exists at least one Γ_B (different from Γ_A) such that $\Gamma_A\Gamma_B = -\Gamma_B\Gamma_A$. Then

$$-Tr(\Gamma_A) = Tr(\Gamma_B\Gamma_A\Gamma_B) = Tr(\Gamma_B^2\Gamma_A) = Tr(\Gamma_A)$$

which would be impossible unless Γ_A were traceless.

theory. If we start introducing derivatives of ψ and $\bar{\psi}$, we can develop an expression like

$$T_{\mu\nu} = -\frac{i\hbar c}{2}\left(\bar{\psi}\gamma_\nu \frac{\partial \psi}{\partial x_\mu} - \frac{\partial \bar{\psi}}{\partial x_\mu}\gamma_\nu \psi\right) + \frac{i\hbar c}{4}\sum_{\substack{\sigma \neq \mu, \nu \\ \mu \neq \nu}} \frac{\partial}{\partial x_\sigma}(\bar{\psi}\gamma_\mu\gamma_\nu\gamma_\sigma\psi), \tag{3.194}$$

which can be shown to be the energy-momentum tensor of the Dirac wave function.

For not too relativistic electrons of positive energies some bilinear covariants are "large" while others are "small." To see this we first recall that if $E \sim mc^2$ and $V \ll mc^2$, then ψ_B is of order (v/c) compared to ψ_A. We can see then that $\bar{\psi}\psi$ and $\bar{\psi}\gamma_4\psi$ given by

$$\begin{aligned} \bar{\psi}\psi &= \psi^\dagger\gamma_4\psi = \psi_A^\dagger\psi_A - \psi_B^\dagger\psi_B, \\ \bar{\psi}\gamma_4\psi &= \psi^\dagger\psi = \psi_A^\dagger\psi_A + \psi_B^\dagger\psi_B \end{aligned} \tag{3.195}$$

are both "large" and in fact equal if terms of order $(v/c)^2$ or higher are ignored. Similarly, since

$$i\gamma_5\gamma_k = \begin{pmatrix} \sigma_k & 0 \\ 0 & -\sigma_k \end{pmatrix}, \qquad \sigma_{ij} = \sum_k = \begin{pmatrix} \sigma_k & 0 \\ 0 & \sigma_k \end{pmatrix}, \tag{3.196}$$

it follows that $i\bar{\psi}\gamma_5\gamma_k\psi$ and $\bar{\psi}\sigma_{ij}\psi$ (ijk cyclic) are "large" and indistinguishable up to order v/c:

$$\bar{\psi}\begin{Bmatrix} i\gamma_5\gamma_k \\ \sigma_{ij} \end{Bmatrix}\psi \approx \psi_A^\dagger\sigma_k\psi_A. \tag{3.197}$$

In contrast to 1, γ_4, $i\gamma_5\gamma_k$, and σ_{ij}, which connect $\psi_A^\dagger$ with ψ_A, the matrices γ_k, $i\gamma_5\gamma_4$, σ_{k4}, and γ_5 connect $\psi_A^\dagger(\psi_B^\dagger)$ with $\psi_B(\psi_A)$. Hence the corresponding bilinear covariants are "small" or, more precisely, of order v/c. For instance,

$$\begin{aligned} \bar{\psi}\gamma_5\psi &= (\psi_A^\dagger, -\psi_B^\dagger)\begin{pmatrix} 0 & -I \\ -I & 0 \end{pmatrix}\begin{pmatrix} \psi_A \\ \psi_B \end{pmatrix} \\ &= -\psi_A^\dagger\psi_B + \psi_B^\dagger\psi_A. \end{aligned} \tag{3.198}$$

Gordon decomposition of the vector current. The remaining part of this section is devoted to a detailed discussion of the vector covariant $\bar{\psi}\gamma_\mu\psi$, which occurs most frequently. We argued earlier that within the framework of the single-particle Dirac theory, $s_\mu = ic\bar{\psi}\gamma_\mu\psi$ is to be regarded as the four-vector probability current. We therefore define

$$j_\mu = es_\mu = iec\bar{\psi}\gamma_\mu\psi, \tag{3.199}$$

which is to be interpreted as the charge-current density. Using steps analogous to (3.45) through (3.48) we can show that j_μ satisfies the continuity equation even in the presence of the electromagnetic interaction. With (3.199) as the charge-current density, the Hamiltonian density for the electromagnetic interaction of the charged Dirac particle is given by

$$\begin{aligned} \mathscr{H}_{\text{int}} &= -j_\mu A_\mu/c = -ie\bar{\psi}\gamma_\mu\psi A_\mu \\ &= -e\psi^\dagger\boldsymbol{\alpha}\psi\cdot\mathbf{A} + e\psi^\dagger\psi A_0. \end{aligned} \tag{3.200}$$

This relationship can also be inferred from the Hamiltonian form of the Dirac equation (cf. Eqs. 3.38 and 3.39):

$$i\hbar(\partial/\partial t)\psi = [(-ic\hbar\nabla - e\mathbf{A})\cdot\boldsymbol{\alpha} + \beta mc^2 + eA_0]\psi. \tag{3.201}$$

To appreciate the physical significance of j_μ we rewrite (3.199) as

$$\begin{aligned} j_\mu &= (iec/2)(\bar{\psi}\gamma_\mu\psi + \bar{\psi}\gamma_\mu\psi) \\ &= \frac{ie\hbar}{2m}\left[-\bar{\psi}\gamma_\mu\gamma_\nu\left(\frac{\partial}{\partial x_\nu} - \frac{ie}{\hbar c}A_\nu\right)\psi + \left\{\left(\frac{\partial}{\partial x_\nu} + \frac{ie}{\hbar c}A_\nu\right)\bar{\psi}\right\}\gamma_\nu\gamma_\mu\psi\right], \end{aligned} \tag{3.202}$$

where we have used (3.60) and its analog for the adjoint wave function $\bar{\psi}$. This encourages us to split j_μ into two parts:

$$j_\mu = j_\mu^{(1)} + j_\mu^{(2)}, \tag{3.203}$$

according to dependency on whether or not the summation index ν in (3.202) coincides or does not coincide with μ. We have

$$j_\mu^{(1)} = \frac{ie\hbar}{2m}\left(\frac{\partial\bar{\psi}}{\partial x_\mu}\psi - \bar{\psi}\frac{\partial\psi}{\partial x_\mu}\right) - \frac{e^2}{mc}A_\mu\bar{\psi}\psi \tag{3.204}$$

and

$$\begin{aligned} j_\mu^{(2)} &= \frac{ie\hbar}{2m}\left[-\bar{\psi}\gamma_\mu\gamma_\nu\frac{\partial}{\partial x_\nu}\psi + \left(\frac{\partial\bar{\psi}}{\partial x_\nu}\right)\gamma_\nu\gamma_\mu\psi \right. \\ &\qquad \left. + \frac{ie}{\hbar c}A_\nu\bar{\psi}\gamma_\mu\gamma_\nu\psi + \frac{ie}{\hbar c}A_\nu\bar{\psi}\gamma_\nu\gamma_\mu\psi\right]_{\mu\neq\nu} \\ &= -\frac{e\hbar}{2m}\frac{\partial}{\partial x_\nu}(\bar{\psi}\sigma_{\nu\mu}\psi). \end{aligned} \tag{3.205}$$

This decomposition is known as the *Gordon decomposition*, named for W. Gordon.

Let us look at each component of (3.204) and (3.205) a little more closely. The four-vector (3.204) does not contain any gamma matrix. In fact $j_k^{(1)}$ would be formally identical to the expression for the three-vector current density in the Schrödinger theory if we could replace the Dirac wave function by the Schrödinger wave function. This is rather gratifying since we know that in the nonrelativistic limit, $(\partial\bar{\psi}/\partial x_k)\psi$, etc., can be legitimately replaced by $(\partial\psi_A^\dagger/\partial x_k)\psi_A$, etc. As for $j_4^{(1)}$ we can easily show that

$$j_4^{(1)} = \frac{ieE}{mc}\bar{\psi}\psi - \frac{ie^2}{mc}\bar{\psi}\psi A_0 \tag{3.206}$$

when the time dependence $e^{-iEt/\hbar}$ is assumed. This reduces to ic times the charge current density $e\psi_A^\dagger\psi_A$ in the Schrödinger theory, provided $E \simeq mc^2$ and $|eA_0| \ll mc^2$. As for (3.205) we recall that in the nonrelativistic limit

$$\bar{\psi}\sigma_{k4}\psi = -\bar{\psi}\sigma_{4k}\psi$$

can be ignored while $\bar{\psi}\sigma_{jk}\psi$ can be interpreted as the spin density $\psi_A^\dagger\sigma_l\psi_A$ (jkl cyclic). In other words, the kth component of (3.205) is $-e\hbar/2m$ times

$$\frac{\partial}{\partial x_j}(\psi_A^\dagger\sigma_l A_A) - \frac{\partial}{\partial x_l}(\psi_A^\dagger\sigma_j\psi_A), \qquad (jkl)\text{ cyclic}, \tag{3.207}$$

which is just the kth component of the curl of the spin density.‡ Thus for slowly moving electrons the Gordon decomposition of j_k can be regarded as the separation of j_k into the convection current due to the moving charge and the current associated with the intrinsic magnetization (magnetic dipole density) of the electron.

When $j_\mu^{(2)}$ interacts with A_μ via (3.200), we have the interaction Hamiltonian density

$$
\begin{aligned}
-\frac{j_\mu^{(2)} A_\mu}{c} &= \frac{e\hbar}{2mc}\left[\frac{\partial}{\partial x_\nu}(\bar{\psi}\sigma_{\nu\mu}\psi)\right]A_\mu \\
&= -\frac{e\hbar}{2mc}\frac{\partial A_\mu}{\partial x_\nu}(\bar{\psi}\sigma_{\nu\mu}\psi) \\
&= -\frac{e\hbar}{2mc}\left[\frac{1}{2}\frac{\partial A_\mu}{\partial x_\nu}(\bar{\psi}\sigma_{\nu\mu}\psi) + \frac{1}{2}\frac{\partial A_\nu}{\partial x_\mu}(\bar{\psi}\sigma_{\mu\nu}\psi)\right] \\
&= -\frac{e\hbar}{2mc}[\tfrac{1}{2}F_{\nu\mu}\bar{\psi}\sigma_{\nu\mu}\psi], \qquad (3.208)
\end{aligned}
$$

where we have dropped $(\partial/\partial x_\nu)(\bar{\psi}\sigma_{\nu\mu}\psi A_\mu)$ which is irrelevant when the interaction density is integrated. Noting that

$$\tfrac{1}{2}F_{\nu\mu}\bar{\psi}\sigma_{\nu\mu}\psi \approx \mathbf{B}\cdot(\psi_A^\dagger\boldsymbol{\sigma}\psi_A) \qquad (3.209)$$

in the nonrelativistic limit, we see that (3.208) can indeed account for the spin magnetic moment interaction with the gyromagnetic ratio $g = 2$, in agreement with our earlier discussion based on (3.67).

Experimentally, as first shown by P. Kusch in 1947, the observed gyromagnetic ratio of the electron is not exactly 2 but rather

$$g = 2\left[1 + \left(\frac{e^2}{4\pi\hbar c}\right)\frac{1}{2\pi} + \cdots\right], \qquad (3.210)$$

which holds also for the muon. The origin of the extra magnetic moment was satisfactorily explained in 1947 by J. Schwinger who took into account the fact that the physical electron can emit or absorb a virtual photon, as we noted in Section 2–8. (We shall come back to this point in Chapter 4.) When the magnetic moment is not correctly given by $g = 2$ we may add a phenomenological term to the interaction Hamiltonian of the form

$$\mathscr{H}_{\text{int}} = -\frac{e\hbar\kappa}{2mc}[\tfrac{1}{2}F_{\nu\mu}\bar{\psi}\sigma_{\nu\mu}\psi], \qquad (3.211)$$

called an *anomalous moment* (Pauli moment) *interaction*.§ The total magnetic

---

‡In this connection recall that according to classical electrodynamics a magnetic dipole density $\mathscr{M}$ gives rise to an effective current density

$$\mathbf{j}_{(\text{eff})} = c\nabla \times \mathscr{M}.$$

See Panofsky and Phillips (1955), p. 120; Jackson (1962), p. 152.

§The interaction (3.211) is to be interpreted as an *effective* Hamiltonian density. To the extent that Schwinger's correction is computable on the basis of the interaction (3.200), there is no need to postulate an additional "fundamental" interaction of the type (3.211), at least for the electron and the muon.

moment computed from the sum of (3.208) and (3.211) is

$$\mu = (e\hbar/2mc)(1 + \kappa). \tag{3.212}$$

For the proton, replacing e and m in (3.208) and (3.211) by $|e|$ and the proton mass m_{p}, we have

$$\mu_{\mathrm{p}} = (|e|\hbar/2m_{\mathrm{p}}c)(1 + \kappa_{\mathrm{p}}), \tag{3.213}$$

where experimentally $\kappa_{\mathrm{p}} = 1.79$, as O. Stern first determined. We thus see that roughly 60% of the observed proton magnetic moment is "anomalous." According to (3.208), spin-$\frac{1}{2}$ particles should have zero magnetic moments as $e \rightarrow 0$, yet experimentally neutral particles, such as the neutron and the Λ-hyperon, are known to possess sizable magnetic moments. They can again be represented phenomenologically by interactions of the type (3.211). In contrast to the anomalous moments of the electron and the muon, the anomalous moments of the proton, the neutron, the Λ-hyperon, and so forth, cannot be accounted for by Schwinger's mechanism. Therefore their existence appears to indicate a failure of the simple prescription $p_\mu \rightarrow p_\mu - qA_\mu/c$. However, if we consider that these particles are complicated objects surrounded by virtual meson clouds, the failure of the simple prescription does not seem surprising. Spin-$\frac{1}{2}$ particles whose electromagnetic properties can be understood on the basis of (3.200) alone are sometimes referred to as *pure Dirac particles*.

Vector covariant for free particles. To investigate further the physical meaning of the gamma matrices, let us consider this time $\bar{\psi}_f\gamma_k\psi_i$, where ψ_i and ψ_f are $E > 0$ plane-wave solutions:

$$\begin{aligned} \psi_i &= \sqrt{\frac{mc^2}{EV}}\, u^{(r)}(\mathbf{p}) \exp\left(\frac{i\mathbf{p}\cdot\mathbf{x}}{\hbar} - \frac{iEt}{\hbar}\right), \\ \psi_f &= \sqrt{\frac{mc^2}{E'V}}\, u^{(r')}(\mathbf{p}') \exp\left(\frac{i\mathbf{p}'\cdot\mathbf{x}}{\hbar} - \frac{iE't}{\hbar}\right). \end{aligned} \tag{3.214}$$

This is of some practical interest in connection with the scattering of a Dirac particle by an external vector potential $\mathbf{A}$, since the transition matrix element which we compute in the Born approximation is essentially $-ie\int \bar{\psi}_f\gamma_k\psi_i A_k d^3x$. Assuming for simplicity that the vector potential is time independent so that $E = E'$ (elastic scattering), we can evaluate $i\gamma_k$ taken between $\bar{u}^{(r')}(\mathbf{p})$ and $u^{(r)}(\mathbf{p})$ as follows:

$$\begin{aligned} i\bar{u}^{(r')}(\mathbf{p}')\gamma_k u^{(r)}(\mathbf{p}) &= u^{(r')\dagger}(\mathbf{p}')\alpha_k u^{(r)}(\mathbf{p}) \\ &= \left(\sqrt{\frac{E+mc^2}{2mc^2}}\right)^2 \left(\chi^{(s')\dagger}, \chi^{(s')\dagger}\frac{\boldsymbol{\sigma}\cdot\mathbf{p}'c}{E+mc^2}\right)\begin{pmatrix} 0 & \sigma_k \\ \sigma_k & 0 \end{pmatrix}\begin{pmatrix} \chi^{(s)} \\ \dfrac{\boldsymbol{\sigma}\cdot\mathbf{p}c}{E+mc^2}\chi^{(s)} \end{pmatrix} \\ &= \chi^{(s')\dagger}\left[\frac{p_k + p_{k'}}{2mc} + \frac{i\boldsymbol{\sigma}\cdot\{(\mathbf{p}'-\mathbf{p})\times\hat{\mathbf{n}}_k\}}{2mc}\right]\chi^{(s)}, \end{aligned} \tag{3.215}$$

where $\chi^{(s)}$ and $\chi^{(s')}$ are the initial and final Pauli two-component spinors. The first term, of course, corresponds to the convection current $j_k^{(1)}$ of (3.201). To see the meaning of the second term, we note that it appears in the transition matrix element as

$$-e\left(\sqrt{\frac{mc^2}{E}}\right)^2 \frac{1}{V}\int_V d^3x\chi^{(s')\dagger}\left[\frac{i\boldsymbol{\sigma}\cdot\{(\mathbf{p}'-\mathbf{p})\times\mathbf{A}\}}{2mc}\right]\chi^{(s)}\exp\left[\frac{i(\mathbf{p}-\mathbf{p}')\cdot\mathbf{x}}{\hbar}\right]$$
$$=\frac{mc^2}{EV}\left(\frac{e\hbar}{2mc}\right)\int_V d^3x\chi^{(s')\dagger}\boldsymbol{\sigma}\cdot\left[\left(\boldsymbol{\nabla}\exp\left[\frac{i(\mathbf{p}-\mathbf{p}')\cdot\mathbf{x}}{\hbar}\right]\right)\times\mathbf{A}\right]\chi^{(s)}$$
$$=-\frac{mc^2}{EV}\frac{e\hbar}{2mc}\chi^{(s')\dagger}\boldsymbol{\sigma}\chi^{(s)}\cdot\int_V d^3x(\boldsymbol{\nabla}\times\mathbf{A})\exp\left[\frac{i(\mathbf{p}-\mathbf{p}')\cdot\mathbf{x}}{\hbar}\right], \tag{3.216}$$

which is recognized as the perturbation matrix element expected from the spin magnetic moment interaction.

Finally, let us consider the case $\psi_i = \psi_f$. Using (3.215) with $\mathbf{p} = \mathbf{p}'$, we have‡

$$i\int_V \bar{\psi}\gamma_k\psi d^3x = \int_V \psi^\dagger\alpha_k\psi d^3x = \left(\frac{mc^2}{EV}\right)\frac{2p_kV}{2mc} = \frac{p_kc}{E}. \tag{3.217}$$

But this is nothing more than the classical particle velocity divided by c. It is interesting to recall in this connection that the analog of $-ie\int\bar{\psi}\gamma_\mu\psi A_\mu\, d^3x$ in the *classical* electrodynamics of a point particle is

$$H_{\text{classical}} = (-e\mathbf{v}\cdot\mathbf{A}/c) + eA_0. \tag{3.218}$$

Since plane-wave solutions of the form (3.214) are orthonormal, the result (3.217) can be generalized to any wave function that can be expressed as a superposition of $E > 0$ free-particle plane-wave solutions:

$$\psi_{E>0} = \sum_{\mathbf{p}}\sum_{r=1,2}\sqrt{\frac{mc^2}{EV}}c_{\mathbf{p},r}u^{(r)}(\mathbf{p})\exp\left(i\frac{\mathbf{p}\cdot\mathbf{x}}{\hbar}-\frac{iEt}{\hbar}\right), \tag{3.219}$$

where $c_{\mathbf{p},r}$ is a Fourier coefficient whose modulus squared directly gives the probability for finding the electron in state $(\mathbf{p}, r)$. With the help of (3.215), we obtain

$$\langle\alpha_k\rangle_+ \equiv \int_V \psi^\dagger_{E>0}\alpha_k\psi_{E>0}\,d^3x$$
$$=\sum_{\mathbf{p}}\sum_{\mathbf{p}'}\sum_{r=1,2}\sum_{r'=1,2}\sqrt{(mc^2)^2/EE'V^2}\,c_{\mathbf{p},r}c^*_{\mathbf{p}',r'}u^{(r')\dagger}(\mathbf{p}')\alpha_k u^{(r)}(\mathbf{p})V\delta_{\mathbf{p},\mathbf{p}'}$$
$$=\sum_{\mathbf{p}}\sum_{r=1,2}|c_{\mathbf{p},r}|^2(p_kc/E) = \langle p_kc/E\rangle, \tag{3.220}$$

where + stands for positive energy. Using similar techniques with negative energy spinors, we can readily obtain

$$\langle\alpha_k\rangle_- = -\langle p_kc/|E|\rangle \tag{3.221}$$

for a wave function made up *exclusively* of negative-energy plane-wave solutions. We shall come back later to these very important relations.

Although we have treated only $i\bar{\psi}\gamma_k\psi = \psi^\dagger\alpha_k\psi$ in detail, the reader may work out the analogs of (3.216) for other matrices γ_4, $i\gamma_5\gamma_\mu$, σ_{k4}, γ_5, and so forth. For instance, the interpretation of $\bar{\psi}\sigma_{k4}\psi$ is of interest in connection with electron-neutron scattering (Problem 3–6).

‡Note that, although we have used the two-component language, we have not made a single nonrelativistic approximation in obtaining (3.215) and (3.217).

3-6. DIRAC OPERATORS IN THE HEISENBERG REPRESENTATION

Heisenberg equation of motion. Up to now we have regarded the Dirac matrices as short-hand devices that rearrange the various components of ψ. In discussing the time development of a matrix element such as $\int \psi'^\dagger(\mathbf{x}, t)\alpha_k\psi(\mathbf{x}, t)d^3x$, we shall find it sometimes more convenient to introduce a time-dependent operator $\alpha_k^{(H)}(t)$ (where H stands for Heisenberg) with the property

$$\int \psi'^\dagger(\mathbf{x}, t)\alpha_k\psi(\mathbf{x}, t)\, d^3x = \int \psi'^\dagger(\mathbf{x}, 0)\alpha_k^{(H)}(t)\psi(\mathbf{x}, 0)\, d^3x. \tag{3.222}$$

The time development of $\int \psi'^\dagger(\mathbf{x}, t)\alpha_k\psi(\mathbf{x}, t)d^3x$ can then be inferred directly from the differential equation that governs the behavior of the operator $\alpha_k^{(H)}$. All this amounts to working in the Heisenberg representation, where the state vector is time independent and the dynamical operator is time dependent. For this reason let us briefly review the connection between the Heisenberg representation and the Schrödinger representation.

By virtue of the Schrödinger equation [or the Dirac equation written in the Hamiltonian form (3.38) and (3.39)], a Schrödinger (or Dirac) wave function, which is not necessarily assumed to be an energy eigenfunction, can be written as

$$\psi(\mathbf{x}, t) = e^{-iHt/\hbar}\psi(\mathbf{x}, 0), \tag{3.223}$$

where H is the Hamiltonian *operator* that acts on the wave function. Associated with a time-independent operator $\Omega^{(S)}$ where S stands for Schrödinger, the corresponding operator in the Heisenberg representation can be defined as

$$\Omega^{(H)}(t) = e^{iHt/\hbar}\Omega^{(S)}e^{-iHt/\hbar}. \tag{3.224}$$

Clearly, at $t = 0$, $\Omega^{(H)}$ coincides with $\Omega^{(S)}$:

$$\Omega^{(H)}(0) = \Omega^{(S)}.$$

With (3.223) and (3.224), the matrix elements of a given operator in the two representations taken between any initial and final state are seen to be the same provided that in evaluating the matrix element in the Heisenberg representation we use the wave function at $t = 0$, that is,

$$\int \psi'^\dagger(\mathbf{x}, t)\Omega^{(S)}\psi(\mathbf{x}, t)\, d^3x = \int \psi'^\dagger(\mathbf{x}, 0)\,\Omega^{(H)}\psi(\mathbf{x}, 0)\, d^3x. \tag{3.225}$$

By considering an infinitesimal displacement in time, we are able to readily deduce the Heisenberg equation of motion:

$$\frac{d\Omega^{(H)}}{dt} = \frac{i}{\hbar}[H, \Omega^{(H)}] + \frac{\partial\Omega^{(H)}}{\partial t}, \tag{3.226}$$

where the last term is nonvanishing only when Ω depends explicitly on time. Note that this equation is equivalent to

$$\frac{d}{dt}\int \psi'^\dagger(\mathbf{x}, t)\Omega^{(S)}\psi(\mathbf{x}, t)\, d^3x$$
$$= \frac{i}{\hbar}\int \psi'^\dagger(\mathbf{x}, t)[H, \Omega^{(S)}]\psi(\mathbf{x}, t)\, d^3x + \int \psi'^\dagger(\mathbf{x}, t)(\partial\Omega^{(S)}/\partial t)\psi(\mathbf{x}, t)\, d^3x. \tag{3.227}$$

With a Dirac matrix, $\alpha_k(\beta)$ we may associate a Heisenberg dynamical operator $\alpha_k^{(H)}(\beta^{(H)})$ with the property that its explicit matrix representation at $t = 0$ is $\alpha_k(\beta)$. Since the operator $e^{-iHt/\hbar}$ that connects $\alpha_k^{(H)}$ with α_k, $\beta_k^{(H)}$ with β, and so forth, is unitary, the anticommutation relations among α_k and β hold also for the corresponding dynamical operators. In this section only, we shall omit the superscript (H) so that α_k will stand for the dynamical operator that corresponds to the matrix α_k discussed in the previous sections. Similarly, the symbols $\dot{\mathbf{x}}$, $\mathbf{p}$, and $\mathbf{L}$ will stand for the velocity operator, the momentum operator, and the orbital angular momentum operator in the Heisenberg representation.

Constants of the motion. With the aid of the Heisenberg equation of motion we can immediately determine whether or not a given observable is a constant of the motion. For instance, for a free particle whose Hamiltonian is‡

$$H = c\alpha_j p_j + \beta mc^2 \tag{3.228}$$

we have the Heisenberg equation for the momentum

$$\frac{dp_k}{dt} = \frac{i}{\hbar}[H, p_k] = 0, \tag{3.229}$$

since p_k commutes with both $c\alpha_j p_j$ and βmc^2. Equation (3.229) tells us that we can find a solution to the Dirac equation which is a simultaneous eigenfunction of the Hamiltonian and the momentum. But we know this already, since the plane wave solutions obtained in Section 3–3 are simultaneous eigenfunctions of H and $\mathbf{p}$.

As a less trivial example, let us consider $\mathbf{L}$ for a free particle. For the x-component of $\mathbf{L}$ we have

$$[H, L_1] = [c\alpha_k p_k, (x_2 p_3 - x_3 p_2)] = -i\hbar c(\alpha_2 p_3 - \alpha_3 p_2), \tag{3.230}$$

and similarly for L_2 and L_3. So

$$d\mathbf{L}/dt = c(\boldsymbol{\alpha} \times \mathbf{p}). \tag{3.231}$$

This means that for a free Dirac particle, $\mathbf{L}$ is *not* a constant of the motion, in sharp contrast with the corresponding operator in the Schrödinger theory.

Next let us consider $\boldsymbol{\Sigma}$. First it is useful to note that

$$\alpha_k = -\Sigma_k \gamma_5 = -\gamma_5 \Sigma_k, \tag{3.232}$$

which is obvious from the explicit forms of α_k, γ_5, and Σ_k.§ Therefore

$$\begin{aligned}[H, \Sigma_1] &= [c\alpha_k p_k, \Sigma_1] = -c[\gamma_5 \Sigma_k p_k, \Sigma_1] \\ &= 2ic(\alpha_2 p_3 - \alpha_3 p_2),\end{aligned} \tag{3.233}$$

---

‡Since it is customary to use α_k and β to write H, we shall, in this section, make exclusive use of α_k and β rather than σ_{k4} and γ_4.

§To prove this without recourse to any particular representation, note, for instance, $\alpha_3 = -i\gamma_3\gamma_4 = i\gamma_1\gamma_2\gamma_1\gamma_2\gamma_3\gamma_4 = -\Sigma_3\gamma_5$, where we have used $(\gamma_1\gamma_2)^2 = -1$.

where we have used $\gamma_5\Sigma_2\Sigma_1 = -i\gamma_5\Sigma_3 = i\alpha_3$, and so forth. Consequently,

$$\frac{d\boldsymbol{\Sigma}}{dt} = -\frac{2c}{\hbar}(\boldsymbol{\alpha} \times \mathbf{p}). \tag{3.234}$$

Thus the spin angular momentum of a free electron is *not* a constant of the motion, either. But taking the dot product of (2.234) and $\mathbf{p}$ and remembering (3.229), we obtain

$$d(\boldsymbol{\Sigma}\cdot\mathbf{p})/dt = 0, \tag{3.235}$$

which means that the helicity $\boldsymbol{\Sigma}\cdot\mathbf{p}/|\mathbf{p}|$ is a constant of the motion, as shown in Section 3–3.

Let us now consider

$$\mathbf{J} = \mathbf{L} + \hbar\boldsymbol{\Sigma}/2. \tag{3.236}$$

Because of (3.231) and (3.234) we have

$$d\mathbf{J}/dt = 0. \tag{3.237}$$

Thus, although $\mathbf{L}$ and $\hbar\boldsymbol{\Sigma}/2$ taken separately are *not* constants of the motion, the sum (3.236) which, according to (3.161), should be identified with the total angular momentum *is* a constant of the motion. As is well known, the constancy of $\mathbf{J}$ is a consequence of invariance under rotation. Hence we may argue that $\mathbf{J}$ must be a constant of the motion even if we add to the free-particle Hamiltonian a central (spherically symmetric) potential $V(r)$. Indeed, the relation (3.237) still holds in the presence of $V(r)$ since both L_k and Σ_k commute with $V(r)$.

Next we shall consider the time derivative of the mechanical momentum

$$\boldsymbol{\pi} = \mathbf{p} - e\mathbf{A}/c \tag{3.238}$$

(as opposed to the *canonical* momentum $\mathbf{p}$) of an electron in the presence of A_μ. Using the Hamiltonian

$$H = c\boldsymbol{\alpha}\cdot\boldsymbol{\pi} + eA_0 + \beta mc^2, \tag{3.239}$$

we obtain

$$\dot{\pi}_k = \frac{i}{\hbar}[H, \pi_k] - \frac{e}{c}\frac{\partial A_k}{\partial t} = \frac{ic}{\hbar}\alpha_j[\pi_j, \pi_k] + \frac{ie}{\hbar}[A_0, \pi_k]. \tag{3.240}$$

But we know that

$$[A_0, \pi_k] = i\hbar\frac{\partial A_0}{\partial x_k},$$

and

$$[\pi_1, \pi_2] = \frac{ie\hbar}{c}\frac{\partial A_2}{\partial x_1} - \frac{ie\hbar}{c}\frac{\partial A_1}{\partial x_2} = \frac{ie\hbar}{c}B_3, \quad \text{etc.} \tag{3.241}$$

Hence

$$\dot{\boldsymbol{\pi}} = e(\mathbf{E} + \boldsymbol{\alpha}\times\mathbf{B}). \tag{3.242}$$

Since $\langle\alpha_k\rangle_+$ has been shown to correspond to the classical particle velocity in units of c (cf. Eq. 3.220), we may be tempted to identify (3.242) with the operator equation for the Lorentz force. However, as we shall see later, we have to be somewhat careful in regarding $c\boldsymbol{\alpha}$ as the operator that corresponds to the particle velocity in the usual sense.

It is also of interest to study the time dependence of $\boldsymbol{\Sigma}\cdot\boldsymbol{\pi}$ for an electron in the electromagnetic field. Assuming that A_μ is time-independent, we obtain

$$\frac{d(\boldsymbol{\Sigma}\cdot\boldsymbol{\pi})}{dt} = \frac{i}{\hbar}[(-c\gamma_5\boldsymbol{\Sigma}\cdot\boldsymbol{\pi} + \beta mc^2 + eA_0), \boldsymbol{\Sigma}\cdot\boldsymbol{\pi}] = \frac{ie}{\hbar}[A_0, \boldsymbol{\Sigma}\cdot\boldsymbol{\pi}] = e\boldsymbol{\Sigma}\cdot\mathbf{E}, \tag{3.243}$$

since both γ_5 and β commute with $\boldsymbol{\Sigma}\cdot\boldsymbol{\pi}$. Let us suppose that there is no electric field. We know that the magnitude of the mechanical momentum of a charged particle is unchanged in a time-independent magnetic field. The constancy of $\boldsymbol{\Sigma}\cdot\boldsymbol{\pi}$ then amounts to the constancy of the helicity. It follows that a longitudinally polarized electron (one whose helicity is $+1$ or -1) entering a region with a magnetic field will remain longitudinally polarized no matter how complicated $\mathbf{B}$ may be. When an electron is injected into a region with a uniform magnetic field $\mathbf{B}$ whose direction is perpendicular to the initial electron velocity, the electron follows a circular path with an angular frequency, known as the *cyclotron frequency*:

$$\omega_L = (|e\mathbf{B}|/mc)\sqrt{1-\beta^2}. \tag{3.244}$$

In this particularly simple case, the constancy of helicity implies that the electron spin precesses in such a way that its precession angular frequency ω_S is equal to ω_L, as pictorially represented in Fig. 3–3.

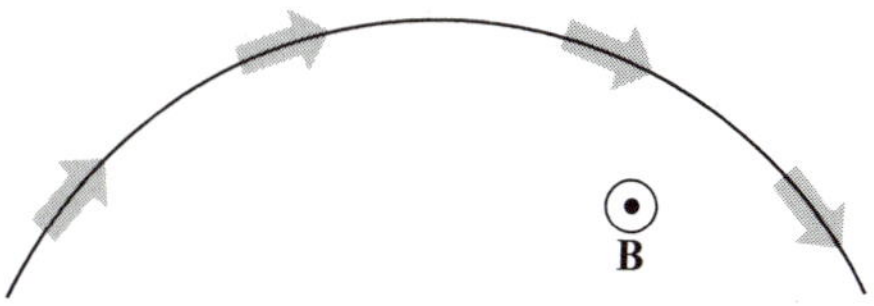

Fig. 3–3. Spin precession of a moving electron in a uniform magnetic field. The gray arrows indicate the spin direction.

In deriving the equality of ω_L and ω_S we implicitly assumed that the gyromagnetic ratio of the electron is strictly two. In reality, because of the anomalous moment of the electron, ω_L and ω_S are not quite equal. It can be shown that the correct relation is given by

$$\frac{\omega_S}{\omega_L} = 1 + \left(\frac{g-2}{2}\right)\frac{1}{\sqrt{1-\beta^2}}. \tag{3.245}$$

This means that as the initially longitudinally polarized electron makes one orbital turn, its spin direction departs from the direction of motion by a very small amount, 1/137 rad if the electron is nonrelativistic (cf. Eq. 3.210). This principle has been used experimentally to make precise measurements on the anomalous moments of the electron and the muon.

"Velocity" in the Dirac theory. Let us now return to the free-particle case. Consider

$$\dot{x}_k = (i/\hbar)[H, x_k] = (ic/\hbar)[\alpha_j p_j, x_k] = c\alpha_k, \tag{3.246}$$

which says that $\boldsymbol{\alpha}$ is the velocity in units of c. (Actually this relation holds even in the presence of A_μ.) At first sight this appears quite reasonable in view of (3.220), which says that $\langle\alpha_k\rangle_+$ is the same as the expectation value of $p_k cH^{-1}$. Note, however, that the eigenvalue of α_k is $+1$ or -1. Hence the eigenvalue of the velocity operator is $\pm c$, as first pointed out by G. Breit in 1928. This is a truly remarkable result, since for a particle of finite mass the classical velocity cannot be equal to $\pm c$. We may also note that because α_k and α_l do not commute when $k \neq l$, a measurement of the x-component of the velocity is incompatible with a measurement of the y-component of the velocity; this may appear strange since we know that p_1 and p_2 commute.

In spite of these peculiarities, there is actually no contradiction with the results derived earlier. The plane-wave solutions (3.114) and (3.115) which are eigenfunctions of $\mathbf{p}$ are not eigenfunctions of α_k (unless the particle is massless), as the reader may verify by directly applying α_k. In fact, since α_k fails to commute with the Hamiltonian (see Eq. 3.248), no energy eigenfunctions are expected to be simultaneous eigenfunctions of α_k.

Let us now look at the time derivative of α_k:

$$\begin{aligned}\dot{\alpha}_k &= (i/\hbar)[H, \alpha_k] \\ &= (i/\hbar)(-2\alpha_k H + \{H, \alpha_k\}) \\ &= (i/\hbar)(-2\alpha_k H + 2c\, p_k),\end{aligned} \tag{3.247}$$

where we have used the fact that α_k anticommutes with every term in H except the term that involves α_k itself. This equation can also be written as

$$\dot{\alpha}_k = -\frac{2c}{\hbar}(\mathbf{\Sigma} \times \mathbf{p})_k - \frac{2i\alpha_k \beta mc^2}{\hbar}. \tag{3.248}$$

We see that the velocity operator $\dot{x}_k = c\alpha_k$ is *not* a constant of the motion despite the fact that the particle is free. Contrast this with Eq. (3.229), which says that the momentum of a free particle *is* a constant of the motion.

The relation (3.247) can be regarded as a differential equation for $\alpha_k(t)$. Keeping in mind that p_k and H are constants of the motion, we easily see by direct substitution that the solution of this differential equation is

$$\alpha_k(t) = c\,p_k H^{-1} + (\alpha_k(0) - c\,p_k H^{-1})\, e^{-2iHt/\hbar}. \tag{3.249}$$

The first term of (3.249) is reasonable since, for an eigenstate of momentum and energy, it gives cp_k/E in agreement with (3.220) and (3.221). But what is the physical significance of the second term? Taken literally, it seems to say that the velocity of the electron has an additional term that fluctuates rapidly about its average value even in the absence of any potential.

As for the coordinate operator, the relation (3.249) can be easily integrated to yield

$$x_k(t) = x_k(0) + c^2 p_k H^{-1} t + (ic\hbar/2)(\alpha_k(0) - c\,p_k H^{-1})H^{-1}\, e^{-2iHt/\hbar}. \tag{3.250}$$

The first and the second terms are understandable because their expectation values are seen to give the trajectory of the wave packet according to the *classical* law:

$$x_k^{(\text{class})}(t) = x_k^{(\text{class})}(0) + t(p_k c^2/E)^{(\text{class})}, \tag{3.251}$$

just as they do in nonrelativistic quantum mechanics. The presence of the third term in (3.250) (which, of course, is a consequence of the second term in Eq. 3.249) appears to imply that the free electron executes very rapid oscillations in addition to the uniform rectilinear motion (3.251). This oscillatory motion, first discussed by E. Schrödinger in 1930, is called *Zitterbewegung* (literally "quivering motion"). We shall say more about this in the next section.

3–7. *ZITTERBEWEGUNG* AND NEGATIVE-ENERGY SOLUTIONS

Expectation values of $\boldsymbol{\alpha}$ and x. The algebraic techniques extensively employed in the previous section are quite powerful when we want to obtain the constants of the motion or establish a correspondence with the classical theory. However, in order to analyze the peculiarities we encountered at the end of the previous section, it is instructive to go back to the Schrödinger representation and reinterpret the operator relations (3.249) and (3.250), using specific wave functions.

Since the positive- and negative-energy plane-wave solutions (3.114) and (3.115) with all possible $\mathbf{p}$ form a complete orthonormal set, the most general free-particle wave function can be written as

$$\psi(\mathbf{x}, t) = \sum_{\mathbf{p}} \sum_{r=1,2} \sqrt{\frac{mc^2}{|E| V}} c_{\mathbf{p},r} u^{(r)}(\mathbf{p}) \exp\left(\frac{i\mathbf{p}\cdot\mathbf{x}}{\hbar} - \frac{i|E|t}{\hbar}\right) + \sum_{\mathbf{p}} \sum_{r=3,4} \sqrt{\frac{mc^2}{|E| V}} c_{\mathbf{p},r} u^{(r)}(\mathbf{p}) \exp\left(\frac{i\mathbf{p}\cdot\mathbf{x}}{\hbar} + \frac{i|E|t}{\hbar}\right), \tag{3.252}$$

where $c_{\mathbf{p},r}$ is to be determined from the Fourier expansion of ψ at $t = 0$. By appropriately choosing $c_{\mathbf{p},r}$, we can write the wave function for an arbitrarily localized free-particle wave packet in the form (3.252). Let us now evaluate $\langle \alpha_k \rangle$. A straightforward calculation using (3.220) and (3.221) gives

$$\begin{aligned}\langle \alpha_k \rangle &= \int_V \psi^\dagger(\mathbf{x}, t) \alpha_k \psi(\mathbf{x}, t) d^3x \\ &= \sum_{\mathbf{p}} \sum_{r=1,2} |c_{\mathbf{p},r}|^2 \frac{p_k c}{|E|} - \sum_{\mathbf{p}} \sum_{r=3,4} |c_{\mathbf{p},r}|^2 \frac{p_k c}{|E|} \\ &+ \sum_{\mathbf{p}} \sum_{r=1,2} \sum_{r'=3,4} \frac{mc^2}{|E|} [c^*_{\mathbf{p},r'} c_{\mathbf{p},r} u^{(r')\dagger}(\mathbf{p}) \alpha_k u^{(r)}(\mathbf{p})\, e^{-2i|E|t/\hbar} \\ &+ c_{\mathbf{p},r'} c^*_{\mathbf{p},r} u^{(r)\dagger}(\mathbf{p}) \alpha_k u^{(r')}(\mathbf{p})\, e^{2i|E|t/\hbar}].\end{aligned} \tag{3.253}$$

The first (second) term which is time independent represents the group velocity of the wave packet made up exclusively of positive- (negative-) energy plane-wave components. The last two terms which are time dependent are more interesting. First note that α_k taken between $u^{(3,4)\dagger}(\mathbf{p})$ and $u^{(1,2)}(\mathbf{p})$ is "large" when $|\mathbf{p}| \ll mc$, in sharp contrast with α_k taken between $u^{(1,2)\dagger}(\mathbf{p})$ and $u^{(1,2)}(\mathbf{p})$ which is of order v/c. Specifically,

$$u^{(3 \text{ or } 4)\dagger}(\mathbf{p}) \alpha_k u^{(1 \text{ or } 2)}(\mathbf{p}) = \chi^{(s_-)\dagger} \sigma_k \chi^{(s_+)} + 0([|\mathbf{p}|/mc]^2), \tag{3.254}$$

where $\chi^{(s_+)}$ and $\chi^{(s_-)}$ are the two-component Pauli spinors corresponding to $u^{(1 \text{ or } 2)}$ and $u^{(3 \text{ or } 4)}$. Therefore the last two terms of (3.253) represent a superposition of violent and rapid oscillations, each with an angular frequency $\sim 2mc^2/\hbar = 1.5 \times 10^{21}$ sec^{-1} and an amplitude $\sim |c^{(1,2)*} c^{(3,4)}|$.

As for $\langle x_k \rangle$ we first observe that the operator relation (3.246) implies

$$(d/dt) \int \psi^\dagger(\mathbf{x}, t) x_k \psi(\mathbf{x}, t)\, d^3x = c \int \psi^\dagger(\mathbf{x}, t) \alpha_k \psi(\mathbf{x}, t)\, d^3x. \tag{3.255}$$

We can therefore obtain $\langle x_k \rangle$ by integrating c times (3.253) with respect to time:

$$\begin{aligned} \langle x_k \rangle = \langle x_k \rangle_{t=0} + \Bigg[&\sum_{\mathbf{p}} \sum_{r=1,2} |c_{\mathbf{p},r}|^2 \frac{p_k c^2}{|E|} t - \sum_{\mathbf{p}} \sum_{r=3,4} |c_{\mathbf{p},r}|^2 \frac{p_k c^2}{|E|} t \\ &+ \sum_{\mathbf{p}} \sum_{r=1,2} \sum_{r'=3,4} \frac{i\hbar}{2mc} \left(\frac{mc^2}{|E|} \right)^2 [c^*_{\mathbf{p},r'} c_{\mathbf{p},r} u^{(r')\dagger}(\mathbf{p}) \alpha_k u^{(r)}(\mathbf{p})\, e^{-2i|E|t/\hbar} \\ &- c_{\mathbf{p}r'} c^*_{\mathbf{p}r} u^{(r)\dagger}(\mathbf{p}) \alpha_k u^{(r')}(\mathbf{p})\, e^{2i|E|t/\hbar}] \Bigg]. \end{aligned} \tag{3.256}$$

Thus on top of the rectilinear motion of the wave packet we again have a superposition of violent oscillations, each with an angular frequency $\sim 2mc^2/\hbar$. If both $c_{\mathbf{p},1 \text{ or } 2}$ and $c_{\mathbf{p},3 \text{ or } 4}$ are appreciable, the fluctuation of the electron coordinate due to these oscillations is of the order of $\hbar/mc = 3.9 \times 10^{-11}$ cm. It is very important to note that the peculiar oscillatory behavior, *Zitterbewegung*, of both $\langle \alpha_k \rangle$ and $\langle x_k \rangle$ is due solely to an interference between the positive- and negative-energy components in the wave packet. The *Zitterbewegung* is completely absent for a wave packet made up exclusively of positive- (negative-) energy plane-wave solutions.

Presence of negative-energy components. At this stage we may naturally ask why we cannot simply forget all about the trouble-making negative-energy components in constructing a wave packet. This can certainly be done for a free-particle wave packet since in a potential-free region a wave packet made up exclusively of positive-energy plane waves at a given time does not develop negative-energy components at later times. On the other hand, we can show that the wave function for a well-localized particle contains, in general, plane-wave components of negative energies. As an illustration, let us consider a harmless-looking four-component wave function at $t = 0$,

$$\psi(\mathbf{x}, 0) = \begin{pmatrix} \phi(\mathbf{x}) \\ 0 \\ 0 \\ 0 \end{pmatrix}, \tag{3.257}$$

where $|\phi(\mathbf{x})|^2$ is assumed to be appreciable only in a region whose linear dimension is Δx_k. From our earlier discussion in Section 3–3, we see that ψ corresponds to the four-component wave function for a nonrelativistic particle with spin-up localized to $\sim \Delta x_k$. We now propose to expand this wave function in various plane-wave components so that $\psi(\mathbf{x}, 0)$ takes the form (3.252), evaluated at $t = 0$. The appropriate Fourier coefficients can be readily found by multiplying (3.257) from the left by $u^{(r)\dagger}(\mathbf{p}) e^{-i\mathbf{p}\cdot\mathbf{x}/\hbar}$ and integrating over the space coordinates. Although the coefficients themselves depend on the detailed form of $\phi(\mathbf{x})$, we easily see that

$$\frac{c_{\mathbf{p},3}}{c_{\mathbf{p},1}} = -\frac{p_3 c}{|E| + mc^2}, \qquad \frac{c_{\mathbf{p},4}}{c_{\mathbf{p},1}} = -\frac{(p_1 - ip_2)c}{|E| + mc^2}. \tag{3.258}$$

This means that a negative-energy component is comparable in importance to the corresponding positive-energy component whenever ϕ has a Fourier component with momentum comparable to mc. On the other hand, we know that the Fourier transform of ϕ is appreciable in a region in momentum-space whose linear dimension is

$$\Delta p_k \sim \hbar/\Delta x_k. \tag{3.259}$$

Suppose the state in question is localized to $\Delta x_k \lesssim \hbar/mc$. The uncertainty relation (3.259) tells us that we need plane-wave components of momenta $|\mathbf{p}| \gtrsim mc$. From (3.258), we then infer that there must be appreciable amounts of negative-energy components.

We have seen that a well-localized state contains, in general, plane-wave components of negative energies. Conversely, we may ask how well localized a state will be which we can form using only plane-wave components of positive energies. A rather careful analysis by T. D. Newton and E. P. Wigner indicates that the best-localized state we can construct in this way is one in which the characteristic linear dimension of the wave packet is $\sim\hbar/mc$ but no smaller. This statement can be shown to be valid also for a Klein-Gordon particle.

It is interesting to note that a *positive*-energy bound-state wave function when expanded in free-particle plane waves usually contains some *negative*-energy components. For instance, let us take the case of the ground-state wave function of the hydrogen atom whose energy is evidently positive, $E = mc^2 - (e^2/8\pi a_0) > 0$. When this bound-state wave function (whose explicit form will be given in the next section) is expanded in free-particle plane waves, we do obtain nonvanishing coefficients for negative-energy plane waves. An immediate consequence of this is that the electron in the hydrogen atom exhibits *Zitterbewegung*. As a result, the effective potential that the electron at $\mathbf{x}$ feels is no longer just $V(\mathbf{x})$ but rather $V(\mathbf{x} + \delta\mathbf{x})$, where $\delta\mathbf{x}$ characterizes the fluctuation of the electron coordinate. Note now that $V(\mathbf{x} + \delta\mathbf{x})$ can be expanded as follows:

$$V(\mathbf{x} + \delta\mathbf{x}) = V(\mathbf{x}) + \delta\mathbf{x}\cdot\nabla V + \frac{1}{2}\delta x_i \delta x_j \frac{\partial^2 V}{\partial x_i \partial x_j} + \cdots \tag{3.260}$$

Assuming that $\mathbf{x}$ fluctuates with magnitude $|\delta\mathbf{x}| \approx \hbar/mc$ without any preferred direction, we obtain for the time average of the difference between $V(\mathbf{x} + \delta\mathbf{x}) - V(\mathbf{x})$:

$$(\Delta V)_{\text{time average}} \approx \frac{1}{2}\frac{1}{3}\left(\frac{\hbar}{mc}\right)^2 \nabla^2 V = \frac{e}{6}\left(\frac{\hbar}{mc}\right)^2 \delta^{(3)}(\mathbf{x}). \tag{3.261}$$

Apart from a numerical factor ($\frac{1}{6}$ instead of $\frac{1}{8}$), this is just the effective potential needed to explain the Darwin term discussed earlier in connection with the approximate treatment of the hydrogen atom (cf. Eqs. 3.83 and 3.87).

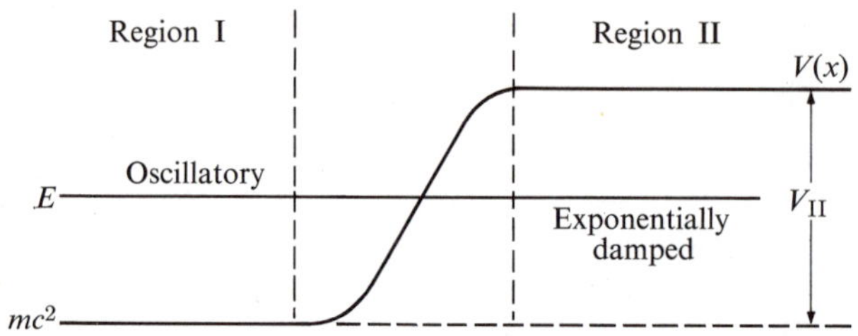

Fig. 3–4. One-dimensional potential with $mc^2 > E - V_{II} > -mc^2$.

Klein's paradox. As a final illustration of the peculiarities attending the negative-energy solutions, let us consider a simple one-dimensional potential (Fig. 3–4). In Region I the particle is free; the height of the potential in Region II is assumed to be V_{II}. When considering a region in which the potential is not varying rapidly, we can proceed directly to obtain the functional form of the wave function. For our purpose it is actually sufficient to look at ψ_A only:

$$(\boldsymbol{\sigma}\cdot\mathbf{p})c\frac{1}{E - V + mc^2}(\boldsymbol{\sigma}\cdot\mathbf{p})c\,\psi_A = \frac{1}{c^2}(E - V - mc^2)\psi_A. \tag{3.262}$$

Wherever V is locally independent of x, we obtain

$$\psi_A \propto \exp\left(\frac{ipx}{\hbar} - \frac{iEt}{\hbar}\right)\chi, \quad \exp\left(-\frac{|p|x}{\hbar} - i\frac{Et}{\hbar}\right)\chi, \tag{3.263}$$

where

$$p^2c^4 = (E - V + mc^2)(E - V - mc^2). \tag{3.264}$$

With $p^2 > 0$ we have an *oscillatory* solution, while with $p^2 < 0$ we have an *exponentially damped* solution. Let us suppose that

$$mc^2 > E - V_{\text{II}} > -mc^2. \tag{3.265}$$

The Region II is a classically forbidden region ($p^2 < 0$), and the free-particle wave function in Region I dies out exponentially as it enters Region II.

So far everything has been straightforward. Let us now consider the potential given by Fig. 3–5. Our experience with nonrelativistic quantum mechanics tells us that the wave function in Region III is even *more strongly* damped. However, when the potential becomes so strongly repulsive that

$$V_{\text{III}} - E > mc^2, \tag{3.266}$$

(3.264) tells us that just the opposite is true. Since both $E-V+mc^2$ and $E-V-mc^2$ are now negative, we have $p^2 > 0$; hence the solution in Region III is *oscillatory* just as is the free-particle solution in Region I. This result is exactly the opposite of the one we set out to find. Semiclassically speaking, a particle initially confined in Region I can tunnel through Region II (just as the α-particle inside an α-emitting nucleus), and behaves in Region III as though it were in an attractive potential instead of the very strong repulsive potential implied by (3.266). This theory is named *Klein's paradox* for O. Klein, who worried about this interesting point in 1930.

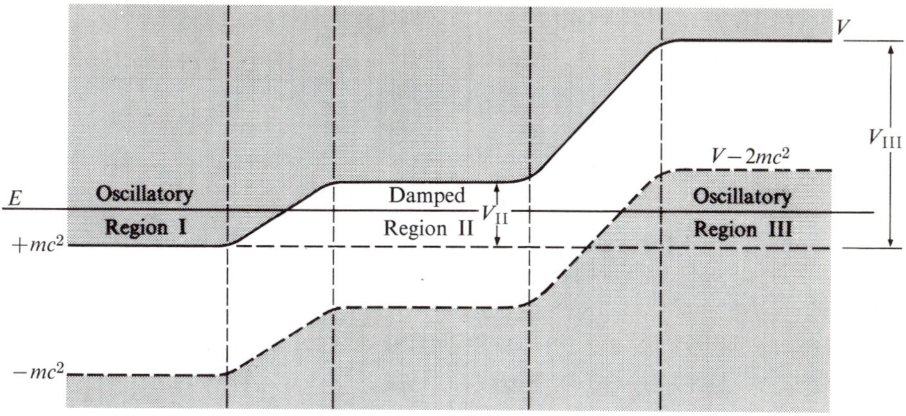

Fig. 3–5. Potential to illustrate Klein's paradox. Oscillatory solutions are expected in shaded regions.

What is the origin of this peculiar behavior? Let us recall that the free-particle solutions to the Dirac equation exhibit an energy spectrum ranging from $-mc^2$ to $-\infty$ as well as from $+mc^2$ to ∞. Now suppose we apply a small positive potential V. The condition that we have negative-energy oscillatory solutions now becomes

$$-\infty < E < -mc^2 + V. \tag{3.267}$$

As V is increased adiabatically, we see that eventually E need not even be negative for (3.267) to be satisfied. Coming back to Fig. 3–5, we see that the oscillatory solution in Region III is essentially a negative-energy solution despite $E > 0$, since it can be obtained from the wave function whose space-time dependence is $\exp[i(px/\hbar) + i(|E|t/\hbar)]$ by simply increasing V adiabatically. Klein's paradox arises because when the potential V is sufficiently positive, an oscillatory negative-energy solution in Region III can have the same positive energy as an oscillatory positive-energy solution in Region I. The tunneling of the electron from Region I into Region III must therefore be viewed as a transition from a positive-energy to a negative-energy state. We shall say more about such a transition in Section 3–9, pp. 131–143. In any case, we find that our intuitive notion that a strong positive potential can repulse the particle breaks down completely when V becomes comparable to $2mc^2$.

Similar peculiarities are present also for strongly attractive potentials. With a moderately attractive finite-ranged potential we can have bound-state solutions ($E < mc^2$) which fall off outside the range of the potential, just as they do in nonrelativistic quantum mechanics so long as the attraction does not exceed a certain critical strength. But when the potential becomes too strong, the Dirac theory starts accommodating solutions with E *less* than mc^2 which are *oscillatory and undamped outside* the range of the potential. The interested reader may verify this point in detail by studying the behavior of the Dirac particle in a deep spherical well (Problem 3–10c).

3–8. CENTRAL FORCE PROBLEMS; THE HYDROGEN ATOM

General considerations. In this section we shall first study some general properties of the wave function for an electron in a spherically symmetric potential. As we have already seen in Section 3–6, the total angular momentum operator $\mathbf{J}$ is a constant of the motion whenever the Hamiltonian is given by

$$H = c\boldsymbol{\alpha}\cdot\mathbf{p} + \beta mc^2 + V(r). \tag{3.268}$$

Let us now look for other constants of the motion. Intuitively, we expect that we must be able to specify whether the electron spin is parallel or antiparallel to the total angular momentum. In nonrelativistic quantum mechanics these two possibilities are distinguished by the eigenvalues of

$$\boldsymbol{\sigma}\cdot\mathbf{J} = \boldsymbol{\sigma}\cdot(\mathbf{L} + \hbar\boldsymbol{\sigma}/2) = (1/\hbar)(\mathbf{J}^2 - \mathbf{L}^2 + \tfrac{3}{4}\hbar^2). \tag{3.269}$$

Alternatively, we may specify l, which can be either $j + \frac{1}{2}$ or $j - \frac{1}{2}$. For a relativistic electron we might try the 4×4 generalization of (3.269), namely $\boldsymbol{\Sigma}\cdot\mathbf{J}$. However, the commutator of H with $\boldsymbol{\Sigma}\cdot\mathbf{J}$ turns out to be rather involved, as the reader may verify. Instead, then, let us try $\beta\boldsymbol{\Sigma}\cdot\mathbf{J}$ which has the same nonrelativistic limit as $\boldsymbol{\Sigma}\cdot\mathbf{J}$:

$$\begin{aligned}[H, \beta\,\boldsymbol{\Sigma}\cdot\mathbf{J}] &= [H, \beta]\,\boldsymbol{\Sigma}\cdot\mathbf{J} + \beta[H, \boldsymbol{\Sigma}]\cdot\mathbf{J} \\ &= -2c\beta(\boldsymbol{\alpha}\cdot\mathbf{p})(\boldsymbol{\Sigma}\cdot\mathbf{J}) + 2ic\beta(\boldsymbol{\alpha}\times\mathbf{p})\cdot\mathbf{J},\end{aligned} \tag{3.270}$$

where we have used (3.233), and

$$\begin{aligned}[H, \beta] &= c\boldsymbol{\alpha}\cdot\mathbf{p}\beta - \beta c\boldsymbol{\alpha}\cdot\mathbf{p} \\ &= -2c\beta\boldsymbol{\alpha}\cdot\mathbf{p}.\end{aligned} \tag{3.271}$$

If we take advantage of

$$\begin{aligned}(\boldsymbol{\alpha}\cdot\mathbf{A})(\boldsymbol{\Sigma}\cdot\mathbf{B}) &= -\gamma_5(\boldsymbol{\Sigma}\cdot\mathbf{A})(\boldsymbol{\Sigma}\cdot\mathbf{B}) \\ &= -\gamma_5\mathbf{A}\cdot\mathbf{B} + i\boldsymbol{\alpha}\cdot(\mathbf{A}\times\mathbf{B}),\end{aligned} \tag{3.272}$$

we can simplify expression (3.270):

$$\begin{aligned}[H, \beta\,\boldsymbol{\Sigma}\cdot\mathbf{J}] &= 2c\beta\gamma_5(\mathbf{p}\cdot\mathbf{J}) \\ &= 2c\beta\gamma_5\mathbf{p}\cdot(\mathbf{L} + \hbar\boldsymbol{\Sigma}/2) \\ &= -c\hbar\beta\boldsymbol{\alpha}\cdot\mathbf{p} = (\hbar/2)[H, \beta],\end{aligned} \tag{3.273}$$

where we have used

$$\mathbf{p}\cdot\mathbf{L} - -i\hbar\boldsymbol{\nabla}\cdot[\mathbf{x}\times(-i\hbar\boldsymbol{\nabla})] = 0 \tag{3.274}$$

and (3.271). Therefore an operator K defined by

$$K = \beta\,\boldsymbol{\Sigma}\cdot\mathbf{J} - \beta\hbar/2 = \beta(\boldsymbol{\Sigma}\cdot\mathbf{L} + \hbar) \tag{3.275}$$

does commute with H:

$$[H, K] = 0. \tag{3.276}$$

Furthermore, using the fact that $\mathbf{J}$ commutes with β and $\boldsymbol{\Sigma}\cdot\mathbf{L}$, we readily see that

$$[\mathbf{J}, K] = 0 \tag{3.277}$$

as well. Thus, for an electron in a central potential, we can construct a simultaneous eigenfunction of H, K, $\mathbf{J}^2$, and J_3. The corresponding eigenvalues are denoted by E, $-\kappa\hbar$, $j(j + 1)\hbar^2$, and $j_3\hbar$.

We shall now derive an important relation between κ and j. First let us consider

$$\begin{aligned} K^2 &= \beta(\mathbf{\Sigma}\cdot\mathbf{L} + \hbar)\beta(\mathbf{\Sigma}\cdot\mathbf{L} + \hbar) \\ &= (\mathbf{\Sigma}\cdot\mathbf{L} + \hbar)^2 \\ &= \mathbf{L}^2 + i\,\mathbf{\Sigma}\cdot(\mathbf{L}\times\mathbf{L}) + 2\hbar\mathbf{\Sigma}\cdot\mathbf{L} + \hbar^2 \\ &= \mathbf{L}^2 + \hbar\mathbf{\Sigma}\cdot\mathbf{L} + \hbar^2. \end{aligned} \tag{3.278}$$

At the same time, since

$$\mathbf{J}^2 = \mathbf{L}^2 + \hbar\mathbf{\Sigma}\cdot\mathbf{L} + 3\hbar^2/4, \tag{3.279}$$

we obtain

$$K^2 = \mathbf{J}^2 + \tfrac{1}{4}\hbar^2, \tag{3.280}$$

which means that the eigenvalues of $\mathbf{J}^2$ and K^2 are related to each other by

$$\kappa^2\hbar^2 = j(j+1)\hbar^2 + \tfrac{1}{4}\hbar^2 = (j+\tfrac{1}{2})^2\hbar^2. \tag{3.281}$$

So we must have

$$\kappa = \pm(j+\tfrac{1}{2}). \tag{3.282}$$

Thus κ is a *nonzero integer* which can be positive or negative. Pictorially speaking, the sign of κ determines whether the spin is antiparallel ($\kappa > 0$) or parallel ($\kappa < 0$) to the total angular momentum in the nonrelativistic limit.

Explicitly, the operator K is given by

$$K = \begin{pmatrix} \boldsymbol{\sigma}\cdot\mathbf{L} + \hbar & 0 \\ 0 & -\boldsymbol{\sigma}\cdot\mathbf{L} - \hbar \end{pmatrix}. \tag{3.283}$$

Thus, if the four-component wave function ψ (assumed to be an energy eigenfunction) is a simultaneous eigenfunction of K, $\mathbf{J}^2$, and J_3, then

$$(\boldsymbol{\sigma}\cdot\mathbf{L} + \hbar)\psi_A = -\kappa\hbar\psi_A, \qquad (\boldsymbol{\sigma}\cdot\mathbf{L} + \hbar)\psi_B = \kappa\hbar\psi_B, \tag{3.284}$$

and

$$\begin{aligned} \mathbf{J}^2\psi_{A,B} &= (\mathbf{L} + \hbar\boldsymbol{\sigma}/2)^2\psi_{A,B} = j(j+1)\hbar^2\psi_{A,B}, \\ J_3\psi_{A,B} &= (L_3 + \hbar\sigma_3/2)\psi_{A,B} = j_3\hbar\psi_{A,B}. \end{aligned} \tag{3.285}$$

The operator $\mathbf{L}^2$ is equal to $\mathbf{J}^2 - \hbar\boldsymbol{\sigma}\cdot\mathbf{L} - \frac{3}{4}\hbar^2$ when it acts on the two-component wave functions ψ_A and ψ_B. This means that any two-component eigenfunction of $\boldsymbol{\sigma}\cdot\mathbf{L} + \hbar$ and $\mathbf{J}^2$ is automatically an eigenfunction of $\mathbf{L}^2$. Thus, although the four-component wave function ψ is not an eigenfunction of $\mathbf{L}^2$ (since H does not commute with $\mathbf{L}^2$), ψ_A and ψ_B *separately* are eigenfunctions of $\mathbf{L}^2$ whose eigenvalues are denoted by $l_A(l_A+1)\hbar^2$ and $l_B(l_B+1)\hbar^2$. From (3.284) and (3.285) we then obtain

$$-\kappa = j(j+1) - l_A(l_A+1) + \tfrac{1}{4}, \qquad \kappa = j(j+1) - l_B(l_B+1) + \tfrac{1}{4}. \tag{3.286}$$

Using (3.282) and (3.386), we can determine l_A and l_B for a given κ. The results are summarized in Table 3–2.

For a given j, l_A can assume two possible values corresponding to the two possible values of κ. This fact is already familiar from our study of nonrelativistic quantum mechanics. For example, for $j = \frac{1}{2}$, l_A can be either 0 or 1 ($s\frac{1}{2}$ or $p\frac{1}{2}$), depending on whether κ is negative or positive. What is new is that for a fixed κ the orbital

Table 3–2

RELATIONS AMONG κ, j, l_A, AND l_B

	l_A	l_B
$\kappa = j + \frac{1}{2}$	$j + \frac{1}{2}$	$j - \frac{1}{2}$
$\kappa = -(j + \frac{1}{2})$	$j - \frac{1}{2}$	$j + \frac{1}{2}$

parities of ψ_A and ψ_B are necessarily opposite. As we showed in Section 3–4, this result can also be derived from the requirement that the four-component wave function ψ have a definite parity (cf. Eqs. 3.172 through 3.175).

We can now write ψ as

$$\psi = \begin{pmatrix} \psi_A \\ \psi_B \end{pmatrix} = \begin{pmatrix} g(r)\mathscr{Y}_{jl_A}^{j_3} \\ if(r)\mathscr{Y}_{jl_B}^{j_3} \end{pmatrix}, \tag{3.287}$$

where $\mathscr{Y}_{jl}^{j_3}$ stands for a normalized spin-angular function (an r-independent eigenfunction of $\mathbf{J}^2$, J_3, $\mathbf{L}^2$, and, of course, $\mathbf{S}^2$) formed by the combination of the Pauli spinor with the spherical harmonics of order l. Explicitly,

$$\mathscr{Y}_{jl}^{j_3} = \sqrt{\frac{l + j_3 + \frac{1}{2}}{2l+1}}\, Y_l^{j_3-1/2}\begin{pmatrix} 1 \\ 0 \end{pmatrix} + \sqrt{\frac{l - j_3 + \frac{1}{2}}{2l+1}}\, Y_l^{j_3+1/2}\begin{pmatrix} 0 \\ 1 \end{pmatrix} \tag{3.288}$$

for $j = l + \frac{1}{2}$, and

$$\mathscr{Y}_{jl}^{j_3} = -\sqrt{\frac{l - j_3 + \frac{1}{2}}{2l+1}}\, Y_l^{j_3-1/2}\begin{pmatrix} 1 \\ 0 \end{pmatrix} + \sqrt{\frac{l + j_3 + \frac{1}{2}}{2l+1}}\, Y_l^{j_3+1/2}\begin{pmatrix} 0 \\ 1 \end{pmatrix} \tag{3.289}$$

for $j = l - \frac{1}{2}$.‡ The radial functions f and g depend, of course, on κ. The factor i multiplying f has been inserted to make f and g real for bound-state (or standing-wave) solutions.

Before we substitute (3.287) in the Dirac equation written in the form

$$c(\boldsymbol{\sigma}\cdot\mathbf{p})\psi_B = (E - V(r) - mc^2)\psi_A, \quad c(\boldsymbol{\sigma}\cdot\mathbf{p})\psi_A = (E - V(r) + mc^2)\psi_B, \tag{3.290}$$

let us note that

$$\begin{aligned} \boldsymbol{\sigma}\cdot\mathbf{p} &= \frac{(\boldsymbol{\sigma}\cdot\mathbf{x})}{r^2}(\boldsymbol{\sigma}\cdot\mathbf{x})(\boldsymbol{\sigma}\cdot\mathbf{p}) \\ &= \frac{(\boldsymbol{\sigma}\cdot\mathbf{x})}{r^2}\left(-i\hbar r\frac{\partial}{\partial r} + i\boldsymbol{\sigma}\cdot\mathbf{L}\right). \end{aligned} \tag{3.291}$$

Moreover, the pseudoscalar operator $(\boldsymbol{\sigma}\cdot\mathbf{x})/r$ acting on $\mathscr{Y}_{jl_A}^{j_3}$ must give an eigenfunction of $\mathbf{J}^2$, J_3, and $\mathbf{L}^2$ with the same j and j_3 but of *opposite orbital parity*. Therefore $[(\boldsymbol{\sigma}\cdot\mathbf{x})/r]\,\mathscr{Y}_{jl_A}^{j_3}$ is equal to $\mathscr{Y}_{jl_B}^{j_3}$ itself up to a multiplicative phase factor. [*Note:* $(\boldsymbol{\sigma}\cdot\mathbf{x})^2/r^2 = 1$.] It is not difficult to show that this phase factor is *minus* one if we

‡See, for example, Merzbacher (1961), p. 402. Throughout this book we shall follow the phase convention used in Condon and Shortley (1951), Rose (1957, 1961), Merzbacher (1961), and Messiah (1962). The phase convention used by Bethe and Salpeter (1957) is slightly different due to an unconventional definition of Y_l^m.

conform to the phase convention used in writing (3.288) and (3.289). In fact, we have already verified this for the special case $j = j_3 = \frac{1}{2}$, $l_A = 0$, as seen from (3.178). Similarly $[(\boldsymbol{\sigma}\cdot\mathbf{x})/r]$ acting on $\mathscr{Y}_{jl_B}^{j_3}$ gives $\mathscr{Y}_{jl_A}^{j_3}$ apart from a minus sign. Thus

$$(\boldsymbol{\sigma}\cdot\mathbf{p})\psi_B = i\frac{(\boldsymbol{\sigma}\cdot\mathbf{x})}{r^2}\left(-i\hbar r\frac{\partial}{\partial r} + i\boldsymbol{\sigma}\cdot\mathbf{L}\right)f\mathscr{Y}_{jl_B}^{j_3}$$

$$= i\frac{(\boldsymbol{\sigma}\cdot\mathbf{x})}{r^2}\left(-i\hbar r\frac{df}{dr} + i(\kappa - 1)\hbar f\right)\mathscr{Y}_{jl_B}^{j_3}$$

$$= -\hbar\frac{df}{dr}\mathscr{Y}_{jl_A}^{j_3} - \frac{(1-\kappa)\hbar}{r}f\mathscr{Y}_{jl_A}^{j_3}. \tag{3.292}$$

Similarly

$$(\boldsymbol{\sigma}\cdot\mathbf{p})\psi_A = i\hbar\frac{dg}{dr}\mathscr{Y}_{jl_B}^{j_3} + i\frac{(1+\kappa)\hbar}{r}g\mathscr{Y}_{jl_B}^{j_3}. \tag{3.293}$$

We now observe that the spin-angular functions completely drop out when we rewrite (3.290), using (3.292) and (3.293):

$$\begin{aligned}-c\hbar\frac{df}{dr} - \frac{(1-\kappa)\hbar c}{r}f &= (E - V - mc^2)g,\\ c\hbar\frac{dg}{dr} + \frac{(1+\kappa)\hbar c}{r}g &= (E - V + mc^2)f.\end{aligned} \tag{3.294}$$

Introducing

$$F(r) = rf(r), \qquad G(r) = rg(r) \tag{3.295}$$

as in nonrelativistic quantum mechanics, we finally get radial equations:

$$\begin{aligned}\hbar c\left(\frac{dF}{dr} - \frac{\kappa}{r}F\right) &= -(E - V - mc^2)G,\\ \hbar c\left(\frac{dG}{dr} + \frac{\kappa}{r}G\right) &= (E - V + mc^2)F.\end{aligned} \tag{3.296}$$

Hydrogen atom. On the basis of the coupled equations (3.296) a variety of problems can be attacked. We shall consider only one problem along this line; the remaining part of this section will be devoted to a discussion of an electron bound to the atomic nucleus by a Coulomb potential. This classical problem (first treated by C. G. Darwin and W. Gordon in 1928) can be solved exactly. The reader who is interested in other central-force problems—the anomalous Zeeman effect, free spherical waves, exact solutions (as opposed to Born approximation solutions to be discussed in Chapter 4) to the Coulomb scattering problem, etc.—may consult Rose's book.‡

In order to simplify (3.296) when V is given by

$$V = -(Ze^2/4\pi r), \tag{3.297}$$

‡Rose (1961), Chapter 5.

we introduce

$$\alpha_1 = (mc^2 + E)/\hbar c, \qquad \alpha_2 = (mc^2 - E)/\hbar c,$$
$$\gamma = (Ze^2/4\pi\hbar c) = Z\alpha \simeq Z/137, \qquad \rho = \sqrt{\alpha_1\alpha_2}\, r. \tag{3.298}$$

Note that $\hbar\sqrt{\alpha_1\alpha_2} = \sqrt{m^2c^4 - E^2}/c$ is just the magnitude of the imaginary momentum of an electron of energy E. The coupled equations we must solve are

$$\left(\frac{d}{d\rho} - \frac{\kappa}{\rho}\right)F - \left(\sqrt{\frac{\alpha_2}{\alpha_1}} - \frac{\gamma}{\rho}\right)G = 0, \qquad \left(\frac{d}{d\rho} + \frac{\kappa}{\rho}\right)G - \left(\sqrt{\frac{\alpha_1}{\alpha_2}} + \frac{\gamma}{\rho}\right)F = 0. \tag{3.299}$$

As in the nonrelativistic treatment of the hydrogen atom, we seek solutions to (3.299) of the form

$$F = e^{-\rho}\rho^s \sum_{m=0} a_m \rho^m, \qquad G = e^{-\rho}\rho^s \sum_{m=0} b_m \rho^m. \tag{3.300}$$

Substituting (3.300) in (3.299), and equating the coefficients of $e^{-\rho}\rho^s\rho^{q-1}$, we obtain the recursion relations

$$(s + q - \kappa)a_q - a_{q-1} + \gamma b_q - \sqrt{\alpha_2/\alpha_1}\, b_{q-1} = 0,$$
$$(s + q + \kappa)b_q - b_{q-1} - \gamma a_q - \sqrt{\alpha_1/\alpha_2}\, a_{q-1} = 0. \tag{3.301}$$

For $q = 0$, we have

$$(s - \kappa)a_0 + \gamma b_0 = 0, \qquad (s + \kappa)b_0 - \gamma a_0 = 0. \tag{3.302}$$

Since a_0 and b_0 are not zero, the secular determinant of (3.302) must vanish; hence

$$s = \pm\sqrt{\kappa^2 - \gamma^2}. \tag{3.303}$$

We must require that $\int \psi^\dagger \psi\, d^3x$ be finite. This requirement amounts to

$$\int |F|^2\, d\rho < \infty, \qquad \int |G|^2\, d\rho < \infty. \tag{3.304}$$

Thus F and G must behave better than $\rho^{-1/2}$ at the origin, which means $s > -\frac{1}{2}$. Since

$$\kappa^2 - \gamma^2 \geq \min(\kappa^2) - \gamma^2 \simeq 1 - (Z/137)^2, \tag{3.305}$$

the above requirement cannot be satisfied if we take the negative root of (3.303). So we are led to take the positive root.‡

It is not difficult to show that F and G would increase exponentially as $\rho \to \infty$ (that is, $F, G \sim e^{+\rho}$ at infinity) if the power series (3.300) did not terminate.§ Assuming that the two series terminate with the same power, there must exist n' with the property

$$a_{n'+1} = b_{n'+1} = 0, \qquad a_{n'} \neq 0, \qquad b_{n'} \neq 0. \tag{3.306}$$

---

‡For $|\kappa| = 1$, $f = F/r$ and $g = G/r$ diverge at the origin (since $s < 1$); yet (3.304) is satisfied.

§The e^ρ behavior for F and G at infinity is allowed if $E > mc^2$, which means a purely imaginary ρ. Indeed the oscillatory behavior of the radial functions at infinity is characteristic of scattering-state solutions which exhibit a continuous energy spectrum.

Setting $q = n' + 1$ in (3.301) we obtain

$$a_{n'} = -\sqrt{\alpha_2/\alpha_1}\, b_{n'} \tag{3.307}$$

from both parts of (3.301) (which incidentally justifies our assumption that the two series terminate with the same power). To write an equation that involves only $a_{n'}$ and $b_{n'}$ whose ratio is now known, we multiply the first of (3.301) by α_1, the second by $\sqrt{\alpha_1\alpha_2}$, set $q = n'$ this time, and subtract:

$$[\alpha_1(s + n' - \kappa) + \sqrt{\alpha_1\alpha_2}\,\gamma]a_{n'} - [\sqrt{\alpha_1\alpha_2}\,(s + n' + \kappa) - \alpha_1\gamma]b_{n'} = 0; \tag{3.308}$$

hence

$$2\sqrt{\alpha_1\alpha_2}\,(s + n') = \gamma(\alpha_1 - \alpha_2), \tag{3.309}$$

or

$$\sqrt{(mc^2)^2 - E^2}\,(s + n') = E\gamma. \tag{3.310}$$

Thus we obtain the energy eigenvalues‡

$$E = \frac{mc^2}{\sqrt{1 + \dfrac{\gamma^2}{(s + n')^2}}} = \frac{mc^2}{\sqrt{1 + \dfrac{Z^2\alpha^2}{(n' + \sqrt{(j + \frac{1}{2})^2 - Z^2\alpha^2})^2}}}\cdot \tag{3.311}$$

Note that E depends only on n' and $j + \frac{1}{2} = |\kappa|$.

In order to compare (3.311) with the corresponding expression obtained in the Schrödinger theory, we define

$$n = n' + (j + \tfrac{1}{2}) = n' + |\kappa|. \tag{3.312}$$

Since the minimum value of n' is zero, we have

$$n > j + \tfrac{1}{2} = |\kappa|, \tag{3.313}$$

which is at least unity. Expanding (3.311) we get

$$E = mc^2\left[1 - \frac{1}{2}\frac{(Z\alpha)^2}{n^2} - \frac{1}{2}\frac{(Z\alpha)^4}{n^3}\left(\frac{1}{j + \frac{1}{2}} - \frac{3}{4n}\right) - \cdots\right]\cdot \tag{3.314}$$

Since§

$$\tfrac{1}{2}\alpha^2 mc^2 = e^2/(8\pi a_{\text{Bohr}}), \tag{3.315}$$

we see that n is indeed identical with the familiar *principal quantum number* in nonrelativistic quantum mechanics. Note also that the leading correction to the Balmer formula is precisely the fine-structure splitting (3.88) which tells us that, for a given n, higher j-states are at higher levels.

In the Dirac theory each state of a hydrogen atom can be completely characterized by n', κ, and j_3. We can translate this classification scheme into the more familiar one based on spectroscopic notation. This is done in Table 3–3 which can be obtained with the help of Table 3–2 and Eqs. (3.312) and (3.313). Note that although $\mathbf{L}^2$ is not "good" in the relativistic theory, it is customary to use the

---

‡Formula (3.311) was first obtained by A. Sommerfeld, using a relativistic version of N. Bohr's old quantum theory.

§In this section we shall use a_{Bohr} rather than a_0 for the Bohr radius to avoid a possible confusion with the coefficient a_0 in (3.300).

Table 3–3

RELATIVISTIC QUANTUM NUMBERS AND SPECTROSCOPIC NOTATION

A pair of states which have the same energy according to (3.311) are denoted by "deg."

n	$n' = n - \|\kappa\| \geq 0$	$\kappa = \pm(j + \frac{1}{2})$	Notation
1	0	-1	$1s\frac{1}{2}$
2	1	-1	$2s\frac{1}{2}$ } deg
2	1	$+1$	$2p\frac{1}{2}$ } deg
2	0	-2	$2p\frac{3}{2}$
3	2	-1	$3s\frac{1}{2}$ } deg
3	2	$+1$	$3p\frac{1}{2}$ } deg
3	1	-2	$3p\frac{3}{2}$ } deg
3	1	$+2$	$3d\frac{3}{2}$ } deg
3	0	-3	$3d\frac{5}{2}$

notation $p_{3/2}$, etc., which actually means $l_A = 1$ with $j = \frac{3}{2}$, etc. In other words the orbital angular momentum of the upper two-component wave function (which becomes the wave function of the Schrödinger-Pauli theory in the nonrelativistic limit) determines the orbital angular momentum in the spectroscopic language. The reader may wonder why we have omitted in Table 3–2 the $\kappa > 0$ states whenever $n' = 0$. The reason for this omission is evident if we go back to the second expression of (3.302) and (3.307) which together imply

$$(s + \kappa)/\gamma = -\sqrt{\alpha_2/\alpha_1}, \qquad n' = 0 \text{ only.} \tag{3.316}$$

This can be satisfied only if κ is negative because s is a positive number smaller than $|\kappa|$ (cf. Eq. 3.303). The absence of the $\kappa > 0$ states for $n' = 0$ corresponds to the familiar rule in nonrelativistic quantum mechanics: The maximum value of l is $n - 1$, not n.

For the ground state ($n' = 0$, $\kappa = -1$), the relation (3.311) simplifies to

$$E_{gd} = mc^2\sqrt{1 - (Z\alpha)^2}. \tag{3.317}$$

So

$$\sqrt{\alpha_1\alpha_2} = Z\alpha mc^2/\hbar c = Z/a_{\text{Bohr}}, \tag{3.318}$$

and

$$\frac{b_0}{a_0} = -\frac{Z\alpha}{1 + \sqrt{1 - (Z\alpha)^2}} = -\frac{(1 - \sqrt{1 - (Z\alpha)^2})}{Z\alpha}. \tag{3.319}$$

Up to an overall multiplicative constant denoted by N we can readily write the ground-state wave function

$$\psi_{gd} = \frac{N}{\sqrt{\pi}}\left(\frac{Z}{a_{\text{Bohr}}}\right)^{3/2} e^{-Zr/a_{\text{Bohr}}}\left(\frac{Zr}{a_{\text{Bohr}}}\right)^{\sqrt{1-(Z\alpha)^2}-1}\begin{pmatrix} \chi^{(s)} \\ \dfrac{i(1 - \sqrt{1 - (Z\alpha)^2})}{Z\alpha}\dfrac{(\boldsymbol{\sigma}\cdot\mathbf{x})}{r}\chi^{(s)} \end{pmatrix}, \tag{3.320}$$

where $\chi^{(s)}$ is the Pauli spinor

$$\begin{pmatrix} 1 \\ 0 \end{pmatrix} \quad \text{or} \quad \begin{pmatrix} 0 \\ 1 \end{pmatrix},$$

depending on whether $j_3 = \frac{1}{2}$ or $-\frac{1}{2}$. A straightforward calculation shows that the normalization requirement for ψ gives

$$N = 2^{\sqrt{1-(Z\alpha)^2}-1} \sqrt{\frac{1+\sqrt{1-(Z\alpha)^2}}{\Gamma(1+2\sqrt{1-(Z\alpha)^2})}}, \tag{3.321}$$

where

$$\Gamma(x) = \int_0^\infty e^{-t} t^{x-1}\, dt,$$
$$\Gamma(m) = (m-1)! \qquad \text{for} \qquad m = \text{positive integer.} \tag{3.322}$$

Note that N approaches 1 as $Z\alpha \longrightarrow 0$. Furthermore $(Zr/a_{\text{Bohr}})^{\sqrt{1-(Z\alpha)^2}-1}$ is essentially unity except at distances of order

$$r \sim \frac{137\hbar}{2mcZ} e^{-2(137)^2/Z^2}. \tag{3.323}$$

As $r \longrightarrow 0$, ψ exhibits a mild singularity. This, however, is of academic interest only, since the wave function at short distances must be modified because of the finite charge distribution of the nucleus. Thus we see that for the ground states of hydrogen-like atoms with low Z, the upper two-component wave function is essentially identical to the Schrödinger wave function multiplied by a Pauli spinor. As for the lower two-component wave function, we merely remark that, apart from $i(\boldsymbol{\sigma}\cdot\mathbf{x})/r$, the ratio of the lower to the upper components is given by (cf. Eq. 3.307)

$$-\frac{a_0}{b_0} = \sqrt{\frac{mc^2 - E_{gd}}{mc^2 + E_{gd}}} \simeq \sqrt{\left(\frac{mv^2}{2}\right)\left(\frac{1}{2mc^2}\right)} = \frac{v}{2c}, \tag{3.324}$$

where v is the "velocity" of the electron in Bohr's circular orbit theory. This result is in agreement with our earlier discussion in Section 3–3 on the ratio of ψ_B to ψ_A.

In 1947 W. E. Lamb and R. C. Retherford observed a splitting between the $2s\frac{1}{2}$- and $2p\frac{1}{2}$-states of the hydrogen atom not given by (3.311). As already discussed in Section 2–8, the main part of this "Lamb shift" can be satisfactorily accounted for when we consider the interaction of the electron with the quantized radiation field.

Another important effect not contained in (3.311) arises from the interaction between the magnetic moment of the nucleus and the magnetic moment of the electron. In the case of the hydrogen atom, for instance, when we compound the electron spin with the proton spin, the net result is $F = 1$ (triplet) or $F = 0$ (singlet), where F is the quantum number corresponding to the *total* spin. Since the magnetic interaction is dependent on the relative orientation of the two magnetic dipole moments, each level of the hydrogen atom characterized by n, j, $l(= l_A)$ is split further into two sublevels corresponding to the two possible values of F

even in the absence of any external magnetic field. This is known as a *hyperfine splitting*. Let us estimate it for the s-states using nonrelativistic quantum mechanics. Classically the magnetic field created by the proton magnetic moment $\mathbf{M}^{(\text{class})}$ is‡

$$\mathbf{B} = \boldsymbol{\nabla} \times \left(\mathbf{M}^{(\text{class})} \times \boldsymbol{\nabla} \frac{1}{4\pi r}\right). \tag{3.325}$$

Quantum-mechanically we replace $\mathbf{M}^{(\text{class})}$ by the magnetic moment operator

$$\mathbf{M} = \frac{|e|\hbar(1+\kappa)}{2m_{\text{p}}c}\boldsymbol{\sigma}_{\text{p}}, \tag{3.326}$$

where $\boldsymbol{\sigma}_{\text{p}}$ is the Pauli matrix for the proton spin and $2(1+\kappa)$ is the g-factor of the proton. (We assume that the magnetic moment distribution of the proton is point-like.) Within the framework of the Schrödinger-Pauli theory we obtain the interaction Hamiltonian operator

$$\begin{aligned} H^{(\text{hf})} &= -\boldsymbol{\mu}\cdot\boldsymbol{\nabla} \times \left(\mathbf{M} \times \boldsymbol{\nabla}\frac{1}{4\pi r}\right) \\ &= (\boldsymbol{\mu}\cdot\mathbf{M})\nabla^2\left(\frac{1}{4\pi r}\right) - [(\boldsymbol{\mu}\cdot\boldsymbol{\nabla})(\mathbf{M}\cdot\boldsymbol{\nabla})]\frac{1}{4\pi r} \\ &= \tfrac{2}{3}(\boldsymbol{\mu}\cdot\mathbf{M})\nabla^2\left(\frac{1}{4\pi r}\right) - \left[(\boldsymbol{\mu}\cdot\boldsymbol{\nabla})(\mathbf{M}\cdot\boldsymbol{\nabla}) - \frac{1}{3}\boldsymbol{\mu}\cdot\mathbf{M}\nabla^2\right]\frac{1}{4\pi r}, \end{aligned} \tag{3.327}$$

where $\boldsymbol{\mu} = (e\hbar/2m_{\text{e}}c)\boldsymbol{\sigma}_{\text{e}}$. The quantity in the brackets transforms like a traceless tensor of rank two; so when it is integrated with a function of $\mathbf{x}$, $f(\mathbf{x})$, it gives a nonvanishing contribution only if the expansion of $f(\mathbf{x})$ in spherical harmonics contains Y_2^m. For the spherically symmetric s states, only the first term of (3.327) is relevant. Using the nonrelativistic wave function ψ_n, we obtain the energy shift

$$\begin{aligned} \Delta E_n &= -\tfrac{2}{3}\boldsymbol{\mu}\cdot\mathbf{M}\int \delta^{(3)}(\mathbf{x})|\psi_n(\mathbf{x})|^2 d^3x \\ &= \left[\frac{(1+\kappa)e^2\hbar^2}{6m_{\text{e}}m_{\text{p}}c^2}\right]\left[\frac{1}{\pi}\left(\frac{1}{a_{\text{Bohr}}n}\right)^3\right]\boldsymbol{\sigma}_{\text{e}}\cdot\boldsymbol{\sigma}_{\text{p}} \\ &= \tfrac{2}{3}\alpha^4(1+\kappa)\left(\frac{m_{\text{e}}}{m_{\text{p}}}\right)\frac{m_{\text{e}}c^2}{n^3}\begin{cases} 1, & F=1, \\ -3, & F=0, \end{cases} \end{aligned} \tag{3.328}$$

as first shown by E. Fermi in 1930. Note that the order of magnitude of this splitting is the fine-structure splitting multiplied by $(m_{\text{e}}/m_{\text{p}})$. For the $1s\frac{1}{2}$-state, the above energy difference corresponds to a radio microwave of 1420 Mc or 21 cm; the radiative transition between these two hyperfine levels is of fundamental importance in radio astronomy. We may parenthetically mention that this energy difference is one of the most accurately measured quantities in modern physics; S. B. Crampton, D. Kleppner, and N. F. Ramsey have determined that the corresponding radiofrequency is $(1420.405751800 \pm 0.000000028)$ Mc.

There are other corrections to formula (3.311). First, we must take into account the motion of the nucleus since the mass of the nucleus is not infinite; a major

‡Panofsky and Phillips (1955), p. 120; Jackson (1962) p. 146.

part of this correction can be taken care of if we use everywhere the reduced mass $m_e m_p/(m_e + m_p)$ in place of m_e. Second, there are other contributions to the Lamb shift not discussed in Chapter 2; especially important is the vacuum polarization effect to be discussed later. Third, the finite size of the nucleus also modifies formula (3.311) especially for the *s* states which are sensitive to small deviations from Coulomb's law at close distances; in the interesting case of the 2*s* state of the hydrogen atom, however, we can estimate the energy shift due to this effect to be only 0.1 Mc, using the observed proton charge radius $\sim 0.7 \times 10^{-13}$ cm.

The utility of the Dirac theory in atomic physics is not limited to light hydrogen-like atoms. For heavy atoms where $(Z\alpha)^2$ is not very small compared with unity (0.45 for uranium), the relativistic effects must be taken into account even for understanding the qualitative features of the energy levels. Although we cannot, in practice, study one-electron ions of heavy atoms, it is actually possible to check the quantitative predictions of the Dirac theory by looking at the energy levels of the innermost (*K*-shell and *L*-shell) electrons of high *Z* atoms which can be inferred experimentally from X-ray spectra. Similar studies have been carried out with muonic atoms (atoms in which one of the electrons is replaced by a negative muon).

Although we shall not discuss the emission and absorption of radiation using the Dirac theory, the results of Section 2–4 are applicable *mutatis mutandis.* All we need to do is make the following replacements:‡

$$-\frac{1}{mc}(\mathbf{p}\cdot\mathbf{A}+\mathbf{A}\cdot\mathbf{p})-\frac{e\hbar}{2mc}\boldsymbol{\sigma}\cdot\mathbf{B}\longrightarrow-\boldsymbol{\alpha}\cdot\mathbf{A}$$
$$\psi^{(\text{Schrödinger-Pauli})}\longrightarrow\psi^{(\text{Dirac})}$$
$$\psi^{(\text{Schrödinger-Pauli})\dagger}\longrightarrow\psi^{(\text{Dirac})\dagger}\quad(\text{not }\bar{\psi}^{(\text{Dirac})}).\tag{3.329}$$

3–9. HOLE THEORY AND CHARGE CONJUGATION

Holes and positrons. Although we have shown that the Dirac theory accommodates negative-energy solutions whose existence should not be ignored altogether, we have as yet not examined their physical significance. This section is devoted to the physical interpretation of the negative-energy states within the framework of a theory in which the electron field is not quantized.

As a simple example to illustrate some of the difficulties with the *original* Dirac theory of 1928, let us consider an atomic electron. According to the quantum theory of radiation developed in the previous chapter, an excited atomic state can lose its energy by spontaneously emitting a photon even in the absence of any external field. This is why all atomic states, with the exception of the ground states, have finite lifetimes. In the Dirac theory, however, the so-called ground state of an atom is not really the lowest state since there exists a continuum of negative-energy states from $-mc^2$ to $-\infty$ for any potential that vanishes at

‡We have to be a little more careful when we treat a process in which the quadratic $\mathbf{A}^2$ term in the nonrelativistic Hamiltonian is important. This point will be discussed in the next section with reference to Thomson scattering.

infinity. We know that an excited atomic state makes a radiative transition to the ground state; similarly, we expect that the atomic electron in the ground state with energy $mc^2 - |E_{BE}|$ can emit spontaneously a photon of energy $\gtrsim 2mc^2$ and fall into a negative-energy state. Furthermore, once it reaches a negative-energy state, it will keep on lowering its energy indefinitely by emitting photons since there is *no lower bound* to the negative-energy spectrum. Since we know that the ground state of an atom is stable, we must somehow prevent such catastrophic transitions.

Faced with this difficulty, Dirac proposed, in 1930, that *all the negative-energy states are completely filled under normal conditions.* The catastrophic transitions mentioned above are then prevented because of the Pauli exclusion principle. What we usually call the vacuum is actually an infinite sea of negative-energy electrons. Occasionally one of the negative-energy electrons in the Dirac sea can absorb a photon of energy $\hbar\omega > 2mc^2$ and become an $E > 0$ state. As a result, a "hole" is created in the Dirac sea. The *observable* energy of the Dirac sea is now the energy of the vacuum *minus* the *negative* energy of the vacated state, hence a *positive* quantity. In this way we expect that the absence of a negative-energy electron appears as the presence of a positive-energy particle. Similarly, when a hole is created in the Dirac sea, the total charge of the Dirac sea becomes

$$Q = Q_{\text{vacuum}} - e = Q_{\text{vacuum}} - (-|e|) = Q_{\text{vacuum}} + |e|; \tag{3.330}$$

hence the *observable* charge of the hole is

$$Q_{\text{obs}} = Q - Q_{\text{vacuum}} = |e|. \tag{3.331}$$

This means that a hole in the sea of negative-energy states looks like a positive-energy particle of charge $|e|$. Thus once we accept (a) that the negative-energy states are completely filled under normal conditions and (b) that a negative-energy electron can absorb a photon of energy $> 2mc^2$ (just as a positive-energy electron can) to become a positive-energy electron, we are unambiguously led to predict the existence of a particle of charge $|e|$ with a positive energy.

When Dirac proposed this "hole theory," there was no good candidate for the predicted positively charged particle. In the beginning Dirac even thought that the hole in the negative-energy state should be identified with the proton. However, it was quickly pointed out by J. R. Oppenheimer that if this interpretation were correct, the hydrogen atom would undergo a self-annihilation into two photons with a lifetime $\sim 10^{-10}$ sec.‡ Moreover, H. Weyl, who looked at the symmetry properties of the Dirac equation, proved that the mass of the particle associated with the hole must be the same as the electron mass. Prior to 1932, because of the experimental absence of the conjectured particle, Dirac's hole theory was not taken seriously. To recapture the prevailing atmosphere of the time, we quote from W. Pauli's *Handbuch* article.§

---

‡This number assumes that the energy released is $2m_ec^2$. If we take the energy released to be $m_ec^2 + m_pc^2$, the lifetime is even shorter.

§The translation from the original German text is the work of J. Alexander, G. F. Chew, W. Selove, and C. N. Yang.

> Recently Dirac attempted the explanation, already discussed by Oppenheimer, of identifying the holes with antielectrons, particles of charge $+|e|$ and the electron mass. Likewise, in addition to protons, there must be antiprotons. The experimental absence of such particles is then traced back to a special initial state in which only one of the two kinds of particles is present. We see that this already appears to be unsatisfactory because the laws of nature in this theory with respect to electrons and antielectrons are exactly symmetrical. Thus γ-ray photons (at least two in order to satisfy the laws of conservation of energy and momentum) must be able to transform, by themselves, into an electron and an antielectron. We do not believe, therefore, that this explanation can be seriously considered.

When the article appeared in print, however, C. D. Anderson had already demonstrated the existence of a positron. Many years later Pauli made the following famous remark on Dirac:

> . . . with his fine instinct for physical realities he started his argument without knowing the end of it.

We shall now examine a little more closely the absorption of a photon by one of the negative-energy electrons in the Dirac sea. As stated earlier, if the photon energy is sufficiently large, an electron in a negative-energy state may be "escalated" to a positive-energy state

$$e^-_{E<0} + \gamma \longrightarrow e^-_{E>0}. \tag{3.332}$$

According to the hole-theoretic interpretation, this appears as

$$\gamma \longrightarrow e^-_{E>0} + e^+_{E>0}, \tag{3.333}$$

since the vacated negative-energy state is observable as a positron state. Although a photon cannot produce an e^-e^+ pair in free space without violating energy and momentum conservation, the process (3.333) can take place in the Coulomb field of a nucleus. As is well known, the production of an electron-positron pair is a very frequent phenomenon when high-energy γ-rays go through matter. We may also consider a closely related process,

$$e^-_{E>0} \longrightarrow e^-_{E<0} + 2\gamma. \tag{3.334}$$

Since all the negative-energy states are supposed to be filled under normal conditions, (3.334) is forbidden except when there is a hole in the normally filled negative-energy states. This means that whenever (3.334) is allowed, we can interpret it as

$$e^-_{E>0} + e^+_{E>0} \longrightarrow 2\gamma. \tag{3.335}$$

This process has also been observed frequently as positrons slow down in solids. We shall present a quantitative treatment of this electron-positron annihilation process in Chapter 4.

At this stage we emphasize again that the electron must obey the Pauli exclusion principle if the hole theory is to make sense. Otherwise we cannot attach much meaning to the notion that the negative-energy states are completely filled. If it were not for the exclusion principle, we could keep on, for millions of years, piling up electrons in the *same* negative-energy state. Even though the energy spectrum

Table 3–4

DYNAMICAL QUANTITIES IN THE HOLE THEORY

	Charge	Energy	Momentum	Spin	Helicity	"Velocity"
$E<0$ Electron state	$-\|e\|$	$-\|E\|$	$\mathbf{p}$	$\frac{\hbar\langle\boldsymbol{\Sigma}\rangle}{2}$	$\boldsymbol{\Sigma}\cdot\hat{\mathbf{p}}$	$\mathbf{v}$
Positron state	$+\|e\|$	$+\|E\|$	$-\mathbf{p}$	$-\frac{\hbar\langle\boldsymbol{\Sigma}\rangle}{2}$	$\boldsymbol{\Sigma}\cdot\hat{\mathbf{p}}$	$\mathbf{v}$

of free Klein-Gordon particles is identical to that of free Dirac particles, it is not possible to construct a sensible hole theory out of Klein-Gordon particles which obey Bose-Einstein statistics.

Let us study the connection between the various dynamical quantities of the positron and those of the negative-energy electron whose absence appears as the presence of the positron in question. We have already seen that both the charge and the energy of the physical positron must be positive. What is the momentum of the positron? Just as in the case of energy, the absence of momentum $\mathbf{p}$ in the Dirac sea appears as the presence of momentum $-\mathbf{p}$. Hence the momentum of the physical ($E>0$) positron state is opposite to that of the corresponding negative-energy electron state. Similarly the absence of a spin-up $E<0$ electron is to be interpreted as the presence of a spin-down $E>0$ positron. Thus we can construct Table 3–4 for a free particle. (We have listed $\langle\boldsymbol{\Sigma}\rangle$ rather than the eigenvalue of Σ_3 since, in general, the plane-wave solutions are not eigenstates of Σ_3).

The entry "velocity" in Table 3–4 requires some explanation. Suppose we consider a wave packet made up of negative-energy solutions whose momenta center around a certain mean value. We can then associate a certain group-velocity with the wave packet. The absence of this $E<0$ wave packet must appear as a wave packet made up of $E>0$ positron states moving in the *same* direction, that is, the velocity of the positron wave packet must be the *same* as that of the corresponding $E<0$ electron wave packet. It is not hard to see that this is possible only if the "velocity" of the negative-energy electron is *opposite* in direction to its momentum. This appears somewhat strange but is completely consistent with (3.221), which says that the expectation value of the velocity operator $c\boldsymbol{\alpha}$ is the negative of the expectation value of $\mathbf{p}c^2/|E|$. The reader who is still not convinced may amuse himself by working out steps analogous to (3.163) through (3.170) for a negative-energy plane wave. If we apply S_{Lor}^{-1} to the wave function $\psi'^{(3,4)}$ for an $E<0$ electron at rest in the primed system, we obtain the wave function $\psi^{(3,4)}$ which corresponds to the negative-energy electron whose momentum (defined as the eigenvalue of $-i\hbar\nabla$) in the unprimed system is *opposite* to the direction of motion of the primed system.

Thomson scattering in the Dirac theory. As a simple calculation that dramatically illustrates the importance of the negative-energy states in an unexpected domain, we shall now compute the cross section for Thomson scattering, that is, the scattering of a low-energy photon ($\hbar\omega \ll mc^2$) by a free electron. As in Section 2–5 (cf. Eqs. 2.158 and 2.168) we expect that the differential cross section is given by $r_0^2|\boldsymbol{\epsilon}^{(\alpha)}\cdot\boldsymbol{\epsilon}^{(\alpha')}|^2$, which is also the same as the classical result. In the Dirac theory there is no analog of the seagull graph Fig. 2–2 (c); we must compute the analogs of Fig. 2–2 (a) and (b) for the free electron. We characterize the initial, final, and intermediate states of the electron by $(\mathbf{p}, r)$, $(\mathbf{p}', r')$, and $(\mathbf{p}'', r'')$ respectively. We then obtain for the transition matrix element,

$$-\frac{e^2c^2\hbar}{2V\sqrt{\omega\omega'}}\sum_{\mathbf{p}''}\sum_{r''=1,\,2}\left(\frac{\langle\mathbf{p}'r'|\boldsymbol{\alpha}\cdot\boldsymbol{\epsilon}^{(\alpha')}e^{-i\mathbf{k}'\cdot\mathbf{x}}|\mathbf{p}''r''\rangle\langle\mathbf{p}''r''|\boldsymbol{\alpha}\cdot\boldsymbol{\epsilon}^{(\alpha)}e^{i\mathbf{k}\cdot\mathbf{x}}|\mathbf{p}r\rangle}{E''-E-\hbar\omega}\right.$$
$$\left.+\frac{\langle\mathbf{p}'r'|\boldsymbol{\alpha}\cdot\boldsymbol{\epsilon}^{(\alpha)}e^{i\mathbf{k}\cdot\mathbf{x}}|\mathbf{p}''r''\rangle\langle\mathbf{p}''r''|\boldsymbol{\alpha}\cdot\boldsymbol{\epsilon}^{(\alpha')}e^{-i\mathbf{k}'\cdot\mathbf{x}}|\mathbf{p}r\rangle}{E''-E+\hbar\omega'}\right), \tag{3.336}$$

where we have used the rule stated in (3.329). Since all the negative-energy states are supposed to be filled, the summation is over positive-energy states only ($r''=1, 2$). The electron is initially at rest. So, as $\mathbf{p}\rightarrow 0$, $\mathbf{k}\rightarrow 0$, a typical matrix element in (3.336) becomes

$$\begin{aligned}\langle\mathbf{p}''r''|\boldsymbol{\alpha}\cdot\boldsymbol{\epsilon}^{(\alpha)}|0r\rangle &= (1/V)\sqrt{mc^2/E''}\int e^{-i\mathbf{p}''\cdot\mathbf{x}/\hbar}u^{(r'')\dagger}(\mathbf{p}'')(\boldsymbol{\alpha}\cdot\boldsymbol{\epsilon}^{(\alpha)})u^{(r)}(0)d^3x\\ &= \delta_{\mathbf{p}''0}u^{(r'')\dagger}(\mathbf{p}'')(\boldsymbol{\alpha}\cdot\boldsymbol{\epsilon}^{(\alpha)})u^{(r)}(0) = 0,\end{aligned} \tag{3.337}$$

since the matrix element of α_k taken between two at-rest $E>0$ spinors vanishes. Because the final electron is also at rest for the scattering of a very soft photon, we easily see that (3.336) is identically zero. This means that the Thomson scattering cross section should vanish, in contradiction to both observation and nonrelativistic quantum mechanics.

What went wrong? In the hole theory we must take into account an additional process which has no analog in nonrelativistic quantum mechanics. Consider a negative-energy electron in the Dirac sea. It can absorb the incident photon (say, at $t=t_1$) and become a positive-energy electron. Even though this virtual transition does not conserve energy (unless $\hbar\omega > 2mc^2$), there is a finite matrix element for it. At a subsequent time ($t=t_2$) the initial electron can fill up the vacated negative-energy state by emitting the outgoing photon. Meanwhile the escalated electron goes on as the positive-energy final-state electron. All this may be visualized physically as follows. The incident photon creates an electron-positron pair at $t=t_1$; subsequently at $t=t_2$ the positron is annihilated by the initial electron, emitting the outgoing photon, as shown in Fig. 3–6(a). Similarly it is possible for the outgoing photon to be emitted first as one of the $E<0$ electrons in the Dirac sea is escalated; subsequently the initial electron fills up the vacated negative-energy state by absorbing the incident photon. This is illustrated in Fig. 3–6(b), which physically represents the creation of an electron-positron pair plus the outgoing photon followed by the annihilation of the positron with the initial electron and the incident photon.

We shall now calculate the matrix elements for the two diagrams. According to the hole theory, for electrons, the initial state is made up of the incident electron and the Dirac sea. In Fig. 3–6(a), one of the negative-energy electrons denoted

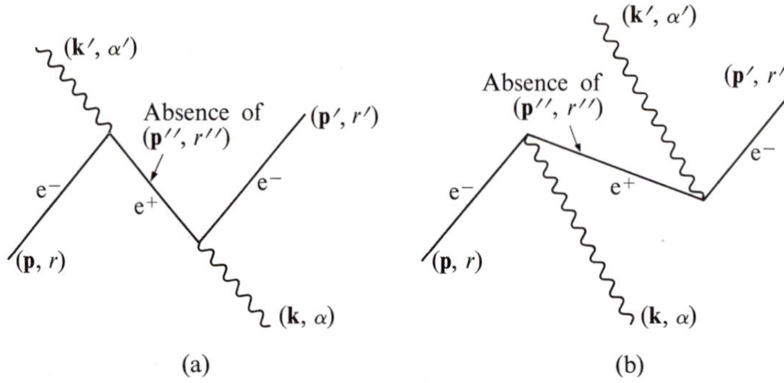

Fig. 3–6. Thomson scattering in the Dirac theory.

by $(\mathbf{p}'', r'')$ makes a transition to a positive-energy state $(\mathbf{p}', r')$ by absorbing the photon $(\mathbf{k}, \alpha)$. The relevant matrix element is $-ec\sqrt{\hbar/2V\omega}\langle \mathbf{p}'r' | \boldsymbol{\alpha}\cdot\boldsymbol{\epsilon}^{(\alpha)} e^{i\mathbf{k}\cdot\mathbf{x}} | \mathbf{p}''r''\rangle$. Since the absorption takes place first, the energy denominator according to the rules of Chapter 2 is $E_I - E_A - \hbar\omega$, where $E_I = E_{\text{vac}} - (-|E''|) + E + E'$ and $E_A = E_{\text{vac}} + E$. For the transition at t_2 we have the matrix element $-ec\sqrt{\hbar/2V\omega}\langle \mathbf{p}''r'' | \boldsymbol{\alpha}\cdot\boldsymbol{\epsilon}^{(\alpha')} e^{-i\mathbf{k}'\cdot\mathbf{x}} | \mathbf{p}r\rangle$. Working out Fig. 3–6(b) in a similar way, we obtain for the transition matrix element for both diagrams combined:

$$-\frac{e^2c^2\hbar}{2V\sqrt{\omega\omega'}}\sum_{\mathbf{p}}\sum_{r''=3,4}\left(\frac{\langle \mathbf{p}''r'' | \boldsymbol{\alpha}\cdot\boldsymbol{\epsilon}^{(\alpha')} e^{-i\mathbf{k}'\cdot\mathbf{x}} | \mathbf{p}r\rangle\langle \mathbf{p}'r' | \boldsymbol{\alpha}\cdot\boldsymbol{\epsilon}^{(\alpha)} e^{i\mathbf{k}\cdot\mathbf{x}} | \mathbf{p}''r''\rangle}{E' + |E''| - \hbar\omega}\right.$$
$$\left. + \frac{\langle \mathbf{p}''r'' | \boldsymbol{\alpha}\cdot\boldsymbol{\epsilon}^{(\alpha)} e^{i\mathbf{k}\cdot\mathbf{x}} | \mathbf{p}r\rangle\langle \mathbf{p}'r' | \boldsymbol{\alpha}\cdot\boldsymbol{\epsilon}^{(\alpha')} e^{-i\mathbf{k}'\cdot\mathbf{x}} | \mathbf{p}''r''\rangle}{E' + |E''| + \hbar\omega'}\right). \tag{3.338}$$

As before, we set $E = E' = mc^2$, $\mathbf{p} = \mathbf{p}' = 0$ as $k \to 0$. Furthermore, because of the space integration that appears in the evaluation of each matrix element, $\mathbf{p}'' = 0$; hence $|E''| = mc^2$. It is now simple to evaluate each of the four matrix elements. For example,

$$\begin{aligned}\langle 0\, r'' | \boldsymbol{\alpha}\cdot\boldsymbol{\epsilon}^{(\alpha)} | 0\, r\rangle &= (0, \chi^{(s'')\dagger})\begin{pmatrix} 0 & \boldsymbol{\sigma}\cdot\boldsymbol{\epsilon}^{(\alpha)} \\ \boldsymbol{\sigma}\cdot\boldsymbol{\epsilon}^{(\alpha)} & 0 \end{pmatrix}\begin{pmatrix} \chi^{(s)} \\ 0 \end{pmatrix} \\ &= \chi^{(s'')}\boldsymbol{\sigma}\cdot\boldsymbol{\epsilon}^{(\alpha)}\chi^{(s)}. \end{aligned} \tag{3.339}$$

As $\hbar\omega, \hbar\omega' \ll mc^2$, the two terms in (3.338) combine. Taking into account the two spin states of the negative-energy electron at rest, we obtain

$$\begin{aligned}&\sum_{r''=3,4}[\langle 0\, r'' | \boldsymbol{\alpha}\cdot\boldsymbol{\epsilon}^{(\alpha')} | 0\, r\rangle\langle 0\, r' | \boldsymbol{\alpha}\cdot\boldsymbol{\epsilon}^{(\alpha)} | 0\, r''\rangle + \langle 0\, r'' | \boldsymbol{\alpha}\cdot\boldsymbol{\epsilon}^{(\alpha)} | 0\, r\rangle\langle 0\, r' | \boldsymbol{\alpha}\cdot\boldsymbol{\epsilon}^{(\alpha')} | 0\, r''\rangle] \\ &= \sum_{s''=1,2}[(\chi^{(s'')\dagger}\boldsymbol{\sigma}\cdot\boldsymbol{\epsilon}^{(\alpha')}\chi^{(s)})(\chi^{(s')\dagger}\boldsymbol{\sigma}\cdot\boldsymbol{\epsilon}^{(\alpha)}\chi^{(s'')}) + (\chi^{(s'')\dagger}\boldsymbol{\sigma}\cdot\boldsymbol{\epsilon}^{(\alpha)}\chi^{(s)})(\chi^{(s')\dagger}\boldsymbol{\sigma}\cdot\boldsymbol{\epsilon}^{(\alpha)}\chi^{(s'')})] \\ &= \chi^{(s')\dagger}[(\boldsymbol{\sigma}\cdot\boldsymbol{\epsilon}^{(\alpha)})(\boldsymbol{\sigma}\cdot\boldsymbol{\epsilon}^{(\alpha')}) + (\boldsymbol{\sigma}\cdot\boldsymbol{\epsilon}^{(\alpha')})(\boldsymbol{\sigma}\cdot\boldsymbol{\epsilon}^{(\alpha)})]\chi^{(s)} \\ &= 2\boldsymbol{\epsilon}^{(\alpha)}\cdot\boldsymbol{\epsilon}^{(\alpha')}\delta_{ss'}, \end{aligned} \tag{3.340}$$

where we have used (3.339) and the closure property of the Pauli spinors

$$\sum_{s''=1,2} \chi^{(s'')}\chi^{(s'')\dagger} = \begin{pmatrix}1\\0\end{pmatrix}(1 \quad 0) + \begin{pmatrix}0\\1\end{pmatrix}(0 \quad 1)$$
$$= \begin{pmatrix}1 & 0\\0 & 1\end{pmatrix}. \tag{3.341}$$

Thus (3.338) is

$$-\frac{e^2c^2\hbar}{2V\sqrt{\omega\omega'}}\frac{1}{mc^2}\delta_{ss'}\cos\Theta, \tag{3.342}$$

where Θ is the angle between $\boldsymbol{\epsilon}^{(\alpha)}$ and $\boldsymbol{\epsilon}^{(\alpha')}$. Apart from the minus sign in front this is exactly the time-independent part of (2.158), which has been shown to be responsible for Thomson scattering.‡ It is amusing to note that the seagull graph (Fig. 2–2c) which is the sole contributor to Thomson scattering in the nonrelativistic theory is replaced in the relativistic theory by the two diagrams of Fig. 3–6 in which the photons are emitted and absorbed one at a time.

The moral we can draw from this calculation is twofold. First, it illustrates (perhaps more vividly than any other example can) that it is absolutely necessary to take into account transitions involving *negative-energy* states if we are to obtain the correct *nonrelativistic* results. It is truly remarkable that only by invoking the concept of a negative-energy state (or a positron state), which is completely foreign to nonrelativistic quantum mechanics, can we arrive at the correct Thomson amplitude. Second, comparing (3.336) with (3.338) and noting that because of energy conservation, the energy denominators in (3.338) can be written as

$$\begin{aligned} E' + |E''| - \hbar\omega &= -(E'' - E + \hbar\omega),\\ E' + |E''| + \hbar\omega' &= -(E'' - E - \hbar\omega), \end{aligned} \tag{3.343}$$

we observe that the negative of (3.338) is formally identical with (3.336) as we replace $r'' = 3,4$ with $r'' = 1,2$ despite the reversal in the time orderings of the matrix elements.§ This kind of observation provides a natural justification of R.P. Feynman's point of view according to which a negative-energy electron going " *backward* in time " is to be regarded as a positron going "*forward* in time." We shall say more about this in the next chapter.

We might add that we can evaluate expressions (3.336) and (3.338) without making the approximation $\hbar\omega \ll mc^2$, $E' = mc^2$. We then obtain the famous

‡If (3.336) is finite (as in the case of the scattering of a high-energy photon), the relative sign of (3.336) and (3.338) is important. Actually the correct amplitude turns out to be the difference of (3.336) and (3.338). This is because for Fig. 3–6 we must take into account the minus sign arising from the fact that the initial $E > 0$ electron is "exchanged" with one of the $E < 0$ electrons in the Dirac sea. The reader need not worry about such subtle sign changes; we shall show in the next chapter that the covariant prescription based on the quantized Dirac theory automatically gives the correct signs for these matrix elements. (See Problem 4–12.)

§In writing (3.336) and (3.338) we have used the usual convention in which the matrix element standing to the right represents the perturbation acting earlier.

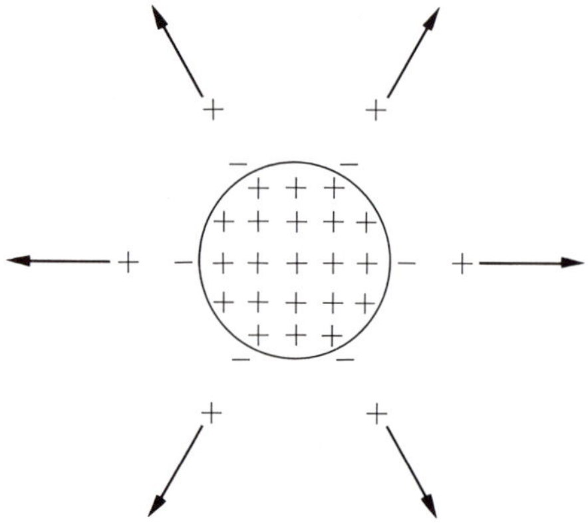

Fig. 3-7. Pictorial representation of vacuum polarization.

formula for Compton scattering

$$\left(\frac{d\sigma}{d\Omega}\right) = \frac{1}{4}r_0^2\left(\frac{\omega'}{\omega}\right)^2\left(\frac{\omega}{\omega'} + \frac{\omega'}{\omega} - 2 + 4\cos^2\Theta\right) \tag{3.344}$$

derived by O. Klein and Y. Nishina in 1929.‡

Virtual electron-positron pairs. As another example illustrating the importance of electron-positron pairs we mention a phenomenon known as *vacuum polarization*. According to the hole theory, the completely filled sea of negative-energy electrons has no observable effects; in particular, what we normally call the vacuum is homogeneous and has no preferred direction. Let us, however, consider a nucleus of charge $Q = Z|e|$ placed in the Dirac sea. There is now a departure from complete homogeneity because the charge distribution of the negative-energy electrons is different from that of the free-field case. In terms of the electron-positron language this is due to the fact that a virtual electron-positron pair created in the Coulomb field behaves in such a way that the electron tends to be attracted to the nucleus while the positron tends to escape from the nucleus. This is pictorially represented in Fig. 3-7. Note that the "induced" negative charges in the vicinity of the nucleus are compensated for by positive charges that "escape to infinity." As a result, the net charge observed at large but finite distances is smaller than the bare charge of the nucleus. In fact what is usually called the observed charge of the nucleus is the original bare charge of the nucleus partially canceled by the charges of the virtual electrons surrounding the nucleus. This situation is rather analogous to that of a charge placed in a dielectric material; the effective charge in a polarized medium is the original charge divided by ϵ, where ϵ is the dielectric constant. In other words, because of virtual electron-positron pairs the vacuum behaves like a polarizable medium.

‡Formula (3.344) can be obtained more readily using covariant perturbation theory (cf. Section 4-4).

On the other hand, at very close distances to the nucleus the "bare charge" itself may be explored; the electron in a hydrogen-like atom should feel at very short distances a stronger attraction than the attraction determined by the Coulomb potential due to the usual observed charge. Since the s state electrons have a greater probability of penetrating the nucleus, we expect that the energy levels of the s states should be displaced to lower levels. Although the argument presented here is rather qualitative, the $2s\frac{1}{2} - 2p\frac{1}{2}$ splitting of the hydrogen atom due to this vacuum polarization effect turns out to be calculable. In 1935, E. A. Uehling predicted that the $2s\frac{1}{2}$-state should lie *lower* by 27 Mc than the $2p\frac{1}{2}$-state.‡ The experimentally observed Lamb shift, as we have seen in Section 2–8, has the opposite sign and is about 40 times larger in magnitude. Although the major part of the Lamb shift is not due to this Uehling effect, for a precise comparison of the experimental value with the theoretical value of the Lamb shift (measured to an accuracy of 0.2 Mc) it has been proved essential to take vacuum polarization seriously. The effect of vacuum polarization is also observable in π mesic and muonic atoms.

The notion that the negative-energy states are completely filled becomes rather treacherous when applied to a particle subject to an external potential. The results of Section 3–7 show that even the (positive-energy) wave function of the hydrogen atom when expanded in plane waves contains small negative-energy components. At first we may be tempted to simply drop the negative-energy components by saying that these states are completely filled. But this cannot be right because, if we do so, we do not even obtain the correct energy levels; in fact, we may recall that the Darwin term (needed for the $2s\frac{1}{2}$–$2p\frac{1}{2}$ degeneracy) can be qualitatively explained by invoking *Zitterbewegung*, which arises from interference of the positive- and negative-energy plane-wave components of the bound-state wave function.

A crude physical argument for the *Zitterbewegung* of an atomic electron within the framework of the hole theory goes as follows. We note that in the Coulomb field of the nucleus a negative-energy electron can make a virtual transition to a positive-energy state (which is equivalent to saying that the Coulomb field can create a virtual electron-positron pair). Now the fact that the wave function in the hydrogen atom contains negative-energy components implies that the atomic electron in the orbit can fill up the hole in the negative-energy state (which means that the atomic electron can annihilate with the positron of the virtual pair). The escalated electron which is left over can now go around the nucleus as the atomic electron. In short, the atomic electron and one of the $E < 0$ electrons in the Dirac sea are visualized as undergoing "exchange scattering." What is the order of magnitude of the distance over which this effect takes place? From the uncertainty principle we expect that the energy violation by an amount $2mc^2$ involved in the escalation of the negative energy electron is allowed only for a time interval $\Delta t \sim \hbar/2mc^2$. (Note incidentally that this is of the order of the reciprocal of the

‡Using covariant perturbation theory, we shall briefly outline the calculation of the Uehling effect in Chapter 5.

Zitterbewegung frequency.) At the time the original atomic electron fills up the vacated negative-energy state, the escalated electron is at most $c(\Delta t) \sim \hbar/2mc$ away from the original electron. This distance is precisely the order of magnitude of the fluctuation of the electron coordinate due to the *Zitterbewegung*.

Charge-conjugate wave function. It is not entirely obvious from the form of the Dirac equation that the space-time development of an electron state in a given potential A_μ is identical to that of the corresponding positron state in the potential $-A_\mu$. For this reason let us cast the Dirac theory into a form which makes the symmetry between the electron and the positron self-evident. This can be best done using a method originally exploited by H. A. Kramers, E. Majorana, and W. Pauli.

We first ask whether the theory based on the Dirac equation with the sign of eA_μ reversed,

$$\left(\frac{\partial}{\partial x_\mu} + \frac{ie}{\hbar c}A_\mu\right)\gamma_\mu \psi^C + \frac{mc}{\hbar}\psi^C = 0, \tag{3.345}$$

is equivalent to the one based on the original Dirac equation (3.60). We assume as usual that there is a definite prescription that relates ψ^C (called the charge-conjugate wave function) and ψ which are respectively solutions to (3.345) and (3.60). Motivated by our experience with the Klein-Gordon theory (cf. Problem 1–3), we try‡

$$\psi^C = S_C \psi^*, \tag{3.346}$$

where S_C is a 4×4 matrix. We must now show that

$$\left[\left(\frac{\partial}{\partial x_k} + \frac{ie}{\hbar c}A_k\right)\gamma_k + \left(\frac{\partial}{\partial x_4} + \frac{ie}{\hbar c}A_4\right)\gamma_4\right] S_C \psi^* + \frac{mc}{\hbar} S_C \psi^* = 0 \tag{3.347}$$

is as good as the original Dirac equation (3.60). Taking the complex conjugate of (3.347) we have

$$\left[\left(\frac{\partial}{\partial x_k} - \frac{ie}{\hbar}A_k\right)\gamma_k^* + \left(-\frac{\partial}{\partial x_4} + \frac{ie}{\hbar c}A_4\right)\gamma_4^*\right] S_C^* \psi + \frac{mc}{\hbar} S_C^* \psi = 0. \tag{3.348}$$

Multiplying (3.348) by $(S_C^*)^{-1}$ from the left and comparing the result with the original Dirac equation, we see that the equivalence of (3.346) and (3.60) can be established if there exists S_C such that

$$\begin{aligned} (S_C^*)^{-1}\gamma_k^* S_C^* &= \gamma_k, \\ (S_C^*)^{-1}\gamma_4^* S_C^* &= -\gamma_4. \end{aligned} \tag{3.349}$$

In the standard (Dirac-Pauli) representation, γ_2 and γ_4 are purely real while γ_1 and γ_3 are purely imaginary. It is easy to see that in this representation,

$$S_C = \gamma_2 = S_C^* = (S_C^*)^{-1} \tag{3.350}$$

‡Note that ψ^C is to be represented by a single *column* matrix, since ψ^* (complex conjugate) rather than $\psi^\dagger$ (Hermitian conjugate) enters.

will do:‡

$$\gamma_2 \begin{Bmatrix} (-\gamma_1) \\ \gamma_2 \\ (-\gamma_3) \end{Bmatrix} \gamma_2 = \begin{Bmatrix} \gamma_1 \\ \gamma_2 \\ \gamma_3 \end{Bmatrix}, \tag{3.351}$$

$$\gamma_2\gamma_4\gamma_2 = -\gamma_4 .$$

Since we have demonstrated the existence of an S_C that satisfies (3.349), we have proved the equivalence of (3.345) and (3.60).

It is very important to note that $S_C = \gamma_2$ is true only in the standard (Dirac-Pauli) representation. In fact the particular forms of S_C depend on the particular representations we happen to use. For instance, in the Majorana representation in which γ_4 is purely imaginary and γ_k is purely real, ψ^C is simply ψ^* itself, as can readily be seen by complex-conjugating (3.60).§ This situation should be contrasted with the parity case where $S_P = \gamma_4$ (up to a phase factor) holds in any representation.

What does (3.346) mean for some of the familiar wave functions we have obtained in the previous sections? Take, for instance, the first of the $E > 0$ plane-wave solutions (3.114). We have

$$\begin{aligned}
&\sqrt{\frac{mc^2}{EV}}\gamma_2 \left[u^{(1)}(\mathbf{p}) \exp\left(\frac{i\mathbf{p}\cdot\mathbf{x}}{\hbar} - \frac{iEt}{\hbar}\right)\right]^* \\
&= \sqrt{\frac{E+mc^2}{2EV}} \begin{pmatrix} 0 & 0 & 0 & -1 \\ 0 & 0 & 1 & 0 \\ 0 & 1 & 0 & 0 \\ -1 & 0 & 0 & 0 \end{pmatrix} \begin{pmatrix} 1 \\ 0 \\ p_3c/(E+mc^2) \\ (p_1+ip_2)c/(E+mc^2) \end{pmatrix}^* \exp\left(-\frac{i\mathbf{p}\cdot\mathbf{x}}{\hbar} + \frac{iEt}{\hbar}\right) \\
&= \sqrt{\frac{E+mc^2}{2EV}} \begin{pmatrix} -(p_1-ip_2)c/(|E|+mc^2) \\ p_3c/(|E|+mc^2) \\ 0 \\ -1 \end{pmatrix} \exp\left(-\frac{i\mathbf{p}\cdot\mathbf{x}}{\hbar} + \frac{i|E|t}{\hbar}\right) \\
&= -\sqrt{\frac{mc^2}{|E|V}} u^{(4)}(-\mathbf{p}) \exp\left(-\frac{i\mathbf{p}\cdot\mathbf{x}}{\hbar} + \frac{i|E|t}{\hbar}\right).
\end{aligned} \tag{3.352}$$

Note that the eigenvalues of $-i\hbar\nabla$ and $i\hbar(\partial/\partial t)$ are $-\mathbf{p}$ and $-|E|$ respectively. Similarly

$$\sqrt{\frac{mc^2}{EV}}\gamma_2 \left[u^{(2)}(\mathbf{p}) \exp\frac{i\mathbf{p}\cdot\mathbf{x}}{\hbar} - \frac{iEt}{\hbar}\right)\right]^* = \sqrt{\frac{mc^2}{|E|V}} u^{(3)}(-\mathbf{p}) \exp\left(-\frac{i\mathbf{p}\cdot\mathbf{x}}{\hbar} + \frac{i|E|t}{\hbar}\right). \tag{3.353}$$

‡More generally we have $S_C = \eta\gamma_2$, where η is an undetermined phase factor; however, this phase factor can be set to 1 by convention.

§In a more advanced treatment of the subject the relations (3.346) and (3.349) are often written as $\psi^C = C\bar{\psi}^T$, $C^{-1}\gamma_\mu C = -\gamma_\mu^T$, where T stands for "transpose." In the standard representation $C = \gamma_2\gamma_4$ up to a phase factor since $\psi^C = \gamma_2\gamma_4(\psi^\dagger\gamma_4)^T = \gamma_2\gamma_4\gamma_4^T\psi^* = \gamma_2\psi^*$.

Thus the charge-conjugate wave function ψ^C obtained from the positive-energy plane-wave solution ψ by means of (3.346) is the wave function for a negative-energy plane wave whose magnitude of the energy is the same and whose momentum is *opposite*. Moreover, the spin direction (or if $\mathbf{p}$ is not along the z-axis, the expectation value of $\boldsymbol{\Sigma}$) is also *reversed* since the index 4(3) goes with the index 1(2). If we now invoke the hole theory, we see that the charge-conjugate wave function describes the dynamical behavior of the negative-energy state whose absence appears as the $E > 0$ *positron* of the *same* $\mathbf{p}$ and *same* $\langle\boldsymbol{\Sigma}\rangle$ (cf. Table 3–4). Likewise, when ψ represents a negative-energy electron state whose absence appears as the positron state of $\mathbf{p}$ and $\langle\boldsymbol{\Sigma}\rangle$, then ψ^C represents the positive-energy electron state of $\mathbf{p}$ and $\langle\boldsymbol{\Sigma}\rangle$.

As another example, let us compare the probability distribution $\psi^\dagger\psi$ with the corresponding $\psi^{C\dagger}\psi^C$ where for the sake of definiteness, ψ may be taken to be the wave function for the ground state of the hydrogen atom. In general, we have

$$\psi^{C\dagger}\psi^C = (\gamma_2\psi^*)^\dagger(\gamma_2\psi^*) = \psi^{*\dagger}\psi^* = \psi^\dagger\psi. \tag{3.354}$$

For the electron in the hydrogen atom, the $A_0(= -iA_4)$ that appears in (3.60) is $|e|/4\pi r$; ψ^C, according to (3.345), is a solution to the Coulomb energy problem in the negative electrostatic potential $-|e|/4\pi r$. Evidently the energy eigenvalue of ψ^C is the negative of that of ψ because of the complex conjugation that appears in (3.346). Thus the relation (3.354) implies that the negative-energy electron going around the negative electrostatic potential has the same probability distribution as the corresponding positive-energy electron going around the positive electrostatic potential. This means that in an electrostatic potential that appears repulsive to the positive-energy electron (for example, in the Coulomb field of the antiproton), the negative-energy *electron* behaves dynamically as though it were in an *attractive* force field. Invoking now the hole-theoretic interpretation, we see that an antiatom in which a positron is bound to the center by $A_0 = -|e|/4\pi r$ looks like the usual atom in which an electron is bound to the center by $A_0 = |e|/4\pi r$.

We define the charge-conjugation operation such that its application on the electron (positron) state of momentum $\mathbf{p}$ and the spin-expectation value $\hbar\langle\boldsymbol{\Sigma}\rangle/2$ results in the positron (electron) state of momentum $\mathbf{p}$ and the spin-expectation value $\hbar\langle\boldsymbol{\Sigma}\rangle/2$. The equivalence of (3.345) and (3.60) implies that if $\psi(\mathbf{x}, t)$ characterizes the space-time behavior of an $E > 0$ electron state in a potential A_μ, then its charge-conjugate wave function $\psi^C(\mathbf{x}, t)$ characterizes the space-time behavior of the *negative-energy electron state whose absence appears as the charge-conjugate (positron) state* in the potential $-A_\mu$.

When we start computing the expectation values of the various dynamical variables using ψ^C, we obtain results which may appear somewhat confusing at first sight. For example, if we naively evaluate the expectation value of $\mathbf{p}$ with respect to ψ and ψ^C, we obtain the result: $\langle\mathbf{p}\rangle$ is opposite to $\langle\mathbf{p}\rangle_C$, as can be seen directly from (3.352) and (3.353) for the free-particle case, and similarly for $\langle\boldsymbol{\Sigma}\rangle$. But we know that the momentum and the spin direction are *unchanged* under charge conjugation which transforms the electron state of momentum $\mathbf{p}$ and

$\langle \mathbf{\Sigma} \rangle$ into the positron state of momentum $\mathbf{p}$ and $\langle \mathbf{\Sigma} \rangle$ (not $-\mathbf{p}$ and $-\langle \mathbf{\Sigma} \rangle$). This peculiarity is due to the fact that in the unquantized Dirac theory the so-called charge-conjugate wave function ψ^C is *not* the wave function of the charge-conjugate state but rather that of the state (subject to the potential whose sign is opposite to the original one) whose *absence* appears as the charge-conjugate state.

What is even more striking, the space integral of the charge density

$$Q = e \int \bar{\psi} \gamma_4 \psi \, d^3x = e \int \psi^\dagger \psi \, d^3x, \tag{3.355}$$

which is the total charge, cannot possibly change its sign when we replace ψ by its charge-conjugate wave function ψ^C, in sharp contrast to the Klein-Gordon case where the substitution $\phi \rightleftarrows \phi^*$ results in the reversal of the charge-current density (cf. Eqs. 1.55 and 3.127).‡ Actually this is expected because when ψ is a positive-energy wave function, ψ^C is the wave function for an $E < 0$ *negatively* charged particle even though it "behaves dynamically" like a positively charged particle in an external electromagnetic field. For a more satisfactory formulation of charge conjugation, it is essential to quantize the electron field according to Fermi-Dirac statistics.

3–10. QUANTIZATION OF THE DIRAC FIELD

Difficulties of the unquantized Dirac theory. One of the great triumphs of relativistic quantum theory is that it has succeeded in providing a theoretical framework within which we can discuss quantitatively a variety of physical phenomena involving the creation and annihilation of various particles. We learned in the last chapter that the "natural language" used to describe the creation and annihilation of photons is that of quantum field theory. In the previous section we did discuss phenomena such as pair creation and pair annihilation. The language used there, however, is very different from that of quantum field theory; instead of saying that the number of electrons is not conserved in pair production, we have argued that the *number of electrons actually is conserved* and that all that happens is just the escalation of a negative-energy electron. In other words, we have tried to describe phenomena such as pair production without abandoning the single-particle interpretation of the Dirac wave function according to which the space integral of $\psi^\dagger \psi$ is a constant of the motion even in the presence of the electromagnetic interaction. In doing so, however, we were forced to depart very radically from the *single*-particle theory itself; in fact, we had to introduce a sea of an *infinite* number of negative-energy particles.

There are essentially two reasons why the hole-theoretic description works. First, as we have already mentioned, crucial to the success of the hole theory is the assumption that the electron obeys the Pauli exclusion principle. Second,

‡In fact $\psi^\dagger \psi$ cannot have its sign changed under any transformation that preserves its positive-definite form.

although electrons and positrons can be created or annihilated, the basic interaction in electrodynamics is such that the *difference* between the number of (positive-energy) electrons and the number of (positive-energy) positrons,

$$N = N(e^-) - N(e^+), \tag{3.356}$$

is conserved. In the hole-theoretic description what we do is just set

$$\begin{aligned} N(e^-_{E>0}) &= N(e^-), \\ N(e^-_{E<0}) &= -N(e^+) + \text{constant background}, \end{aligned} \tag{3.357}$$

so that the newly defined electron number, given by the *sum*

$$N' = N(e^-_{E>0}) + N(e^-_{E<0}), \tag{3.358}$$

is necessarily conserved whenever (3.356) is conserved.

In the "real world" there are nonelectromagnetic phenomena which do not conserve (3.356). Take, for instance, a beta (plus) decay

$$p \longrightarrow n + e^+ + \nu. \tag{3.359}$$

Although the free proton cannot undergo this disintegration process because of energy conservation, a proton bound in a nucleus can emit a positron and neutrino and turns itself into a neutron. In the hole-theoretic interpretation we may try to attribute the presence of the e^+ in the final state to the absence of a negative-energy electron in the Dirac sea. But where is the electron which used to occupy the now vacated negative-energy state? It is apparent that the probability of finding the electron is no longer conserved.‡

Second quantization. Our beta-decay example reveals that it is actually much more sensible to construct a formalism in which we allow electrons and positrons to be destroyed or created more freely. Guided by the success of the quantum theory of radiation, we are tempted to follow, as much as possible, the quantization procedure we used in the photon case. We shall first construct a "classical" theory of the Dirac field using the standard Lagrangian formalism of Chapter 1 and then quantize the dynamical excitations of the Dirac field by replacing the Fourier coefficients by creation and annihilation operators. At this stage it is not completely clear whether this method is a legitimate one. As we emphasized in Section 2–3, the classical field theory is a limit of the quantum field theory where the occupation number goes to infinity, but we know that the occupation number of a particular electron state is *at most one*. However, let us go ahead with the Lagrangian formulation of the classical Dirac field.§

---

‡We might argue that in this β^+ process a negative-energy electron gives up its charge to the proton and gets escalated to a (positive-energy) neutrino state. But note that the electron and the neutrino are different particles which must be described by different wave equations.

§The reader who is unhappy with our procedure may study an alternative, more axiomatic, approach based on J. Schwinger's action principle, discussed, for instance, in Chapter 1 of Jauch and Rohrlich (1955). In Schwinger's formalism the field variables are treated as operators from the very beginning.

The basic free-field Lagrangian density from which the field equation may be derived is taken to be

$$\begin{aligned}\mathscr{L} &= -c\hbar\bar{\psi}\gamma_\mu(\partial/\partial x_\mu)\psi - mc^2\bar{\psi}\psi \\ &= -c\hbar\bar{\psi}_\alpha(\gamma_\mu)_{\alpha\beta}(\partial/\partial x_\mu)\psi_\beta - mc^2\,\delta_{\alpha\beta}\bar{\psi}_\alpha\psi_\beta,\end{aligned} \tag{3.360}$$

which is a Lorentz invariant scalar density. In the Lagrangian formulation each of the four components of ψ and $\bar{\psi}$ is to be regarded as an independent field variable. Varying $\bar{\psi}_\alpha$ [which actually stands for $(\psi^\dagger)_\gamma(\gamma_4)_{\gamma\alpha}$] we obtain four Euler-Lagrange equations of the form $\partial\mathscr{L}/\partial\bar{\psi}_\alpha = 0$ which can be summarized as the single Dirac equation (3.31). To obtain the field equation for $\bar{\psi}$, we first make the replacement

$$-c\hbar\bar{\psi}_\alpha(\gamma_\mu)_{\alpha\beta}\frac{\partial}{\partial x_\mu}\psi_\beta \longrightarrow c\hbar\left(\frac{\partial}{\partial x_\mu}\bar{\psi}\right)_\alpha(\gamma_\mu)_{\alpha\beta}\psi_\beta, \tag{3.361}$$

which is justified since the difference is just a four-divergence. Varying ψ_β we then get the adjoint equation (3.46). The "canonical momentum" π conjugate to ψ is‡

$$\pi_\beta = \frac{\partial\mathscr{L}}{\partial(\partial\psi_\beta/\partial t)} = i\hbar\,\bar{\psi}_\alpha(\gamma_4)_{\alpha\beta} = i\hbar\psi_\beta^\dagger. \tag{3.362}$$

The Hamiltonian density is then obtainable by the standard prescription (1.4):

$$\begin{aligned}\mathscr{H} &= c\pi_\beta\frac{\partial\psi_\beta}{\partial x_0} - \mathscr{L} \\ &= c\hbar\left(i\psi^\dagger\frac{\partial\psi}{\partial x_0} - i\bar{\psi}\gamma_4\frac{\partial\psi}{\partial x_0} + \bar{\psi}\gamma_k\frac{\partial}{\partial x_k}\psi\right) + mc^2\bar{\psi}\psi \\ &= \psi^\dagger(-i\hbar c\boldsymbol{\alpha}\cdot\boldsymbol{\nabla} + \beta mc^2)\psi.\end{aligned} \tag{3.363}$$

Thus the total Hamiltonian of the free Dirac field is

$$H = \int \psi^\dagger(-i\hbar c\boldsymbol{\alpha}\cdot\boldsymbol{\nabla} + \beta mc^2)\psi\, d^3x. \tag{3.364}$$

Since the plane-wave solutions (3.114) and (3.115) taken at $t = 0$ form a complete orthonormal set, an arbitrary four-component field at $t = 0$ can be expanded in free-particle plane waves. The Dirac field ψ becomes a quantized field if we replace the Fourier coefficients in the plane-wave expansion by operators of the type considered at the end of Section 2–2. We have

$$\psi(\mathbf{x}, t) = \frac{1}{\sqrt{V}}\sum_{\mathbf{p}}\sum_{r=1}^{4}\sqrt{\frac{mc^2}{|E|}}\,b_{\mathbf{p}}^{(r)}(t)u^{(r)}(\mathbf{p})\,e^{i\mathbf{p}\cdot x/\hbar}, \tag{3.365}$$

where ψ is now an operator assumed to act on state vectors in occupation number space. We interpret $b_{\mathbf{p}}^{(r)}$ and $b_{\mathbf{p}}^{(r)\dagger}$ as respectively the annihilation and the creation operators for state $(\mathbf{p}, r)$. A single electron state characterized by $(\mathbf{p}, r)$ is represented by $b_{\mathbf{p}}^{(r)\dagger}(0)|0\rangle$. As we have already seen in Chapter 2, the Pauli exclusion prin-

‡As we have written the Lagrangian density, the canonical momentum conjugate to $\bar{\psi}$ vanishes.

ciple is guaranteed if we use the Jordan-Wigner anticommutation relations (cf. Eq. 2.49)

$$\begin{aligned} \{b_{\mathbf{p}}^{(r)}, b_{\mathbf{p}'}^{(r')\dagger}\} &= \delta_{rr'}\delta_{\mathbf{pp}'}, \\ \{b_{\mathbf{p}}^{(r)}, b_{\mathbf{p}'}^{(r')}\} &= 0, \\ \{b_{\mathbf{p}}^{(r)\dagger}, b_{\mathbf{p}'}^{(r')\dagger}\} &= 0, \end{aligned} \tag{3.366}$$

from which it follows that the eigenvalue of the number operator defined by

$$N_{\mathbf{p}}^{(r)} = b_{\mathbf{p}}^{(r)\dagger} b_{\mathbf{p}}^{(r)} \tag{3.367}$$

is zero or one.

We assume that the Hamiltonian operator of the quantized Dirac field has the same form as the classical expression (3.363). We then have‡

$$\begin{aligned} H &= \frac{1}{V}\int \sum_{\mathbf{p}}\sum_{\mathbf{p}'}\sum_{r=1}^{4}\sum_{r'=1}^{4}\left(\sqrt{\frac{mc^2}{|E|}}\, b_{\mathbf{p}}^{(r)\dagger}u^{(r)\dagger}(\mathbf{p})\, e^{-i\mathbf{p}\cdot\mathbf{x}/\hbar}\right)(-i\hbar c\boldsymbol{\alpha}\cdot\nabla + \beta mc^2) \\ &\quad\times\left(\sqrt{\frac{mc^2}{|E'|}}\, b_{\mathbf{p}'}^{(r')}u^{(r')}(\mathbf{p}')\, e^{i\mathbf{p}'\cdot\mathbf{x}/\hbar}\right) d^3x \\ &= \sum_{\mathbf{p}}\sum_{\mathbf{p}'}\sum_{r}\sum_{r'}\delta_{\mathbf{pp}'}\frac{mc^2 E'}{\sqrt{|EE'|}}\, b_{\mathbf{p}}^{(r)\dagger}b_{\mathbf{p}'}^{(r')}u^{(r)\dagger}(\mathbf{p})u^{(r')}(\mathbf{p}') \\ &= \sum_{\mathbf{p}}\sum_{r=1,2}|E|\, b_{\mathbf{p}}^{(r)\dagger}b_{\mathbf{p}}^{(r)} - \sum_{\mathbf{p}}\sum_{r=3,4}|E|\, b_{\mathbf{p}}^{(r)\dagger}b_{\mathbf{p}}^{(r)}, \end{aligned} \tag{3.368}$$

because of the orthogonality and normalization relations (3.106) and (3.110), and the energy momentum relation

$$E' = \pm\sqrt{|\mathbf{p}'|^2c^2 + m^2c^4} \qquad \text{for} \quad r' = \begin{cases}1, 2,\\ 3, 4.\end{cases} \tag{3.369}$$

Recall that by our definition, the creation and annihilation operators $b_{\mathbf{p}}^{(r)\dagger}$ and $b_{\mathbf{p}}^{(r)}$ are *time-dependent* operators. Their time dependence can be inferred from the Heisenberg equation of motion,

$$\begin{aligned} \dot{b}_{\mathbf{p}}^{(r)} &= \frac{i}{\hbar}[H, b_{\mathbf{p}}^{(r)}] = \mp\frac{i}{\hbar}b_{\mathbf{p}}^{(r)}|E| \qquad \text{for} \quad r = \begin{cases}1, 2,\\ 3, 4,\end{cases} \\ \dot{b}_{\mathbf{p}}^{(r)\dagger} &= \frac{i}{\hbar}[H, b_{\mathbf{p}}^{(r)\dagger}] = \pm\frac{i}{\hbar}b_{\mathbf{p}}^{(r)}|E| \qquad \text{for} \quad r = \begin{cases}1, 2,\\ 3, 4,\end{cases} \end{aligned} \tag{3.370}$$

where we have taken advantage of the very useful relation

$$[AB, C] = A\{B, C\} - \{A, C\}B. \tag{3.371}$$

Thus we have

$$b_{\mathbf{p}}^{(r)}(t) = b_{\mathbf{p}}^{(r)}(0)\, e^{\mp i|E|t/\hbar} \qquad \text{for} \quad r = \begin{cases}1, 2,\\ 3, 4,\end{cases} \tag{3.372}$$

‡Previously we used the symbol H for the Hamiltonian operator $(-i\hbar c\boldsymbol{\alpha}\cdot\nabla + \beta mc^2)$ acting on the Dirac wave function. In this section, H stands for the total Hamiltonian operator of the free Dirac field acting on state vectors in occupation-number space.

and a similar relation for $b_{\mathbf{p}}^{(r)\dagger}(t)$. The expansion (3.365) now becomes

$$\psi(\mathbf{x}, t) = \frac{1}{V}\sum_{\mathbf{p}}\sqrt{\frac{mc^2}{|E|}}\left(\sum_{r=1,2} b_{\mathbf{p}}^{(r)}(0)\, u^{(r)}(\mathbf{p}) \exp\left[\frac{i\mathbf{p}\cdot\mathbf{x}}{\hbar} - \frac{i|E|t}{\hbar}\right] + \sum_{r=3,4} b_{\mathbf{p}}^{(r)}(0)\, u^{(r)}(\mathbf{p}) \exp\left[\frac{i\mathbf{p}\cdot\mathbf{x}}{\hbar} + \frac{i|E|t}{\hbar}\right]\right). \tag{3.373}$$

The *quantized* free field is now seen to satisfy the same field equation (viz. the Dirac equation) as the one derived from the *classical* variational principle. Note that this is not *a priori* self-evident; recall, in particular, that b and $b^\dagger$ can no longer be written as linear combinations of P and Q satisfying $[Q, P] = i\hbar$. What is even more striking, the form of the field equation obtainable from the Heisenberg equation of motion is the same whether the creation and annihilation operators satisfy anticommutation relations or commutation relations, as the reader may readily verify.‡ Mathematically this remarkable feature is a consequence of the fact that the commutator $[AB, C]$ that appears in (3.371) can also be written as

$$[AB, C] = A[B, C] + [A, C]B. \tag{3.374}$$

From the fact that we can get the same field equation whether the electron satisfies Fermi-Dirac statistics or Bose-Einstein statistics, we may be tempted to infer that quantum field theory does not "know" which statistics the electron is supposed to satisfy. This inference, however, is *not* correct, as we shall see shortly.

At this point it turns out to be more convenient to redefine the b and $b^\dagger$ so that they now become *time-independent* operators. We do this because we would like to exhibit the time dependence of the field operator ψ more explicitly. We set

$$b_{\mathbf{p}}^{(r)(\text{new})} = b_{\mathbf{p}}^{(r)(\text{old})}(0),$$
$$b_{\mathbf{p}}^{(r)(\text{old})}(t) = b_{\mathbf{p}}^{(r)(\text{new})}\, e^{\mp i|E|t/\hbar} \qquad \text{for} \quad r = \begin{cases} 1, 2, \\ 3, 4. \end{cases} \tag{3.375}$$

The form of the anticommutation relations is unchanged under this replacement.

In the previous sections the symbol ψ stood for a single-particle wave function while in the present section the same symbol ψ is used to denote the quantized Dirac field which is an operator that can act on state vectors in occupation-number space. To distinguish between the two possibilities the wave function ψ is often called a *c*-number field; the field operator ψ, a *q*-number field. We naturally ask: what is the connection between the *c*-number ψ and the *q*-number ψ? For a single-particle plane-wave state the desired connection is easily established:

$$\psi^{(c\text{-number})} = \langle 0 | \psi^{(q\text{-number})} | b_{\mathbf{p}}^{(r)\dagger}\Phi_0\rangle, \tag{3.376}$$

where $\psi^{(q\text{-number})}$ is given by (3.373) with $b_{\mathbf{p}}^{(r)}$ replacing $b_{\mathbf{p}}^{(r)}(0)$. To see this, just note

$$\langle 0 | b_{\mathbf{p}'}^{(r')} b_{\mathbf{p}}^{(r)\dagger} | 0\rangle = \delta_{rr'}\delta_{\mathbf{pp}'}. \tag{3.377}$$

‡Note that (3.368) would still be valid even if the electron satisfied Bose-Einstein statistics.

Therefore, for $r = 1, 2$, we have

$$\langle 0 | \psi^{(q\text{-number})} | b_{\mathbf{p}}^{(r)\dagger}\Phi_0\rangle = \sqrt{\frac{mc^2}{EV}}\,u^{(r)}(\mathbf{p}) \exp\left(i\frac{\mathbf{p}\cdot\mathbf{x}}{\hbar} - \frac{iEt}{\hbar}\right), \tag{3.378}$$

which is indeed the wave function for a positive-energy plane wave characterized by $(\mathbf{p}, r)$. The transition from $\psi^{(c\text{-number})}$ to $\psi^{(q\text{-number})}$ is sometimes called *second quantization*.‡ Although in the above example we considered just a single-particle state, it is important to keep in mind that, in general, $\psi^{(q\text{-number})}$ can actually operate on the state vector for an *assembly* of electrons (and positrons). The Dirac equation in the quantized theory should be regarded as a differential equation that determines the dynamical behavior of the entire aggregate of electrons (and positrons).

Let us now go back to the expression for the Hamiltonian operator (3.368), which is unchanged under the replacement (3.375). This expression for the total Hamiltonian makes good sense since we showed in Chapter 2 that $b_{\mathbf{p}}^{(r)\dagger}\,b_{\mathbf{p}}^{(r)}$ (whose eigenvalue is zero or one) is to be interpreted as the number operator. The energy of a one-electron state characterized by $(\mathbf{p}, r)$ is just $|E|$ or $-|E|$ depending on whether $r = 1, 2$ or $r = 3, 4$. The total energy of an ensemble of $E > 0$ electrons is just the sum of the energies of the individual electrons.

We can also compute the total charge operator. Following steps analogous to (1.51) through (1.54), we can readily show that $i\bar{\psi}\gamma_\mu\psi$ satisfies the continuity equation. Assuming that the charge density operator is given by $e\psi^\dagger\psi$ even in the q-number theory, we get for the total charge operator,

$$\begin{aligned} Q &= e\int \psi^\dagger\psi\, d^3x \\ &= e\sum_{\mathbf{p}}\sum_{\mathbf{p}'}\sum_{r}\sum_{r'} (mc^2/\sqrt{|EE'|})\,\delta_{\mathbf{pp}'} b_{\mathbf{p}}^{(r)\dagger} b_{\mathbf{p}'}^{(r')} u^{(r)\dagger}(\mathbf{p})u^{(r')}(\mathbf{p}') \\ &= e\sum_{\mathbf{p}}\sum_{r=1}^{4} b_{\mathbf{p}}^{(r)\dagger}b_{\mathbf{p}}^{(r)}. \end{aligned} \tag{3.379}$$

This is again expected.§ As for the total momentum of the Dirac field we may start with (cf. Problem 1–1)

$$P_k = -i\int \mathscr{T}_{4k}\, d^3x, \tag{3.380}$$

where

$$\mathscr{T}_{4k} = -\frac{\partial\mathscr{L}}{\partial(\partial\psi_\alpha/\partial x_4)}\frac{\partial\psi_\alpha}{\partial x_k} - \frac{\partial\bar{\psi}_\alpha}{\partial x_k}\frac{\partial\mathscr{L}}{\partial(\partial\bar{\psi}_\alpha/\partial x_4)}. \tag{3.381}$$

We then obtain

$$\mathbf{P} = -i\hbar\int \psi^\dagger\nabla\psi\, d^3x = \sum_{\mathbf{p}}\sum_{r} \mathbf{p}\, b_{\mathbf{p}}^{(r)\dagger}b_{\mathbf{p}}^{(r)}. \tag{3.382}$$

---

‡By *first quantization* one simply means $\mathbf{p}^{(\text{classical})} \longrightarrow -i\hbar\nabla$, etc., for the dynamical variables of a single particle.

§Recall that the negative-energy electron has electric charge $e = -|e|$ even though it "behaves dynamically" like a positively charged particle.

Positron operators and positron spinors. Although (3.368), (3.379), and (3.382) are satisfactory from the hole-theoretic point of view, the persistent appearance of negative energies seems somewhat distasteful. It is much better to have a formalism in which the free-particle energy is always positive while the total charge is positive or negative (depending on whether there are more positrons or electrons). With this aim in mind let us define $b_{\mathbf{p}}^{(s)}$, $d_{\mathbf{p}}^{(s)}$, $u^{(s)}(\mathbf{p})$, and $v^{(s)}(\mathbf{p})$ with $s = 1, 2$ such that

$$b_{\mathbf{p}}^{(s)} = b_{\mathbf{p}}^{(r)}, \qquad (r = s) \quad \text{for } r = 1, 2,$$

$$d_{\mathbf{p}}^{(s)\dagger} = \mp b_{-\mathbf{p}}^{(r)} \qquad \begin{cases} s = 1 & \text{for } r = 4, \\ s = 2 & \text{for } r = 3; \end{cases} \tag{3.383}$$

and

$$u^{(s)}(\mathbf{p}) = u^{(r)}(\mathbf{p}), \qquad (r = s) \quad \text{for } r = 1, 2,$$

$$v^{(s)}(\mathbf{p}) = \mp u^{(r)}(-\mathbf{p}) \qquad \begin{cases} s = 1 & \text{for } r = 4, \\ s = 2 & \text{for } r = 3. \end{cases} \tag{3.384}$$

The basic motivation for all this stems from the fact that the annihilation of a negative-energy electron of momentum $-\mathbf{p}$ and spin-*down* appears as the creation of a positron with momentum $+\mathbf{p}$ and spin-*up*. We later see that $d^\dagger$ (d) can indeed be interpreted as the creation (annihilation) operator of a positron. We have reshuffled the order of the r- and s-indices and inserted minus signs in such a way that

$$\begin{aligned} S_C u^{(s)*}(\mathbf{p}) &= \gamma_2 u^{(s)*}(\mathbf{p}) = v^{(s)}(\mathbf{p}), \\ S_C v^{(s)*}(\mathbf{p}) &= \gamma_2 v^{(s)*}(\mathbf{p}) = u^{(s)}(\mathbf{p}), \end{aligned} \tag{3.385}$$

with the *same* s $(= 1, 2)$ (cf. 3.352). Note that the d and $d^\dagger$ satisfy the same anticommutation relations as the b and the $b^\dagger$:

$$\begin{aligned} \{d_{\mathbf{p}}^{(s)}, d_{\mathbf{p}'}^{(s')\dagger}\} &= \delta_{ss'}\, \delta_{\mathbf{pp}'}, \\ \{d_{\mathbf{p}}^{(s)}, d_{\mathbf{p}'}^{(s')}\} &= \{d_{\mathbf{p}}^{(s)\dagger}, d_{\mathbf{p}'}^{(s')\dagger}\} = 0. \end{aligned} \tag{3.386}$$

We also have

$$\begin{aligned} \{b_{\mathbf{p}}^{(s)}, d_{\mathbf{p}'}^{(s')}\} = \{b_{\mathbf{p}}^{(s)}, d_{\mathbf{p}'}^{(s')\dagger}\} &= \{b_{\mathbf{p}}^{(s)\dagger}, d_{\mathbf{p}'}^{(s')\dagger}\} \\ &= \{b_{\mathbf{p}}^{(s)\dagger}, d_{\mathbf{p}'}^{(s')}\} = 0. \end{aligned} \tag{3.387}$$

For later purposes it turns out to be useful to collect formulas for u and v. First, (3.105) now becomes

$$\begin{aligned} (i\gamma \cdot p + mc)u^{(s)}(\mathbf{p}) &= 0, \\ (-i\gamma \cdot p + mc)v^{(s)}(\mathbf{p}) &= 0, \end{aligned} \tag{3.388}$$

where $p = (\mathbf{p}, iE/c)$ with E *positive*, even in the equation for $v^{(s)}(\mathbf{p})$. The orthogonality and normalization relations (3.106) and (3.110) become

$$\begin{gathered} u^{(s)\dagger}(\mathbf{p})u^{(s)}(\mathbf{p}) = \delta_{ss'}(E/mc^2), \qquad v^{(s')\dagger}(\mathbf{p})v^{(s)}(\mathbf{p}) = \delta_{ss'}(E/mc^2), \\ v^{(s')\dagger}(-\mathbf{p})u^{(s)}(\mathbf{p}) = u^{(s')\dagger}(-\mathbf{p})v^{(s)}(\mathbf{p}) = 0, \end{gathered} \tag{3.389}$$

where E is again understood to be positive. In terms of $\bar{u}^{(s)}(\mathbf{p})$ and $\bar{v}^{(s)}(\mathbf{p})$ we obtain from the Hermitian conjugate of (3.388),

$$\begin{aligned}\bar{u}^{(s)}(\mathbf{p})(i\gamma\cdot p + mc) &= 0,\\ \bar{v}^{(s)}(\mathbf{p})(-i\gamma\cdot p + mc) &= 0,\end{aligned} \tag{3.390}$$

where we have used $(\pm i\gamma\cdot p + mc)^{\dagger}\gamma_4 = \gamma_4(\pm i\gamma\cdot p + mc)$. It is also straightforward to prove with the aid of (3.389)

$$\begin{aligned}\bar{u}^{(s')}(\mathbf{p})u^{(s)}(\mathbf{p}) = \delta_{ss'}, \qquad \bar{v}^{(s')}(\mathbf{p})v^{(s)}(\mathbf{p}) = -\delta_{ss'},\\ \bar{u}^{(s')}(\mathbf{p})v^{(s)}(\mathbf{p}) = \bar{v}^{(s')}(\mathbf{p})u^{(s)}(\mathbf{p}) = 0,\end{aligned} \tag{3.391}$$

for example, by multiplying the first expression of (3.388) from the left by $\bar{u}^{(s')}(\mathbf{p})\gamma_4$, multiplying the first expression of (3.390) (with s replaced by s') from the right by $\gamma_4 u^{(s)}(\mathbf{p})$, and adding the two. The expansions for ψ and $\bar{\psi}$ now take the form

$$\begin{aligned}\psi(\mathbf{x}, t) &= \frac{1}{\sqrt{V}}\sum_{\mathbf{p}}\sum_{s=1,2}\sqrt{\frac{mc^2}{E}}\left(b_{\mathbf{p}}^{(s)}u^{(s)}(\mathbf{p})\exp\left[\frac{i\mathbf{p}\cdot\mathbf{x}}{\hbar} - \frac{iEt}{\hbar}\right]\right.\\ &\qquad\left. + d_{\mathbf{p}}^{(s)\dagger}v^{(s)}(\mathbf{p})\exp\left[\frac{-i\mathbf{p}\cdot\mathbf{x}}{\hbar} + \frac{iEt}{\hbar}\right]\right),\\ \bar{\psi}(\mathbf{x}, t) &= \frac{1}{\sqrt{V}}\sum_{\mathbf{p}}\sum_{s=1,2}\sqrt{\frac{mc^2}{E}}\left(d_{\mathbf{p}}^{(s)}\bar{v}^{(s)}(\mathbf{p})\exp\left[\frac{i\mathbf{p}\cdot\mathbf{x}}{\hbar} - \frac{iEt}{\hbar}\right]\right.\\ &\qquad\left. + b_{\mathbf{p}}^{(s)\dagger}\bar{u}^{(s)}(\mathbf{p})\exp\left[\frac{-i\mathbf{p}\cdot\mathbf{x}}{\hbar} + \frac{iEt}{\hbar}\right]\right),\end{aligned} \tag{3.392}$$

where from now on it is understood that E *shall always stand for the positive square root* $\sqrt{|\mathbf{p}^2|c^2 + m^2c^4}$. Note that in obtaining (3.392) from (3.373) we used the fact that the sum over $\mathbf{p}$ runs over *all* directions ($-\mathbf{p}$ as well as $\mathbf{p}$).

Going back to the Hamiltonian operator and the total charge operator, we can now rewrite (3.368) and (3.379) as follows:

$$\begin{aligned}H &= \sum_{\mathbf{p}}\sum_{s} E(b_{\mathbf{p}}^{(s)\dagger}b_{\mathbf{p}}^{(s)} - d_{-\mathbf{p}}^{(s)}d_{-\mathbf{p}}^{(s)\dagger})\\ &= \sum_{\mathbf{p}}\sum_{s} E(b_{\mathbf{p}}^{(s)\dagger}b_{\mathbf{p}}^{(s)} + d_{\mathbf{p}}^{(s)\dagger}d_{\mathbf{p}}^{(s)} - 1),\end{aligned} \tag{3.393}$$

and

$$\begin{aligned}Q &= e\sum_{\mathbf{p}}\sum_{s}(b_{\mathbf{p}}^{(s)\dagger}b_{\mathbf{p}}^{(s)} + d_{-\mathbf{p}}^{(s)}d_{-\mathbf{p}}^{(s)\dagger})\\ &= e\sum_{\mathbf{p}}\sum_{s}(b_{\mathbf{p}}^{(s)\dagger}b_{\mathbf{p}}^{(s)} - d_{\mathbf{p}}^{(s)\dagger}d_{\mathbf{p}}^{(s)} + 1).\end{aligned} \tag{3.394}$$

We recall that the anticommutation relations for the d and $d^{\dagger}$ are completely identical in from with those of the b and $b^{\dagger}$. This means, among other things, that the eigenvalue of $d_{\mathbf{p}}^{(s)\dagger}d_{\mathbf{p}}^{(s)}$ is one or zero. From (3.393) and (3.394) we see that if we interpret $d_{\mathbf{p}}^{(s)\dagger}d_{\mathbf{p}}^{(s)}$ as the number operator for a positive-energy positron, then we have the following satisfactory result: *A state with an extra positron has the expected extra positive energy and positive* $(-e = |e|)$ *charge.* Thus we take

$$N_{\mathbf{p}}^{(e^-, s)} = b_{\mathbf{p}}^{(s)\dagger}b_{\mathbf{p}}^{(s)}, \qquad N_{\mathbf{p}}^{(e^+, s)} = d_{\mathbf{p}}^{(s)\dagger}d_{\mathbf{p}}^{(s)} \tag{3.395}$$

to be the occupation-number operators for the electron and the positron state characterized by $(\mathbf{p}, s)$. In hole-theory language, where r runs from 1 to 4, the application of an $E < 0$ electron creation operator $b_{\mathbf{p}}^{(3,4)\dagger}$ to the "physical vacuum" must result in a null state since all the negative-energy states are already filled. In our new notation this means that the application of $d_{\mathbf{p}}^{(1,2)}$ to the vacuum must result in a null state. This is reasonable if $d_{\mathbf{p}}^{(1,2)}$ is to be understood as the annihilation operator for a positron with $(\mathbf{p}, s)$; in the vacuum there is no positron to be annihilated. So for the vacuum state we require

$$b_{\mathbf{p}}^{(s)}|0\rangle = 0, \qquad d_{\mathbf{p}}^{(s)}|0\rangle = 0. \tag{3.396}$$

From the anticommutation relation between $N_{\mathbf{p}}^{(e^+,s)}$, $d_{\mathbf{p}}^{(s)}$, and $d_{\mathbf{p}}^{(s)\dagger}$, it follows that the eigenvalue of $N_{\mathbf{p}}^{(e^+,s)}$ is zero for the vacuum state and one for a single positron state $d_{\mathbf{p}}^{(s)\dagger}|0\rangle$ (cf. Eqs. 2.55 through 2.57). In other words, $d_{\mathbf{p}}^{(s)\dagger}$ is the creation operator for a positron.

The expressions (3.393) and (3.394) are still not completely satisfactory. It is true that according to (3.393) the vacuum is the state with the lowest possible energy; however, if we apply H to the vacuum state, we get $-\Sigma_{\mathbf{p}}\Sigma_s E$ which is $-\infty$. Physically this means that the infinite negative energy of the Dirac sea has not yet been properly subtracted. We can redefine the energy scale so that H applied to $|0\rangle$ gives a zero eigenvalue. We then have

$$H = \sum_{\mathbf{p}} \sum_{s} E(N_{\mathbf{p}}^{(e^-,s)} + N_{\mathbf{p}}^{(e^+,s)}) \tag{3.397}$$

whose eigenvalue is necessarily positive semidefinite. Similarly subtracting the infinite negative charge of the Dirac sea, we obtain

$$\begin{aligned} Q &= e \sum_{\mathbf{p}} \sum_{s} (N_{\mathbf{p}}^{(e^-,s)} - N_{\mathbf{p}}^{(e^+,s)}) \\ &= -|e| \sum_{\mathbf{p}} \sum_{s} (N_{\mathbf{p}}^{(e^-,s)} - N_{\mathbf{p}}^{(e^+,s)}). \end{aligned} \tag{3.398}$$

This subtraction procedure amounts to starting with the charge density‡

$$\rho = e\psi^\dagger\psi - e\langle\psi^\dagger\psi\rangle_0. \tag{3.399}$$

Note that, unlike the total charge in the c-number theory (3.355) which is necessarily negative, the eigenvalue of (3.398) can be negative or positive. Now at last we can forget completely about negative energy electrons, the picturesque Dirac sea, the negatively charged particles with $E < 0$ that behave like positively charged particles, the absence of $\mathbf{p}$ appearing as the presence of $-\mathbf{p}$, and all that. From now on we can work with electrons and positrons of *positive energies only*.

We have seen that once we define both the energy and the charge of the vacuum state to be zero, then the total energy of the Dirac field is necessarily positive

---

‡The expression (3.399) can be shown to be equal to

$$(e/2)(\psi^\dagger\psi - \psi^T\psi^{\dagger T})$$

which is not zero in the q-number theory. This method of eliminating the undesirable vacuum expectation value is due to W. Heisenberg.

semidefinite, while the total charge can be negative or positive. We emphasize that in obtaining this satisfactory result the *anticommutation* relations for the creation and annihilation operators have played a crucial role. Had we used commutation relations instead, we would have ended up with an expression for the Hamiltonian operator whose eigenvalue has no lower bound [since $b_{\mathbf{p}}^{(r)\dagger} b_{\mathbf{p}}^{(r)}$ with $r = 3, 4$ in (3.368) can take on an arbitrarily large positive number]. Thus the Dirac field must be quantized according to Fermi-Dirac statistics if we demand that there be a state with the lowest energy. We have actually illustrated a special case of a very general theorem which states that half-integral spin fields must be quantized according to Fermi-Dirac statistics while integer spin fields must be quantized according to Bose-Einstein statistics. This spin-statistics theorem, first proved by W. Pauli in 1940, is one of the crowning achievements of relativistic quantum theory.

Let us now look at the total momentum operator (3.382). We get

$$\begin{aligned}\mathbf{P} &= \sum_{\mathbf{p}} \sum_{s=1,2} \mathbf{p}(b_{\mathbf{p}}^{(s)\dagger} b_{\mathbf{p}}^{(s)} + d_{-\mathbf{p}}^{(s)} d_{-\mathbf{p}}^{(s)\dagger}) \\ &= \sum_{\mathbf{p}} \sum_{s=1,2} \mathbf{p} b_{\mathbf{p}}^{(s)\dagger} b_{\mathbf{p}}^{(s)} + \sum_{\mathbf{p}} \sum_{s=1,2} (-\mathbf{p})(-d_{\mathbf{p}}^{(s)\dagger} d_{\mathbf{p}}^{(s)} + 1) \\ &= \sum_{\mathbf{p}} \sum_{s=1,2} \mathbf{p}(N_{\mathbf{p}}^{(e^-,s)} + N_{\mathbf{p}}^{(e^+,s)}) \end{aligned} \tag{3.400}$$

(since $\sum_{\mathbf{p}} \mathbf{p} = 0$). This justifies our earlier assertion that the physical momentum of the positron state $d_{\mathbf{p}}^{(s)\dagger}|0\rangle$ is $\mathbf{p}$, not $-\mathbf{p}$ (cf. Eq. 3.384).

In order to convince ourselves that $b_{\mathbf{p}}^{(s)\dagger}|0\rangle$ and $d_{\mathbf{p}}^{(s)\dagger}|0\rangle$ with the same $(\mathbf{p}, s)$ really are an electron and a positron state with the *same* spin direction, it is instructive to work out the effect of applying the spin operator to these states. Taking the spin density to be $(\hbar/2)\psi^\dagger \mathbf{\Sigma} \psi$ we obtain

$$S_3 = (\hbar/2) \int \psi^\dagger \Sigma_3 \psi \, d^3x \tag{3.401}$$

for the z-component of the spin operator.‡ For an electron,

$$\begin{aligned} S_3 b_{\mathbf{p}}^{(s)\dagger}|0\rangle &= [S_3, b_{\mathbf{p}}^{(s)\dagger}]|0\rangle \\ &= \frac{\hbar}{2} \int \psi^\dagger \Sigma_3 \{\psi, b_{\mathbf{p}}^{(s)\dagger}\} \, d^3x |0\rangle \\ &= \frac{\hbar}{2} \frac{mc^2}{E} u^{(s)\dagger}(\mathbf{p}) \Sigma_3 u^{(s)}(\mathbf{p}) b_{\mathbf{p}}^{(s)\dagger}|0\rangle. \end{aligned} \tag{3.402}$$

‡Within the framework of the Lagrangian formalism the ultimate justification for interpreting $(\hbar/2)\psi^\dagger \mathbf{\Sigma} \psi$ as the spin density rests on the fact that the constancy of

$$\int \psi^\dagger[-i\hbar(\mathbf{x} \times \nabla)_3 + (\hbar/2)\Sigma_3]\psi \, d^3x$$

is guaranteed by the invariance of the Lagrangian density under an infinitesimal rotation around the z-axis [see, for example, Bjorken and Drell (1965), pp. 17–19, 55]. Note that no additive constant is needed because, with S_3 given by (3.401), $S_3|0\rangle = 0$ is automatically satisfied by rotational invariance.

In contrast,

$$\begin{aligned} S_3 d_{\mathbf{p}}^{(s)\dagger}|0\rangle &= [S_3, d_{\mathbf{p}}^{(s)\dagger}]|0\rangle \\ &= -\frac{\hbar}{2}\int \{\psi^\dagger, d_{\mathbf{p}}^{(s)\dagger}\}\Sigma_3\psi\, d^3x|0\rangle \\ &= -\frac{\hbar}{2}\frac{mc^2}{E}v^{(s)\dagger}(\mathbf{p})\Sigma_3 v^{(s)}(\mathbf{p})\, d_{\mathbf{p}}^{(s)\dagger}|0\rangle. \end{aligned} \tag{3.403}$$

If the free-particle spinor corresponding to an electron state has an eigenvalue $\Sigma_3 = +1$, then (3.402) tells us that $S_3 = \hbar/2$, as expected. On the other hand, if we want to describe a positron with spin-*up*, (3.403) demands that we use a free-particle spinor v for which the eigenvalue of Σ_3 is *minus* one. As an example, take the electron at rest with spin-up. The corresponding free particle spinor is $u^{(1)}(0)$ with $\Sigma_3 = +1$. For the positron state at rest with the same spin direction we are supposed to use $v^{(1)}(0) = -u^{(4)}(0)$ which indeed has $\Sigma_3 = -1$. This, of course, is expected on the basis of the hole theory.

Under the charge conjugation operation defined in the previous section, a single electron state $b_{\mathbf{p}}^{(s)\dagger}|0\rangle$ goes into the corresponding positron state $d_{\mathbf{p}}^{(s)\dagger}|0\rangle$ with the same $(\mathbf{p}, s)$. Thus

$$b_{\mathbf{p}}^{(s)} \rightleftarrows d_{\mathbf{p}}^{(s)}, \qquad b_{\mathbf{p}}^{(s)\dagger} \rightleftarrows d_{\mathbf{p}}^{(s)\dagger} \tag{3.404}$$

under charge conjugation. We see that the total charge operator (3.398) indeed changes its sign under charge conjugation in contrast to its c-number analog (3.355), which always has the same sign. The Hamiltonian operator (3.397) does remain invariant under charge conjugation, as it should.

In the previous section we also examined the invariance property of the Dirac equation under $eA_\mu \longrightarrow -eA_\mu$. In the quantized theory the steps (3.345) through (3.351) go through just as before if we replace ψ^* by $\psi^{\dagger T}$ (where the transpose operation brings the free-particle spinor back to the column matrix form without affecting the creation and annihilation operators); ψ^C, given by

$$\psi^C = \gamma_2 \psi^{\dagger T}, \tag{3.405}$$

is now called the charge-conjugate field (rather than the charge-conjugate wave function). Using (3.385), we find that ψ^C in the free-field case is given by

$$\psi^C = \frac{1}{\sqrt{V}}\sum_{\mathbf{p}}\sum_{s}\sqrt{\frac{mc^2}{E}}\left(b_{\mathbf{p}}^{(s)\dagger}v_{\mathbf{p}}^{(s)}\exp\left[\frac{-i\mathbf{p}\cdot\mathbf{x}}{\hbar}+\frac{iEt}{\hbar}\right] + d_{\mathbf{p}}^{(s)}u^{(s)}(\mathbf{p})\exp\left[\frac{i\mathbf{p}\cdot\mathbf{x}}{\hbar}-\frac{iEt}{\hbar}\right]\right). \tag{3.406}$$

Comparing this with (3.392), we see that the transformation

$$\psi \longrightarrow \psi^C \tag{3.407}$$

is achieved if we make the substitution (3.404) which is precisely the charge conjugation operation; ψ^C annihilates positrons and creates electrons just as ψ annihilates electrons and creates positrons.

We shall now briefly mention the anticommutation relations among ψ, $\psi^\dagger$, and $\bar{\psi}$. First, it is evident that

$$\{\psi_\alpha(x), \psi_\beta(x')\} = \{\psi_\alpha^\dagger(x), \psi_\beta^\dagger(x')\} = \{\bar{\psi}_\alpha(x), \bar{\psi}_\beta(x')\} = 0, \tag{3.408}$$

where x now stands for the four-vector $(\mathbf{x}, ict)$. For $\psi(x)$ and $\psi^\dagger(x')$, we have the equal-time anticommutation relation first written by W. Heisenberg and W. Pauli:

$$\{\psi_\alpha(\mathbf{x}, t), \psi_\beta^\dagger(\mathbf{x}', t')\}_{t=t'} = \delta_{\alpha\beta}\delta^{(3)}(\mathbf{x} - \mathbf{x}'), \tag{3.409}$$

the proof of which is left as an exercise (Problem 3–13). This also implies

$$\{\psi_\alpha(\mathbf{x}, t), \bar{\psi}_\beta(\mathbf{x}', t')\}_{t=t'} = (\gamma_4)_{\alpha\beta}\delta^{(3)}(\mathbf{x} - \mathbf{x}'). \tag{3.410}$$

As for the anticommutation relation between ψ and $\bar{\psi}$ at different times, for our purpose it suffices to remark that $\{\psi_\alpha(x), \bar{\psi}_\beta(x')\}$ is a function of the four-vector $x - x'$ such that it vanishes when x and x' are separated by a space-like distance:‡

$$\{\psi_\alpha(x), \bar{\psi}_\beta(x')\} = 0$$

if

$$(x - x')^2 = (\mathbf{x} - \mathbf{x}')^2 - c^2(t - t')^2 > 0. \tag{3.411}$$

Because of the *anticommutation* relation, it is clear that $\psi(x)$ and $\bar{\psi}(x')$ *do not commute* when $x - x'$ is spacelike. This is not disturbing since ψ and $\bar{\psi}$, having no classical analog, are not "measurable" in the same sense as **E** and **B** are measurable. On the other hand, for the charge-current density

$$j_\mu(x) = ie\bar{\psi}\gamma_\mu\psi - ie\langle 0|\bar{\psi}\gamma_\mu\psi|0\rangle, \tag{3.412}$$

which *is* "measurable," we obtain from (3.408) and (3.410),

$$[j_\mu(x), j_\nu(x')] = 0 \qquad \text{if} \quad (x - x')^2 > 0, \tag{3.413}$$

where we have used

$$[AB, CD] = -AC\{D, B\} + A\{C, B\}D - C\{D, A\}B + \{C, A\}DB. \tag{3.414}$$

Thus measurements of charge-current densities performed at two different points separated by a spacelike distance cannot influence each other, in agreement with the causality principle.

Electromagnetic and Yukawa couplings. Let us now talk about the interaction of electrons and positrons with the electromagnetic field. The Hamiltonian density for the basic interaction is taken to be

$$\mathscr{H}_{\text{int}} = -ie\bar{\psi}\gamma_\mu\psi A_\mu, \tag{3.415}$$

where ψ is now the quantized electron field; A_μ can be either classical or quantized. This interaction can be derived from the Lagrangian density

$$\mathscr{L}_{\text{int}} = ie\bar{\psi}\gamma_\mu\psi A_\mu, \tag{3.416}$$

since $\mathscr{H}_{\text{int}}$ is just the negative of $\mathscr{L}_{\text{int}}$ whenever $\mathscr{L}_{\text{int}}$ does not contain time derivatives of field operators. Strictly speaking, $ie\bar{\psi}\gamma_\mu\psi$ should be replaced by (3.412), but in practice the form (3.415) is sufficient since a constant (c-number) interaction cannot cause a transition between different states.

---

‡The explicit form of $\{\psi_\alpha(x), \bar{\psi}_\beta(x')\}$ known as $-iS_{\alpha\beta}(x - x')$ may be found in more advanced textbooks, for example, Mandl (1959), pp. 30–35, pp. 54–55; Schweber (1961), pp. 180–182, pp. 225–227.

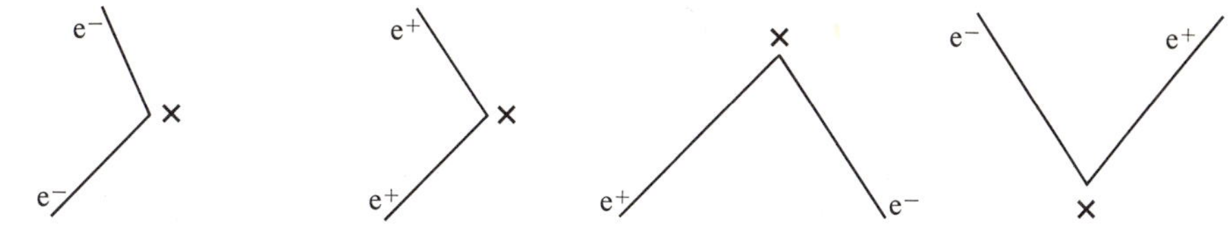

(a) $ie\bar{\psi}^{(-)}\gamma_\mu\psi^{(+)}A_\mu$ (b) $ie\bar{\psi}^{(+)}\gamma_\mu\psi^{(-)}A_\mu$ (c) $ie\bar{\psi}^{(+)}\gamma_\mu\psi^{(+)}A_\mu$ (d) $ie\bar{\psi}^{(-)}\gamma_\mu\psi^{(-)}A_\mu$

Fig. 3–8. Four possibilities realized by the interaction (3.415).

In lowest-order perturbation theory (and also in covariant perturbation based on the interaction representation to be discussed in the next chapter), we may regard ψ in (3.415) as the free field ψ given by (3.392) and look at the effect of $\mathscr{H}_{\text{int}}$ on a given initial state. It is convenient to decompose each of ψ and $\bar{\psi}$ into two parts as follows:

$$\begin{aligned} \psi &= \psi^{(+)} + \psi^{(-)}, \\ \bar{\psi} &= \bar{\psi}^{(+)} + \bar{\psi}^{(-)}, \end{aligned} \tag{3.417}$$

where

$$\begin{aligned} \psi^{(+)} &= \frac{1}{\sqrt{V}} \sum_{\mathbf{p}} \sum_{s} \sqrt{\frac{mc^2}{E}}\, b_{\mathbf{p}}^{(s)} u^{(s)}(\mathbf{p}) \exp\left(\frac{i\mathbf{p}\cdot\mathbf{x}}{\hbar} - \frac{iEt}{\hbar}\right), \\ \psi^{(-)} &= \frac{1}{\sqrt{V}} \sum_{\mathbf{p}} \sum_{s} \sqrt{\frac{mc^2}{E}}\, d_{\mathbf{p}}^{(s)\dagger} v^{(s)}(\mathbf{p}) \exp\left(\frac{-i\mathbf{p}\cdot\mathbf{x}}{\hbar} + \frac{iEt}{\hbar}\right), \\ \bar{\psi}^{(+)} &= \frac{1}{\sqrt{V}} \sum_{\mathbf{p}} \sum_{s} \sqrt{\frac{mc^2}{E}}\, d_{\mathbf{p}}^{(s)} \bar{v}^{(s)}(\mathbf{p}) \exp\left(\frac{i\mathbf{p}\cdot\mathbf{x}}{\hbar} - \frac{iEt}{\hbar}\right), \\ \bar{\psi}^{(-)} &= \frac{1}{\sqrt{V}} \sum_{\mathbf{p}} \sum_{s} \sqrt{\frac{mc^2}{E}}\, b_{\mathbf{p}}^{(s)\dagger} \bar{u}^{(s)}(\mathbf{p}) \exp\left(\frac{-i\mathbf{p}\cdot\mathbf{x}}{\hbar} + \frac{iEt}{\hbar}\right), \end{aligned} \tag{3.418}$$

Note that $\psi^{(+)}$ (which is called the positive *frequency* part of ψ) is linear in the electron annihilation operators; it can therefore annihilate electrons but cannot do anything else. Likewise $\psi^{(-)}$, $\bar{\psi}^{(+)}$, and $\bar{\psi}^{(-)}$ respectively create positrons, annihilate positrons, and create electrons. So we see that the interaction (3.415) can do four different types of things, as shown in Fig. 3–8 where the symbol $\times$ may stand for an interaction with an external classical potential, the emission of a photon, or the absorption of a photon. The question of which of the four possibilities given in Fig. 3–8 is actually realized in a particular physical process depends entirely on what kind of initial and final state are present. For instance, if initially there is only a positron, the action of (3.415) may result in the scattering of the positron, represented by Fig. 3–8(b). On the other hand, if initially there is an electron-positron pair, and if we also know that there is neither an electron nor a positron in the final state, the matrix element of $\mathscr{H}_{\text{int}}$ is that of a pair annihilation process represented by Fig. 3–8(c). Note that in all cases the difference between the number of electrons and the number of positrons (3.356) is conserved. In Chapter 4 we shall work out a number of problems in quantum electrodynamics based on (3.415).

The meson-nucleon interaction proposed by H. Yukawa in 1935 is patterned after (3.415).‡ We have

$$\mathscr{H}_{\text{int}} = G\bar{\psi}\psi\phi, \tag{3.419}$$

where ψ and ϕ are now the nucleon and a neutral spin-zero meson field. Note that ψ can annihilate nucleons and create antinucleons while $\bar{\psi}$ can create nucleons and annihilate antinucleons. The constant, G, characterizes the strength of the coupling of the meson field to the nucleon. Alternatively we may have§

$$\mathscr{H}_{\text{int}} = iG\bar{\psi}\gamma_5\psi\phi. \tag{3.420}$$

Both (3.419) and (3.420) are invariant under proper orthochronous Lorentz transformations, as is evident from Table 3–1. The form of the interaction (3.419) can be made invariant under space inversion if ϕ in the space-inverted system is given by

$$\phi'(x') = \phi(x), \qquad x' = (-\mathbf{x}, ict). \tag{3.421}$$

On the other hand (3.420) can also be made invariant under space inversion if ϕ in the space-inverted system is given by

$$\phi'(x') = -\phi(x), \qquad x' = (-\mathbf{x}, ict), \tag{3.422}$$

since $\bar{\psi}\gamma_5\psi$ changes its sign. If we require that parity be conserved, we may have *either* (3.419) *only or* (3.420) *only* but not both simultaneously. In the next section we shall point out that a Yukawa interaction of the type (3.419) leads to an s-wave emission (absorption) of a meson by the nucleon while an interaction of the type (3.420) leads to a p-wave emission (absorption). The meson fields that transform like (3.421) and (3.422) under space inversion are called respectively a *scalar* and a *pseudoscalar* field; the couplings (3.419) and (3.420) are respectively called a *scalar* coupling and a *pseudoscalar* (*pseudoscalar*) coupling.|| The meson fields corresponding to the observed neutral spin-zero mesons, π^0 and η, turn out to be both pseudoscalar. We shall say more about the meson-nucleon interaction when we discuss the one-pion exchange potential in the next chapter.

3–11. WEAK INTERACTIONS AND PARITY NONCONSERVATION; THE TWO-COMPONENT NEUTRINO

Classification of interactions. To illustrate the power of the field-theoretic formalism we have developed in the last section, we shall discuss in this section some simple examples taken from the physics of so-called "weak interactions." As is well

---

‡We emphasize that the field-theoretic description of the meson-nucleon interaction is not as firmly founded as that of the electromagnetic interaction of electrons and positrons. Quantum field theory does, however, provide a convenient language to describe some phenomena involving mesons and nucleons.

§The factor i is necessary to make the interaction density Hermitian. Note $(\bar{\psi}\gamma_5\psi)^\dagger = \psi^\dagger\gamma_5\gamma_4\psi = -\bar{\psi}\gamma_5\psi$.

||So called to distinguish it from a pseudoscalar (pseudovector) coupling of the type $(iF\hbar/m_\pi c)(\partial\phi/\partial x_\mu)\,\bar{\psi}\gamma_5\gamma_\mu\psi$.

known, the basic interactions of elementary particles can be classified into three categories:

a) *strong* interactions, for example, $n + p \rightarrow n + p$, $\pi^- + p \rightarrow \Lambda + K^0$, $\rho^0 \rightarrow \pi^+ + \pi^-$;

b) *electromagnetic* interactions, for example, $e^+ + e^- \rightarrow 2\gamma$, $\gamma + p \rightarrow p + \pi^0$, $\Sigma^0 \rightarrow \Lambda + \gamma$;

c) *weak* interactions, for example, $n \rightarrow p + e^- + \bar{\nu}$, $\bar{\nu}' + p \rightarrow \mu^+ + \Lambda$, $K^+ \rightarrow \pi^+ + \pi^0$.

To this list we should add a fourth (and the oldest) class of interactions, the *gravitational* interactions. But gravity turns out to be of little interest in our present-day understanding of elementary particles. What is remarkable is that all four classes of interactions are characterized by *dimensionless coupling constants* that differ by many orders of magnitude; the dimensionless coupling constant that characterizes electromagnetic interaction is $(e^2/4\pi\hbar c) = \alpha \simeq \frac{1}{137}$. As we shall show in Section 4–6, for the analogous coupling constant in pion-nucleon physics, defined as in (3.420), we have $(G^2/4\pi\hbar c) \simeq 14$, which is a typical strong-interaction coupling constant. In contrast, the constant that characterizes the strength of a typical weak-interaction process, when defined in a similar manner, turns out to be as small as 10^{-12} to 10^{-14}; this will be illustrated in a moment when we discuss Λ decay and pion decay. Gravitational interactions are even weaker; at the same separation distance the gravitational attraction between two protons is about 10^{-37} times weaker than the electrostatic repulsion between them.

Another striking feature of the elementary particle interactions is that some conservation laws which are obeyed to high degrees of accuracy by the strong and electromagnetic interactions are known to be violated by the weak interactions. In particular, weak-interaction processes *in general* do not conserve parity. In this section we shall illustrate this point using the language of quantum field theory.

Parity. One learns in nonrelativistic quantum mechanics that a state with momentum $\mathbf{p}$ goes into a state with momentum $-\mathbf{p}$ under parity, as is evident from the operator form of momentum, $\mathbf{p} = -i\hbar\boldsymbol{\nabla}$. For instance, in the Schrödinger theory the plane-wave solution $\psi(\mathbf{x}, t) = \exp(i\mathbf{p}\cdot\mathbf{x}/\hbar - iEt/\hbar)$ goes into $\psi(-\mathbf{x}, t) = \exp[i(-\mathbf{p})\cdot\mathbf{x}/\hbar - iEt/\hbar]$, which we recognize as the plane-wave solution with momentum *opposite* to the original one. The orbital angular momentum $\mathbf{L}$ is unchanged under parity since it is given by $\mathbf{x} \times \mathbf{p}$. Furthermore, we argue that the spin angular momentum is also even under parity since space inversion commutes with an infinitesimal rotation. As a result the magnetic moment interaction $-(e\hbar/2mc)\boldsymbol{\sigma}\cdot\mathbf{B}$, for instance, is invariant under parity since $\mathbf{B}$ does not change under parity either. From these considerations we expect that in field theory an electron state $b_{\mathbf{p}}^{(s)\dagger}|0\rangle$ goes into $b_{-\mathbf{p}}^{(s)\dagger}|0\rangle$ (with the same s), etc. We shall see, in a moment, that this is essentially the case except that we have to be careful with the relative transformation properties of the electron and positron states.

In Section 3–4 we proved that the Dirac wave function in the space-inverted system $\mathbf{x}' = -\mathbf{x}$, $t' = t$ is related to the wave function in the original system by $\psi'(\mathbf{x}', t) = \gamma_4\psi(\mathbf{x}, t)$ (up to a phase factor). The proof given there goes through just as well even if ψ stands for the quantized Dirac field. If we now consider the space inversion operation (commonly called the parity operation) that transforms $\mathbf{x}$ into $-\mathbf{x}$, we see that the functional form of the field operator changes as

$$\psi(\mathbf{x}, t) \longrightarrow \psi'(\mathbf{x}, t) = \gamma_4\psi(-\mathbf{x}, t), \tag{3.423}$$

since what used to be called $\mathbf{x}' = -\mathbf{x}$ is now $\mathbf{x}$. What does this mean? Going back to the plane-wave expansion of the filed operator, we have

$$\gamma_4\psi(-\mathbf{x}, t) = \sum_{\mathbf{p}}\sum_{s}\sqrt{\frac{mc^2}{E}}\left[b_{\mathbf{p}}^{(s)}\gamma_4 u^{(s)}(\mathbf{p})\exp\left(-\frac{i\mathbf{p}\cdot\mathbf{x}}{\hbar} - \frac{iEt}{\hbar}\right)\right.$$
$$\left. + d_{\mathbf{p}}^{(s)\dagger}\gamma_4 v^{(s)}(\mathbf{p})\exp\left(\frac{i\mathbf{p}\cdot\mathbf{x}}{\hbar} + \frac{iEt}{\hbar}\right)\right]. \tag{3.424}$$

But it is easy to show (cf. Problem 3–4) that

$$\gamma_4 u^{(s)}(\mathbf{p}) = u^{(s)}(-\mathbf{p}), \qquad \gamma_4 v^{(s)}(\mathbf{p}) = -v^{(s)}(-\mathbf{p}). \tag{3.425}$$

Note the minus sign preceding the positron spinor, $v^{(s)}(-\mathbf{p})$; a special case of this has already been worked out in Section 3–4, where we learned that a positive-energy at-rest spinor and a negative-energy at-rest spinor behave oppositely under parity. Thus

$$\gamma_4\psi(-\mathbf{x}, t) = \sum_{\mathbf{p}}\sum_{s}\sqrt{\frac{mc^2}{E}}\left[b_{-\mathbf{p}}^{(s)}u^{(s)}(\mathbf{p})\exp\left(\frac{i\mathbf{p}\cdot\mathbf{x}}{\hbar} - \frac{iEt}{\hbar}\right)\right.$$
$$\left. - d_{-\mathbf{p}}^{(s)\dagger}v^{(s)}(\mathbf{p})\exp\left(\frac{-i\mathbf{p}\cdot\mathbf{x}}{\hbar} + \frac{iEt}{\hbar}\right)\right]. \tag{3.426}$$

Comparing this with (3.392) we deduce that the transformation (3.423) is accomplished if the creation and annihilation operators change in the following manner:

$$\begin{aligned} b_{\mathbf{p}}^{(s)} &\longrightarrow b_{-\mathbf{p}}^{(s)}, & d_{\mathbf{p}}^{(s)} &\longrightarrow -d_{-\mathbf{p}}^{(s)}, \\ b_{\mathbf{p}}^{(s)\dagger} &\longrightarrow b_{-\mathbf{p}}^{(s)\dagger}, & d_{+\mathbf{p}}^{(s)\dagger} &\longrightarrow -d_{-\mathbf{p}}^{(s)\dagger}. \end{aligned} \tag{3.427}$$

An immediate consequence of this is that a single-particle state with $(\mathbf{p}, s)$ goes into a state with $(-\mathbf{p}, s)$, as expected from nonrelativistic quantum mechanics, except that a positron state acquires a minus sign,

$$\begin{aligned} b_{\mathbf{p}}^{(s)\dagger}|0\rangle &\longrightarrow \Pi\, b_{\mathbf{p}}^{(s)\dagger}|0\rangle = b_{-\mathbf{p}}^{(s)\dagger}|0\rangle, \\ d_{\mathbf{p}}^{(s)\dagger}|0\rangle &\longrightarrow \Pi\, d_{\mathbf{p}}^{(s)\dagger}|0\rangle = -d_{-\mathbf{p}}^{(s)\dagger}|0\rangle, \end{aligned} \tag{3.428}$$

where Π is the parity operator that acts on state vectors.‡

We shall now discuss the physical significance of the minus sign in the second part of (3.428). Consider, for simplicity, an electron-positron system in which the electron and the positron are both at rest, hence are in a relative s-state. According

‡Using the operator Π, we can write (3.427) as $\Pi b_{\mathbf{p}}^{(s)}\Pi^{-1} = b_{-\mathbf{p}}^{(s)}$, etc. We assume that the vacuum state is even by convention: $\Pi|0\rangle = |0\rangle$.

to (3.428), such a system transforms as

$$b_{\mathbf{p}=0}^{(s)\dagger} d_{\mathbf{p}=0}^{(s')\dagger} |0\rangle \longrightarrow -b_{\mathbf{p}=0}^{(s)\dagger} d_{\mathbf{p}=0}^{(s')\dagger} |0\rangle \tag{3.429}$$

under parity. This means that the parity of an s-state e^-e^+ system is *odd*.‡ Actually we could have anticipated this from the hole theory. According to the hole-theoretic interpretation, what is physically observable is the relative parity between the completely filled Dirac sea and the Dirac sea with one negative-energy electron missing. Remembering that parity is a multiplicative concept in the sense that the parity of a composite system is the product of the parities of the constituent systems, we infer that the observable parity of the positron is the *same* as that of the missing negative-energy electron. But, according to our earlier discussion following (3.179), the parity of a negative-energy electron at rest is odd (when the phase factor η is so chosen that the parity of a positive-energy electron at rest is even). Hence the parity of a positron at rest is odd (relative to that of an electron at rest).

If we follow steps similar to (3.423) through (3.429) with a non-Hermitian (charged) field $\phi_{ch}(\mathbf{x}, t)$ which transforms as

$$\phi_{ch}(\mathbf{x}, t) \longrightarrow \pm\phi_{ch}(-\mathbf{x}, t) \tag{3.430}$$

under parity, it is not difficult to show that a π^- state and a π^+ state transform in the *same* way under parity and that a $\pi^+\pi^-$ system in a relative s-state is *even* (in sharp contrast to the e^+e^- case). Quite generally, the "intrinsic parity" of the "antiparticle" is opposite to that of the corresponding "particle" in the fermion case and is the same in the boson case. This is one of the most important results of relativistic quantum theory. In Section 4–4, we shall present experimental evidence in favor of the odd parity of an s-state e^+e^- system.

Hyperon decay. So much for the transformation properties of the free-particle states. Let us now examine how we may describe parity-nonconserving decay processes using the language of quantum field theory. As a particularly simple example, we shall consider the decay of a free Λ hyperon (known to be a spin-$\frac{1}{2}$ particle):

$$\Lambda \longrightarrow p + \pi^-. \tag{3.431}$$

A simple interaction density (operator) that can account for this process is

$$\mathscr{H}_{\text{int}} = \phi_\pi^\dagger \bar{\psi}_p (g + g'\gamma_5)\psi_\Lambda + \text{Hc}, \tag{3.432}$$

where ψ_Λ can annihilate Λ hyperons and create anti-Λ hyperons,§ $\bar{\psi}_p$ can annihilate antiprotons and create protons, and $\phi_\pi^\dagger$ can create π^- and annihilate π^+. We have added Hc which stands for the "Hermitian conjugate" because, without

‡The skeptical reader may demonstrate that this conclusion (which can be verified experimentally) is independent of our choice $\eta = 1$ in $\psi'(x') = \eta\gamma_4\psi(x)$.

§The anti-Λ hyperon is to be distinguished from the Λ hyperon. First, even though they have the same charge (namely, zero), the same mass, and the same lifetime, their magnetic moments are opposite. Second, annihilations of a $\bar{\Lambda}p$ system into mesons, for example, $\bar{\Lambda} + p \longrightarrow K^+ + \pi^+ + \pi^-$, have been observed, whereas reactions of the type $\Lambda + p \longrightarrow$ mesons are strictly forbidden.

it, the interaction density would not be Hermitian.‡ Explicitly stated,

$$\begin{aligned} \text{Hc} &= \psi_\Lambda^\dagger(g^* + g'^*\gamma_5)\gamma_4\psi_p\phi_\pi \\ &= \bar{\psi}_\Lambda(g^* - g'^*\gamma_5)\psi_p\phi_\pi. \end{aligned} \tag{3.433}$$

We see that the Hc can give rise to processes like

$$\bar{\Lambda} \longrightarrow \bar{p} + \pi^+, \tag{3.434}$$

where $\bar{\Lambda}$ and $\bar{p}$ stand for an anti-Λ hyperon and an antiproton. It is a characteristic feature of the quantum-field-theoretic description that if (3.431) takes place, then (3.434) must also take place. At the end of this section, we shall show that this feature is a consequence of what is known as CPT invariance. Experimentally the decay process (3.434) has indeed been observed to be as follows:

$$\begin{aligned} &p + \bar{p} \longrightarrow \Lambda + \bar{\Lambda} \\ &\qquad\qquad\qquad\quad \llcorner\!\!\longrightarrow \bar{p} + \pi^+ \\ &\qquad\qquad \llcorner\!\!\longrightarrow p + \pi^-. \end{aligned} \tag{3.435}$$

We do not actually believe that the Hamiltonian density (3.432) is a "fundamental" interaction in nature in the same way as we believe that $-ie\bar{\psi}_e\gamma_\mu\psi_e A_\mu$ is "fundamental." So many new particles and new decay processes are observed nowadays in high-energy nuclear physics that if we were to introduce a new fundamental interaction every time a new decay process was discovered, a complete list of the fundamental interactions would become ridiculously long. Anybody in his right mind would then say that most (or perhaps all) of the interactions in the list could not possibly be "fundamental." Unfortunately, as yet we do not know what the basic interaction mechanisms are which give rise to a phenomenological interaction of the kind described by (3.432). In any case, for computational purposes, let us go along with (3.432).

We shall now investigate the transformation properties of (3.432). First, we can easily see that (3.432) is invariant under a proper orthochronous Lorentz transformation,

$$\begin{aligned} \bar{\psi}'_p \begin{Bmatrix} 1 \\ \gamma_5 \end{Bmatrix} \psi'_\Lambda\phi'_\pi &= \bar{\psi}_p S_{\text{Lor}}^{-1} \begin{Bmatrix} 1 \\ \gamma_5 \end{Bmatrix} S_{\text{Lor}}\psi_\Lambda\phi_\pi \\ &= \bar{\psi}_p \begin{Bmatrix} 1 \\ \gamma_5 \end{Bmatrix} \psi_\Lambda\phi_\pi, \end{aligned} \tag{3.436}$$

since the form of $S_{\text{Lor}}(S_{\text{Lor}}^{-1})$ is independent of the type of spin-$\frac{1}{2}$ field. Next, we shall show that the interaction (3.432) is not invariant under parity unless $g = 0$ or $g' = 0$. Since ψ_p, ψ_Λ, and ϕ_π transform under parity as

$$\psi_p \longrightarrow \eta_p\gamma_4\psi_p(-\mathbf{x}, t), \qquad \psi_\Lambda \longrightarrow \eta_\Lambda\gamma_4\psi_\Lambda(-\mathbf{x}, t), \qquad \phi_\pi \longrightarrow \eta_\pi\phi_\pi(-\mathbf{x}, t), \tag{3.437}$$

‡The use of a Hermitian Hamiltonian density is required since the expectation value of the Hamiltonian operator must be real. Furthermore, in Section 4–2 we shall show that the use of a non-Hermitian Hamiltonian violates probability conservation.

the interaction density (3.432) changes under the parity operation as follows:

$$\mathscr{H}_{\text{int}}(\mathbf{x}, t) \longrightarrow \eta_{\text{p}}^* \eta_\pi^* \eta_\Lambda \phi_\pi^\dagger(-\mathbf{x}, t)\bar{\psi}_{\text{p}}(-\mathbf{x}, t)(g - g'\gamma_5)\psi_\Lambda(-\mathbf{x}, t) + \text{Hc}. \tag{3.438}$$

Suppose $g' = 0$; then $\mathscr{H}_{\text{int}}$ transforms like a scalar density,

$$\mathscr{H}_{\text{int}}(\mathbf{x}, t) \longrightarrow \mathscr{H}_{\text{int}}(-\mathbf{x}, t), \tag{3.439}$$

provided we choose the phase factors in (3.437) in such a way that $\eta_{\text{p}}^*\eta_\pi^*\eta_\Lambda = 1$. Similarly for $g = 0$, $\mathscr{H}_{\text{int}}$ again transforms like a scalar density provided we set $\eta_{\text{p}}^*\eta_\pi^*\eta_\Lambda = -1$. However, if both g and g' are nonvanishing, then $\mathscr{H}_{\text{int}}$ cannot be made to transform like a scalar density no matter how we choose η_{p}, η_π, and η_Λ. This is what we mean by saying that *the interaction is not invariant under parity.*

All this appears somewhat formal. We now wish to exhibit some of the physical manifestations of the parity-nonconserving interaction (3.432). For this purpose let us obtain the transition matrix element for this process. First, recall that the state vectors we have been using are time independent, while the interaction Hamiltonian density is made up of time-dependent field operators. In the language of time-dependent perturbation theory, discussed in Section 2–4, a state vector in our occupation-number space corresponds to a time-independent wave function $u_n(\mathbf{x})$ rather than to a time-dependent one $u_n(\mathbf{x})e^{-iE_nt/\hbar}$. Recalling the connection between the operators in the Heisenberg and the Schrödinger representations, we see that the matrix element of $\int \mathscr{H}_{\text{int}} d^3x$ taken between the initial Λ state and the final π^-p state corresponds, in the language of Chapter 2, to the matrix element of $e^{iH_0t/\hbar}H_Ie^{-iH_0t/\hbar}$ taken between $u_i(\mathbf{x})$ and $u_f(\mathbf{x})$; the only new point is that our Hamiltonian can now change the nature of the particles. Using this fact and recalling the expression for $c^{(1)}$ given by (2.109) we see that $c^{(1)}(t)$ in our case is given by

$$c^{(1)}(t) = (-i/\hbar)\left\langle f\left|\int_0^t dt' \int \mathscr{H}_{\text{int}}(\mathbf{x}, t')\, d^3x\right|i\right\rangle, \tag{3.440}$$

where the perturbation is assumed to be turned on at $t = 0$.‡

In the language of quantum field theory the initial and final states in Λ decay are given by

$$\begin{aligned} |i\rangle &= b_{\mathbf{p}}^{(\Lambda, s)\dagger}|0\rangle, \\ |f\rangle &= a^\dagger(\mathbf{p}_\pi)b_{\mathbf{p}'}^{(\text{p}, s')\dagger}|0\rangle, \end{aligned} \tag{3.441}$$

where $b_{\mathbf{p}}^{(\Lambda, s)\dagger}$, $b_{\mathbf{p}'}^{(\text{p}, s')\dagger}$, and $a^\dagger(\mathbf{p}_\pi)$ are the creation operators for the Λ hyperon, the proton, and the π^- meson. It is legitimate to replace ψ_Λ by its positive frequency part $\psi_\Lambda^{(+)}$ since the negative frequency part of ψ_Λ acting on the initial state would result in a $(\Lambda + \bar{\Lambda})$ state which, because of orthogonality, gives zero when multiplied by $\langle f|\phi_\pi^\dagger\bar{\psi}_{\text{p}}$ from the left. Furthermore, the only part of $\psi_\Lambda^{(+)}$ which contributes is just $(1/\sqrt{V})\, b_{\mathbf{p}}^{(\Lambda, s)}u_\Lambda^{(s)}(\mathbf{p}) \exp(i\mathbf{p}\cdot\mathbf{x}/\hbar - iEt/\hbar)$, since $b_{\mathbf{p}''}^{(\Lambda, s'')}b_{\mathbf{p}}^{(\Lambda, s)\dagger}|0\rangle = 0$, unless $\mathbf{p}'' = \mathbf{p}$, $s'' = s$. All this amounts to saying that it is all right to replace the

‡We shall present a more formal discussion of the connection between the transition matrix and the interaction Hamiltonian in Section 4–2, when we shall discuss the S-matrix expansion in the interaction representation.

field operator ψ_Λ by the wave function of the initial Λ particle multiplied by the corresponding annihilation operator. Likewise it is easy to see that $\phi_\pi^\dagger$ can be replaced by just

$$a^\dagger(\mathbf{p}_\pi)c\sqrt{\hbar/2\omega_\pi V}\exp(-i\mathbf{p}_\pi\cdot\mathbf{x}/\hbar + i\omega_\pi t)$$

(with $\hbar\omega_\pi = \sqrt{|\mathbf{p}_\pi|^2c^2 + m_\pi^2c^4}$), and $\bar{\psi}_\mathrm{p}$ can be replaced by

$$\sqrt{m_\mathrm{p}c^2/E_\mathrm{p}V}\, b_{\mathbf{p}'}^{(\mathrm{p},s')\dagger}\bar{u}_\mathrm{p}^{(s')}(\mathbf{p}')\exp(-i\mathbf{p}'\cdot\mathbf{x}/\hbar + iE_\mathrm{p}t/\hbar).$$

The result of all this gives

$$\begin{aligned}\left\langle f\middle|\int\mathscr{H}_{\mathrm{int}}d^3x\middle|i\right\rangle \\ &= \langle f|a^\dagger(\mathbf{p}_\pi)b_{\mathbf{p}'}^{(\mathrm{p},s')\dagger}b_{\mathbf{p}=0}^{(\Lambda,s)}|i\rangle(c\sqrt{\hbar/2\omega_\pi})(\sqrt{m_\mathrm{p}c^2/E_\mathrm{p}})\bar{u}_\mathrm{p}^{(s')}(\mathbf{p}')(g+g'\gamma_5)u_\Lambda^{(s)}(0) \\ &\quad\times\left[\frac{1}{\sqrt{V^3}}\int\exp\left(-\frac{i\mathbf{p}_\pi\cdot\mathbf{x}}{\hbar}-\frac{i\mathbf{p}\cdot\mathbf{x}}{\hbar}\right)d^3x\right]\exp\left(i\omega_\pi t+\frac{iE_\mathrm{p}t}{\hbar}-\frac{im_\Lambda c^2t}{\hbar}\right),\end{aligned} \tag{3.442}$$

where we have assumed that the initial Λ particle is at rest. Using

$$b_{\mathbf{p}=0}^{(\Lambda,s)}b_{\mathbf{p}=0}^{(\Lambda,s)\dagger}|0\rangle = (1 - b_{\mathbf{p}=0}^{(\Lambda,s)\dagger}b_{\mathbf{p}=0}^{(\Lambda,s)})|0\rangle, \quad \text{etc.},$$

it is not difficult to see that $\langle f|a^\dagger(\mathbf{p}_\pi)b_{\mathbf{p}'}^{(\mathrm{p},s')\dagger}b_{\mathbf{p}=0}^{(\Lambda,s)}|i\rangle$ can eventually be reduced to $\langle 0|0\rangle = 1$. The exponential time dependence in (3.442) is precisely the kind that appears in the derivation of the Golden Rule in time-dependent perturbation theory (cf. Eq. 2.113); if we assume that the perturbation is switched on at $t = 0$ and acts for a long time, the modulus squared of $\int_0^t \exp(i\omega_\pi t + iE_\mathrm{p}t'/\hbar - im_\Lambda c^2t'/\hbar)dt'$ leads to $2\pi\hbar t$ times the usual δ function that expresses energy conservation. Note also that the space integral in (3.442) simply tells us that the transition matrix element is zero unless momentum is conserved.

To sum up, the time-dependent matrix element that appears in the Golden Rule can be obtained immediately from the Hamiltonian operator $\int\mathscr{H}_{\mathrm{int}}d^3x$ just by replacing the quantized field operators in $\mathscr{H}_{\mathrm{int}}$ by the appropriate initial and final wave functions with their time dependence omitted. In other words, we get the correct results by pretending that the q-number density (3.432) made up of the field operators is a c-number density made up of the initial- and final-state wave functions.‡

Let us now simplify the spinor product in (3.442). We have

$$\begin{aligned}&\bar{u}_\mathrm{p}^{(s')}(\mathbf{p}')(g+g'\gamma_5)u_\Lambda^{(s)}(0) \\ &= \sqrt{\frac{m_\mathrm{p}c^2+E_\mathrm{p}}{2m_\mathrm{p}c^2}}\left(\chi^{(s')\dagger}, -\chi^{(s')\dagger}\frac{(\boldsymbol{\sigma}\cdot\mathbf{p}'c)}{E_\mathrm{p}+m_\mathrm{p}c^2}\right)\begin{pmatrix} g & -g' \\ -g' & g\end{pmatrix}\begin{pmatrix}\chi^{(s)} \\ 0\end{pmatrix} \\ &= \sqrt{\frac{m_\mathrm{p}c^2+E_\mathrm{p}}{2m_\mathrm{p}c^2}}\chi^{(s')\dagger}\left(g + g'\frac{\boldsymbol{\sigma}\cdot\mathbf{p}'c}{E_\mathrm{p}+m_\mathrm{p}c^2}\right)\chi^{(s)}.\end{aligned} \tag{3.443}$$

‡In fact this kind of replacement is implicit in the discussions of beta decay that appear in most textbooks on nuclear physics, for example, Segrè (1964), Chapter 9, and Preston (1964), Chapter 15. We now understand why beta decay can be discussed at an elementary level without using the language of quantum field theory.

Assuming that the initial Λ is polarized with spin along the positive z-axis, we obtain

$$\left(g + g' \frac{\boldsymbol{\sigma}\cdot\mathbf{p}'c}{E_p + mc^2}\right)\chi^{(s)} = (a_s + a_p \cos\theta)\begin{pmatrix}1\\0\end{pmatrix} + a_p \sin\theta\, e^{i\phi}\begin{pmatrix}0\\1\end{pmatrix}, \tag{3.444}$$

where

$$a_s = g, \qquad a_p = g'(|\mathbf{p}'|c/(E_p + m_p c^2)) \tag{3.445}$$

and the angles θ and ϕ characterize the orientation of $\mathbf{p}'$ relative to the Λ spin direction. The physical meaning of g and g' can now be seen as follows. If $g \neq 0$, $g' = 0$, the final state proton can be described by an $s\frac{1}{2}$, $j_z = \frac{1}{2}$ wave function $\binom{1}{0}$. On the other hand, if $g = 0$, $g' \neq 0$, the final proton is in a pure $p\frac{1}{2}$, $j_z = \frac{1}{2}$ state described by $\cos\theta\binom{1}{0} + \sin\theta\, e^{i\phi}\binom{0}{1}$. In other words, the scalar coupling (the g-term) gives rise to an $s\frac{1}{2}\ \pi^-$p system, whereas the pseudoscalar (γ_5) coupling (the g'-term) gives rise to a $p\frac{1}{2}\ \pi^-$p system. If g and g' are both nonvanishing, both $s\frac{1}{2}$ and $p\frac{1}{2}$ are allowed; in other words, the same initial state can go into final states of *opposite* parities. Recalling that for $g \neq 0$, $g' \neq 0$ the interaction density (3.432) is not invariant under parity, we see that an $\mathscr{H}_{\text{int}}$ that does not transform like (3.439) under parity indeed gives rise to final states of opposite parities.

We shall digress here and examine the pseudoscalar (pseudoscalar) coupling of the pion to the nucleon (3.420) which gives rise to processes like $\text{p} \longrightarrow \text{p} + \pi^0$. We may argue that when the proton dissociates itself into a π^0 and a proton, the final πp system must be in $p\frac{1}{2}$ state. Unfortunately, such dissociation processes are forbidden by energy momentum conservation if all the particles are free. However, taking advantage of a reaction in which the nucleon is bound (specifically $\pi^+ + \text{d} \longrightarrow \text{p} + \text{p}$) it has been proved possible to show that the $\pi^\pm$ is pseudoscalar (with the convention that the proton and the neutron are both even), that is, $\phi(\mathbf{x}, t) \longrightarrow -\phi(-\mathbf{x}, t)$. Note that if there is just a γ_5 type coupling, parity is conserved; the "intrinsic" odd parity of the pion is compensated for by the odd orbital parity.‡

Coming back to Λ decay, we can now compute the decay-angular distribution. The relative probabilities of observing the proton with spin-up and spin-down can be obtained immediately from (3.444):

$$\begin{aligned} &\text{spin-up:} && |a_s + a_p \cos\theta|^2, \\ &\text{spin-down:} && |a_p|^2 \sin^2\theta, \end{aligned} \tag{3.446}$$

which results in the decay-angular distribution of the proton

$$1 - \alpha\cos\theta, \tag{3.447}$$

where§

$$\alpha = -\frac{2\,\text{Re}(a_s a_p^*)}{|a_s|^2 + |a_p|^2}. \tag{3.448}$$

‡An alternative way of understanding what is meant by the intrinsic odd parity of the pion is to visualize the pion as a very tightly bound state of a nucleon (intrinsically even) and an antinucleon (intrinsically odd) in a relative s-state.

§The minus sign in (3.447) arises from the fact that the experimentalists usually talk about the decay-angular distribution of the pion relative to the Λ spin: $1 + \alpha\cos\theta^{(\pi)}$ where $\cos\theta^{(\pi)} = -\cos\theta$.

Recalling that θ is measured from the Λ spin direction, we see that the angular distribution (3.447) implies that whenever $\mathrm{Re}(a_s a_p^*) \neq 0$ or equivalently $\mathrm{Re}(gg'^*) \neq 0$, there exists an observable effect that depends on $\langle \boldsymbol{\sigma}_\Lambda \rangle \cdot \mathbf{p}'$, a pseudoscalar quantity that changes its sign under parity.

To really understand the meaning of parity nonconservation in this decay process, it is instructive to work out the special decay configuration $\theta = 0$. This is shown in Fig. 3–9(a). The transition probability for this process is, according to (3.446), $|a_s + a_p|^2$ apart from kinematical factors. If we apply the parity operation to the decay configuration shown in Fig. 3–9(a), we obtain the decay configuration shown in Fig. 3–9(b) since, under parity, momentum changes but spin does not. However, according to (3.446), the transition probability for the physical situation described by Fig. 3–9(b) is $|a_s - a_p|^2$ since $\theta = \pi$. Thus the transition probability for $|i\rangle \to |f\rangle$ is not the same as that for $\Pi|i\rangle \to \Pi|f\rangle$ unless $\mathrm{Re}(a_s a_p^*) = 0$. Since the configuration in Fig. 3–9(b) is the mirror image of the configuration in Fig. 3–9(a) apart from a 180° rotation about an axis perpendicular to $\mathbf{p}'$, we conclude that the mirror image of our world looks different from our world if $\mathrm{Re}(gg'^*) \neq 0$.

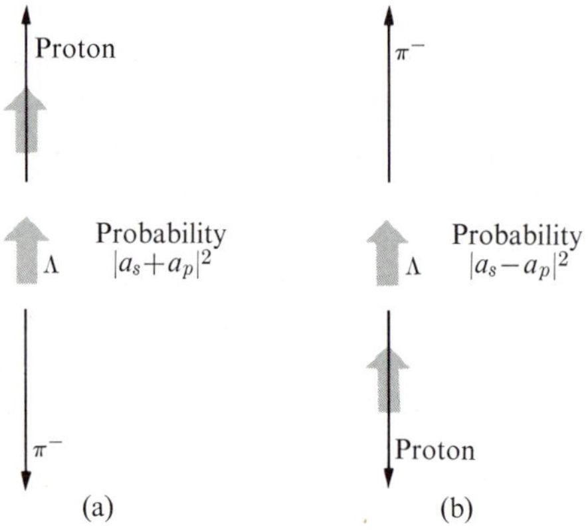

Fig. 3–9. Λ decay. Parity conservation would require that the two decay configurations (which go into each other under space inversion) be physically realizable with the same transition probability. The gray arrows indicate the spin direction.

Although we cannot prepare a polarized sample of Λ hyperons using a magnet, it turns out that the Λ hyperons produced in

$$\pi^- + \mathrm{p} \longrightarrow \Lambda + \mathrm{K}^0 \tag{3.449}$$

are strongly polarized in the direction $\mathbf{p}_{\pi\ \text{incident}} \times \mathbf{p}_\Lambda$. Parity nonconservation in Λ decay was unambiguously established, in 1957, by a Berkeley group and by a Columbia-Michigan-Pisa-Bologna collaboration group who showed that there are more decay pions emitted with $(\mathbf{p}_{\pi\ \text{incident}} \times \mathbf{p}_\Lambda)\cdot\mathbf{p}_{\pi\ \text{decay}} > 0$ than with $(\mathbf{p}_{\pi\ \text{incident}} \times \mathbf{p}_\Lambda)\cdot\mathbf{p}_{\pi\ \text{decay}} < 0$. Now, at last, it is possible to communicate even to intelligent beings in outer space that the incident pion direction, the Λ direction, and the preferential direction of the decay pion, taken in that order, form the three axes of what we mean by a right-handed system.

Prior to 1956 practically everybody tacitly assumed that it was "illegal" to write parity-nonconserving interactions like (3.432). There was a good reason for this; the success of the parity selection rules in atomic and nuclear physics shows that the principle of parity conservation holds to a high degree of accuracy in both electromagnetic and strong interactions. In the years from 1954 to 1956, as various experimental groups studied the properties of "strange" mesons called τ^+ and θ^+ which decay via weak interactions as

$$\tau^+ \longrightarrow 2\pi^+ + \pi^-, \qquad \theta^+ \longrightarrow \pi^+ + \pi^0, \tag{3.450}$$

it soon became evident that τ^+ and θ^+ have the same mass and the same lifetime; it therefore appeared natural to assume that a τ-like decay event and a θ-like decay event simply represent different decay modes of the same parent particle (now called K^+). However, using an ingenious argument based only on parity and angular momentum conservation, R. H. Dalitz was able to show that the experimental energy and angular distributions of the pions from τ-decay strongly suggest that the τ and the θ could not possibly have the same spin parity. Since, at that time, people believed in parity conservation, this led to the famous τ-θ puzzle.‡ Faced with this dilemma, in the spring of 1956 T. D. Lee and C. N. Yang systematically investigated the validity of parity conservation in elementary particle interactions. Their conclusion was that in the realm of weak interactions parity conservation (which holds extremely well for the strong and electromagnetic interactions) was "only an extrapolated hypothesis unsupported by experimental evidence." Furthermore, they suggested a number of experiments that are really sensitive to the question of whether or not parity is conserved. (Their list of suggested experiments included the decay angular distribution of a polarized Λ hyperon which we have been discussing.) As is well known, subsequent experiments (beginning with the historic Co^{60} experiment of C. S. Wu and coworkers and the π-μ-e experiments of J. I. Friedman and V. L. Telegdi and of R. L. Garwin, L. M. Lederman, and M. Weinrich) have unequivocally supported the idea that weak interactions in general do not conserve parity.

Coming back to Λ decay, let us work out the decay rate using the Golden Rule and (3.442) through (3.446). The $\cos\theta$ term drops out as we integrate over all angles. For the reciprocal of the partial lifetime, we get

$$\begin{aligned}
\frac{1}{\tau_{\Lambda\to p\pi^-}} &= \frac{\Gamma(\Lambda \longrightarrow p\pi^-)}{\hbar} \\
&= \frac{2\pi}{\hbar}\frac{c^2\hbar}{2\omega_\pi}\frac{m_pc^2}{E_pV}\frac{m_pc^2+E_p}{2m_pc^2}\left(|g|^2 + \frac{|g'|^2|\mathbf{p}|^2c^2}{E_p+m_pc^2}\right)\frac{4\pi V}{(2\pi\hbar)^3}\frac{|\mathbf{p}'|^2 d|\mathbf{p}'|}{d(E_\pi+E_p)} \\
&= \left(\frac{|g|^2}{4\pi\hbar c} + \frac{|g'|^2}{4\pi\hbar c}\frac{(E_p - m_pc^2)}{(E_p+m_pc^2)}\right)\frac{|\mathbf{p}'|(E_p+m_pc^2)}{\hbar m_\Lambda c},
\end{aligned} \tag{3.451}$$

‡For detailed discussions of the τ-θ puzzle see, for example, Nishijima (1964), pp. 315–323; Sakurai (1964), pp. 47–51.

where we have used

$$\frac{d|\mathbf{p}'|}{d(E_p + E_\pi)} = \frac{d|\mathbf{p}'|}{[(|\mathbf{p}'|c^2/E_p) + (|\mathbf{p}'|c^2/E_\pi)]\,d|\mathbf{p}'|}$$
$$= \frac{\hbar\omega_\pi E_p}{m_\Lambda|\mathbf{p}'|c^4}. \tag{3.452}$$

If we insert the experimentally measured mean life of the Λ particle (2.6×10^{-10} sec) and the branching ratio into the $\pi^- p$ decay mode (known to be about $\frac{2}{3}$), we obtain

$$\frac{|g|^2}{4\pi\hbar c} + 0.003\frac{|g'|^2}{4\pi\hbar c} \simeq 2 \times 10^{-14}. \tag{3.453}$$

Thus the dimensionless coupling constants

$$|g|^2/(4\pi\hbar c) \qquad \text{and} \qquad |g'|^2/(4\pi\hbar c)$$

for Λ decay are seen to be small compared to $e^2/(4\pi\hbar c) \simeq \frac{1}{137}$ by many orders of magnitude. The interaction responsible for Λ decay is indeed "weak."

Fermi theory of beta decay. Historically the theory of weak interactions started when E. Fermi wrote, in 1932, a Hamiltonian density that involves the proton, neutron, electron, and neutrino fields to account for nuclear beta decay:

$$\mathrm{n} \longrightarrow \mathrm{p} + \mathrm{e}^- + \bar{\nu}. \tag{3.454}$$

Fermi assumed for simplicity that the derivatives of the field operators do not appear. With this hypothesis the most general interaction density invariant under proper orthochronous Lorentz transformations has the form

$$\mathscr{H}_{\text{int}} = \sum_i (\bar{\psi}_p\Gamma_i\psi_n)[\bar{\psi}_e\Gamma_i(C_i + C_i'\gamma_5)\psi_\nu] + \text{Hc}, \tag{3.455}$$

where‡

$$\Gamma_i = 1,\ \gamma_\lambda,\ \sigma_{\lambda\sigma},\ i\gamma_5\gamma_\lambda,\ \gamma_5. \tag{3.456}$$

We have subscribed to the usual convention according to which the light neutral particle emitted together with the e^- in (3.454) is an "antineutrino" ($\bar{\nu}$), not a "neutrino" (ν), and the field operator ψ_ν annihilates neutrinos and creates antineutrinos. Evidently the explicitly written part of (3.455) can account for the neutrino induced reaction

$$\nu + \mathrm{n} \longrightarrow \mathrm{e}^- + \mathrm{p} \tag{3.457}$$

as well as for β^- decay (3.454). The Hc in (3.455) can describe β^+ decay, K (electron) capture, etc.:

$$\begin{aligned} \mathrm{p} &\longrightarrow \mathrm{n} + \mathrm{e}^+ + \nu, \\ \mathrm{e}^- + \mathrm{p} &\longrightarrow \mathrm{n} + \nu, \end{aligned} \tag{3.458}$$

since, when explicitly written, it contains the annihilation operators for protons and electrons and the creation operators for neutrons, positrons, and neutrinos.

‡We avoid the indices μ and ν to prevent possible confusions with muon and neutrino.

The constants C_i and C_i' characterize the strength of the interactions of type i (scalar, vector, tensor, axial vector, and pseudoscalar); they have the dimension of energy times volume. From our earlier discussion on Λ decay, it is evident that parity conservation requires either

$$C_i' = 0 \qquad \text{for all } i \quad (\eta_p^* \eta_n \eta_e^* \eta_\nu = 1),$$

or (3.459)

$$C_i = 0 \qquad \text{for all } i \quad (\eta_p^* \eta_n \eta_e^* \eta_\nu = -1).$$

If neither of the two possibilities is satisfied, then the interaction density (3.455) is not invariant under parity.

As it stands, (3.455) contains 10 arbitrary constants (which need not be purely real). About a quarter-century after the appearance of Fermi's paper, it finally became evident that the correct Hamiltonian density that phenomenologically described nuclear beta decay was

$$\mathscr{H}_{\text{int}} = C_V(\bar{\psi}_p \gamma_\lambda \psi_n)[\bar{\psi}_e \gamma_\lambda (1 + \gamma_5)\psi_\nu] + C_A(\bar{\psi}_p i\gamma_5 \gamma_\lambda \psi_n)[\bar{\psi}_e i\gamma_5 \gamma_\lambda (1 + \gamma_5)\psi_\nu] + \text{Hc} \tag{3.460}$$

with

$$C_V = 6.2 \times 10^{-44}\ \text{MeV cm}^3 \simeq (10^{-5}/\sqrt{2})\, m_p c^2 (\hbar/m_p c)^3, \qquad C_A/C_V \simeq -1.2. \tag{3.461}$$

The interaction (3.460) with $C_V \simeq -C_A$ is known as the $V - A$ interaction; it was written, on aesthetic grounds, by E. C. G. Sudarshan and R. E. Marshak, by R. P. Feynman and M. Gell-Mann, and by J. J. Sakurai in advance of the confirming experiments. Since the nucleon can be assumed to be nonrelativistic, only the time component of the vector covariant and the space components of the axial vector covariant contribute (γ_k and $i\gamma_5\gamma_4$ are "small") unless the symmetry of the initial and final nuclear states is such that the expectation values of 1 (the nonrelativistic limit of $\bar{\psi}_p \gamma_4 \psi_n$) and σ_k (the nonrelativistic limit of $\psi_p i\gamma_5\gamma_k\psi_n$) are both zero. The vector interaction gives the Fermi selection rule $\Delta J = 0$, no parity change, while the axial-vector interaction gives the Gamow-Teller selection rule $\Delta J = 0$, ± 1, no parity change, for the *nuclear* states.

We shall not discuss in detail the various aspects of nuclear beta decay: the electron spectrum, the *ft*-values, forbidden transitions, the electron-neutrino angular correlation, the angular distributions of electrons from polarized nuclei, etc. They are treated in standard textbooks on nuclear physics.‡ We concentrate on just one aspect of (3.460), namely the physical meaning of $(1 + \gamma_5)\psi_\nu$.

Two-component neutrino. The neutrino field ψ_ν can be expanded just as in (3.392). The only difference is that its mass is consistent with zero, $m_\nu < 200\ \text{eV}/c^2$ experimentally. The positive frequency part of ψ_ν is linear in the free-particle spinor for an annihilated neutrino. So let us investigate the effect of $(1 + \gamma_5)$ on $u^{(s)}(\mathbf{p})$.

---

‡See, for example, Preston (1962), Chapter 15; Källén (1964), Chapter 13.

As $m_\nu \longrightarrow 0$, we have

$$\gamma_5 \begin{pmatrix} \chi^{(s)} \\ (\boldsymbol{\sigma}\cdot\mathbf{p}c/E)\chi^{(s)} \end{pmatrix} = -\begin{pmatrix} 0 & I \\ I & 0 \end{pmatrix}\begin{pmatrix} \chi^{(s)} \\ \boldsymbol{\sigma}\cdot\hat{\mathbf{p}}\chi^{(s)} \end{pmatrix} = -\begin{pmatrix} \boldsymbol{\sigma}\cdot\hat{\mathbf{p}}\chi^{(s)} \\ \chi^{(s)} \end{pmatrix}. \tag{3.462}$$

Meanwhile

$$-\boldsymbol{\Sigma}\cdot\hat{\mathbf{p}} \begin{pmatrix} \chi^{(s)} \\ (\boldsymbol{\sigma}\cdot\mathbf{p}c/E)\chi^{(s)} \end{pmatrix} = -\begin{pmatrix} \boldsymbol{\sigma}\cdot\hat{\mathbf{p}} & 0 \\ 0 & \boldsymbol{\sigma}\cdot\hat{\mathbf{p}} \end{pmatrix}\begin{pmatrix} \chi^{(s)} \\ \boldsymbol{\sigma}\cdot\hat{\mathbf{p}} \ \chi^{(s)} \end{pmatrix} = -\begin{pmatrix} \boldsymbol{\sigma}\cdot\hat{\mathbf{p}}\chi^{(s)} \\ \chi^{(s)} \end{pmatrix}. \tag{3.463}$$

This means that as $m_\nu \longrightarrow 0$, the γ_5 operator (sometimes called the *chirality* operator‡) and the negative of the helicity operator have the same effect on a free-particle spinor $u^{(s)}(\mathbf{p})$. In particular, eigenspinors of $\boldsymbol{\Sigma}\cdot\hat{\mathbf{p}}$ are also eigenspinors of γ_5 with opposite eigenvalues, and $(1+\gamma_5)$ acting on the free-particle spinor for a right-handed particle gives zero as $m_\nu \longrightarrow 0$. It then follows that $(1+\gamma_5)\psi_\nu^{(+)}$ annihilates only left-handed (helicity $= -1$) neutrinos, denoted by ν_L. Likewise it is easy to see that

$$\begin{aligned} &(1+\gamma_5)\psi_\nu^{(-)} && \text{creates} && \bar{\nu}_R, \\ &\bar{\psi}_\nu^{(+)}(1-\gamma_5) && \text{annihilates} && \bar{\nu}_R, \\ &\bar{\psi}_\nu^{(-)}(1-\gamma_5) && \text{creates} && \nu_L, \end{aligned}$$

where $\bar{\nu}_R$ stands for a right-handed antineutrino. Now the Hc in (3.460) contains $\bar{\psi}_\nu(1-\gamma_5)$ since

$$\left[\bar{\psi}_e \begin{Bmatrix} \gamma_\lambda \\ i\gamma_5\gamma_\lambda \end{Bmatrix}(1+\gamma_5)\psi_\nu\right]^\dagger = \psi_\nu^\dagger(1+\gamma_5)\begin{Bmatrix} \gamma_\lambda \\ -i\gamma_\lambda\gamma_5 \end{Bmatrix}\gamma_4\psi_e = \bar{\psi}_\nu(1-\gamma_5)\begin{Bmatrix} \mp\gamma_\lambda \\ \pm i\gamma_5\gamma_\lambda \end{Bmatrix}\psi_e, \tag{3.464}$$

where the upper signs are for the space components, the lower for the time components. An immediate consequence of this is that the helicity of the neutrino emitted in K (electron) capture [the second of (3.458)] is -1. This has indeed been shown to be the case by a beautiful experiment of M. Goldhaber, L. Grodzins, and A. W. Sunyar. The positive (negative) helicity of the antineutrino (neutrino) emitted in β^- (β^+) decay has also been inferred from the electron (positron) polarization and the $e^-\bar{\nu}(e^+\nu)$ angular correlation (cf. Problem 3–14). Parity conservation would require that in a physically realizable process the emission probability for a right-handed particle be the same as that for a left-handed particle since the helicity changes sign under parity. It is therefore evident that the interaction (3.460) which produces only $\nu_L(\bar{\nu}_R)$ in β^+ (β^-) decay is incompatible with the principle of parity conservation.

---

‡This expression is derived from the Greek word χειρ meaning "hand." The term "chirality" was first used by Lord Kelvin in a somewhat different context.

There is another (simpler) way to see the connection between helicity and chirality for a massless particle. Let us go back to (3.26) which was obtained by linearizing the Waerden equation. When the fermion mass is zero, the two equations are completely decoupled,‡

$$\left(i\boldsymbol{\sigma}\cdot\nabla - i\frac{\partial}{\partial x_0}\right)\phi^{(L)} = 0, \qquad \left(-i\boldsymbol{\sigma}\cdot\nabla - i\frac{\partial}{\partial x_0}\right)\phi^{(R)} = 0. \tag{3.465}$$

Let us now postulate that only the first of (3.465) has to do with physical reality. Evidently a free-particle solution of the $\phi^{(L)}$ equation in the c-number (wave-function) language satisfies

$$\boldsymbol{\sigma}\cdot\mathbf{p} = -E/c, \tag{3.466}$$

where E can be positive or negative. This means that the helicity of a positive-energy neutrino is negative, while the helicity of a negative-energy neutrino is positive. Using the hole theory, we infer that the helicity of a (positive-energy) antineutrino is also positive (cf. Table 3–4). We should emphasize that the assertion that the neutrino (antineutrino) is always left-handed (right-handed) makes sense only if the mass is *strictly* zero. Otherwise we can easily perform a Lorentz transformation that changes a left-handed particle into a right-handed one.

Meanwhile, according to Problem 3–5, in the Weyl representation of the Dirac equation we have

$$\psi = \begin{pmatrix} \phi^{(R)} \\ \phi^{(L)} \end{pmatrix}, \tag{3.467}$$

and

$$\gamma_5 = \begin{pmatrix} -I & 0 \\ 0 & I \end{pmatrix}. \tag{3.468}$$

It is then evident that $(1+\gamma_5)\psi_\nu$ selects just $\phi^{(L)}$, that is, only the lower two of the four components of ψ in the Weyl representation:

$$(1+\gamma_5)\psi = 2\begin{pmatrix} 0 \\ \phi^{(L)} \end{pmatrix}. \tag{3.469}$$

‡It is amusing that the Maxwell equations applying to empty space can also be written in a form similar to (3.465). The reader may show that (1.57) and (1.58) with $\rho = 0$, $\mathbf{j} = 0$ can be combined to give

$$\left(-i\mathbf{S}\cdot\nabla - i\frac{\partial}{\partial x_0}\right)\Phi = 0,$$

where

$$S_1 = i\begin{pmatrix} 0 & 0 & 0 \\ 0 & 0 & -1 \\ 0 & 1 & 0 \end{pmatrix}, \quad S_2 = i\begin{pmatrix} 0 & 0 & 1 \\ 0 & 0 & 0 \\ -1 & 0 & 0 \end{pmatrix}, \quad S_3 = i\begin{pmatrix} 0 & -1 & 0 \\ 1 & 0 & 0 \\ 0 & 0 & 0 \end{pmatrix};$$

and

$$\Phi = \begin{pmatrix} B_1 - iE_1 \\ B_2 - iE_2 \\ B_3 - iE_3 \end{pmatrix}.$$

Suppose we demand that only

$$(1+\gamma_5)\psi_\nu \qquad \text{and} \qquad \bar{\psi}_\nu(1-\gamma_5) = [(1+\gamma_5)\psi_\nu]^\dagger\gamma_4$$

appear in the beta-decay interaction. Because of (3.469) this requirement amounts to saying that only $\phi^{(L)}$ and $\phi^{(L)\dagger}$ appear, hence the neutrinos (antineutrinos) emitted in β^+ (β^-) decay are necessarily left- (right-) handed. We may argue that the appearance of only $\phi^{(L)}$ is due to either (a) an intrinsic property of the *free* neutrino itself, or (b) a property of the parity-nonconserving beta-decay *interaction* that just happens to select $\phi^{(L)}$ (and also of other interactions in which the neutrino participates). If the neutrino mass is strictly zero, the two points of view are, in practice, completely indistinguishable.‡ A theory of neutrinos based on $\phi^{(L)} \neq 0$, $\phi^{(R)} = 0$ (or vice versa) is called the two-component theory of the neutrino.

Historically, the idea that the neutrino is described by $\phi^{(L)}$ only (or $\phi^{(R)}$ only) was advanced by H. Weyl in 1929; it was rejected by W. Pauli in his *Handbuch* article on the grounds that the wave equation for $\phi^{(L)}$ only (called the Weyl equation) is not manifestly covariant under space inversion (cf. Problem 3–5). With the advent of parity nonconservation in 1957 the Weyl equation was revived by A. Salam, by L. D. Landau, and by T. D. Lee and C. N. Yang.

Muon capture,

$$\mu^- + \mathrm{p} \longrightarrow \mathrm{n} + \nu', \tag{3.470}$$

can also be described by an interaction of the form (3.455); ψ_e and ψ_ν are now replaced by ψ_μ and $\psi_{\nu'}$. The particle ν' whose mass is also consistent with zero turns out again to be left-handed (its antiparticle $\bar{\nu}'$ is right-handed). For some years it was generally believed that ν and ν' were identical; however, motivated by the experimental absence of $\mu^+ \longrightarrow e^+ + \gamma$ (which can be best understood if $\nu \neq \nu'$), B. Pontecorvo and others proposed an experiment to test the assumption that ν is the same as ν'. Noting that ν' also appears in

$$\pi^+ \longrightarrow \mu^+ + \nu' \tag{3.471}$$

(because π^+ can virtually disintegrate into a proton and an antineutron), we can settle the question of the identity or the nonidentity of ν and ν' by examining whether or not the neutral particle from pion decay (3.471) can induce a high-energy neutrino reaction, that is,

$$\nu' + \mathrm{n} \longrightarrow e^- + \mathrm{p}. \tag{3.472}$$

In 1962 M. Schwartz and collaborators established experimentally that (3.472) is forbidden, while

$$\nu' + \mathrm{n} \longrightarrow \mu^- + \mathrm{p} \tag{3.473}$$

is fully allowed for ν' from pion decay; these experimental facts are in agreement with the idea that ν and ν' are different.

---

‡In general, you may introduce as many fields as you like without changing the physical content, so long as the fields introduced are coupled to nothing.

Pion decay and the CPT theorem. As a final example, let us make a comparison between

$$\pi^- \longrightarrow \mu^- + \bar{\nu}' \tag{3.474}$$

and

$$\pi^- \longrightarrow e^- + \bar{\nu}. \tag{3.475}$$

It is appealing to assume that the interactions responsible for (3.474) and (3.475) are the same in form as well as in strength, that is, the interactions are invariant under

$$e^{\pm} \longleftrightarrow \mu^{\pm}, \qquad \nu \longleftrightarrow \nu', \qquad \bar{\nu} \longleftrightarrow \bar{\nu}'. \tag{3.476}$$

In addition it is reasonable to postulate that in (3.475) the field operators ψ_e and ψ_ν enter in the same combination as in nuclear beta decay, viz. in the combination $i\bar{\psi}_e\gamma_5\gamma_\lambda(1+\gamma_5)\psi_\nu$ [which is also equal to $-i\bar{\psi}_e\gamma_\lambda(1+\gamma_5)\psi_\nu$]. With these two hypotheses, the interaction densities may be taken as

$$\mathscr{H}_{\text{int}} = if\frac{\hbar}{m_\pi c}\frac{\partial\phi_\pi}{\partial x_\lambda}[\bar{\psi}_e\gamma_5\gamma_\lambda(1+\gamma_5)\psi_\nu] + \text{Hc}, \tag{3.477}$$

for (3.475) and

$$\mathscr{H}_{\text{int}} = if\frac{\hbar}{m_\pi c}\frac{\partial\phi_\pi}{\partial x_\lambda}[\bar{\psi}_\mu\gamma_5\gamma_\lambda(1+\gamma_5)\psi_{\nu'}] + \text{Hc}, \tag{3.478}$$

for (3.474). The constant f is assumed to be universal, that is, the same for both (3.474) and (3.475); we have inserted the factor $\hbar/m_\pi c$ just to make $(f^2/4\pi\hbar c)$ dimensionless. Using the same argument as in Λ decay, we obtain the following for the π-e process:

$$c^{(1)} = -\frac{i}{\hbar}\frac{f\hbar}{m_\pi c}\left(c\sqrt{\frac{\hbar}{2\omega_\pi}}\right)\left(\sqrt{\frac{m_\nu c^2}{E_\nu}}\right)\left(\sqrt{\frac{m_e c^2}{E_e}}\right)[i\bar{u}_e\gamma_5\gamma_\lambda(1+\gamma_5)v_\nu]$$
$$\times\frac{1}{V^{3/2}}\int_0^t dt'\int d^3x'\left[\left(\frac{\partial}{\partial x'_\lambda}\exp\left[i\frac{p^{(\pi)}\cdot x'}{\hbar}\right]\right)\left(\exp\left[-\frac{ip^{(\bar{\nu})}\cdot x'}{\hbar}-\frac{ip^{(e)}\cdot x'}{\hbar}\right]\right)\right], \tag{3.479}$$

where we have suppressed the momentum and spin indices for the free-particle spinors. The four-gradient acting on the pion plane wave just brings down $ip_\lambda^{(\pi)}/\hbar$; because of energy-momentum conservation we get

$$\begin{aligned} ip_\lambda^{(\pi)}\bar{u}_e\gamma_5\gamma_\lambda(1+\gamma_5)v_\nu &= -ip_\lambda^{(\pi)}\bar{u}_e\gamma_\lambda(1+\gamma_5)v_\nu \\ &= -i\bar{u}_e(\gamma\cdot p^{(e)} + \gamma\cdot p^{(\bar{\nu})})(1+\gamma_5)v_\nu \\ &= m_e c\bar{u}_e(1+\gamma_5)v_\nu, \end{aligned} \tag{3.480}$$

where we have used (3.388) and (3.390) with $m_\nu = 0$. Let us now work in the rest system of the decaying pion with the z-axis along the antineutrino momentum denoted by $\mathbf{p}$. Using the explicit forms of free-particle spinors (cf. Eqs. 3.115 and 3.384), we can easily verify that for $\mathbf{p}$ along the quantization axis,

$$(1+\gamma_5)v_\nu^{(1)}(\mathbf{p}) = 2v_\nu^{(1)}(\mathbf{p}), \qquad (1+\gamma_5)v_\nu^{(2)}(\mathbf{p}) = 0 \tag{3.481}$$

as $m_\nu \longrightarrow 0$, which supports our earlier assertion that $(1+\gamma_5)\psi_\nu^{(-)}$ creates right-handed antineutrinos only. The amplitude for the production of a right-handed (spin-*down*) electron with momentum $\mathbf{p}_e = -\mathbf{p}$ is proportional to

$$\sqrt{\frac{m_\nu c^2}{E_\nu}}\sqrt{\frac{m_e c^2}{E_e}}\,\bar{u}_e^{(2)}(-\mathbf{p})(1+\gamma_5)v_\nu^{(1)}(\mathbf{p})$$

$$= -2\sqrt{\frac{m_\nu c^2 + E_\nu}{2E_\nu}}\sqrt{\frac{m_e c^2 + E_e}{2E_e}}\left(0, 1, 0, -\frac{|\mathbf{p}|c}{E_e + m_e c^2}\right)\begin{pmatrix}0\\-1\\0\\1\end{pmatrix}$$

$$= \frac{E_e + m_e c^2 + |\mathbf{p}|c}{\sqrt{E_e(E_e + m_e c^2)}} = \sqrt{\frac{m_\pi c^2}{2E_e}}, \tag{3.482}$$

as $m_\nu \longrightarrow 0$, where we have used

$$|\mathbf{p}| = \frac{E_\nu}{c} = \frac{(m_\pi^2 - m_e^2)c}{2m_\pi},$$

$$E_e + |\mathbf{p}|c = m_\pi c^2. \tag{3.483}$$

Note that the use of the normalization convention (3.389) does not cause any trouble in the $m_\nu \longrightarrow 0$ limit. For the other (left-handed) spin state of the electron we can readily show that

$$\bar{u}_e^{(1)}(-\mathbf{p})(1+\gamma_5)v_\nu^{(1)}(\mathbf{p}) = 0, \tag{3.484}$$

which we could actually have guessed from angular momentum conservation. (The pion is spinless and the orbital angular momentum cannot have a nonvanishing component along $\mathbf{p}$.) The only other thing we must evaluate is the phase-space factor

$$\frac{|\mathbf{p}|^2 d|\mathbf{p}|}{d(E_\nu + E_e)} = \frac{E_e|\mathbf{p}|^2}{m_\pi c^3} = \frac{E_e(m_\pi^2 - m_e^2)^2}{4m_\pi^3 c}. \tag{3.485}$$

Collecting all the factors and using the Golden Rule, we finally get

$$\frac{1}{\tau(\pi^- \longrightarrow e^- + \bar{\nu})} = \left(\frac{f^2}{4\pi\hbar c}\right)\frac{m_e^2(m_\pi^2 - m_e^2)^2 c^2}{\hbar m_\pi^5}. \tag{3.486}$$

Similarly

$$\frac{1}{\tau(\pi^- \longrightarrow \mu^- + \bar{\nu}')} = \left(\frac{f^2}{4\pi\hbar c}\right)\frac{m_\mu^2(m_\pi^2 - m_\mu^2)^2 c^2}{\hbar m_\pi^5}. \tag{3.487}$$

What is more significant, the ratio of (3.475) to (3.474) is

$$\frac{\Gamma(\pi^- \longrightarrow e^- + \bar{\nu})}{\Gamma(\pi^- \longrightarrow \mu^- + \bar{\nu}')} = \left(\frac{m_e}{m_\mu}\right)^2\frac{(m_\pi^2 - m_e^2)^2}{(m_\pi^2 - m_\mu^2)^2} \simeq 1.3 \times 10^{-4}. \tag{3.488}$$

Note that the muonic decay mode is much more frequent despite the smaller Q-value available; it is hard to beat the factor $(m_e/m_\mu)^2 \simeq 2.3 \times 10^{-5}$.

Actually the fact that the transition probability computed with (3.477) is proportional to m_e^2 is not surprising. To see this we first note that, if $m_e = 0$, then

$$\bar{\psi}_e\gamma_5\gamma_\lambda(1+\gamma_5)\psi_\nu = \bar{\psi}_e(1-\gamma_5)\gamma_5\gamma_\lambda\psi_\nu$$

would create only left-handed electrons. But the simultaneous creation of a left-handed electron and a right-handed $\bar{\nu}$ in pion decay is strictly forbidden by angular momentum conservation. It is therefore expected that the amplitude for (3.473) is proportional to m_e.

Although the result (3.488) was first obtained by M. Ruderman and R. Finkelstein‡ as early as 1949, only two years after the discovery of the pion, for a very long time the experimentalists failed to find the electronic decay mode; the quoted upper limit was an order of magnitude *lower* than (3.488). When it became evident that the electron and neutrino enter in the combination $\bar{\psi}_e\gamma_\lambda(1+\gamma_5)\psi_\nu$ in beta decay and also in muon decay, the idea of describing pion decay by (3.477) and (3.478) became so attractive that some theoreticians even said, "The experiments must be wrong." Truly enough, a number of experiments performed since 1958 at CERN and other places brilliantly confirmed the predicted ratio (3.488). We mention this example just to emphasize that the power of relativistic quantum mechanics in making quantitative predictions is not limited to the domain of electromagnetic interactions.

As a by-product of our calculation we obtain

$$(f^2/4\pi\hbar c) \simeq 1.8 \times 10^{-15}, \tag{3.489}$$

when the observed pion lifetime is inserted in (3.487). So this is again an example of "weak" interactions. We may also mention that a similar calculation with Hc in (3.478) gives the following result:

$$\begin{aligned}\Gamma(\pi^+ \longrightarrow \mu^+ + \nu'_L) &= \Gamma(\pi^- \longrightarrow \mu^- + \bar{\nu}'_R) \neq 0,\\ \Gamma(\pi^+ \longrightarrow \mu^+ + \nu'_R) &= \Gamma(\pi^- \longrightarrow \mu^- + \bar{\nu}'_L) = 0.\end{aligned} \tag{3.490}$$

This means that charge conjugation invariance [which demands $\Gamma(\pi^+ \longrightarrow \mu^+ + \nu'_L) = \Gamma(\pi^- \longrightarrow \mu^- + \bar{\nu}'_L)$, etc.] is violated as well as invariance under parity.

It is very important to note that the equality of the π^+ and the π^- lifetime holds despite the breakdown of charge conjugation invariance.

Following E. P. Wigner, we shall define the time-reversal operation which reverses both momentum and spin. This definition of time reversal agrees with our intuitive notion of the "reversal of motion."§ Consider now the product

---

‡The original calculation of Ruderman and Finkelstein was based on the parity conserving interaction $if(\hbar/m_\pi c)(\partial\phi_\pi/\partial x_\lambda)\bar{\psi}_e\gamma_5\gamma_\lambda\psi_\nu$. The numerical result for the ratio of the two decay modes is unchanged since

$$\bar{\psi}_e\gamma_5\gamma_\lambda\psi_\nu = \tfrac{1}{2}[\bar{\psi}_e\gamma_5\gamma_\lambda(1+\gamma_5)\psi_\nu + \bar{\psi}_e\gamma_5\gamma_\lambda(1-\gamma_5)\psi_\nu],$$

and the calculation for the emission of a right-handed ν proceeds in the same way as that for the emission of a left-handed ν.

§If the final state is not described by a plane wave, a much more careful discussion of time reversal is needed. This is because time reversal interchanges the roles of an incoming and an outgoing state. For a detailed discussion of time reversal in both nonrelativistic and relativistic quantum mechanics, see Sakurai (1964), Chapter 4.

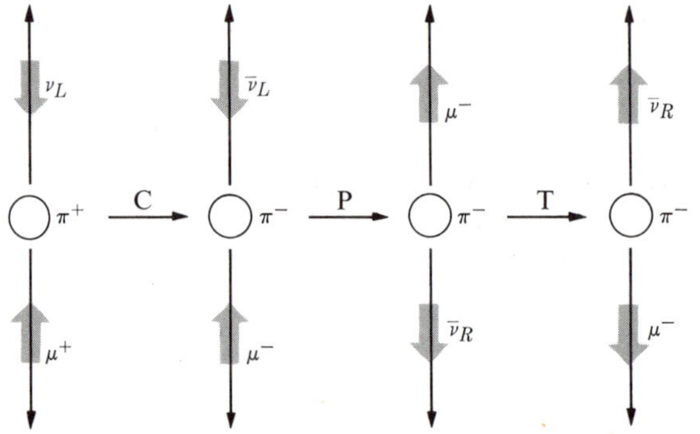

Fig. 3–10. CPT operation applied to pion decay. The gray arrows indicate the spin direction.

Table 3–5

TRANSFORMATION PROPERTIES UNDER CHARGE CONJUGATION, PARITY, TIME REVERSAL, AND CPT

	Q	**p**	**J**	Helicity
Charge conjugation‡	−	+	+	+
Parity	+	−	+	−
Time reversal	+	−	−	+
CPT	−	+	−	−

‡By charge conjugation we mean "particle-antiparticle conjugation."

of charge conjugation, parity, and time reversal, denoted by CPT. The relations (3.490) are completely consistent with invariance under CPT; the decay configuration obtained by applying CPT is seen to be a physically realizable configuration with the same transition probability (Fig. 3–10 and Table 3–5).

In the formalism developed in this chapter, the invariance under CPT can be regarded as a consequence of the use of a Hermitian Hamiltonian. (Actually the pion decay interaction we have written turns out also to be invariant under CP, the product of charge conjugation and parity, but it is easy to write down a Hermitian Hamiltonian that violates CP invariance; for example, the electric-dipole-moment interaction of Problem 3–16.) In a more axiomatic formulation of quantum field theory it can be shown that it is impossible to violate CPT invariance without drastically altering the structure of the theory. The fact that a wide class of quantum field theories is invariant under CPT was first demonstrated by G. Lüders and W. Pauli; its physical implications have been extensively discussed by T. D. Lee, R. Oehme, and C. N. Yang.§

§For a more complete discussion of CPT invariance see Nishijima (1964), pp. 329–339; Sakurai (1964), Chapter 6; Streater and Wightman (1963), pp. 142–146.

PROBLEMS

3-1. Prove Eq. 3.13.

3-2. Consider an electron in a uniform and constant magnetic field **B** along the z-axis. Obtain the most general four-component positive-energy eigenfunctions. Show that the energy eigenvalues are given by

$$E = \sqrt{m^2c^4 + c^2p_3^2 + 2ne\hbar c\,|\mathbf{B}|}$$

with $n = 0, 1, 2, \ldots$ List all the constants of the motion.

3-3. (a) Construct the normalized wave functions for $E > 0$ plane waves which are eigenstates of the helicity operator. (The momentum $\mathbf{p}$ is not necessarily assumed to be in the z-direction.) Evaluate the expectation values of $\boldsymbol{\Sigma}\cdot\hat{\mathbf{p}}$ and $i\gamma_5\boldsymbol{\gamma}\cdot\hat{\mathbf{p}} = \beta\boldsymbol{\Sigma}\cdot\hat{\mathbf{p}}$, that is, $\int \psi^\dagger\boldsymbol{\Sigma}\cdot\hat{\mathbf{p}}\psi\, d^3x = \int \bar{\psi}i\gamma_5\boldsymbol{\gamma}\cdot\hat{\mathbf{p}}\psi\, d^3x$ and $\int \psi^\dagger i\gamma_5\boldsymbol{\gamma}\cdot\hat{\mathbf{p}}\psi\, d^3x = \int \bar{\psi}\boldsymbol{\Sigma}\cdot\mathbf{p}\psi\, d^3x$.

(b) Construct the normalized wave function for an $E > 0$ transversely polarized plane wave whose propagation and spin ($\langle\boldsymbol{\Sigma}\rangle$) directions are along the positive z- and the positive x-axes respectively. Evaluate the expectation values of Σ_1 and $i\gamma_5\gamma_1 = \beta\Sigma_1$.

3-4. Let $\psi(\mathbf{x}, t)$ be the wave function for an $E > 0$ plane wave.

(a) Show that $\gamma_4\psi(-\mathbf{x}, t)$ is indeed the wave function for the $E > 0$ plane wave with momentum reversed and the spin direction unchanged.

(b) Show that $i\Sigma_2\psi^*(\mathbf{x}, -t)$ is the wave function for the "time-reversed state" in the sense that the momentum and spin directions are both reversed while the sign of the energy is unchanged.

3-5. Consider the coupled two-component equations (cf. Eq. 3.26)

$$-i\frac{\partial}{\partial x_\mu}\sigma_\mu^{(R)}\phi^{(R)} = -\frac{mc}{\hbar}\phi^{(L)},$$

$$i\frac{\partial}{\partial x_\mu}\sigma_\mu^{(L)}\phi^{(L)} = -\frac{mc}{\hbar}\phi^{(R)},$$

$$\sigma_\mu^{(R)} = (\boldsymbol{\sigma}, i), \qquad \sigma_\mu^{(L)} = (\boldsymbol{\sigma}, -i).$$

(a) Write the above wave equations in the form

$$\left(\gamma'_\mu\frac{\partial}{\partial x_\mu} + \frac{mc}{\hbar}\right)\psi' = 0,$$

where

$$\psi' = \begin{pmatrix}\phi^{(R)}\\ \phi^{(L)}\end{pmatrix}.$$

Obtain the explicit forms of γ'_μ and $\gamma'_5 = \gamma'_1\gamma'_2\gamma'_3\gamma'_4$ and check $\{\gamma'_\mu, \gamma'_\nu\} = 2\delta_{\mu\nu}$ for $\mu, \nu = 1, \ldots, 5$. Find a unitary S that relates the new (Weyl) set $\{\gamma'_\mu\}$ and the standard (Dirac-Pauli) set $\{\gamma_\mu\}$ via $\gamma'_\mu = S\gamma_\mu S^{-1}$.

(b) Without using the invariance properties of the Dirac theory, discussed in Section 3-4, establish directly the relativistic covariance of the coupled two-component equations. In particular, find 2×2 matrices $S^{(R)}$ and $S^{(L)}$ such that

$$\phi^{(R,L)\prime}(x') = S^{(R,L)}\phi^{(R,L)}(x),$$

where x and x' are related by a Lorentz transformation along the x_1-axis. Discuss also the covariance of the coupled wave equations under parity; $\phi^{(R,L)\prime}(x') = ?$, where $x' = (-\mathbf{x}, ict)$.

3–6. (a) Reduce

$$\bar{u}^{(r')}(\mathbf{p}')\sigma_{k4}u^{(r)}(\mathbf{p}) = iu^{(r')\dagger}(\mathbf{p}')\gamma_k u^{(r)}(\mathbf{p})$$

to the form $\chi^{(s')\dagger}\mathcal{O}\chi^{(s)}$ where $u^{(r)}(\mathbf{p})$ and $u^{(r')}(\mathbf{p}')$ are positive-energy free particle spinors and $\chi^{(s)}$ and $\chi^{(s')}$ are the corresponding two-component Pauli spinors. Assume $|\mathbf{p}| = |\mathbf{p}'|$.

(b) The interaction of a neutron with the electromagnetic field can be represented at low energies by the phenomenological Hamiltonian density

$$\mathscr{H}_{\text{int}} = -(\kappa |e| \hbar/2m_n c)[\tfrac{1}{2}F_{\mu\nu}\bar{\psi}\sigma_{\mu\nu}\psi], \qquad \kappa = -1.91.$$

Using the expression obtained in (a), compute the differential cross section for the scattering of a slow neutron by an electrostatic field in Born approximation. Interpret your results physically. Show, in particular, that even a very slow neutron can have an attractive short-ranged ($\delta^3(\mathbf{x})$-like) interaction (known as the *Foldy interaction*) with an electron. Show also that if this interaction is represented by an equivalent spherical potential well of radius $r_0 = (e^2/4\pi m_e c^2)$, then the depth of the potential is 4.08 keV.‡

3–7. Using the uncertainty principle, N. Bohr argued that it is impossible to prepare a beam of free electrons with all spins pointing in the same direction by means of a Stern-Gerlach type experiment or, more generally, by a selection mechanism that takes advantage of the classical concept of a particle trajectory.§ Justify Bohr's thesis from the point of view of the Dirac theory of the electron. Would this still be true even if the electron had a large anomalous magnetic moment?

3–8. Consider the unitary operator

$$U = \sqrt{\frac{mc^2 + |E|}{2|E|}} + \frac{\beta\boldsymbol{\alpha}\cdot\mathbf{p}c}{\sqrt{2|E|(mc^2 + |E|)}}$$

where $|E|$ is to be understood as a "square root operator," $\sqrt{c^2\mathbf{p}^2 + m^2c^4}$, which just gives $\sqrt{|\mathbf{p}|^2c^2 + m^2c^4}$ as its eigenvalue when it acts on the wave function for a free-particle plane wave.

(a) Show that the application of U to the wave function for a positive- (negative-) energy plane wave results in a wave function whose lower (upper) two components are missing.

(b) The above transformation (first considered in 1948 by M. H. Pryce‖) can be regarded as a change in the representation of the Dirac matrices. Show that the operator in the usual representation that corresponds to $\mathbf{x}$ in the new representation is

$$\mathbf{X} = \mathbf{x} + \frac{ic\hbar\beta\boldsymbol{\alpha}}{2|E|} - \frac{ic^3\hbar\beta(\boldsymbol{\alpha}\times\mathbf{p})\mathbf{p}}{2|E|^2(|E| + mc^2)} - \frac{c^2\hbar(\boldsymbol{\sigma}\times\mathbf{p})}{2|E|(|E| + mc^2)}.$$

(The operator $\mathbf{X}$ is sometimes called the *mean position operator*.)

---

‡Although the Foldy interaction is much weaker than the nuclear interaction, it can be detected experimentally as thermal neutrons are scattered coherently by high-Z atoms.

§The original argument of Bohr can be found in Mott and Massey (1949), p. 61.

‖This transformation is a special case of a class of transformations considered extensively by L. L. Foldy and S. A. Wouthuysen, and by S. Tani.

(c) Prove

$$\dot{\mathbf{X}} = \pm c^2\mathbf{p}/|E|.$$

(Note that the expression for $\dot{\mathbf{X}}$ is free from *Zitterbewegung*, but the nonlocality of $\mathbf{X}$ is the price we must pay.)

3–9. At some instant of time (say, $t = 0$) the normalized wave function for an electron is known to be

$$\psi(\mathbf{x}, 0) = \frac{1}{\sqrt{V}} \begin{pmatrix} a \\ b \\ c \\ d \end{pmatrix} e^{ip_3x_3/\hbar}$$

where a, b, c, and d are independent of the space-time coordinates and satisfy $|a|^2 + |b|^2 + |c|^2 + |d|^2 = 1$. Find the probabilities of observing the electron with (i) $E > 0$ spin-up, (ii) $E > 0$ spin-down, (iii) $E < 0$ spin-up, and (iv) $E < 0$ spin-down.

3–10. Consider a Dirac particle subject to a (three-dimensional) spherical well potential

$$V(r) = -V_0 < 0 \qquad \text{for} \quad r < r_0,$$
$$V(r) = 0 \qquad \text{for} \quad r > r_0.$$

(a) Obtain the exact four-component energy eigenfunctions for $j = \frac{1}{2}$ ("even") bound states, where "even" means even orbital parity for the upper two components.

(b) Set up an equation that determines the energy eigenvalues.

(c) What happens if the strength of the potential is increased so that V_0 becomes comparable to or larger than $2mc^2$?

3–11. Discuss how the numbers of nodes of the radial functions $G(r)$ and $F(r)$ of the hydrogen atom are related to the quantum numbers n, j, and l.

3–12. Consider a positive-energy electron at rest with spin-up. Suppose at $t = 0$ we apply an external (classical vector potential) represented by

$$\mathbf{A} = \hat{\mathbf{n}}_3 a \cos \omega t$$

(where a is space-time independent, and $\hat{\mathbf{n}}_3$ stands for a unit vector in the positive z-direction). Show that for $t > 0$ there is a finite probability of finding the electron in a negative-energy state if the negative-energy states are assumed to be initially empty. In particular work out quantitatively the following two cases: $\hbar\omega \ll 2mc^2$ and $\hbar\omega \approx 2mc^2$.

3–13. (a) Prove

$$\sum_{r=1}^{4} u_\alpha^{(r)}(\mathbf{p}) u_\beta^{(r)\dagger}(\mathbf{p}) = \sum_{s=1}^{2} [u_\alpha^{(s)}(\mathbf{p}) u_\beta^{(s)\dagger}(\mathbf{p}) + v_\alpha^{(s)}(-\mathbf{p}) v_\beta^{(s)\dagger}(-\mathbf{p})]$$
$$= \delta_{\alpha\beta}(|E|/mc^2).$$

(b) Using the above relation, prove the equal-time anticommutation relation between $\psi_\alpha(x)$ and $\psi_\beta^\dagger(x)$ (3.409).

3–14. Consider an allowed pure Fermi β^+ decay. Only the vector interaction contributes, and it is legitimate to replace γ_μ by γ_4. Without using the trace techniques to be introduced in the next chapter show that the positron-neutrino angular correlation is given by

$$1 + (v/c)_{e^+} \cos\theta,$$

where θ is the angle between the momenta of e^+ and ν.‡ Compute also the expectation value of the positron helicity, and interpret your result, using angular momentum conservation.

3–15. We showed in Section 2–3 that in atomic physics, favored radiative transitions take place between states of opposite parities, that is, parity must *change*. Meanwhile, we know that the fundamental electromagnetic interaction *conserves* parity. Resolve this paradox.

3–16. (a) Suppose the electron had a static electric dipole moment analogous to the magnetic moment. Write a Hamiltonian density that represents the interaction of the electric dipole moment with the electromagnetic field and prove that it is not invariant under parity.

(b) Show that the electric dipole moment interaction would lead to a mixing (in the quantum-mechanical sense) between the $2s\frac{1}{2}$ and $2p\frac{1}{2}$ states of the hydrogen atom. From the fact that the observed and the calculated Lamb shift agree within 0.5 Mc, obtain an upper limit on the magnitude of the electric dipole moment of the electron. (*Caution:* the relevant matrix element vanishes if the nonrelativistic wave functions for $2s\frac{1}{2}$ and $2p\frac{1}{2}$ are used.)

---

‡Because $\gamma_4 u(\mathbf{p}) = u(-\mathbf{p})$, it is easy to see that the angular coefficient would be just opposite in sign if the interaction were scalar. Historically the vector nature of Fermi transitions was first established by J. S. Allen and collaborators, who showed that the positron and the neutrino tend to be emitted in the same direction in the β^+ decay of A^{35} (almost pure Fermi).

CHAPTER 4

COVARIANT PERTURBATION THEORY

4-1. NATURAL UNITS AND DIMENSIONS

In relativistic quantum theory it is most convenient to use units in which action (energy times time) is measured in $\hbar$, and length divided by time is measured in c. This system of units is referred to as *natural units*. When we work in natural units (as we shall for the remaining part of this book) the symbol m_e may mean not only the electron mass but also any one of the following:

a) reciprocal length

$$m_e\left(=\frac{1}{\hbar/m_e c}\right)=\frac{1}{3.86\times 10^{-11}\,\text{cm}}, \tag{4.1}$$

b) reciprocal time

$$m_e\left(=\frac{1}{\hbar/m_e c^2}\right)=\frac{1}{1.29\times 10^{-21}\,\text{sec}}, \tag{4.2}$$

c) energy

$$m_e(=m_e c^2)=0.511\text{ MeV}, \tag{4.3}$$

d) momentum

$$m_e(=m_e c)=0.511\text{ MeV}/c. \tag{4.4}$$

In natural units the fine-structure constant is simply

$$\alpha=\frac{e^2}{4\pi}\simeq\frac{1}{137.04}. \tag{4.5}$$

In the "pure" electrodynamics of electrons and photons, apart from m_e there is no other constant that has the dimension of mass or reciprocal length. So, from purely dimensional considerations, we may argue that the cross section for any electrodynamic process which is of order e^2 in the amplitude must be of the order of

$$\frac{\alpha^2}{m_e^2}=\left(\frac{e^2}{4\pi}\right)^2\frac{1}{m_e^2}=r_0^2, \tag{4.6}$$

r_0 being the classical radius of the electron. Indeed, apart from the numerical factor $8\pi/3$, (4.6) is precisely the Thomson cross section computed in Chapter 2. Working with α and m_e, we can form other constants which are familiar from atomic physics. For instance, the Bohr radius is given by

$$a_0=1/\alpha m_e, \tag{4.7}$$

so that

$$a_0:\frac{1}{m_e}:r_0\simeq 137:1:\frac{1}{137}. \tag{4.8}$$

The Rydberg energy is simply

$$\text{Ry}_\infty=\alpha^2\,m_e/2=13.6\text{ eV}. \tag{4.9}$$

In high-energy physics a typical energy scale is set by the rest energy of the pion,

$$m_\pi = \begin{cases} 139.6 \text{ MeV} & \text{for } \pi^\pm \\ 135.0 \text{ MeV} & \text{for } \pi^0. \end{cases} \tag{4.10}$$

A typical length is then given by

$$1/m_\pi \simeq \sqrt{2} \times 10^{-13} \text{ cm}, \tag{4.11}$$

which is accurate to 0.07 % if the charged pion mass is used. Since the dimensionless constant analogous to α is of the order of unity in strong-interaction physics, a typical cross section in collisions of strong-interacting particles is expected to be

$$1/m_\pi^2 \simeq 2 \times 10^{-26} \text{ cm}^2 = 20 \text{ mb}, \tag{4.12}$$

which is, in fact, roughly equal to the total cross section for $\pi^\pm$-p collisions at high energies. Suppose there is a short-lived state that decays via strong interactions. The decay width of such a short-lived state must be of the order of the pion rest energy. As an example, we may mention that the decay width of the 750-MeV ρ meson is about 100 MeV.‡ A decaying state whose decay width is 100 MeV has a mean lifetime of 6.58×10^{-24} sec, which is a typical time scale of strong interactions. As we mentioned in Section 3–11, the square of a weak-interaction coupling constant is smaller than that of a strong-interaction coupling constant by a factor of about 10^{14}. As a result, the time scale characteristic of weak interactions is about $10^{14} \times 6.6 \times 10^{-24}$ sec. Indeed the mean lifetimes of particles like K mesons and Λ hyperons are in the vicinity of 10^{-8} to 10^{-10} sec.

In any practical calculation we can work with $\hbar = 1$, $c = 1$ throughout. At the very end of the calculation we may insert, if we wish, the numerical values for $\hbar$ and c in the appropriate places to make the dimension right. In going from natural units to the usual units all we need to remember are:

$$\hbar = 6.58 \times 10^{-22} \text{ MeV-sec}, \tag{4.13}$$

and

$$\hbar c = 1.973 \times 10^{-11} \text{ MeV-cm}. \tag{4.14}$$

For instance, take the example of Σ^0 decay ($\Sigma^0 \longrightarrow \Lambda + \gamma$), which the reader may have worked out in Problem 2-5. The mean lifetime in the system of units used in Chapter 2 turns out to be

$$\tau = \frac{\pi(m_\Lambda + m_\Sigma)^2 c^2}{e^2 \hbar k^3}. \tag{4.15}$$

---

‡More quantitatively, if the ρ decay interaction ($\rho^0 \longrightarrow \pi^+ + \pi^-$) is represented by

$$\mathscr{H}_{\text{int}} = g\phi_\mu\left(\phi_{ch}^\dagger \frac{\partial}{\partial x_\mu}\phi_{ch} - \frac{\partial \phi_{ch}^\dagger}{\partial x_\mu}\phi_{ch}\right)$$

where ϕ_μ, ϕ_{ch}, and $\phi_{ch}^\dagger$ are the field operators for ρ^0, π^-, and π^+, then we can show

$$\Gamma(\rho^0 \longrightarrow \pi^+ + \pi^-) = \left(\frac{g^2}{4\pi}\right)\frac{(m_\rho^2 - 4m_\pi^2)^{3/2}}{12m_\rho^2}$$

which gives $(g^2/4\pi) = 2$ when $\Gamma = 100$ MeV.

If we compute the same quantity using natural units, we, of course, get

$$\tau = \frac{\pi(m_\Lambda + m_\Sigma)^2}{e^2 k^3}. \tag{4.16}$$

Now k may mean the wave number of the photon $E_\gamma/\hbar c$, the photon energy E_γ, the photon momentum E_γ/c or the photon angular frequency $\omega = E_\gamma/\hbar$; likewise m_Λ may mean $m_\Lambda c^2$, $(\hbar/m_\Lambda c)^{-1}$ or $(\hbar/m_\Lambda c^2)^{-1}$. But we need not worry about all this. All we should note is that when m_Λ, m_Σ, and k in the formula (4.16) are expressed in MeV, we must multiply the right-hand side of (4.16) by $\hbar = 6.58 \times 10^{-22}$ MeV-sec to make the expression dimensionally correct. Noting that $(m_\Lambda + m_\Sigma)c^2 = 2307$ MeV, and $E_\gamma = 74.5$ MeV, we obtain

$$\tau = \frac{137}{4}\left(\frac{2307 \text{ MeV}}{74.5 \text{ MeV}}\right)^2 \frac{6.58 \times 10^{-22} \text{ MeV-sec}}{74.5 \text{ MeV}} \simeq 2.9 \times 10^{-19} \text{ sec}. \tag{4.17}$$

It is a complete waste of time to keep track of all the $\hbar$ and c at each stage of the calculation, to convert the value of the hyperon mass in grams, or to express e in terms of $(\text{gram})^{1/2}(\text{cm})^{3/2}(\text{sec})^{-1}$, etc. It is so easy to get numbers if we just work in natural units.

To conclude this section we shall make a few remarks on the dimensions of the field operators. The Hamiltonian density $\mathscr{H}$ or the Lagrangian density $\mathscr{L}$ must have the dimension of energy divided by volume:

$$[\mathscr{L}] = \text{energy}/(\text{length})^3 \xrightarrow{\text{natural units}} (\text{mass})^4. \tag{4.18}$$

In the free-field Lagrangian density the boson field ϕ (or $\mathbf{A}$) appears in the form $(\partial\phi/\partial x_\mu)^2$, $(mc/\hbar)^2\phi^2$, $[(1/c)\partial\mathbf{A}/\partial t]^2$, etc. This means that

$$[\phi] = \sqrt{\text{energy}/\text{length}} \xrightarrow{\text{natural units}} \text{mass}. \tag{4.19}$$

Note that $c\sqrt{\hbar/2\omega V}$ which appears in the plane-wave expansion (2.60) indeed has the dimension of $\sqrt{\text{energy}/\text{length}}$ which is expected since the creation and annihilation operators defined in Chapter 2 are dimensionless. According to (3.360) the Dirac field ψ appears in the Lagrangian density as $\hbar\bar{\psi}\gamma_\mu(\partial\psi/\partial x_\mu)$ and $mc^2\bar{\psi}\psi$. Hence

$$[\psi] = \sqrt{1/(\text{length})^3} \xrightarrow{\text{natural units}} (\text{mass})^{3/2}. \tag{4.20}$$

Note again that $\sqrt{mc^2/EV}$ which appears in the plane-wave expansion (3.392) is dimensionally correct since the creation and annihilation operators as well as the free-particle spinor $u^{(s)}(\mathbf{p})$ are dimensionless. In constructing interaction densities it is worth remembering that expressions like $i\bar{\psi}\gamma_\mu\psi A_\mu$ and ϕ^4 automatically have the dimension of the Hamiltonian density in natural units, while an expression like ϕ^3 must be multiplied by a constant that has the dimension of mass.

4-2. S-MATRIX EXPANSION IN THE INTERACTION REPRESENTATION

Interaction representation. In this chapter we wish to discuss a variety of physical processes assuming the validity of perturbation theory. The usual time-dependent perturbation theory used in Chapter 2 is not too convenient for treating processes involving relativistic particles because neither the space-time coordinates nor the energy-momentum variables appear in a covariant manner. It turns out that within the framework of the Hamiltonian formalism manifestly covariant expressions for transition matrices can be most readily obtained if we use a representation known as the *interaction representation*. The utility of this representation in quantum field theory was first recognized by S. Tomonaga and by J. Schwinger in 1947.

In the usual Schrödinger representation the time development of a state vector $\Phi^{(S)}(t)$ is given by the Schrödinger equation

$$i(\partial\Phi^{(S)}/\partial t) = H^{(S)}\Phi^{(S)}, \tag{4.21}$$

where the superscript (S) stands for Schrödinger. We can split $H^{(S)}$ into two parts,

$$H^{(S)} = H_0^{(S)} + H_I^{(S)}, \tag{4.22}$$

where $H_0^{(S)}$ is the Hamiltonian in the absence of perturbation, and $H_I^{(S)}$ stands for the perturbation Hamiltonian. In field theory $H_0^{(S)}$ is taken to be the free-field Hamiltonian while $H_I^{(S)}$ is the space integral of the interaction Hamiltonian density.

Consider now the transformation

$$\begin{aligned} \Phi(t) &= e^{iH_0^{(S)}t}\Phi^{(S)}, \\ O(t) &= e^{iH_0^{(S)}t}O^{(S)}e^{-iH_0^{(S)}t}, \end{aligned} \tag{4.23}$$

where $O^{(S)}$ is an operator in the Schrödinger representation. The above transformation can be regarded as a change in the representation—from the Schrödinger representation to what is known as the interaction representation. The symbols Φ and O which have no superscripts stand for the state vector and the operator in the interaction representation which correspond respectively to $\Phi^{(S)}$ and $O^{(S)}$ in the Schrödinger representation. The time development of $\Phi(t)$ is given by

$$\begin{aligned} i\frac{\partial\Phi}{\partial t} &= i\left[iH_0^{(S)}e^{iH_0^{(S)}t}\Phi^{(S)} + e^{iH_0^{(S)}t}\frac{\partial\Phi^{(S)}}{\partial t}\right] \\ &= -H_0^{(S)}e^{iH_0^{(S)}t}\Phi^{(S)} + e^{iH_0^{(S)}t}(H_0^{(S)} + H_I^{(S)})e^{-iH_0^{(S)}t}e^{iH_0^{(S)}t}\Phi^{(S)} \\ &= H_I\Phi, \end{aligned} \tag{4.24}$$

where H_I is the perturbation Hamiltonian in the interaction representation. Meanwhile, noting that $O^{(S)}$ and O coincide at $t = 0$, we have

$$O(t) = e^{iH_0^{(S)}t}O(0)e^{-iH_0^{(S)}t}, \tag{4.25}$$

from which follows the equation of motion for O,

$$\dot{O} = i[H_0^{(S)}, O] = i[H_0, O], \tag{4.26}$$

since $H_0^{(S)}$ and H_0 are identical. It is amusing to note that (4.24) would give a constant state vector just as in the Heisenberg representation if H_I were zero and that (4.26) resembles the Heisenberg equation with H_0 replacing the total Hamiltonian. We could have anticipated all this, since the transformation (4.23) would be identical with the familiar transformation that relates the Schrödinger representation to the Heisenberg representation if $H_0^{(S)}$ were replaced by $H^{(S)}$.

The differences among the three representations are summarized more systematically as follows:

a) In the *Heisenberg representation* the state vector is completely time independent, whereas the operator which is time dependent satisfies the Heisenberg equation.

b) In the *Schrödinger representation* it is the state vector that carries the entire time dependence.

c) In the *interaction representation* the time dependence is split into two parts: the time dependence of the noninteracting systems is carried by the operators, while the remaining part of the time dependence due to the interaction is taken up by the state vector. For this reason the interaction representation is sometimes referred to as the *intermediate representation.*

In quantum field theory the field operator in the interaction representation satisfies the equation of motion (4.26) which involves H_0 only. This means that the field operator satisfies the *free-field* equation even in the presence of the interaction provided that we use the interaction representation. This is indeed gratifying since from the previous chapters we already know a great deal about the explicit forms of the free-field operators. We can, for instance, use the plane-wave expansion (3.392) for the electron field.

As for the interaction Hamiltonian, H_I differs from $H_I^{(S)}$ in that the field operators occurring in H_I are time-dependent free-field operators.

***U*-matrix and *S*-matrix.** We now propose to obtain a formal solution to the differential equation (4.24). First, let us define an operator $U(t, t_0)$ by

$$\Phi(t) = U(t, t_0)\Phi(t_0), \tag{4.27}$$

where $\Phi(t_0)$ is the state vector that characterizes the system at some fixed time t_0. Clearly we have

$$U(t_0, t_0) = 1. \tag{4.28}$$

Equation (4.24) is equivalent to

$$i(\partial/\partial t)U(t, t_0) = H_I(t)U(t, t_0). \tag{4.29}$$

But the differential equation (4.29) and the boundary condition (4.28) can be combined to give a single integral equation

$$U(t, t_0) = 1 - i\int_{t_0}^{t} dt\, H_I(t)U(t, t_0). \tag{4.30}$$

Let us solve this by the well-known iteration procedure. We have

$$\begin{aligned} U(t, t_0) &= 1 - i\int_{t_0}^{t} dt_1 H_I(t_1)\left[1 - i\int_{t_0}^{t_1} dt_2 H_I(t_2)U(t_2, t_0)\right] \\ &= 1 - i\int_{t_0}^{t} dt_1 H_I(t_1) + (-i)^2\int_{t_0}^{t} dt_1\int_{t_0}^{t_1} dt_2 H_I(t_1)H_I(t_2) + \cdots \\ &\quad + (-i)^n\int_{t_0}^{t} dt_1\int_{t_0}^{t_1} dt_2\cdots\int_{t_0}^{t_{n-1}} dt_n H_I(t_1)H_I(t_2)\cdots H_I(t_n) + \cdots \end{aligned} \tag{4.31}$$

They physical meaning of U is as follows. Suppose the system is known to be in state i at t_0. The probability of finding the system in state f at some later time t

is given by

$$|\langle\Phi_f|\,U(t,t_0)\Phi_i\rangle|^2 = |\,U_{fi}(t,t_0)|^2. \tag{4.32}$$

The average transition probability per unit time for $i \longrightarrow f$ is

$$w = \frac{1}{t-t_0}|\,U_{fi}(t,t_0) - \delta_{fi}|^2, \tag{4.33}$$

since the "1" in (4.31) would be present even if the interaction were absent. Although this ratio has a rather complicated time dependence when the time interval $t - t_0$ is small, it approaches a definite limit as $t_0 \longrightarrow -\infty$ and $t \longrightarrow +\infty$, as we shall see explicitly in working out actual physical problems in the next section. This motivates us to define what is known as the S-matrix by

$$S = U(\infty, -\infty). \tag{4.34}$$

From (4.27) we see that the S-matrix connects $\Phi(\infty)$ and $\Phi(-\infty)$ according to

$$\Phi(\infty) = S\Phi(-\infty). \tag{4.35}$$

The explicit form of the S-matrix can easily be read from (4.31):

$$\begin{aligned} S &= S^{(0)} + S^{(1)} + S^{(2)} + S^{(3)} + \cdots \\ &= 1 - i\int_{-\infty}^{\infty} dt_1 H_I(t_1) + (-i)^2 \int_{-\infty}^{\infty} dt_1 \int_{-\infty}^{t_1} dt_2 H_I(t_1)H_I(t_2) + \cdots \\ &\quad + (-i)^n \int_{-\infty}^{\infty} dt_1 \int_{-\infty}^{t_1} dt_2 \cdots \int_{-\infty}^{t_{n-1}} dt_n H_I(t_1)H_I(t_2)\cdots H_I(t_n) + \cdots. \end{aligned} \tag{4.36}$$

Whether or not the expansion (4.36) is useful depends on the strength of the interaction. In quantum electrodynamics the S-matrix expansion turns out to converge rapidly so that we can obtain results that agree extremely well with observation just by considering the first few terms; this, of course, is because the dimensionless constant α is small compared to unity. In the realm of strong-interaction physics, however, the situation is not so fortunate because the relevant expansion parameter is of the order of unity.

At this stage it is perhaps of some interest to discuss the connection between the formalism developed here and Dirac's time-dependent perturbation theory extensively used in Chapter 2. First, let us look at the formula (2.107), which describes the time development of the coefficient $c_j(t)$. We may imagine that the c_j form a column matrix that corresponds to the state vector Φ in the language of this section; the differential equation (2.107) is then seen to correspond precisely to (4.24). The coefficient $c_j(t)$ is nothing more than the matrix element of the expansion (4.31) taken between the initial (k) and the final (j) state; $c_j^{(0)} = \delta_{jk}$ corresponds to the leading term, 1, of (4.31); $c_j^{(1)}(t)$ given by (2.109) is readily seen to correspond to the second term in the expansion (4.31) with $t_0 = 0$ if we recall that

$$\begin{aligned} \langle j|\,H_I(t)\,|k\rangle &= \langle j|\,e^{iH_0t}H_I^{(S)}e^{-iH_0t}\,|k\rangle \\ &= \langle j|\,H_I^{(S)}\,|k\rangle e^{i(E_j-E_k)t}. \end{aligned} \tag{4.37}$$

A similar remark holds for higher-order terms. From this point of view the basic idea of interaction representation is as old as Dirac's time-dependent perturbation theory.

There is, however, an important difference between the perturbation method we learn in nonrelativistic quantum mechanics and the perturbation method we shall develop in this chapter. In the nonrelativistic theory, using (4.37), we first integrate over t_1, t_2, etc. in (4.36) to obtain a *transition matrix* T_{fi} that goes as follows:

$$S_{fi} = \delta_{fi} - 2\pi i\delta(E_f - E_i)T_{fi}, \tag{4.38}$$

$$T_{fi} = H^{(S)}_{fi} + \sum_n \frac{H^{(S)}_{fn}H^{(S)}_{ni}}{E_i - E_n + i\epsilon} + \cdots \tag{4.39}$$

(where we have assumed for simplicity that $H^{(S)}$ is time independent) and subsequently perform a space integration of the type

$$H^{(S)}_{fn} = (1/V)\int e^{-i\mathbf{p}_f\cdot\mathbf{x}}H^{(S)}e^{i\mathbf{p}_n\cdot\mathbf{x}}\,d^3x, \tag{4.40}$$

etc. This is not how we shall proceed in this chapter. Instead, we propose to perform the space and time integrations *simultaneously* in a covariant manner. As a result, it becomes possible to obtain a manifestly covariant expression for the transition matrix even though neither the basic equation (4.24) nor the expression for the interaction Hamiltonian

$$H_I = \int \mathscr{H}_{\text{int}}\,d^3x \tag{4.41}$$

is covariant. We shall see how this method works in the following sections.

Unitarity. We shall finish this section by discussing the unitarity property of the S-matrix:

$$SS^\dagger = 1, \tag{4.42a}$$

$$S^\dagger S = 1, \tag{4.42b}$$

or equivalently

$$\sum_n S_{fn}S^*_{in} = \delta_{fi}, \tag{4.43a}$$

$$\sum_n S^*_{nf}S_{ni} = \delta_{fi}. \tag{4.43b}$$

Although there are several ways of proving this important property, we present here a straightforward proof (to second order) based on the explicit form of the S-matrix expansion (4.36). We assume that H_I is linear in some coupling parameter λ; for instance, in the case of

$$H_I = -ie\int \bar{\psi}\gamma_\mu\psi A_\mu\,d^3x$$

the parameter λ is just e itself. The expansion (4.36) can then be regarded as a power series in λ; $S^{(0)}$ varies as $\lambda^0 = 1$, $S^{(1)}$ as λ, $S^{(2)}$ as λ^2, etc. Meanwhile the unitarity equation (4.43a) can be written as

$$\sum_n (S^{(0)}_{fn} + S^{(1)}_{fn} + S^{(2)}_{fn} + \cdots)(S^{(0)*}_{in} + S^{(1)*}_{in} + S^{(2)*}_{in} + \cdots) = \delta_{fi}. \tag{4.44}$$

This must hold as an identity relation for any value of λ. So the coefficient of *each* power of λ on the left-hand side of (4.44) must vanish if the unitarity relation is to be satisfied. Thus

$$\sum_n S^{(0)}_{fn} S^{(0)*}_{in} - \delta_{fi} = 0, \tag{4.45a}$$

$$\sum_n (S^{(0)}_{fn} S^{(1)*}_{in} + S^{(1)}_{fn} S^{(0)*}_{in}) = 0, \tag{4.45b}$$

$$\sum_n (S^{(0)}_{fn} S^{(2)*}_{in} + S^{(2)}_{fn} S^{(0)*}_{in}) + \sum_n S^{(1)}_{fn} S^{(1)*}_{in} = 0, \tag{4.45c}$$

etc. The first expression (4.45a) is trivially satisfied, since

$$S^{(0)}_{fn} = \delta_{fn}, \qquad S^{(0)*}_{in} = \delta_{in}. \tag{4.46}$$

Because of (4.46) the unitarity relation to order λ (4.45b) now becomes

$$i \int_{-\infty}^{\infty} dt \langle i | H_I(t) | f \rangle^* - i \int_{-\infty}^{\infty} dt \langle f | H_I(t) | i \rangle = 0. \tag{4.47}$$

This is satisfied provided that H_I is Hermitian.

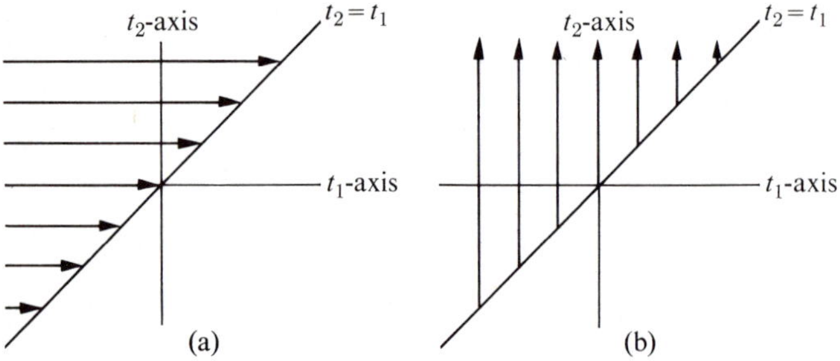

Fig. 4-1. Region of integration in (4.49) and (4.50).

Before we proceed to prove (4.45c) we must prove a very important identity for $S^{(2)}$:

$$\begin{aligned} S^{(2)} &= (-i)^2 \int_{-\infty}^{\infty} dt_1 \int_{-\infty}^{t_1} dt_2 H_I(t_1) H_I(t_2) \\ &= (-i)^2 \int_{-\infty}^{\infty} dt_1 \int_{t_1}^{\infty} dt_2 H_I(t_2) H_I(t_1). \end{aligned} \tag{4.48}$$

To prove this we rewrite the first form by interchanging the variables of integration,

$$S^{(2)} = (-i)^2 \int_{-\infty}^{\infty} dt_2 \int_{-\infty}^{t_2} dt_1 H_I(t_2) H_I(t_1). \tag{4.49}$$

We are supposed to fix t_2 and integrate over t_1 from $-\infty$ to t_2; we then integrate over t_2 from $-\infty$ to ∞. This means that in a two-dimensional $t_1 t_2$ plane we must integrate $H_I(t_2)H_I(t_1)$ over the region above the line $t_2 = t_1$ as indicated in Fig. 4–1(a). But it is clear from Fig. 4–1(b) that we can write (4.49) equally well as

$$S^{(2)} = (-i)^2 \int_{-\infty}^{\infty} dt_1 \int_{t_1}^{\infty} dt_2 H_I(t_2) H_I(t_1), \tag{4.50}$$

which proves (4.48). We shall call the first and the second forms of (4.48) respectively "Form 1" and "Form 2" for $S^{(2)}$.

Using this important identity for $S^{(2)}$ and (4.47) already demonstrated, we see from (4.45c) that in order to prove the unitarity relation to second order in λ, it is sufficient to prove that

$$\sum_n \int_{-\infty}^{\infty} dt \langle f | H_I(t) | n \rangle \int_{-\infty}^{\infty} dt \langle n | H_I(t) | i \rangle$$
$$= -(-i)^2 \left[\int_{-\infty}^{\infty} dt_1 \int_{t_1}^{\infty} dt_2 \langle i | H_I(t_2) H_I(t_1) | f \rangle^* + \int_{-\infty}^{\infty} dt_1 \int_{-\infty}^{t_1} dt_2 \langle f | H_I(t_1) H_I(t_2) | i \rangle \right], \tag{4.51}$$

where we have "Form 2" for $S_{in}^{(2)*}$ in (4.45c) and "Form 1" for $S_{fn}^{(2)}$ in (4.45c). Using the hermiticity of $H_I(t)$ and inserting a complete set of intermediate states $\sum_n | n \rangle \langle n | = 1$ between $H_I(t_1)$ and $H_I(t_2)$, we readily see that the right-hand side of (4.51) is reduced to

$$\sum_n \int_{-\infty}^{\infty} dt_1 \left[\int_{t_1}^{\infty} dt_2 + \int_{-\infty}^{t_1} dt_2 \right] \langle f | H_I(t_1) | n \rangle \langle n | H_I(t_2) | i \rangle, \tag{4.52}$$

which is the same as the left-hand side of (4.51). This method can be generalized to prove that the S-matrix is unitary, in the sense of (4.43a), to all (finite) orders. A similar proof can be carried out for (4.43b).

To understand the physical significance of the unitarity property we have just verified to second order let us go back to the probabilistic interpretation of the U-matrix given earlier. Suppose the system is certain to be in state i at $t = -\infty$. According to (4.32) the probability of finding the system in some particular final state f at $t = \infty$ is then given by $| S_{fi} |^2$. When we consider *all possible* final states, the sum of the probabilities at $t = \infty$ must add up to one:

$$\sum_f | S_{fi} |^2 = 1. \tag{4.53}$$

This requirement is guaranteed if the S-matrix is unitary. As we emphasized in Chapter 3, in relativistic quantum mechanics the probability density of observing a single given particle integrated over all space is no longer constant since particles are freely created or annihilated. For instance, when we start with $e^+ + e^-$ with momenta $\mathbf{p}$ and $-\mathbf{p}$, a possible final state may be not only $e^+ + e^-$ with $\mathbf{p}'$ and $-\mathbf{p}'$ but also 2γ, 3γ, $e^+ + e^- + \gamma$, etc. Yet probability conservation still holds in the sense of (4.53) provided we include in the sum all possible states into which the initial state can make energy-conserving transitions.

We would like to emphasize the role played by the hermiticity of the interaction Hamiltonian in proving the unitarity relation for the S-matrix. Conversely, we might argue on the basis of (4.47) that probability conservation in the quantum theory demands that the interaction density be taken to be Hermitian. In fact, we can take probability conservation or the unitarity requirement as the starting point. Assuming that $S^{(1)}$ is given by $-i$ times the time integral of H_I, we may ask how $S^{(2)}$ must look if unitarity is to be satisfied. In this way, instead of deriving the S-matrix expansion from the Hamiltonian formalism, we may derive the S-matrix expansion from the unitarity requirement for the S-matrix and a few other general principles. This approach, advocated particularly by N. N. Bogoliubov and collaborators is an interesting alternative point of view, but we shall not discuss it any further.‡

In terms of the T-matrix defined by (4.38) the unitarity relation reads

$$i(T_{fi} - T^*_{if}) = 2\pi \sum_n \delta(E_f - E_n) T^*_{nf} T_{ni}. \tag{4.54}$$

From this relation it is possible to derive the optical theorem which relates the total cross section to the imaginary part of the forward scattering amplitude (cf. Eq. 2.219). Since the derivation of this important theorem from (4.54) is treated in most modern textbooks on nonrelativistic quantum mechanics, we shall not give it here.§

A somewhat different application of the unitarity principle will be discussed in Section 4–7 where we shall show how (4.45c) can be used to relate vacuum polarization to pair production by an external field.

4–3. FIRST-ORDER PROCESSES; MOTT SCATTERING AND HYPERON DECAY

Matrix element for electron scattering. We are now in a position to apply the S-matrix formalism developed in the previous section to concrete physical problems. Consider first the simplest problem in relativistic quantum mechanics, the scattering of a relativistic electron by an external (unquantized) potential. Although this problem can be treated in an elementary manner just as well by the usual Born approximation method (cf. Section 3–5), we shall deliberately use the language of quantum field theory to illustrate some of the techniques and formalisms which will be useful for tackling more complex problems.

As in the unquantized Dirac theory, the basic interaction density is taken as

$$\mathscr{H}_{\text{int}} = -ie\bar{\psi}\gamma_\mu\psi A^{(e)}_\mu, \tag{4.55}$$

where $A^{(e)}_\mu$ stands for an external (unquantized) potential. For example, $A^{(e)}_\mu(x)$ may be a time-independent Coulomb potential

$$\mathbf{A} = 0, \qquad A_4 = -ieZ/4\pi r \tag{4.56}$$

generated by an infinitely heavy nucleus (assumed to be fixed in space) or a time-dependent three-vector potential

$$\mathbf{A} = (0, 0, a\cos\omega t), \qquad A_4 = 0, \tag{4.57}$$

where a is space-time independent. Although the form (4.55) is familiar, we have to be a little careful with what is meant by the expression $\bar{\psi}\gamma_\mu\psi$. First, let us rewrite it, using the decomposition (3.417):

$$\begin{aligned}\bar{\psi}\gamma_\mu\psi &= (\bar{\psi}^{(+)} + \bar{\psi}^{(-)})_\alpha(\gamma_\mu)_{\alpha\beta}(\psi^{(+)} + \psi^{(-)})_\beta \\ &= (\gamma_\mu)_{\alpha\beta}(\bar{\psi}^{(+)}_\alpha\psi^{(+)}_\beta + \bar{\psi}^{(-)}_\alpha\psi^{(+)}_\beta + \bar{\psi}^{(+)}_\alpha\psi^{(-)}_\beta + \bar{\psi}^{(-)}_\alpha\psi^{(-)}_\beta).\end{aligned} \tag{4.58}$$

‡See Bogoliubov and Shirkov (1959), pp. 206–226.

§Merzbacher (1961), pp. 495–499; Messiah (1962), p. 866.

As we explained in Section 3–10, the current operator must be so constructed that its vacuum expectation value vanishes. This requirement is not satisfied by (4.58) because $\langle \bar{\psi}_\alpha^{(+)} \psi_\beta^{(-)} \rangle_0$ is infinite. However, the expression

$$(\gamma_\mu)_{\alpha\beta}(\bar{\psi}_\alpha^{(+)}\psi_\beta^{(+)} + \bar{\psi}_\alpha^{(-)}\psi_\beta^{(+)} - \psi_\beta^{(-)}\bar{\psi}_\alpha^{(+)} + \bar{\psi}_\alpha^{(-)}\psi_\beta^{(-)}), \tag{4.59}$$

which differs from (4.58) only by a c-number function $(\gamma_\mu)_{\alpha\beta}\{\bar{\psi}_\alpha^{(+)}, \psi_\beta^{(-)}\}$, does satisfy the required property since $\bar{\psi}^{(+)}$ acting on the vacuum state vanishes. In the literature, to distinguish (4.59) from (4.58) one often denotes (4.59) by $:\bar{\psi}\gamma_\mu\psi:$. We shall not use this notation but always understand $\bar{\psi}\gamma_\mu\psi$ to be (4.59) rather than (4.58) whenever ambiguities arise. The physical meaning of each term of (4.59) has already been discussed in Section 3–10. We merely recall that the first term destroys an electron-positron pair; the second destroys an electron and creates an electron, hence it is appropriate for describing electron scattering; the third destroys a positron and creates a positron (positron scattering); and the fourth creates an electron-positron pair. The four possibilities are schematically shown in Fig. 3–8.

To compute the transition amplitude from state i to a particular final state f, we must consider $S_{fi} - \delta_{fi}$ since the leading term $S^{(0)} = 1$ simply represents the amplitude for a physical situation in which nothing *new* happens. Since $S_{fi} - \delta_{fi}$ is just $S^{(1)}$ to lowest nonvanishing order, we consider

$$S_{fi}^{(1)} = -i\left\langle \Phi_f \middle| -ie \int d^4x\, \bar{\psi}\gamma_\mu\psi A_\mu^{(e)} \middle| \Phi_i \right\rangle, \tag{4.60}$$

where the four-dimensional integration is over all space and time. In our particular problem, Φ_i is an electron state with $p = (\mathbf{p}, iE)$ and s; Φ_f is also an electron state with $p' = (\mathbf{p}', iE')$ and s'. Denoting by b^+ and b'^+ the corresponding creation operators, we have

$$\Phi_i = b^+ |0\rangle, \qquad \Phi_f = b'^+ |0\rangle. \tag{4.61}$$

It is intuitively obvious that out of the four terms in (4.59) only $\bar{\psi}_\alpha^{(-)}\psi_\beta^{(+)}$ contributes. This can be verified rigorously. Take, for instance, $\bar{\psi}_\alpha^{(+)}\psi_\beta^{(+)}$. A typical matrix element will look like $\langle 0 | b' d b'' b^+ | 0 \rangle$ so far as the creation and annihilation operators are concerned, where we have denoted a typical electron annihilation operator contained in $\psi_\beta^{(+)}$ by b'', a typical positron annihilation operator in $\bar{\psi}_\alpha^{(+)}$ by d. But

$$\langle 0 | b' d b'' b^+ | 0 \rangle = \langle 0 | b' b'' b^+ d | 0 \rangle = 0, \tag{4.62}$$

since the positron annihilation operator d acting on the vacuum vanishes. Thus it is sufficient to consider the $\bar{\psi}_\alpha^{(-)}\psi_\beta^{(+)}$ term only. Furthermore, we claim that it is legitimate to replace $\psi^{(+)}$ by $\sqrt{m/EV}\, b u^{(s)}(\mathbf{p}) e^{ip\cdot x}$, $\bar{\psi}^{(-)}$ by $\sqrt{m/E'V}\, b'^+ \bar{u}^{(s')}(\mathbf{p}') e^{-ip'\cdot x}$. To prove this rigorously, let b'' and b'''^+ be a typical annihilation and a creation operator contained in $\psi^{(+)}$ and $\bar{\psi}^{(-)}$ respectively. We must consider

$$\langle 0 | b' b'''^+ b'' b^+ | 0 \rangle = \langle 0 | \{b', b'''^+\}\{b'', b^+\} | 0 \rangle, \tag{4.63}$$

which vanishes if the momentum and spin indices of an annihilation operator (creation operator) in $\psi^{(+)}(\bar{\psi}^{(-)})$ do not coincide with those of the initial (final) electron. Thus, in the language of quantum field theory, the lowest-order scattering is viewed as the simultaneous annihilation of the incident electron and creation of the final outgoing electron at the same space-time point at which the potential $A_\mu^{(e)}$ is operative. This is shown schematically in Fig. 4–2(a). In contrast the second-order amplitude, which we shall not consider here, is represented by Fig. 4–2(b).

The anticommutator $\{b'', b^\dagger\}$ in (4.63) is unity when the momentum and spin indices of b'' coincide with those of $b^\dagger$. The case is similar for $\{b', b'''^\dagger\}$. So we finally get

$$S_{fi}^{(1)} = -e\int d^4x\left[\sqrt{\frac{m}{E'V}}\bar{u}^{(s')}(\mathbf{p}')e^{-ip'\cdot x}\right] \times \gamma_\mu\left[\sqrt{\frac{m}{EV}}u^{(s)}(\mathbf{p})e^{ip\cdot x}\right]A_\mu^{(e)}(x). \tag{4.64}$$

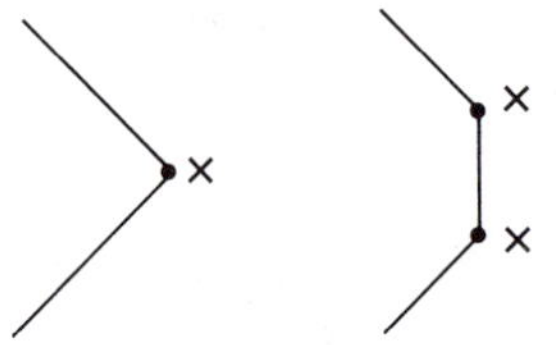

Fig. 4-2. Scattering of an electron by an external potential.

Specializing now to the Coulomb potential case (4.56) and abbreviating $\bar{u}^{(s')}(\mathbf{p}')$ and $u^{(s)}(\mathbf{p})$ by $\bar{u}'$ and u, we obtain

$$S_{fi}^{(1)} = -2\pi i\delta(E-E')\left[-\frac{m}{EV}\frac{Ze^2}{|\mathbf{p}'-p|^2}\bar{u}'\gamma_4 u\right], \tag{4.65}$$

where we have used

$$\begin{aligned}\int d^4x\,A_4^{(e)}e^{i(p-p')\cdot x} &= 2\pi\delta(E-E')\int d^3x(-iZe/4\pi r)e^{-i(\mathbf{p}-\mathbf{p}')\cdot\mathbf{x}}\\ &= -2\pi i\delta(E-E')(Ze/|\mathbf{p}-\mathbf{p}'|^2).\end{aligned} \tag{4.66}$$

The results (4.64) and (4.66) are precisely what we expect when we solve the problem in an elementary manner using the Born approximation method (cf. Section 3–5).

The transition probability is $|S_{fi}^{(1)}|^2$ divided by the time interval $t - t_0$ that appears in (4.33). If we naively square (4.65), we get the square of the δ-function that expresses energy conservation. The meaning of this can be understood if we set $t = T/2$, $t_0 = -T/2$, and let $T \longrightarrow \infty$. By the δ-function in (4.65) we actually mean

$$\delta(E'-E) = \lim_{T\to\infty}(1/2\pi)\int_{-T/2}^{T/2} e^{i(E-E')t_1}\,dt_1. \tag{4.67}$$

But this evaluated at $E' - E = 0$ is just $T/2\pi$. We therefore set‡

$$|2\pi\delta(E'-E)|^2 = 2\pi T\delta(E'-E). \tag{4.68}$$

‡A more "rigorous" way of proving (4.68) is given below:

$$\begin{aligned}|\delta(E'-E)|^2 &= \lim_{T\to\infty}\left|\left(1/2\pi\right)\int_{-T/2}^{T/2} e^{i(E'-E)t}\,dt\right|^2\\ &= \lim_{T\to\infty}\left|\frac{\sin[(E'-E)T/2]}{\pi(E'-E)}\right|^2\\ &= (T/2\pi)\delta(E'-E),\end{aligned}$$

where we have used (2.115).

Because of this, $|S_{fi}|^2/T$ is seen to be independent of T. To get the differential cross-section element into $d\Omega$, we must multiply $|S_{fi}|^2/T$ by the phase space factor and divide by the incident flux density which, in this case, is just $|\mathbf{p}|/EV$. Thus

$$\begin{aligned} d\sigma &= \frac{|S_{fi}|^2}{T}\left[\frac{Vd^3p'}{(2\pi)^3}\right]\left(\frac{EV}{|\mathbf{p}|}\right) \\ &= \frac{Z^2e^4}{4\pi^2}\frac{m^2}{E|\mathbf{p}|}\frac{|\bar{u}'\gamma_4 u|^2}{|\mathbf{p}-\mathbf{p}'|^4}\delta(E'-E)d^3p' \\ &= \frac{Z^2e^4m^2}{4\pi^2}\frac{|\bar{u}'\gamma_4 u|^2}{|\mathbf{p}-\mathbf{p}'|^4}\,d\Omega, \end{aligned} \tag{4.69}$$

where we have used

$$\begin{aligned} d^3p'\delta(E'-E) &= \frac{|\mathbf{p}|^2}{(\partial E/\partial|\mathbf{p}|)}\,d\Omega \\ &= |\mathbf{p}|E\,d\Omega, \end{aligned} \tag{4.70}$$

with $|\mathbf{p}'| = |\mathbf{p}|$.

Spin sum and projection operator. The problem is now reduced to that of evaluating $|\bar{u}'\gamma_4 u|^2$. For a transition from a particular momentum-spin state to some other momentum-spin state we may use the explicit forms of the free-particle spinors, just as we did in discussing weak interaction phenomena (cf. Section 3–11). Quite often, however, we are not interested in the transition probability into a particular spin state, since what is readily observable is the angular distribution of scattered electrons without any regard to their spin orientations. We must then sum over the final spin states. This spin sum can most readily be performed using a trick originally introduced by H. Casimir. First, recall that $|\bar{u}'\gamma_4 u|^2$ really means

$$\begin{aligned} |\bar{u}'\gamma_4 u|^2 &= u^\dagger\gamma_4\gamma_4 u'\bar{u}'\gamma_4 u \\ &= \bar{u}_\delta(\gamma_4)_{\delta\gamma}u'_\gamma\bar{u}'_\beta(\gamma_4)_{\beta\alpha}u_\alpha. \end{aligned} \tag{4.71}$$

So we must evaluate the spin sum $\sum_{s'=1}^{2} u_\gamma^{(s')}(\mathbf{p}')\bar{u}_\beta^{(s')}(\mathbf{p}')$. To do this we take advantage of the following matrix relation:

$$\sum_{s=1}^{2} u_\alpha^{(s)}(\mathbf{p})\bar{u}_\beta^{(s)}(\mathbf{p}) = [(-i\gamma\cdot p + m)/2m]_{\alpha\beta}. \tag{4.72}$$

The proof of this relation goes as follows. First, for the free-particle spinors that represent the electron at rest, we have

$$\begin{aligned} \sum_{s=1}^{2} u^{(s)}(0)\bar{u}^{(s)}(0) &= \begin{pmatrix}1\\0\\0\\0\end{pmatrix}(1\quad 0\quad 0\quad 0) + \begin{pmatrix}0\\1\\0\\0\end{pmatrix}(0\quad 1\quad 0\quad 0) \\ &= \begin{pmatrix}1 & & & \\ & 1 & & \\ & & 0 & \\ & & & 0\end{pmatrix} = \frac{\gamma_4+1}{2} = \frac{m\gamma_4+m}{2m}. \end{aligned} \tag{4.73}$$

Now consider the "inverse" Lorentz transformation of Section 3–4 (cf. Eq. 3.165) which "boosts" the electron at rest.

$$\begin{aligned} u^{(s)}(0) &\longrightarrow u^{(s)}(\mathbf{p}) = S_{\text{Lor}}^{-1} u^{(s)}(0), \\ \bar{u}^{(s)}(0) &\longrightarrow \bar{u}^{(s)}(\mathbf{p}) = \bar{u}^{(s)}(0) S_{\text{Lor}}. \end{aligned} \tag{4.74}$$

Multiplying (4.73) by S_{Lor}^{-1} from the left and by S_{Lor} from the right, we readily obtain (4.72) with the aid of

$$\begin{aligned} S_{\text{Lor}}^{-1} \gamma_4 S_{\text{Lor}} &= \gamma_4 \cosh \chi - i\gamma \cdot \hat{\mathbf{p}} \sinh \chi \\ &= -i\gamma \cdot p/m \end{aligned} \tag{4.75}$$

(cf. Eq. 3.139). The matrix (4.72) can be regarded as the projection operator for the electron spinor $u^{(s)}(\mathbf{p})$ in the sense that when it acts on $u^{(s)}(\mathbf{p})$, we merely obtain $u^{(s)}(\mathbf{p})$ itself, whereas when it acts on the positron spinor $v^{(s)}(\mathbf{p})$, it gives zero. This is evident from (3.391). It is not difficult to show that the analog of (4.71) for the positron spinor is

$$\sum_{s=1}^{2} v_\alpha^{(s)}(\mathbf{p}) \bar{v}_\beta^{(s)}(\mathbf{p}) = -\left(\frac{i\gamma \cdot p + m}{2m}\right)_{\alpha\beta}, \tag{4.76}$$

where p is the momentum-energy four-vector of the *physical* positron state (*not* that of the negative-energy electron state whose absence appears as the positron state). Note that

$$\sum_{s=1}^{2} [u_\alpha^{(s)}(\mathbf{p}) \bar{u}_\beta^{(s)}(\mathbf{p}) - v_\alpha^{(s)}(\mathbf{p}) \bar{v}_\beta^{(s)}(\mathbf{p})] = \delta_{\alpha\beta}. \tag{4.77}$$

(Contrast this with Problem 3–13.)

Cross section for Mott scattering. Let us go back to electron scattering. We may further suppose that the incident electron beam is unpolarized. We can then sum over the initial spin states using Casimir's trick once again and divide by two. So we are led to consider

$$\begin{aligned} \frac{1}{2} \sum_s \sum_{s'} |\bar{u}' \gamma_4 u|^2 &= \frac{1}{2} \sum_s \sum_{s'} \bar{u}_\delta^{(s)}(\mathbf{p}) (\gamma_4)_{\delta\gamma} u_\gamma^{(s')}(\mathbf{p}') \bar{u}_\beta^{(s')}(\mathbf{p}') (\gamma_4)_{\beta\alpha} u_\alpha^{(s)}(\mathbf{p}) \\ &= \frac{1}{2} (\gamma_4)_{\delta\gamma} \left(\frac{-i\gamma \cdot p' + m}{2m}\right)_{\gamma\beta} (\gamma_4)_{\beta\alpha} \left(\frac{-i\gamma \cdot p + m}{2m}\right)_{\alpha\delta} \\ &= \frac{1}{2} \text{Tr}\left[\gamma_4 \left(\frac{-i\gamma \cdot p' + m}{2m}\right) \gamma_4 \left(\frac{-i\gamma \cdot p + m}{2m}\right)\right] \\ &= (1/8m^2) \text{Tr}[(-i\gamma \cdot p'' + m)(-i\gamma \cdot p + m)], \end{aligned} \tag{4.78}$$

where the symbol Tr stands for the "trace" (the sum of the diagonal elements) and p'' is defined by

$$p_k'' = -p_k', \qquad p_4'' = p_4'. \tag{4.79}$$

To evaluate a trace of this kind it is useful to note that for any pair of four-vectors a_μ and b_μ, we have

$$\text{Tr}[(\gamma \cdot a)(\gamma \cdot b)] = \text{Tr}[(\gamma \cdot b)(\gamma \cdot a)] = 4a \cdot b. \tag{4.80}$$

This can easily be proved since

$$\gamma_\mu a_\mu \gamma_\nu b_\nu + \gamma_\nu b_\nu \gamma_\mu a_\mu = 2\delta_{\mu\nu} a_\nu b_\mu = 2a \cdot b, \tag{4.81}$$

and

$$\mathrm{Tr}(1) = 4. \tag{4.82}$$

Recall also that the γ-matrices are traceless so that a term like $\mathrm{Tr}(\gamma \cdot p)$ vanishes. Quite generally the trace of the product of an odd number of γ-matrices vanishes as we can prove most elegantly, using γ_5 which anticommutes with every γ_μ ($\mu = 1, 2, 3, 4$):

$$\begin{aligned}\mathrm{Tr}(\underbrace{\gamma_\mu \gamma_\nu \cdots \gamma_\sigma \gamma_\rho}_{\text{odd number}}) &= \mathrm{Tr}(\gamma_5^2 \gamma_\mu \gamma_\nu \cdots \gamma_\sigma \gamma_\rho) \\ &= \mathrm{Tr}(\gamma_5 \gamma_\mu \gamma_\nu \cdots \gamma_\sigma \gamma_\rho \gamma_5) \\ &= -\mathrm{Tr}(\gamma_\mu \gamma_\nu \cdots \gamma_\sigma \gamma_\rho).\end{aligned} \tag{4.83}$$

Thus

$$\begin{aligned}\frac{1}{2}\sum_s \sum_{s'} |\bar{u}'\gamma_4 u|^2 &= \frac{1}{8m^2}(-4p'' \cdot p + 4m^2) \\ &= \frac{1}{8m^2}(4\mathbf{p}\cdot\mathbf{p}' + 4E^2 + 4m^2) \\ &= \frac{1}{m^2}\left(E^2 - |\mathbf{p}|^2 \sin^2\frac{\theta}{2}\right).\end{aligned} \tag{4.84}$$

Collecting everything we obtain the differential cross section for the scattering of unpolarized electrons:

$$\frac{d\bar{\sigma}}{d\Omega} = \frac{1}{4}(Z\alpha)^2 \frac{E^2}{|\mathbf{p}|^4 \sin^4(\theta/2)}\left(1 - \beta^2 \sin^2\frac{\theta}{2}\right), \tag{4.85}$$

where we have used

$$|\mathbf{p} - \mathbf{p}'|^2 = 4|\mathbf{p}|^2 \sin^2(\theta/2), \qquad \beta = |\mathbf{p}|/E. \tag{4.86}$$

It is seen that, as $v \longrightarrow 0$, Eq. (4.85) is reduced to the famous Rutherford cross section‡

$$\left(\frac{d\bar{\sigma}}{d\Omega}\right) = \frac{1}{4}\frac{(Z\alpha)^2}{m^2 v^4 \sin^4(\theta/2)}. \tag{4.87}$$

It is also interesting to note that for the scattering of the charged Klein-Gordon particle we simply have (4.85), with the $\beta^2 \sin^2(\theta/2)$ term omitted, as the reader may easily verify.

The problem we have been treating was discussed, in 1929, without recourse to perturbation theory by N. F. Mott using an $E > m$ Coulomb wave function (involving hypergeometric functions) obtained earlier by W. Gordon and C. G. Darwin. For this reason the scattering of a relativistic electron by a Coulomb field is referred to as Mott scattering.§ Our perturbative treatment is expected to be valid at sufficiently high energies with relatively low values of Z, or more quantitatively for $Z/(137\beta) \ll 1$.

---

‡Goldstein (1951), p. 84; Merzbacher (1961), p. 228.

§The reader who is interested in the nonperturbative treatment of Mott scattering may consult Mott and Massey (1949), pp. 78–84.

Finite charge distribution. In representing the potential by $A_4 = -iZe/4\pi r$ we have tacitly assumed that the nucleus is pointlike. We shall now show that at sufficiently high energies a large-angle Mott scattering may be used to probe the finite size of the nucleus. Let $\rho(\mathbf{x})$ be the charge density of the nucleus. The $1/r$ potential characteristic of a point nucleus is now modified as follows:

$$-\frac{Ze}{4\pi|\mathbf{x}|} \longrightarrow \frac{1}{4\pi}\int d^3x' \frac{\rho(\mathbf{x}')}{|\mathbf{x}-\mathbf{x}'|}. \tag{4.88}$$

Hence, calling the momentum transfer $\mathbf{q}$

$$\mathbf{q} = \mathbf{p}' - \mathbf{p}, \tag{4.89}$$

we see that the analog of (4.66) is

$$\begin{aligned}\int d^4x\, A_4 e^{i(p-p')\cdot x} &= 2\pi i\delta(E'-E)\int d^3x \int d^3x' \frac{\rho(\mathbf{x}')e^{-i\mathbf{q}\cdot\mathbf{x}}}{4\pi|\mathbf{x}-\mathbf{x}'|}\\ &= 2\pi i\delta(E'-E)\int d^3x' e^{-i\mathbf{q}\cdot\mathbf{x}'}\rho(\mathbf{x}')\int d^3x \frac{e^{-i\mathbf{q}\cdot(\mathbf{x}-\mathbf{x}')}}{4\pi|\mathbf{x}-\mathbf{x}'|}\\ &= -2\pi i\delta(E'-E)\frac{Ze}{|\mathbf{q}|^2}F(\mathbf{q}),\end{aligned} \tag{4.90}$$

where the function $F(\mathbf{q})$, called the *form factor* of the nucleus, is essentially the Fourier transform of the charge distribution

$$F(\mathbf{q}) = \frac{\int \rho(\mathbf{x})e^{-i\mathbf{q}\cdot\mathbf{x}}d^3x}{-Ze}. \tag{4.91}$$

For a point nucleus where we have $\rho(x) = -Ze\delta^{(3)}(x)$, $F(\mathbf{q})$ would be identically equal to unity in agreement with our earlier result. Evidently the cross section is given by

$$(d\bar{\sigma}/d\Omega) = (d\bar{\sigma}/d\Omega)_{\text{point}}|F(\mathbf{q})|^2, \tag{4.92}$$

where $(d\bar{\sigma}/d\Omega)_{\text{point}}$ stands for the expression (4.85). If $|\mathbf{q}|^2 = 4|\mathbf{p}|^2\sin^2(\theta/2)$ is not too large, we can expand the exponential in (4.91) as follows:

$$\begin{aligned}F(\mathbf{q}) &= \frac{1}{-Ze}\int\left(1 - i\mathbf{q}\cdot\mathbf{x} - \frac{(\mathbf{q}\cdot\mathbf{x})^2}{2} + \cdots\right)\rho(\mathbf{x})\,d^3x\\ &= 1 - \frac{\langle r^2\rangle}{6}|\mathbf{q}|^2 + \cdots,\end{aligned} \tag{4.93}$$

where we have taken advantage of the fact that $\rho(\mathbf{x})$ is spherically symmetric. [Clearly $F(\mathbf{q})$ depends only on $|\mathbf{q}|^2$ whenever the charge density is spherically symmetric.] The quantity

$$\langle r^2\rangle = \int r^2\rho\, d^3x\Big/\int \rho\, d^3x \tag{4.94}$$

that appears in the expansion (4.93) is known as the *mean square radius*. The relations (4.92) and (4.93) have actually been used experimentally to determine the size of the nucleus, particularly by R. Hofstadter and coworkers.

Helicity change; spin projection operator. In deriving the differential cross section (4.85) we assumed that the incident electron beam is unpolarized. We also know that the final expression (4.85) is easy to arrive at, once we master Casimir's trace method. However, it is sometimes instructive to work out how the spin direction changes as the electron undergoes scattering. For this reason let us evaluate the S-matrix element between definite helicity states. Clearly factors like m/E and $1/|\mathbf{p} - \mathbf{p}|^2$ are the same as before. We are concerned with $\bar{u}'\gamma_4 u$, where u and u' are now eigenspinors of helicity. Let $u^{(+)}(\mathbf{p})$ and $u^{(-)}(\mathbf{p})$ stand for the free-particle spinors for electrons with momentum $\mathbf{p}$ and helicities $+1$ and -1 respectively. Choosing the z-axis along the initial momentum $\mathbf{p}$, we find that $u^{(+)}(\mathbf{p})$ and $u^{(-)}(\mathbf{p})$ coincide, of course, with $u^{(1)}(\mathbf{p})$ and $u^{(2)}(\mathbf{p})$. We must now construct the helicity eigenspinors $u^{(\pm)}(\mathbf{p}')$, where $\mathbf{p}'$ is no longer along the z-axis. We can do this most easily by using the rotation operator of Section 3–4 (cf. also Problem 3–3a). If the scattering plane (the plane determined by $\mathbf{p}$ and $\mathbf{p}'$) is chosen to coincide with the xz-plane, all we need to do is rotate the electron state of helicity $+1$ (-1) moving in the positive z-direction by the scattering angle θ around the y-axis, that is,‡

$$u^{(+)}(\mathbf{p}') = S_{\text{rot}}^{-1}u^{(1)}(\mathbf{p}) = \left(\cos\frac{\theta}{2} - i\Sigma_2 \sin\frac{\theta}{2}\right)u^{(1)}(\mathbf{p})$$

$$= \sqrt{\frac{E+m}{2m}}\left(\begin{array}{c|c} \cos\frac{\theta}{2} - i\sigma_2\sin\frac{\theta}{2} & 0 \\ \hline 0 & \cos\frac{\theta}{2} - i\sigma_2\sin\frac{\theta}{2}\end{array}\right)\begin{pmatrix} 1 \\ 0 \\ \frac{|\mathbf{p}|}{E+m} \\ 0\end{pmatrix}$$

$$= \sqrt{\frac{E+m}{2m}}\begin{pmatrix} \cos\frac{\theta}{2} \\ \sin\frac{\theta}{2} \\ \frac{|\mathbf{p}|}{E+m}\cos\frac{\theta}{2} \\ \frac{|\mathbf{p}|}{E+m}\sin\frac{\theta}{2}\end{pmatrix}. \tag{4.95}$$

Similarly

$$u^{(-)}(\mathbf{p}') = \sqrt{\frac{E+m}{2m}}\begin{pmatrix} -\sin\frac{\theta}{2} \\ \cos\frac{\theta}{2} \\ \frac{|\mathbf{p}|}{E+m}\sin\frac{\theta}{2} \\ -\frac{|\mathbf{p}|}{E+m}\cos\frac{\theta}{2}\end{pmatrix}. \tag{4.96}$$

With these helicity eigenspinors it is now straightforward to evaluate the amplitudes for "helicity-nonflip" scattering and "helicity-flip" scattering, illustrated in

‡The term S_{rot}^{-1} rather than S_{rot} appears because we are rotating the *physical state* rather than the *coordinate system.*

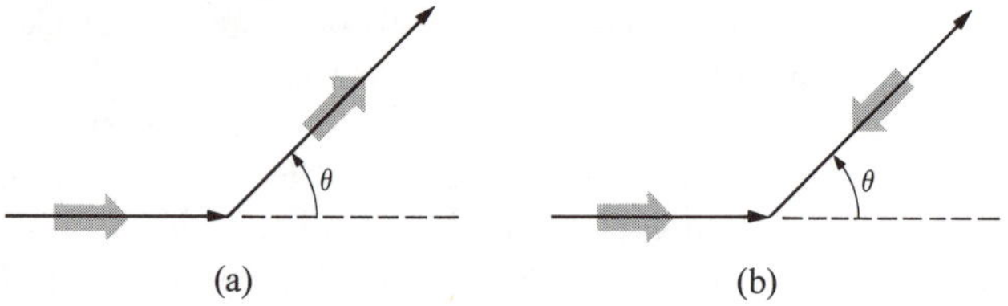

Fig. 4–3. (a) Helicity nonflip scattering, (b) helicity flip scattering. The gray arrows indicate the spin direction.

Fig. 4–3(a) and (b). Apart from a common proportionality factor, they are simply

$$\begin{aligned}\bar{u}^{(+)}(\mathbf{p}')\gamma_4 u^{(+)}(\mathbf{p}) &= \frac{E+m}{2m}\left[\cos\frac{\theta}{2} + \left(\frac{|\mathbf{p}|}{E+m}\right)^2 \cos\frac{\theta}{2}\right] \\ &= \frac{E}{m}\cos\frac{\theta}{2}, \end{aligned} \tag{4.97a}$$

and

$$\begin{aligned}\bar{u}^{(-)}(\mathbf{p}')\gamma_4 u^{(+)}(\mathbf{p}) &= \frac{E+m}{2m}\left[-\sin\frac{\theta}{2} + \left(\frac{|\mathbf{p}|}{E+m}\right)^2 \sin\frac{\theta}{2}\right] \\ &= -\sin\frac{\theta}{2}. \end{aligned} \tag{4.97b}$$

Two limiting cases of (4.97) are of interest. First, if the electron is nonrelativistic, the ratio of the probability of helicity nonflip to that of helicity flip is $\cot^2(\theta/2)$ which is recognized to be the ratio of the probabilities of observing an $s_3 = \frac{1}{2}$ electron with spin in the direction $\mathbf{p}'$ and in the direction $-\mathbf{p}'$‡. This is expected since the electrostatic field of the nucleus cannot change the spin direction of a very slowly moving electron. On the other hand, for extreme relativistic electrons the probability of helicity flip to that of helicity nonflip goes to zero, as $(m/E)^2$ for a fixed θ. This is again not surprising if we recall from Section 3–11 that (a) apart from a minus sign, chirality coincides with helicity for extreme relativistic fermions and (b) the γ_4 interaction connects states of the same chirality. To summarize, the spin direction does not change in the scattering of slowly moving electrons, while it is the helicity that is unchanged in the scattering of extreme relativistic electrons. As is evident from the derivation, this statement holds quite generally for the lowest-order scattering of a spin-$\frac{1}{2}$ particle by any potential that transforms like the fourth component of a vector field.

Summing the two transition probabilities for helicity flip and helicity nonflip, we see that the lowest-order cross section for Mott scattering of a *longitudinally polarized* electron is the same as that of an *unpolarized* electron since

$$\begin{aligned}|\bar{u}^{(+)}(\mathbf{p}')\gamma_4 u^{(+)}(\mathbf{p})|^2 + |\bar{u}^{(-)}(\mathbf{p}')\gamma_4 u^{(+)}(\mathbf{p})|^2 &= \left(\frac{E}{m}\right)^2 \cos^2\frac{\theta}{2} + \sin^2\frac{\theta}{2} \\ &= \frac{1}{m^2}\left(E^2 - |\mathbf{p}|^2 \sin^2\frac{\theta}{2}\right), \end{aligned} \tag{4.98}$$

‡See, for example, Merzbacher (1961), pp. 276–277.

which is completely identical to (4.84). This does not mean, however, that we can never detect electron polarization by Mott scattering. It can be shown that if the scattering of a *transversely* polarized electron is computed to *higher* orders, the differential cross section does depend on $\sin\phi$, where ϕ is the azimuthal angle measured from the polarization direction of the incident electron.

When we obtained (4.84) using the trace method, it was unnecessary to use the explicit form of the free-particle spinor. In computing the transition probability between particular helicity states, it is also possible to use the trace method without recourse to the explicit forms of the helicity eigenspinors. All we need to do is construct the projection operator for a particular spin state just as we can compute the sum taken over only the electron states (as distinguished from positron states) once we have the electron projection operator (4.72).

In the Pauli spin formalism, it is well known that

$$\frac{1+\boldsymbol{\sigma}\cdot\hat{\mathbf{n}}}{2} \tag{4.99}$$

is the projection operator for a two-component spinor whose spin direction is $\hat{\mathbf{n}}$. Motivated by this, we may try

$$\frac{1+\boldsymbol{\Sigma}\cdot\hat{\mathbf{n}}}{2} \tag{4.100}$$

in the relativistic theory. However, as we emphasized in Section 3–3, a free-particle spinor is, in general, not an eigenspinor of $\boldsymbol{\Sigma}\cdot\hat{\mathbf{n}}$. Moreover, this form is not manifestly covariant. So, remembering that Σ_k and $\beta\Sigma_k = i\gamma_5\gamma_k$ have the same nonrelativistic limit, we may try instead

$\beta = \gamma_4$

$$\frac{1+i\gamma_5\gamma\cdot w}{2}, \tag{4.101}$$

where w is a four-vector such that in the electron-rest system it is a unit (three-) vector with a vanishing fourth component:

$$w\,|_{\text{rest frame}} = (\hat{\mathbf{n}}, 0). \tag{4.102}$$

More generally,

$$w^2 = |\mathbf{w}|^2 - |w_0|^2 = 1, \qquad w\cdot p = 0 \tag{4.103}$$

in *any* frame. The physical meaning of w is such that when we consider a Lorentz transformation which brings the electron to rest, then in that rest frame, (4.101) is the projection operator for the electron spin state whose spin points in the direction $\hat{\mathbf{n}}$. If we are interested in constructing the spin projection operator for a transversely polarized electron, then the four-vector w is again given by

$$w^{(\text{trans})} = (\hat{\mathbf{n}}, 0). \tag{4.104}$$

This is clear because, quite generally, a three-vector is unchanged under a Lorentz transformation in a direction perpendicular to the direction of the three-vector. The reader who has worked out Problem 3–3(b) will recognize that the use of $i\gamma_5\boldsymbol{\gamma}\cdot\hat{\mathbf{n}}$ is completely reasonable for a transversely polarized electron. On the other hand, if we want the projection operator for a longitudinally polarized state with helicity $+1$, we must use

$$w^{(\text{long})} = \left(\frac{E\hat{\mathbf{p}}}{m}, \frac{i|\mathbf{p}|}{m}\right), \tag{4.105}$$

which evidently satisfies (4.103). In fact, it is not difficult to verify that $\boldsymbol{\Sigma}\cdot\hat{\mathbf{p}}$ and $(i\gamma_5\boldsymbol{\gamma}\cdot\hat{\mathbf{p}}E - \gamma_5\gamma_4|\mathbf{p}|)/m$ have the *same* effect when they act on helicity eigenspinors.*

It is important to note that $i\gamma\cdot p$ and $i\gamma_5\gamma\cdot w$ commute:

$$\begin{aligned}[\gamma\cdot p, \gamma_5\gamma\cdot w] &= -\gamma_5(\gamma\cdot p\,\gamma\cdot w + \gamma\cdot w\,\gamma\cdot p) \\ &= -2\gamma_5 p\cdot w = 0,\end{aligned} \tag{4.106}$$

where we have used (4.81) and (4.103). It is therefore possible to construct a simultaneous eigenspinor of $-i\gamma\cdot p$ and $i\gamma_5\gamma\cdot w$ with eigenvalues m and 1 respectively. We shall denote such an eigenspinor by $u^{(w)}(\mathbf{p})$. To summarize the various free-particle spinors defined in this text, the description of momentum eigenstates based on $u^{(1)}$ and $u^{(2)}$ makes use of the eigenvalue of the 2×2 Pauli matrix σ_3 for the upper two components in the Dirac-Pauli representation. The description in terms of $u^{(+)}$ and $u^{(-)}$ is characterized by the eigenvalue of the helicity operator $\boldsymbol{\Sigma}\cdot\hat{\mathbf{p}}$. Finally the description in terms of $u^{(w)}(\mathbf{p})$ makes use of the eigenvalue of $i\gamma_5\gamma\cdot w$.

In practical calculations we often need to evaluate $u_\alpha^{(w)}(\mathbf{p})\bar{u}_\beta^{(w)}(\mathbf{p})$ *without* the spin sum (cf. Eq. 4.71). In analogy with (4.27) we expect that this is given by the product of the electron- (as distinguished from the positron-) projection operator multiplied by the covariant spin-projection operator (4.101):

$$u_\alpha^{(w)}(\mathbf{p})\bar{u}_\beta^{(w)}(\mathbf{p}) = \left(\frac{-i\gamma\cdot p + m}{2m}\right)_{\alpha\gamma}\left(\frac{1 + i\gamma_5\gamma\cdot w}{2}\right)_{\gamma\beta}. \tag{4.107}$$

This assertion can be proved in a simple manner as follows: First, (4.107) is evidently true for the spin-up electron at rest,

$$\begin{aligned}\left(\frac{\gamma_4 m + m}{2m}\right)\left(\frac{1 + i\gamma_5\gamma_3}{2}\right) &= \begin{pmatrix}1&0&0&0\\0&0&0&0\\0&0&0&0\\0&0&0&0\end{pmatrix}\begin{pmatrix}1&0&0&0\\0&0&0&0\\0&0&0&0\\0&0&0&1\end{pmatrix} \\ &= \begin{pmatrix}1\\0\\0\\0\end{pmatrix}(1\quad 0\quad 0\quad 0).\end{aligned} \tag{4.108}$$

Considering a rotation which transforms a unit vector along the z-axis into an arbitrary unit vector $\hat{\mathbf{n}}$, we obtain

$$u^{(w)}(0)\bar{u}^{(w)}(0) = \left(\frac{\gamma_4 m + m}{2m}\right)\left(\frac{1 + i\gamma_5\gamma\cdot\hat{\mathbf{n}}}{2}\right) \tag{4.109}$$

for any at-rest spinor with spin along $\hat{\mathbf{n}}$. We can then "Lorentz-boost" (4.109) to get (4.107) in the most general case just as we obtained (4.72) from (4.73). As an

* The general form of w appears on p. 314.

example, we may mention that we can evaluate $|\bar{u}^{(+)}(\mathbf{p}')\gamma_4 u^{(+)}(\mathbf{p})|^2$ without recourse to the particular form of the helicity eigenspinors

$$
\begin{aligned}
|\bar{u}^{(+)}(\mathbf{p}')\gamma_4 u^{(+)}(\mathbf{p})|^2 &= \bar{u}_\delta^{(w)}(\mathbf{p})(\gamma_4)_{\delta\gamma} u_\gamma^{(w')}(\mathbf{p}')\bar{u}_\beta^{(w')}(\mathbf{p}')(\gamma_4)_{\beta\alpha} u_\alpha^{(w)}(\mathbf{p}) \\
&= \mathrm{Tr}\left[\gamma_4 \frac{(-i\gamma\cdot p' + m)}{2m}\frac{(1 + i\gamma_5\gamma\cdot w')}{2}\gamma_4 \frac{(-i\gamma\cdot p + m)}{2m}\frac{(1 + i\gamma_5\gamma\cdot w)}{2}\right], \qquad (4.110)
\end{aligned}
$$

where, since we are considering longitudinally polarized states with helicity plus one, w and w' are given by (4.105). A straightforward (but tedious) calculation shows, as it should, that this is nothing more than (4.97a) squared.‡

Pair annihilation and pair creation. So far we have just talked about electron scattering by an external potential. It is important to note that an external potential is capable of not only scattering electrons (or positrons) but also annihilating or creating electron-positron pairs. Take, for instance,

$$
e^- + e^+ \longrightarrow \text{vacuum}. \qquad (4.111)
$$

Although this cannot take place in the absence of a potential because of energy momentum conservation, if there is a time-dependent potential whose angular frequency is equal to the sum of the electron and the positron energies, we expect, in general, a finite transition probability for (4.111), as we shall see in a moment. The initial state Φ_i is an electron-positron state represented by

$$
\Phi_i = |e^- e^+\rangle = b^{(s_-)\dagger}(\mathbf{p}_-)d^{(s_+)\dagger}(\mathbf{p}_+)|0\rangle, \qquad (4.112)
$$

where $(\mathbf{p}_-, s_-)$ and $(\mathbf{p}_+, s_+)$ characterize the electron and the positron state respectively. We have, to lowest order,

$$
S_{fi}^{(1)} = -i\left\langle 0 \left| -ie\int d^4x \bar{\psi}\gamma_\mu\psi A_\mu \right| e^- e^+\right\rangle. \qquad (4.113)
$$

This time the only term in (4.58) that contributes is $\bar{\psi}_\alpha^{(+)}(\gamma_\mu)_{\alpha\beta}\psi_\beta^{(+)}$, which annihilates an electron-positron pair. Using the now familiar argument, we readily get

$$
S_{fi}^{(1)} = -e\int d^4x \left[\sqrt{\frac{m}{E_+V}}\bar{v}^{(s_+)}(\mathbf{p}_+)e^{ip_+\cdot x}\right]\gamma_\mu\left[\sqrt{\frac{m}{E_-V}}u^{(s_-)}(\mathbf{p}_-)e^{ip_-\cdot x}\right]A_\mu^{(e)}(x) \qquad (4.114)
$$

with $p_\pm = (\mathbf{p}_\pm, iE_\pm) = (\mathbf{p}_\pm, i\sqrt{|\mathbf{p}_\pm|^2 + m^2})$. This expression is strikingly similar to the first-order S-matrix for electron scattering given by (4.64); the only difference is that the adjoint wave function for the outgoing electron is replaced by the adjoint wave function that represents the annihilated positron. This similarity is *not* a coincidence. After all, in the hole-theoretic description pair annihilation is nothing more than the "scattering" of a positive-energy electron into an empty negative-energy state. We shall come back to this similarity between (4.64) and (4.114) when we discuss Feynman's space-time approach in Section 4–5.

It is clear that $S_{fi}^{(1)}$ for pair annihilation vanishes for a time-independent potential (for example, the static Coulomb potential) because of $\delta(E_+ + E_-)$ that arises

‡In this particular example, it does not pay to use the covariant spin-projection operator. It is much faster to work with (4.95) as we have done.

as a result of the t integration. This, however, need not be the case for a time-dependent potential such as (4.57) with $\omega = E_+ + E_-$, since the integrand now contains a term whose time dependence goes as $\exp[-i(E_+ + E_- - \omega)t]$, which results in $\delta(E_+ + E_- - \omega)$ when integrated. Physically this difference arises because a time-dependent potential can transfer energy to, or absorb energy from, the charged particles, whereas a time-independent potential (such as the static Coulomb potential) can transfer momentum to the charged particles but cannot exchange energy with them.

We may remark that a process inverse to (4.111),

$$\text{vacuum} \longrightarrow e^+ + e^-, \tag{4.115}$$

can be treated in a completely analogous manner. In Problem 4–5 the reader is asked to compute the probability per unit time per unit volume that the time-dependent potential (4.57) will create an electron-positron pair in vacuum. We have to admit that the potential (4.57) is not a very realistic one. In this connection we should mention that the electromagnetic excitation and the de-excitation of spin-zero nuclei of the same Z (for example, $0^{16} \rightleftarrows 0^{16*}$ where 0^{16*} stands for an excited $J = 0$ state of 0^{16} with an excitation energy ΔE of 6.05 MeV) can be represented by a phenomenological potential of the type

$$\begin{aligned} \mathbf{A} &= 0, \qquad A_0 = f(r)\cos(\Delta E t), \\ f(r) &= 0 \quad \text{for} \quad r > \text{nuclear radius} \end{aligned} \tag{4.116}$$

so far as their effect on the electron-positron system is concerned. This effective time-dependent potential offers a convenient framework within which to discuss a process like

$$0^{16*} \longrightarrow 0^{16} + e^+ + e^-, \tag{4.117}$$

which, from the point of view of the electron-positron system, is just pair creation (4.115). In nuclear physics a process such as (4.117) is known as *internal conversion* with pair creation, a theory of which was first formulated by H. Yukawa and S. Sakata in 1935.‡

Hyperon decay. Mott scattering, which we have treated in detail (or more generally any problem that involves an external potential), is actually not such a good example for illustrating the covariant perturbation method. This is because the very presence of an infinitely heavy nucleus prompts us to select a particular Lorentz frame in which the nucleus is situated at the origin. The lack of covariance is reflected by the fact that the δ-function that expresses energy conservation is not accompanied by the $\delta^{(3)}$-function that expresses momentum conservation. Physically this arises from the fact that an infinitely heavy nucleus can accommodate any amount of momentum. If, instead, we treat the electron *plus* the nucleus *as a whole* as a quantum-mechanical system, the formalism automatically gives both the conservation of total energy and the conservation of total momentum (cf.

‡We shall not discuss this interesting phenomenon any further. The reader may consult Akhietzer and Berestetzkii (1965), pp. 567–569.

Problem 4–13, where electron scattering by a proton is treated without the assumption that the proton is infinitely heavy.)

To illustrate the techniques encountered in covariant perturbation theory let us now reconsider the decay of a Λ hyperon (treated earlier in Section 3–11). Using the interaction density (3.432), we see that the first-order S-matrix element for this decay process is

$$\begin{aligned}S_{fi}^{(1)} &= -i\left\langle \pi^- \mathrm{p}\left| \int d^4x \phi_{ch}^{\dagger} \bar{\psi}_{\mathrm{p}}(g + g'\gamma_5)\psi_{\Lambda} \right| \Lambda \right\rangle \\ &= -i \int d^4x \frac{e^{-ip_\pi \cdot x}}{\sqrt{2E_\pi V}} \left[\sqrt{\frac{m_{\mathrm{p}}}{E_{\mathrm{p}}V}} \bar{u}_{\mathrm{p}} e^{-ip_{\mathrm{p}}\cdot x}\right](g + g'\gamma_5)\left[\sqrt{\frac{m_\Lambda}{E_\Lambda V}} u_\Lambda e^{ip_\Lambda \cdot x}\right] \\ &= -i(2\pi)^4 \delta^{(4)}(p_\Lambda - p_\pi - p_{\mathrm{p}}) \sqrt{\frac{1}{2E_\pi V}} \sqrt{\frac{m_{\mathrm{p}}}{E_{\mathrm{p}}V}} \sqrt{\frac{m_\Lambda}{E_\Lambda V}} \bar{u}_{\mathrm{p}}(g + g'\gamma_5)u_\Lambda. \end{aligned} \tag{4.118}$$

Note the appearance of a four-dimensional delta function that expresses both energy and momentum conservation. It is convenient to define a covariant matrix element denoted by $\mathscr{M}_{fi}$ with the kinematical factors $\delta^{(4)}(p_\Lambda - p_\pi - p_{\mathrm{p}})$, $\sqrt{m_{\mathrm{p}}/E_{\mathrm{p}}V}$, etc., eliminated:

$$S_{fi} = \delta_{fi} - i(2\pi)^4 \delta^{(4)}(p_\Lambda - p_\pi - p_{\mathrm{p}}) \sqrt{\left(\frac{1}{2E_\pi V}\right)\left(\frac{m_{\mathrm{p}}}{E_{\mathrm{p}}V}\right)\left(\frac{m_\Lambda}{E_\Lambda V}\right)} \mathscr{M}_{fi}, \tag{4.119}$$

where δ_{fi} (which comes from $S^{(0)}$) is zero in this case and $\mathscr{M}_{fi}$ is to be computed to only first order in g and g'. With this definition,

$$\mathscr{M}_{fi} = \bar{u}_{\mathrm{p}}(g + g'\gamma_5)u_\Lambda, \tag{4.120}$$

which is invariant under Lorentz transformation.

We may be interested in computing the decay distribution of the final π^-p system arising from a Λ hyperon whose spin state is characterized by the four-vector w. If we are not interested in observing the polarization of the decay proton, we must sum over the proton spin states. So we are led to consider

$$\begin{aligned}\sum_{\text{proton spin}} |\mathscr{M}_{fi}|^2 &= \sum_{\text{proton spin}} [u_\Lambda^\dagger(g^* + g'^*\gamma_5)\gamma_4 u_{\mathrm{p}}][\bar{u}_{\mathrm{p}}(g + g'\gamma_5)u_\Lambda] \\ &= \mathrm{Tr}\left[(g^* - g^{*\prime}\gamma_5)\left(\frac{-i\gamma\cdot p_{\mathrm{p}} + m_{\mathrm{p}}}{2m_{\mathrm{p}}}\right)\right. \\ &\quad \left.\times (g + g'\gamma_5)\left(\frac{-i\gamma\cdot p_\Lambda + m_\Lambda}{2m_\Lambda}\right)\left(\frac{1 + i\gamma_5\gamma\cdot w}{2}\right)\right]. \end{aligned} \tag{4.121}$$

The evaluation of the trace is facilitated if we note that

$$\mathrm{Tr}(\gamma_5 \underbrace{\gamma_\mu \gamma_\nu \cdots}_{\text{less than four gamma matrices } (\mu,\nu \neq 5)}) = 0 \tag{4.122}$$

The only terms that survive in $4m_{\mathrm{p}}m_\Lambda$ times (4.121) are

$$\begin{aligned}&\tfrac{1}{2}|g|^2 \mathrm{Tr}[(-i\gamma\cdot p_{\mathrm{p}})(-i\gamma\cdot p_\Lambda)] + \tfrac{1}{2}|g|^2 \mathrm{Tr}(m_{\mathrm{p}}m_\Lambda), \\ &+\tfrac{1}{2}|g'|^2 \mathrm{Tr}[(-\gamma_5)(-i\gamma\cdot p_{\mathrm{p}})\gamma_5(-i\gamma\cdot p_\Lambda)], \\ &+\tfrac{1}{2}|g'|^2 \mathrm{Tr}[(-\gamma_5)m_{\mathrm{p}}\gamma_5 m_\Lambda] + \tfrac{1}{2}g^*g' \mathrm{Tr}[(-i\gamma\cdot p_{\mathrm{p}})\gamma_5 i\gamma_5\gamma\cdot w m_\Lambda], \\ &+\tfrac{1}{2}g'^*g \mathrm{Tr}[(-\gamma_5)(-i\gamma\cdot p_{\mathrm{p}})i\gamma_5\gamma\cdot w m_\Lambda], \end{aligned} \tag{4.123}$$

where we note that a term like $\mathrm{Tr}[(-\gamma_5)m_p(i\gamma_5\gamma\cdot w)(-i\gamma\cdot p_\Lambda)]$ vanishes because $w\cdot p_\Lambda = 0$. We can "kill" the γ_5 in (4.123) by noting that $\gamma_5\gamma_\mu\gamma_5 = -\gamma_\mu$ and $\gamma_5^2 = 1$. We then use (4.80) and (4.82) to obtain

$$4m_p m_\Lambda \sum_{\text{proton spin}} |\mathcal{M}_{fi}|^2 = |g|^2(-2p_p\cdot p_\Lambda + 2m_p m_\Lambda) + |g'|^2(-2p_p\cdot p_\Lambda - 2m_p m_\Lambda) + 2\mathrm{Re}(gg'^*)(2p_p\cdot w m_\Lambda). \tag{4.124}$$

Without committing ourselves to any particular Lorentz frame we can compute the differential decay rates as follows:

$$dw = \frac{1}{T}|(2\pi)^4\delta^{(4)}(p_\Lambda - p_\pi - p_p)|^2\left(\frac{1}{2E_\pi V}\right)\left(\frac{m_p}{E_p V}\right)\left(\frac{m_\Lambda}{E_\Lambda V}\right)\frac{Vd^3p_p}{(2\pi)^3}\frac{Vd^3p_\pi}{(2\pi)^3}\sum_{\text{proton spin}}|\mathcal{M}_{fi}|^2. \tag{4.125}$$

The appearance of both $V\,d^3p_p/(2\pi)^3$ and $V\,d^3p_\pi/(2\pi)^3$ can be understood if we recall that we have not yet used the constraint imposed by the momentum conservation implied by the delta function. We can handle the square of the four-dimensional delta function just as in (4.67) and (4.68), remembering that

$$\delta^{(3)}(\mathbf{p}_\Lambda - \mathbf{p}_\pi - \mathbf{p}_p) = \lim_{V\to\infty}[1/(2\pi)^3]\int_V d^3x \exp[i(\mathbf{p}_\Lambda - \mathbf{p}_\pi - \mathbf{p}_p)\cdot\mathbf{x}], \tag{4.126}$$

which is $V/(2\pi)^3$ if $\mathbf{p}_\Lambda - \mathbf{p}_\pi - \mathbf{p}_p = 0$, hence

$$|\delta^{(4)}(p_\Lambda - p_\pi - p_p)|^2 = \frac{VT}{(2\pi)^4}\delta^{(4)}(p_\Lambda - p_\pi - p_p). \tag{4.127}$$

The expression (4.125) is now reduced to‡

$$dw = (2\pi)^4\left(\frac{1}{2E_\Lambda}\right)\left[\frac{d^3p_p}{(2\pi)^3 2E_p}\right]\left[\frac{d^3p_\pi}{(2\pi)^3 2E_\pi}\right](4m_p m_\Lambda)\sum_{\text{proton spin}}|\mathcal{M}_{fi}|^2\,\delta^{(4)}(p_\Lambda - p_\pi - p_p). \tag{4.128}$$

Note that the normalization volume V no longer appears. We note that apart from $1/E_\Lambda$ the expression (4.128) is completely relativistically invariant. To prove this assertion it is sufficient to note that

$$\begin{aligned}\frac{d^3p}{2E} &= \int d^4p\,\frac{\delta(p_0 - \sqrt{|\mathbf{p}|^2 + m^2})}{2\sqrt{|\mathbf{p}|^2 + m^2}}\\ &= \int_{p_0>0} d^4p\,\frac{\delta(p_0 - \sqrt{|\mathbf{p}|^2 + m^2}) + \delta(p_0 + \sqrt{|\mathbf{p}|^2 + m^2})}{2\sqrt{|\mathbf{p}|^2 + m^2}}\\ &= \int_{p_0>0}\delta(p^2 + m^2)\,d^4p,\end{aligned} \tag{4.129}$$

where p_0 is understood to be an integration variable in contrast to $E = \sqrt{|\mathbf{p}|^2 + m^2}$ which is fixed once $\mathbf{p}$ is given.§ Furthermore, the dependence of dw on $1/E_\Lambda$ is readily understandable from the point of view of the special theory of relativity which requires that a faster moving Λ must decay more slowly.

---

‡A generalization of this formula, to any decay process, can be found in Appendix D.

§For any timelike vector (for example, the four-momentum vector of a free particle) the sign of the zeroth component has a Lorentz invariant meaning.

The expression (4.128) with $\sum |\mathcal{M}_{fi}|^2$ given by (4.124) gives the differential decay rate for a polarized Λ hyperon in any Lorentz frame. We can now work in the rest system of the Λ hyperon so that $\mathbf{p} = \mathbf{p}_p = -\mathbf{p}_\pi$, $\mathbf{p}_\Lambda = 0$. Taking advantage of the delta function that expresses momentum conservation, we can immediately integrate over the pion momentum to obtain

$$dw = 2\pi \frac{|\mathbf{p}|^2 d|\mathbf{p}|\, d\Omega}{(2\pi)^3}\left(\frac{m_p}{E_p}\right)\left(\frac{1}{2E_\pi}\right) \sum |\mathcal{M}_{fi}|^2\, \delta(m_\Lambda - E_\pi - E_p), \tag{4.130}$$

which is in complete agreement with the Golden Rule when we realize that m_p/E_p and $1/2E_\pi$ combine with $|\mathcal{M}_{fi}|^2$ to give $V|(H'_I)_{fi}|^2$. Meanwhile, remembering that w is simply a unit (three-) vector in the Λ rest system, we have

$$\begin{aligned} 4m_p m_\Lambda \sum |\mathcal{M}_{fi}|^2 = 2[&|g|^2 (E_p m_\Lambda + m_p m_\Lambda) + |g'|^2 (E_p m_\Lambda - m_p m_\Lambda) \\ &+ 2\mathrm{Re}(gg'^*)(\mathbf{p}_p \cdot \hat{\mathbf{n}})], \end{aligned} \tag{4.131}$$

which leads to the decay angular distribution (3.447) obtained earlier. If we are interested in the decay width we must integrate (4.130) over the directions of $\mathbf{p}$. Thus

$$\Gamma(\Lambda \longrightarrow \pi^- + p) = \frac{1}{2}\left(\frac{1}{4\pi}\right)^2 \frac{|\mathbf{p}|}{m_\Lambda^2} \int d\Omega \sum_{\text{proton spin}} |\mathcal{M}_{fi}|^2\, 4m_p m_\Lambda, \tag{4.132}$$

where we have used

$$\begin{aligned} d|\mathbf{p}|\,\delta(m_\Lambda - E_\pi - E_p) &= \frac{1}{\partial(E_\pi + E_p)/\partial|\mathbf{p}|} = \frac{1}{(|\mathbf{p}|/E_\pi) + (|\mathbf{p}|/E_p)} \\ &= \frac{E_\pi E_p}{|\mathbf{p}|\, m_\Lambda}. \end{aligned} \tag{4.133}$$

We may parenthetically remark that (4.132) is a special case of a very useful formula applicable to any two-body decay process:‡

$$\Gamma(1 \longrightarrow 2 + 3) = \frac{1}{2}\left(\frac{1}{4\pi}\right)^2 \frac{|\mathbf{p}|}{m_1^2} \int d\Omega \sum_{\text{final spin}} |\mathcal{M}_{fi}|^2 \prod_{\substack{\text{external}\\ \text{fermion}}} (2m_{\text{fermi}}). \tag{4.134}$$

Combining (4.131) with (4.132), we finally get

$$\Gamma(\Lambda \longrightarrow \pi^- + p) = \left[\frac{|g|^2}{4\pi} + \frac{|g'|^2}{4\pi}\left(\frac{E_p - m_p}{E_p + m_p}\right)\right] \frac{|\mathbf{p}|(E_p + m_p)}{m_\Lambda} \tag{4.135}$$

in complete agreement with (3.451).

The reader is now in a position to work out the angular distribution and the decay rate of almost any decay process that can be described adequately by a first-order S-matrix. As exercises, he may try $\omega \longrightarrow e^+ + e^-$ and $\Sigma^0 \longrightarrow \Lambda + \gamma$ (given as Problems 4–6 and 4–7).

‡The product

$$\prod_{\text{external fermion}} (2m_{\text{fermi}})$$

arises because the normalization constant for a boson differs from that for a fermion [$\sqrt{1/2EV}$ and $\sqrt{m/EV}$ respectively; cf Eq. (4.119)]. This difference would have been absent had we normalized the free-particle spinor by $\bar{u}^{(s)}(\mathbf{p})u^{(s')}(\mathbf{p}) = 2m\delta_{ss'}$.

4–4. TWO-PHOTON ANNIHILATION AND COMPTON SCATTERING; THE ELECTRON PROPAGATOR

S-matrix for two-photon annihilation. We shall now proceed to discuss processes that involve second-order matrix elements. The fundamental relations on which everything will be based are the two alternative forms of the $S^{(2)}$-matrix derived in Section 4–2 (cf. Eq. 4.48). If we use "Form 1," we have

$$S^{(2)} = (-i)^2 \int_{-\infty}^{\infty} dt_1 \int_{-\infty}^{t_1} dt_2\, H_I(t_1) H_I(t_2)$$
$$= (-i)^2 \int d^4x_1 \int_{t_2<t_1} d^4x_2\, \mathscr{H}_{\text{int}}(x_1)\mathscr{H}_{\text{int}}(x_2), \tag{4.136}$$

where the four-dimensional integration over x_2 is to be performed only for $t_2 < t_1$. With "Form 2" we have

$$S^{(2)} = (-i)^2 \int_{-\infty}^{\infty} dt_1 \int_{t_1}^{\infty} dt_2\, H_I(t_2) H_I(t_1)$$
$$= (-i)^2 \int d^4x_1 \int_{t_2>t_1} d^4x_2\, \mathscr{H}_{\text{int}}(x_2)\mathscr{H}_{\text{int}}(x_1) \tag{4.137}$$

with t_2 *later* than t_1.‡

Without further preliminary remarks we wish to consider a specific physical process that can be described by considering $S^{(2)}$. We shall first treat the problem of two-photon annihilation in detail and show how the notion of the electron propagator arises in a concrete physical situation. The discussion of second-order processes which involve the propagation of virtual photons or mesons will be deferred until Section 4–6.

By two-photon annihilation we simply mean that

$$e^+ + e^- \longrightarrow 2\gamma. \tag{4.138}$$

The initial and the final states are written as

$$\Phi_i = |e^-e^+\rangle = b^{(s_-)}(\mathbf{p}_-)^\dagger d^{(s_+)}(\mathbf{p}_+)^\dagger |0\rangle,$$
$$\Phi_f = |2\gamma\rangle = a^\dagger_{\mathbf{k}_1\alpha_1} a^\dagger_{\mathbf{k}_2\alpha_2} |0\rangle. \tag{4.139}$$

Needless to say, $S^{(0)}$ and $S^{(1)}$ give no contributions. So, assuming that S can be approximated by $S^{(2)}$, we write S_{fi} in "Form 1" as follows:

$$S_{fi} = (-e)^2 \int d^4x_1 \int_{t_2<t_1} d^4x_2 \langle 2\gamma | \bar{\psi}(x_1)\gamma_\mu \psi(x_1) A_\mu \bar{\psi}(x_2)\gamma_\nu \psi(x_2) A_\nu(x_2) | e^-e^+\rangle. \tag{4.140}$$

---

‡In the literature "Form 1" and Form 2" are often combined as follows:

$$S^{(2)} = [(-i)^2/2] \int_{-\infty}^{\infty} dt_1 \int_{-\infty}^{\infty} dt_2\, P[H_I(t_1) H_I(t_2)],$$

where the symbol P stands for an operator that arranges the product of factors as follows:

$$P[A(t_1)B(t_2)] = \begin{cases} A(t_1)B(t_2) & \text{if} \quad t_1 > t_2, \\ B(t_2)A(t_1) & \text{if} \quad t_2 > t_1. \end{cases}$$

This P-product (*Dyson chronological product*) should not be confused with the T-product (*Wick time-ordered product*), to be introduced later.

This particular problem (treated to this order) involves only real photons, and it is therefore legitimate to regard A_μ as the (quantized) transverse electromagnetic field discussed in Chapter 2.‡ So far as the electron field operators are concerned, the final state is just the vacuum state; likewise, so far as the photon field is concerned, the initial state is the vacuum state. Since the free-electron field and the free-photon field operate in different subspaces, we can write (4.140) as

$$S_{fi} = (-e)^2 \int d^4x_1 \int_{t_2<t_1} d^4x_2 \langle 0 | \bar{\psi}(x_1)\gamma_\mu\psi(x_1)\bar{\psi}(x_2)\gamma_\nu\psi(x_2) | \mathrm{e}^-\mathrm{e}^+\rangle \times \langle 2\gamma | A_\mu(x_1)A_\nu(x_2) | 0\rangle. \tag{4.141}$$

From Chapter 2 we already know how to handle $\langle 2\gamma | A_\mu(x_1)A_\nu(x_2) | 0\rangle$. Noting that $A_\mu(x_1)$ can create either photon 1 or photon 2 at x_1, we get

$$\langle 2\gamma | A_\mu(x_1)A_\nu(x_2) | 0\rangle = \left[\left(\frac{1}{\sqrt{2\omega_1 V}}\epsilon_\mu^{(\alpha_1)}e^{-ik_1\cdot x_1}\right)\left(\frac{1}{\sqrt{2\omega_2 V}}\epsilon_\nu^{(\alpha_2)}e^{-ik_2\cdot x_2}\right) + \left(\frac{1}{\sqrt{2\omega_2 V}}\epsilon_\mu^{(\alpha_2)}e^{-ik_2\cdot x_1}\right)\left(\frac{1}{\sqrt{2\omega_1 V}}\epsilon_\nu^{(\alpha_1)}e^{-ik_1\cdot x_2}\right)\right], \tag{4.142}$$

where $\epsilon_\mu^{(\alpha_1)}$ simply stands for

$$\epsilon_\mu^{(\alpha_1)} = (\boldsymbol{\epsilon}^{(\alpha_1)}, 0) \tag{4.143}$$

with $\boldsymbol{\epsilon}^{(\alpha_1)}$ perpendicular to $\mathbf{k}_1$. We have, of course, $|\mathbf{k}_1| = \omega_1$, since we are talking about a real photon.

Let us now concentrate on the matrix element of the electron field operators. If we rewrite $\bar{\psi}\gamma_\mu\psi$ in (4.141) using (4.59), it may appear at first sight that there are as many as sixteen terms. Fortunately most of them give rise to no contribution when taken between the $\mathrm{e}^-\mathrm{e}^+$ state and the vacuum state. For instance, take $\bar{\psi}_\alpha^{(+)}(x_1)\psi_\beta^{(+)}(x_1)\bar{\psi}_\gamma^{(+)}(x_2)\psi_\delta^{(+)}(x_2)$, which annihilates two electrons and two positrons. So far as the electron-positron system is concerned, there are only two particles in the initial state, none in the final state. Hence the matrix element of such a product does not contribute to this particular process. More generally, since we have to go from a two-particle state to a no-particle state, there must be three $(+)$ and one $(-)$. Furthermore, it is legitimate to replace $\bar{\psi}_\alpha(x_1)\psi_\beta(x_1)$ by $\bar{\psi}_\alpha^{(+)}(x_1)\psi_\beta^{(+)}(x_1)$ since $\bar{\psi}_\alpha^{(-)}(x_1)$ and/or $\psi_\beta^{(-)}(x_1)$ acting on $\langle 0|$ from the *right* give zero. (Physically this is reasonable since $\bar{\psi}_\alpha^{(-)}(x_1)$ and/or $\psi_\beta^{(-)}(x_1)$ would give rise to at least one electron and/or one positron created at t_1 which survive(s) at $t = +\infty$.) So out of the sixteen terms we need to consider only two terms:

$$(\gamma_\mu)_{\alpha\beta}(\gamma_\nu)_{\gamma\delta}\bar{\psi}_\alpha^{(+)}(x_1)\psi_\beta^{(+)}(x_1)\bar{\psi}_\gamma^{(-)}(x_2)\psi_\delta^{(+)}(x_2) \tag{4.144}$$

and

$$-(\gamma_\mu)_{\alpha\beta}(\gamma_\nu)_{\gamma\delta}\bar{\psi}_\alpha^{(+)}(x_1)\psi_\beta^{(+)}(x_1)\psi_\delta^{(-)}\bar{\psi}_\gamma^{(+)}(x_2). \tag{4.145}$$

Note that the ordering of $\bar{\psi}_\gamma^{(+)}(x_2)$ and $\psi_\gamma^{(-)}(x_2)$ and the associated minus sign in (4.145) are chosen to agree with those of the third term of (4.59). We propose to

‡One may argue on the basis of Appendix A that the complete interaction Hamiltonian must also contain the instantaneous Coulomb interaction. However, the Coulomb interaction is irrelevant if we are computing two-photon annihilation to order e^2 in amplitude.

move the positron annihilation operator $\bar{\psi}_\alpha^{(+)}$ in the first term (4.144) so that the product $\bar{\psi}_\alpha^{(+)}(x_1)\psi_\delta^{(+)}(x_2)$ may immediately annihilate the electron-positron pair in the initial state:

$$(\gamma_\mu)_{\alpha\beta}(\gamma_\nu)_{\gamma\delta}\bar{\psi}_\alpha^{(+)}(x_1)\psi_\beta^{(+)}(x_1)\bar{\psi}_\gamma^{(-)}(x_2)\psi_\delta^{(+)}(x_2)$$
$$= (\gamma_\mu)_{\alpha\beta}(\gamma_\nu)_{\gamma\delta}\psi_\beta^{(+)}(x_1)\bar{\psi}_\gamma^{(-)}(x_2)\bar{\psi}_\alpha^{(+)}(x_1)\psi_\delta^{(+)}(x_2). \tag{4.146}$$

As for the second term (4.145), we interchange x_1 and x_2 and make the following changes in the summation indices $\mu \longleftrightarrow \nu$, $\alpha\beta \longleftrightarrow \gamma\delta$ so that the second term may look as similar as possible to the first one. The product (4.145) now becomes

$$-(\gamma_\nu)_{\gamma\delta}(\gamma_\mu)_{\alpha\beta}\bar{\psi}_\gamma^{(+)}(x_2)\psi_\delta^{(+)}(x_2)\psi_\beta^{(-)}(x_1)\bar{\psi}_\alpha^{(+)}(x_1)$$
$$= -(\gamma_\mu)_{\alpha\beta}(\gamma_\nu)_{\gamma\delta}\bar{\psi}_\gamma^{(+)}(x_2)\psi_\beta^{(-)}(x_1)\bar{\psi}_\alpha^{(+)}(x_1)\psi_\delta^{(+)}(x_2). \tag{4.147}$$

It is legitimate to interchange freely the integration variables x_1 and x_2 in this manner provided that we keep in mind that t_2 is now *later* than t_1. *This forces us to use* "Form 2," which after all was originally obtained from "Form 1" by interchanging the integration variables [cf. (4.49) and (4.50)]. The photon-field matrix element is unaffected by these changes because (4.142) is symmetric under $x_1 \longleftrightarrow x_2$, $\mu \longleftrightarrow \nu$. Thus we have

$$S_{fi} = (-e)^2 \int d^4x_1 \int d^4x_2 \, \langle 2\gamma \,|\, A_\mu(x_1)A_\nu(x_2) \,|\, 0\rangle(\gamma_\mu)_{\alpha\beta}(\gamma_\nu)_{\gamma\delta}$$
$$\times [\langle 0 \,|\, \psi_\beta^{(+)}(x_1)\bar{\psi}_\gamma^{(-)}(x_2)\bar{\psi}_\alpha^{(+)}(x_1)\psi_\delta^{(+)}(x_2) \,|\, e^-e^+\rangle\theta(t_1 - t_2)$$
$$- \langle 0 \,|\, \bar{\psi}_\gamma^{(+)}(x_2)\psi_\beta^{(-)}(x_1)\bar{\psi}_\alpha^{(+)}(x_1)\psi_\delta^{(+)}(x_2) \,|\, e^-e^+\rangle\theta(t_2 - t_1)], \tag{4.148}$$

where the θ function is defined by

$$\theta(t) = \begin{cases} 1 & t > 0, \\ 0 & t < 0. \end{cases} \tag{4.149}$$

There still remains the task of simplifying the electron-field matrix elements. This can be done by inserting a complete set of intermediate states $\sum_n |n\rangle\langle n|$ between $\psi_\beta^{(+)}(x_1)\bar{\psi}_\gamma^{(-)}(x_2)$ [or $\bar{\psi}_\gamma^{(+)}(x_2)\psi_\beta^{(-)}(x_1)$] and $\bar{\psi}_\alpha^{(+)}(x_1)\psi_\delta^{(+)}(x_2)$. Evidently the only contribution comes from the vacuum state. We have

$$\langle 0 \,|\, \psi_\beta^{(+)}(x_1)\bar{\psi}_\gamma^{(-)}(x_2)\bar{\psi}_\alpha^{(+)}(x_1)\psi_\delta^{(+)}(x_2) \,|\, e^-e^+\rangle$$
$$= \langle 0 \,|\, \psi_\beta^{(+)}(x_1)\bar{\psi}_\gamma^{(-)}(x_2) \,|\, 0\rangle\langle 0 \,|\, \bar{\psi}_\alpha^{(+)}(x_1)\psi_\delta^{(+)}(x_2) \,|\, e^-e^+\rangle$$
$$= \langle 0 \,|\, \psi_\beta^{(+)}(x_1)\bar{\psi}_\gamma^{(-)}(x_2) \,|\, 0\rangle\left[\sqrt{\frac{m}{E_+V}}\,\bar{v}_\alpha^{(s_+)}(\mathbf{p}_+)e^{ip_+\cdot x_1}\right]\left[\sqrt{\frac{m}{E_-V}}\,u_\delta^{(s_-)}(\mathbf{p}_-)e^{ip_-\cdot x_2}\right]. \tag{4.150}$$

An analogous reduction can be performed for the second term. Putting everything together we finally obtain

$$S_{fi} = (-e)^2 \int d^4x_1 \int d^4x_2 \left[\left(\frac{1}{\sqrt{2\omega_1 V}}\epsilon_\mu^{(\alpha_1)}e^{-ik_1\cdot x_1}\right)\left(\frac{1}{\sqrt{2\omega_2 V}}\epsilon_\nu^{(\alpha_2)}e^{-ik_2\cdot x_2}\right)\right.$$
$$\left. + \begin{Bmatrix} k_1 \longleftrightarrow k_2 \\ \alpha_1 \longleftrightarrow \alpha_2 \end{Bmatrix}\right]\left[\sqrt{\frac{m}{E_+V}}\,\bar{v}_\alpha^{(s_+)}(\mathbf{p}_+)e^{ip_+\cdot x_1}\right]$$
$$\times (\gamma_\mu)_{\alpha\beta}[\langle 0|\psi_\beta^{(+)}(x_1)\bar{\psi}_\gamma^{(-)}(x_2)|0\rangle\theta(t_1 - t_2) - \langle 0|\bar{\psi}_\gamma^{(+)}(x_2)\psi_\beta^{(-)}(x_1)|0\rangle\theta(t_2 - t_1)]$$
$$\times (\gamma_\nu)_{\gamma\delta}\left[\sqrt{\frac{m}{E_-V}}\,u_\delta^{(s_-)}(\mathbf{p}_-)e^{ip_-\cdot x_2}\right], \tag{4.151}$$

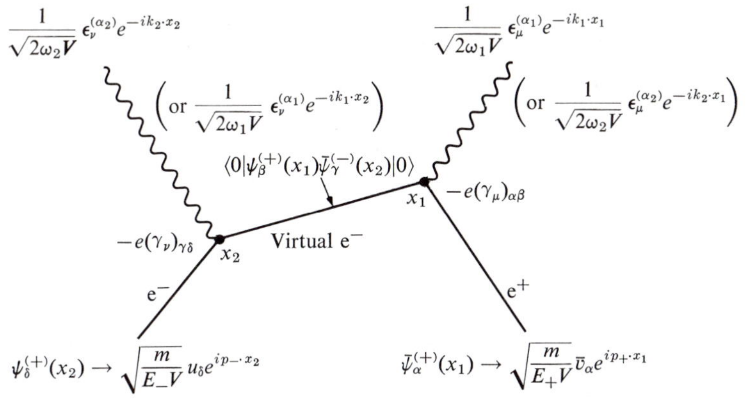

(a) $t_1 > t_2$

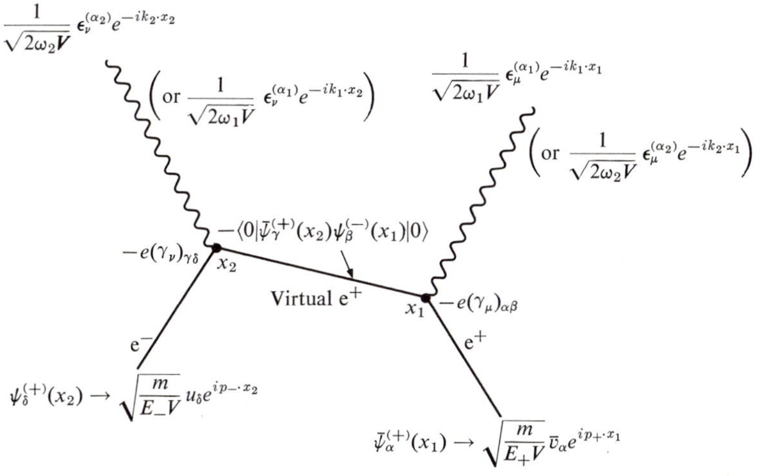

(b) $t_2 > t_1$

Fig. 4–4. Time-ordered graphs for two-photon annihilation.

where we have abbreviated the second term of Eq. (4.142) by the expression $\{k_1 \leftrightarrow k_2, \alpha_1 \leftrightarrow \alpha_2\}$.

Expression (4.151) has both familiar and unfamiliar aspects. First, the appearance of $(1/\sqrt{2\omega_1 V})\epsilon_\mu^{(\alpha_1)}e^{-ik_1\cdot x_1}$, etc., is familiar from Chapter 2, where we showed that this is expected whenever a real photon characterized by $(\mathbf{k}_1, \alpha_1)$ is emitted at x_1. Similarly $\sqrt{m/E_- V}\, u e^{ip_-\cdot x_2}$ and $\sqrt{m/E_+ V}\, \bar v e^{ip_+\cdot x_1}$ simply represent the annihilation of the incident electron at x_2 and the incident positron at x_1 just as in (4.114). It is clear that of the four electron-positron operators we started with, we have used up two $[\bar\psi_\alpha^{(+)}(x_2)$ and $\psi_\delta^{(+)}(x_2)]$ to annihilate the incident positron and the incident electron. The remaining two operators $\psi_\beta(x_1)$ and $\bar\psi_\gamma(x_2)$ (not used in annihilating the incident particles) are seen to be *paired* and *sandwiched between the vacuum states;* this is really new. When t_1 is later than t_2, we are instructed to use the first vacuum product, in which $\bar{\psi}_\gamma^{(-)}(x_2)$ first creates an electron at exactly the same space-time point (x_2) at which the incident electron is annihilated and one of the photons is emitted. The appearance of $\psi_\beta^{(+)}(x_1)$ in the same vacuum product tells us that the electron created at x_2 is subsequently annihilated at x_1, the point at which the incident positron is annihilated and the other photon is emitted. Thus for $t_1 > t_2$, the vacuum expectation value $\langle 0|\psi_\beta^{(+)}(x_1)\bar{\psi}_\gamma^{(-)}(x_2)|0\rangle$ characterizes the propagation of a *virtual electron* from x_2 to x_1. This is shown in Fig. 4–4(a), where the "internal line" that joins x_2 and x_1 represents the virtual electron going from x_2 to x_1. Similarly, when $t_2 > t_1$, then $\langle 0|\bar{\psi}_\gamma^{(+)}(x_2)\psi_\beta^{(-)}(x_1)|0\rangle$ represents the propagation of a *virtual positron* from x_1 to x_2, as shown schematically in Fig. 4–4(b). Note that in both cases the action of the creation operator [$\psi_\beta^{(-)}(x_1)$ or $\bar{\psi}_\gamma^{(-)}(x_2)$] precedes that of the annihilation operator [$\psi_\beta^{(+)}(x_1)$ or $\bar{\psi}_\gamma^{(+)}(x_2)$] as demanded by the "causality principle."

Electron propagator. The major breakthrough in the evaluation of the matrix element is achieved when we *combine* the two vacuum-expectation-value terms in (4.151) into a single expression. We propose to consider *together* the propagation of a virtual electron from x_2 to x_1 for $t_1 > t_2$ and the propagation of a virtual positron from x_1 to x_2 for $t_2 > t_1$ (*not separately* as in the old-fashioned perturbation method). With this in mind we define what is known as the time-ordered product (*Wick time-ordered product*) of the fermion operators as follows:

$$T(\psi_\alpha(x)\bar{\psi}_\beta(x')) = \begin{cases} \psi_\alpha(x)\bar{\psi}_\beta(x') & \text{for} \quad t > t' \\ -\bar{\psi}_\beta(x')\psi_\alpha(x) & \text{for} \quad t' > t. \end{cases} \tag{4.152}$$

Since the (+) part [the (−) part] of the field operator gives no contribution when it acts on the vacuum state from the left (from the right), we have

$$\langle 0|T(\psi_\alpha(x)\bar{\psi}_\beta(x'))|0\rangle = \begin{cases} \langle 0|\psi_\alpha^{(+)}(x)\bar{\psi}_\beta^{(-)}(x')|0\rangle & \text{for} \quad t > t', \\ -\langle 0|\bar{\psi}_\beta^{(+)}(x')\psi_\alpha^{(-)}(x)|0\rangle & \text{for} \quad t' > t, \end{cases} \tag{4.153}$$

which is precisely the form that appears in (4.151). Let us now compute (4.153). For $t > t'$ we must consider

$$\begin{aligned} &\langle 0|\psi_\alpha^{(+)}(x)\bar{\psi}_\beta^{(-)}(x')|0\rangle \\ &= \frac{1}{V}\left\langle 0\left| \sum_{\mathbf{p},s}\sqrt{\frac{m}{E}}\, b^{(s)}(\mathbf{p})u_\alpha^{(s)}(\mathbf{p})\exp(ip\cdot x) \sum_{\mathbf{p}',s'}\sqrt{\frac{m}{E'}}\, b^{(s')+}(\mathbf{p}')\bar{u}_\beta^{(s')}(\mathbf{p}')\exp(-ip'\cdot x')\right| 0\right\rangle \\ &= \frac{1}{(2\pi)^3} m \int \frac{d^3p}{E} \sum_s u_\alpha^{(s)}(\mathbf{p})\bar{u}_\beta^{(s)}(\mathbf{p})\exp[ip\cdot(x-x')] \\ &= \frac{1}{(2\pi)^3}\int \frac{d^3p}{2E}[-i\boldsymbol{\gamma}\cdot\mathbf{p} - i\gamma_4(iE) + m]_{\alpha\beta}\exp[i\mathbf{p}\cdot(\mathbf{x}-\mathbf{x}') - iE(t-t')], \end{aligned} \tag{4.154}$$

where we have used

$$\langle 0|b^{(s)}(\mathbf{p})b^{(s')+}(\mathbf{p}')|0\rangle = \delta_{\mathbf{p}\mathbf{p}'}\delta_{ss'}, \tag{4.155}$$

the well-known rule

$$\lim_{V\to\infty}\frac{1}{V}\sum_{\mathbf{p}} = \int\frac{d^3p}{(2\pi)^3}, \tag{4.156}$$

and (4.72). Similarly for $t' > t$ we get with the aid of (4.76),

$$
\begin{aligned}
&-\langle 0 | \bar{\psi}_\beta^{(+)}(x') \psi_\alpha^{(-)}(x) | 0 \rangle \\
&= -\frac{1}{(2\pi)^3} m \int \frac{d^3p}{E} \sum_s v_\alpha(\mathbf{p}) \bar{v}_\beta(\mathbf{p}) \exp[i\mathbf{p}\cdot(\mathbf{x}' - \mathbf{x}) - iE(t' - t)] \\
&= -\frac{1}{(2\pi)^3} m \int \frac{d^3p}{E} \left[-\left(\frac{i\gamma\cdot p + m}{2m} \right)_{\alpha\beta} \right] \exp[i\mathbf{p}\cdot(\mathbf{x}' - \mathbf{x}) - iE(t' - t)] \\
&= \frac{1}{(2\pi)^3} \int \frac{d^3p}{2E} [-i\boldsymbol{\gamma}\cdot\mathbf{p} - i\gamma_4(-iE) + m]_{\alpha\beta} \exp[i\mathbf{p}\cdot(\mathbf{x} - \mathbf{x}') - i(-E)(t - t')],
\end{aligned}
\tag{4.157}
$$

where in obtaining the last line we have changed the sign of the integration variable ($\mathbf{p} \rightarrow -\mathbf{p}$) so that the whole expression may look similar to (4.154). We shall now demonstrate that the vacuum expectation value (4.153) is equal to the four-dimensional integral

$$
-\frac{i}{(2\pi)^4} \int d^4p \frac{(-i\gamma\cdot p + m)_{\alpha\beta} \exp[ip\cdot(x - x')]}{p^2 + m^2 - i\epsilon}, \tag{4.158}
$$

where ϵ is an infinitesimal positive real quantity and $p^2 = |\mathbf{p}|^2 - p_0^2$ with p_0 *not*, in general, equal to $\sqrt{|\mathbf{p}|^2 + m^2}$. We first integrate over p_0:

$$
-\frac{i}{(2\pi)^4} \int d^3p \int_{-\infty}^{\infty} dp_0 \frac{(-i\gamma\cdot p + m)_{\alpha\beta} \exp[ip\cdot(x - x')]}{[-(p_0 - \sqrt{|\mathbf{p}|^2 + m^2})](p_0 + \sqrt{|\mathbf{p}|^2 + m^2}) - i\epsilon} \tag{4.159}
$$

Regarding the integrand as a function of the complex variable p_0, we note that it has poles at two points,

$$
\begin{aligned}
p_0 &= \sqrt{|\mathbf{p}|^2 + m^2} - i\delta = E - i\delta, \\
p_0 &= -\sqrt{|\mathbf{p}|^2 + m^2} + i\delta = -E + i\delta,
\end{aligned}
\tag{4.160}
$$

with $\delta = (\epsilon/2E)$ where E, as usual, is understood to be $\sqrt{|\mathbf{p}|^2 + m^2}$, in contrast to p_0, which is variable. We are, of course, supposed to integrate along the real axis in the p_0-plane. This can be done using the contour-integral method. If $t > t'$, we must use the semicircular path indicated in Fig. 4–5(a) since the p_0 that appears in $\exp[-ip_0(t - t')]$ cannot be positive imaginary. This means that for $t > t'$ we just

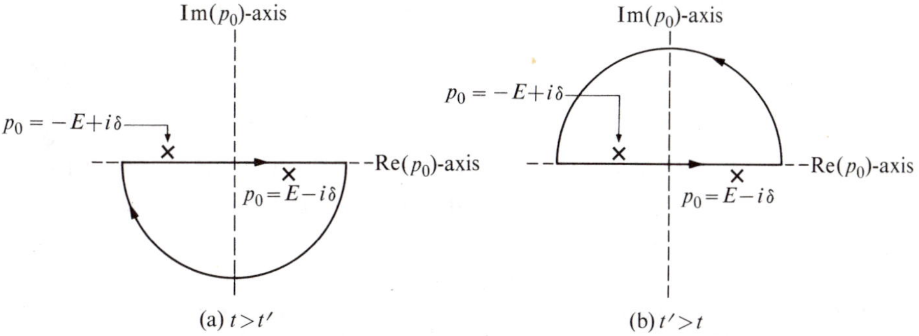

Fig. 4–5. The p_0 integration in (4.159).

pick up the pole contribution from $p_0 = E - i\delta$. Hence (4.159) is equal to

$$\frac{(2\pi i)(-i)}{(2\pi)^4}\int \frac{d^3p}{2E}[-i\boldsymbol{\gamma}\cdot\mathbf{p} - i\gamma_4(iE) + m]_{\alpha\beta}\exp[i\mathbf{p}\cdot(\mathbf{x}-\mathbf{x}') - iE(t-t')] \quad (4.161)$$

for $t > t'$, in complete agreement with (4.154). Similarly, by considering the contour indicated in Fig. 4–5(b) we readily see that the four-dimensional integral is equal to (4.157) for $t' > t$.

To summarize, we have proved that

$$\langle 0|T(\psi_\alpha(x)\bar{\psi}_\beta(x'))|0\rangle = -\frac{i}{(2\pi)^4}\int d^4p\,\frac{(-i\gamma\cdot p + m)_{\alpha\beta}}{p^2+m^2-i\epsilon}\exp[ip\cdot(x-x')] \quad (4.162)$$

regardless of whether t is later or earlier than t′. Note that (4.162) called the "electron propagator" is a 4×4 matrix which depends on the four-vector difference $x - x'$. In the literature (4.162) is often denoted by $-\frac{1}{2}(S_F(x-x'))_{\alpha\beta}$ (where F stands for Feynman).‡ The *electron propagator* is undoubtedly one of the most important functions in twentieth-century physics; together with the *covariant photon propagator* to be introduced in Section 4–6, it has enabled us to atttack a variety of physical problems which we could not manage earlier. Although its explicit form in the coordinate (x) space can be obtained in terms of a Hankel function of the second kind, it is more useful to leave it in the Fourier integral form (4.162). In the next section we shall show that it is one of the Green's functions associated with the Dirac equation.

Coming back to two-photon annihilation, we find that the expression for the S-matrix element (4.151) now becomes

$$\begin{aligned}
S_{fi} &= (-e)^2\int d^4x_1\int d^4x_2\left(\frac{1}{\sqrt{2\omega_1V}}\epsilon_\mu^{(\alpha_1)}e^{-ik_1\cdot x_1}\right)\left(\frac{1}{\sqrt{2\omega_2V}}\epsilon_\nu^{(\alpha_2)}e^{-ik_2\cdot x_2}\right)\\
&\quad\times\left(\sqrt{\frac{m}{E_+V}}\bar{v}_\alpha e^{ip_+\cdot x_1}\right)(\gamma_\mu)_{\alpha\beta}\langle 0|T(\psi_\beta(x_1)\bar{\psi}_\gamma(x_2))|0\rangle(\gamma_\nu)_{\gamma\delta}\left(\sqrt{\frac{m}{E_-V}}u_\delta e^{ip_-\cdot x_2}\right)\\
&\quad+\begin{Bmatrix}k_1\longleftrightarrow k_2\\ \alpha_1\longleftrightarrow\alpha_2\end{Bmatrix}\\
&= (-e)^2\sqrt{\left(\frac{m}{E_-V}\right)\left(\frac{m}{E_+V}\right)\left(\frac{1}{2\omega_1V}\right)\left(\frac{1}{2\omega_2V}\right)}\int d^4x_1\int d^4x_2\,e^{-ik_1\cdot x_1-ik_2\cdot x_2}\epsilon_\mu^{(\alpha_1)}\epsilon_\nu^{(\alpha_2)}\\
&\quad\times e^{-ip_+\cdot x_1+ip_-\cdot x_2}\bar{v}\gamma_\mu\left(-\frac{i}{(2\pi)^4}\int d^4q\,\frac{e^{iq\cdot(x_1-x_2)}(-i\gamma\cdot q+m)}{q^2+m^2-i\epsilon}\right)\gamma_\nu u\\
&\quad+\begin{Bmatrix}k_1\longleftrightarrow k_2\\ \alpha_1\longleftrightarrow\alpha_2\end{Bmatrix},
\end{aligned} \quad (4.163)$$

where in next to the last line the usual rule for matrix multiplication is understood. It is indeed gratifying that the vacuum-expectation value that appears in (4.163) is manifestly covariant. We may come to appreciate this point even more if we note that (a) the notion of t_2 being earlier or later than t_1 is not a Lorentz invariant

‡Sometimes the symbol $(S_F(x-x'))_{\alpha\beta}$ is used for (4.162) itself. Our definition of S_F agrees with that given in F. J. Dyson's original paper on this subject.

concept when $x_1 - x_2$ is spacelike; (b) unlike the anticommutator $\{\psi_\alpha(x), \psi_\beta(x')\}$, the electron propagator does not vanish for a spacelike separation‡ (as can be shown by examining the explicit form of the electron propagator in the coordinate (x) space). In other words, the separation shown in Fig. 4–4(a) and (b) is not a Lorentz invariant procedure. To put it in a more pictorial way, a physical process in which we visualize a virtual *electron* as propagating from x_2 to x_1 may appear in some other Lorentz frame as a process in which we visualize a virtual *positron* as propagating from x_1 to x_2.§ Yet when Fig. 4–4(a) and (b) are considered *together*, the two vacuum-expectation-value terms combine into a manifestly covariant form.

The fact that we get a mathematically simple Lorentz covariant expression when we consider Fig. 4–4(a) and (b) together suggests that we should take the following viewpoint. Instead of saying that the virtual particle is an electron that goes from x_2 to x_1 when $t_1 > t_2$ and a positron that goes from x_1 to x_2 when $t_2 > t_1$, we might as well say the following: (a) It is always an electron that goes from x_2 to x_1 regardless of whether t_1 is later or earlier than t_2, (b) if t_2 happens to be later than t_1, the electron simply goes "backward in time." (We say this with the understanding that the electron wave function for the virtual electron is that of a negative-energy electron because of the appearance of $v^{(s)}(\mathbf{p})$ [$= \pm u^{(3,4)}(-\mathbf{p})$].) From this point of view the positron that propagates forward in time [e.g. the virtual positron in Fig. 4–4(b)] can be considered as being equivalent to a negative-energy electron that propagates "backward in time." This interpretation (proposed originally by E. C. G. Stueckelberg in 1941) forms the basis of Feynman's space-time approach (to be discussed in the next section).||

Given an internal line segment that joins x_1 and x_2, we must still specify whether the electron goes from x_1 to x_2 or from x_2 to x_1. This can be done simply by drawing an "arrow." If the arrow points from x_2 to x_1, we understand it to be one of the following two situations: either the electron goes forward in time from x_2 to x_1 ($t_1 > t_2$), or the positron goes forward in time from x_1 to x_2 ($t_2 > t_1$). If we do not ask whether t_1 is earlier or later than t_2, Figs. 4–4(a) and (b) are topologically equivalent. So they are collectively represented by a single graph Fig. 4–6(a) where t_2 can now be earlier or later than t_1. It is also convenient to draw an arrow for an external line which represents a real (as opposed to a virtual) electron or positron. If the direction of the arrow coincides with the direction of increasing t (upward), then the external line represents a real electron. If the arrow direction is opposite to the increasing t direction, then the external line represents a real positron. All this

‡Feynman (1961b), pp. 83–86, gives an enlightening discussion as to why the electron propagator need not vanish for a space-like separation.

§In this sense we agree with "purists" such as G. Källén who remarks, "We should like to warn the reader against too pictorial an interpretation of the diagrams in terms of particles propagating from one space-time point to another since such pictures sometimes lead to serious misunderstanding of the physics involved."

||The first statement in the literature proposing that a virtual positron appearing in the intermediate states can be treated as a negative-energy electron except for the reversal of the time ordering is found in a 1930 paper of P. A. M. Dirac. See also our earlier discussion of Thomson scattering given in Section 3–9.

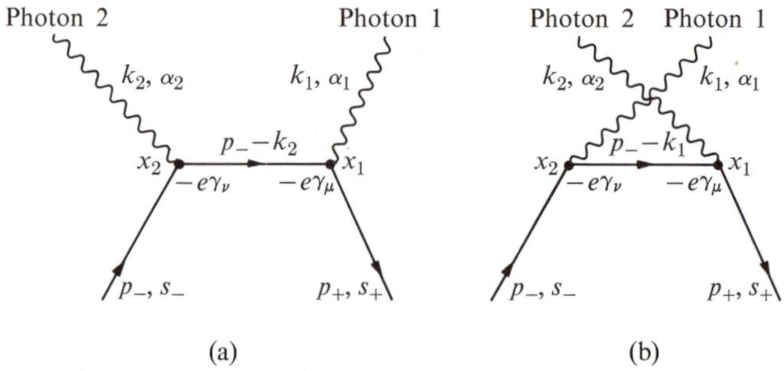

Fig. 4–6. Feynman graphs for two-photon annihilation, where t_2 can be earlier or later than t_1.

is clear from Fig. 4–6(a). The "arrow rule" can be concisely summarized by saying that we associate with the action of $\psi(x)[\bar{\psi}(x)]$ an arrow toward (leaving) the point x regardless of whether the action of the field operator represents annihilation or creation, whether the particle created or annihilated is real or virtual, and whether the particle is an electron or a positron.

The introduction of the arrows makes it simpler to keep track of the ordering of the various factors in the S-matrix element (4.163). All we need to remember is "follow the arrow." We start with the wave function for the incident electron to be annihilated at x_2, $-e\gamma_\nu$ to represent the interaction with the radiation field at x_2, the electron propagator that takes care of the propagation of the virtual electron from x_2 to x_1 (with the understanding that the virtual particle is actually a positron if $t_2 > t_1$), $-e\gamma_\mu$ to represent the second interaction with the radiation field at x_1, and finally the (adjoint) wave function to represent the incident positron to be annihilated at x_1. If we subscribe to the "follow-the-arrow" prescription, there is no longer any need to write the matrix indices explicitly.

We have argued that Figs. 4–4(a) and (b) can be combined into a single graph, Fig. 4–6(a). However, we must still make a distinction as to whether the interaction of the electron-positron system with the radiation field at x_1 represents the emission of photon 1 or of photon 2. For this reason there are actually two topologically inequivalent graphs we must consider, as is clear from Fig. 4–6(a) and (b). In terms of the S-matrix element (4.163), the term denoted by $k_1 \leftrightarrow k_2$, $\alpha_1 \leftrightarrow \alpha_2$ corresponds to Fig. 4–6(b). Figure 4–6(b) is sometimes referred to as a *crossed diagram.*

Covariant matrix element for two-photon annihilation. Let us now perform the indicated integrations in (4.163). We propose to perform the space-time integrations first. Omitting those factors which do not depend on x_1 and x_2, we have

$$\int d^4x_1 \int d^4x_2 \exp[i(-k_1 + p_+ + q)\cdot x_1] \exp[i(-k_2 + p_- - q)\cdot x_2]$$
$$= (2\pi)^4 \delta^4[q - (-p_+ + k_1)](2\pi)^4 \delta^4[q - (p_- - k_2)] \qquad (4.164)$$

for Fig. 4–5(a). One of the $(2\pi)^4$ will be absorbed in the defining equation of the covariant $\mathscr{M}$-matrix to be given below, and the other $(2\pi)^4$ will be made to cancel with $1/(2\pi)^4$ in the Fourier integral expression for the electron propagator. We now perform the indicated energy-momentum (q-space) integration which simply kills one of the $\delta^{(4)}$ functions (say, the second one) in (4.164) with the result that q is replaced everywhere by $p_- - k_2$. This amounts to saying that *both* energy and momentum are conserved at each vertex, in sharp contrast to the analogous situation in the old-fashioned perturbation method according to which energy is *not* conserved in the intermediate states.‡ There survives a four-dimensional δ-function that expresses *overall* energy momentum conservation. The crossed diagram Fig. 4–6(b) can be treated in a similar manner; clearly q is replaced by $p_- - k_1$ in this case. As in the Λ-decay case we define the covariant matrix element $\mathscr{M}_{fi}$ by

$$S_{fi} = \delta_{fi} - i(2\pi)^4\delta^{(4)}(p_- + p_+ - k_1 - k_2)\sqrt{\left(\frac{m}{E_-V}\right)\left(\frac{m}{E_+V}\right)\left(\frac{1}{2\omega_1 V}\right)\left(\frac{1}{2\omega_2 V}\right)}\mathscr{M}_{fi}, \tag{4.165}$$

where δ_{fi} is again zero for this particular process. We end up with the remarkably simple expression

$$\begin{aligned}-i\mathscr{M}_{fi} = (-e)^2\bar{v}^{(s_+)}(p_+)\Bigg[&\gamma\cdot\epsilon^{(\alpha_1)}\frac{-i\gamma\cdot(p_- - k_2) + m}{i[(p_- - k_2)^2 + m^2 - i\epsilon]}\gamma\cdot\epsilon^{(\alpha_2)} \\ &+ \gamma\cdot\epsilon^{(\alpha_2)}\frac{-i\gamma\cdot(p_- - k_1) + m}{i[(p_- - k_1)^2 + m^2 - i\epsilon]}\gamma\cdot\epsilon^{(\alpha_1)}\Bigg]u^{(s_-)}(\mathbf{p}_-).\end{aligned} \tag{4.166}$$

The first and second terms correspond to Fig. 4–6(a) and (b) respectively. Note that the intermediate momentum of the virtual electron is $q = p_- - k_2$ or $p_- - k_1$. The relative sign of the two terms is positive as expected from the Bose-Einstein nature of the photon; we get the correct result automatically since we have used a formalism in which the radiation fields commute.

Matrix element for Compton scattering. From this single example of two-photon annihilation, it is perhaps not completely obvious that the various electron field operators in the S-matrix necessarily combine in such a miraculous way that we always end up with an elegant expression involving the covariant electron propagator. For this reason, before we proceed to obtain the cross section for two-photon annihilation, let us work out the second-order matrix element for another process, Compton scattering. This time the S-matrix element analogous to (4.140) is

$$S_{fi} = (-e)^2\int d^4x_1\int_{t_2<t_1} d^4x_2\langle\gamma' e^{-\prime}|\bar{\psi}(x_1)\gamma_\mu\psi(x_1)A_\mu(x_1)\bar{\psi}(x_2)\gamma_\nu\psi(x_2)A_\nu(x_2)|\gamma e^-\rangle, \tag{4.167}$$

where $|\gamma e^-\rangle$ and $|\gamma' e^{-\prime}\rangle$ symbolically represent the initial and final $\gamma + e^-$ states. Remembering that t_1 is later than t_2, we argue that $\psi(x_1)$ cannot create a positron (since there is no positron in the final state) and that $\bar{\psi}(x_2)$ cannot annihilate a positron (since there is no positron in the initial state). Thus $\psi(x_1)$ and $\bar{\psi}(x_2)$ can be

‡We shall come back to this important difference between the old-fashioned method and the modern (covariant) method in Section 4–6.

replaced by $\psi^{(+)}(x_1)$ and $\bar{\psi}^{(-)}(x_2)$. Furthermore there must be two (+) and two (−) since we start with one electron and end up with one electron. The products of the electron-positron operators that give rise to nonvanishing contributions are then seen to be

$$(\gamma_\mu)_{\alpha\beta}(\gamma_\nu)_{\gamma\delta}\bar{\psi}_\alpha^{(-)}(x_1)\psi_\beta^{(+)}(x_1)\bar{\psi}_\gamma^{(-)}(x_2)\psi_\delta^{(+)}(x_2) \tag{4.168}$$

and

$$(\gamma_\mu)_{\alpha\beta}(\gamma_\nu)_{\gamma\delta}\bar{\psi}_\alpha^{(+)}(x_1)\psi_\beta^{(+)}(x_1)\bar{\psi}_\gamma^{(-)}(x_2)\psi_\delta^{(-)}(x_2). \tag{4.169}$$

When sandwiched between the initial and the final electron states, (4.168) becomes

$$\begin{aligned}
&(\gamma_\mu)_{\alpha\beta}(\gamma_\nu)_{\gamma\delta}\langle e^{-\prime}|\bar{\psi}_\alpha^{(-)}(x_1)\psi_\beta^{(+)}(x_1)\bar{\psi}_\gamma^{(-)}(x_2)\psi_\delta^{(+)}(x_2)|e^-\rangle \\
&= (\gamma_\mu)_{\alpha\beta}(\gamma_\nu)_{\gamma\delta}\langle e^{-\prime}|\bar{\psi}_\alpha^{(-)}(x_1)|0\rangle\langle 0|\psi_\beta^{(+)}(x_1)\bar{\psi}_\gamma^{(-)}(x_2)|0\rangle\langle 0|\psi_\delta^{(+)}(x_2)|e^-\rangle \\
&= \left[\sqrt{\frac{m}{E'V}}\bar{u}'_\alpha e^{-ip'\cdot x_1}\right](\gamma_\mu)_{\alpha\beta}\langle 0|\psi_\beta^{(+)}(x_1)\bar{\psi}_\gamma^{(-)}(x_2)|0\rangle(\gamma_\nu)_{\gamma\delta}\left[\sqrt{\frac{m}{EV}}u_\delta e^{ip\cdot x_2}\right].
\end{aligned} \tag{4.170}$$

As for (4.169) we interchange $x_1 \leftrightarrow x_2$, $\mu \leftrightarrow \nu$, and $\alpha\beta \leftrightarrow \gamma\delta$. Clearly

$$\begin{aligned}
&(\gamma_\mu)_{\alpha\beta}(\gamma_\nu)_{\gamma\delta}\bar{\psi}_\gamma^{(+)}(x_2)\psi_\delta^{(+)}(x_2)\bar{\psi}_\alpha^{(-)}(x_1)\psi_\beta^{(-)}(x_1) \\
&\quad = (\gamma_\mu)_{\alpha\beta}(\gamma_\nu)_{\gamma\delta}[-\bar{\psi}_\gamma^{(+)}(x_2)\bar{\psi}_\alpha^{(-)}(x_1)\psi_\delta^{(+)}(x_2)\psi_\beta^{(-)}(x_1) \\
&\qquad + \bar{\psi}_\gamma^{(+)}(x_2)\{\psi_\delta^{(+)}(x_2), \bar{\psi}_\alpha^{(-)}(x_1)\}\psi_\beta^{(-)}(x_1)] \\
&\quad = (\gamma_\mu)_{\alpha\beta}(\gamma_\nu)_{\gamma\delta}[-\bar{\psi}_\alpha^{(-)}(x_1)\bar{\psi}_\gamma^{(+)}(x_2)\psi_\beta^{(-)}(x_1)\psi_\delta^{(+)}(x_2) \\
&\qquad + \bar{\psi}_\gamma^{(+)}(x_2)\{\psi_\delta^{(+)}(x_2), \bar{\psi}_\alpha^{(-)}(x_1)\}\psi_\beta^{(-)}(x_1)].
\end{aligned} \tag{4.171}$$

Recall now that $\{\psi_\delta^{(+)}(x_2), \bar{\psi}_\alpha^{(-)}(x_1)\}$ is just a c-number function of $x_1 - x_2$ and that $\bar{\psi}_\gamma^{(+)}$ and $\psi_\beta^{(-)}$ respectively annihilate and create positrons. This means that the second term of (4.171) gives no contribution to Compton scattering.‡ Sandwiching (4.171) between the initial and the final electron states, we obtain

$$\begin{aligned}
&(\gamma_\mu)_{\alpha\beta}(\gamma_\nu)_{\gamma\delta}\langle e^{-\prime}|-\bar{\psi}_\alpha^{(-)}(x_1)\bar{\psi}_\gamma^{(+)}(x_2)\psi_\beta^{(-)}(x_1)\psi_\delta^{(+)}(x_2)|e^-\rangle \\
&= \left[\sqrt{\frac{m}{E'V}}\bar{u}'_\alpha e^{-ip'\cdot x_1}\right](\gamma_\mu)_{\alpha\beta}\langle 0|-\bar{\psi}_\gamma^{(+)}(x_2)\psi_\beta^{(-)}(x_1)|0\rangle(\gamma_\nu)_{\gamma\delta}\left[\sqrt{\frac{m}{EV}}u_\delta e^{ip\cdot x_2}\right],
\end{aligned} \tag{4.172}$$

where we should keep in mind that t_2 is now later than t_1. Note that just as in the annihilation problem some of the field operators are used to annihilate and create the incident and the outgoing electron while the remaining operators are used to create and annihilate a virtual electron (in the case $t_1 > t_2$) and a virtual positron (in the case $t_2 > t_1$). Considering the two terms (4.170) and (4.172) together, and taking advantage of the fact that the photon-field matrix element is symmetric under $\mu \leftrightarrow \nu$, $x_1 \leftrightarrow x_2$, we see that the two terms combine to form a single expression valid regardless of whether t_1 is later or earlier than t_2:

$$\left[\sqrt{\frac{m}{E'V}}\bar{u}_\alpha e^{-ip'\cdot x_1}\right](\gamma_\mu)_{\alpha\beta}\langle 0|T(\psi_\beta(x_1)\bar{\psi}_\gamma(x_2))|0\rangle(\gamma_\nu)_{\gamma\delta}\left[\sqrt{\frac{m}{EV}}u_\delta e^{ip\cdot x_2}\right]. \tag{4.173}$$

‡To prove this rigorously, note that $\langle 0|b'd'd^\dagger b^\dagger|0\rangle = \langle 0|\{d', d^\dagger\}\{b', b^\dagger\}|0\rangle = 0$ since the initial and the final electron states are in general different.

The photon-field matrix element is well known (from Section 2–5, where we discussed Rayleigh scattering in detail). We therefore get

$$
\begin{aligned}
S_{fi} = (-e)^2 \int d^4x_1 \int d^4x_2 & \left[\frac{1}{\sqrt{2\omega_1 V}}\epsilon_\mu^{(\alpha)} e^{ik\cdot x_1}\right]\left[\frac{1}{\sqrt{2\omega_2 V}}\epsilon_\nu^{(\alpha')} e^{-ik'\cdot x_2}\right] \\
\times & \left[\sqrt{\frac{m}{E'V}}\bar{u}'_\alpha e^{-ip'\cdot x_1}\right](\gamma_\mu)_{\alpha\beta}\langle 0 | T(\psi_\beta(x_1)\bar{\psi}_\gamma(x_2)) | 0\rangle (\gamma_\nu)_{\gamma\delta}\left[\sqrt{\frac{m}{EV}}u_\delta e^{ip\cdot x_2}\right] \\
+ & \begin{Bmatrix} k \longleftrightarrow k' \\ \alpha \longleftrightarrow \alpha' \end{Bmatrix}.
\end{aligned}
\tag{4.174}
$$

The final expression involves the covariant electron propagator just as advertised earlier. The remaining steps are very similar to (4.164) through (4.166). When the indicated integrations are performed, we end up with the $\mathscr{M}$-matrix element [defined as in (4.165)]:

$$
\begin{aligned}
-i\mathscr{M}_{fi} = (-e)^2 \bar{u}^{(s')}(\mathbf{p}') & \left[\gamma\cdot\epsilon^{(\alpha)} \frac{-i\gamma\cdot(p-k')+m}{i[(p-k')^2+m^2-i\epsilon]}\gamma\cdot\epsilon^{(\alpha')}\right. \\
& \left. + \gamma\cdot\epsilon^{(\alpha')}\frac{-i\gamma\cdot(p+k)+m}{i[(p+k)^2+m^2-i\epsilon]}\gamma\cdot\epsilon^{(\alpha)}\right] u^{(s)}(\mathbf{p}).
\end{aligned}
\tag{4.175}
$$

This expression can be visualized by drawing Feynman diagrams (Fig. 4–7). Again, as Feynman put it, "the result is much easier to understand than the derivation."

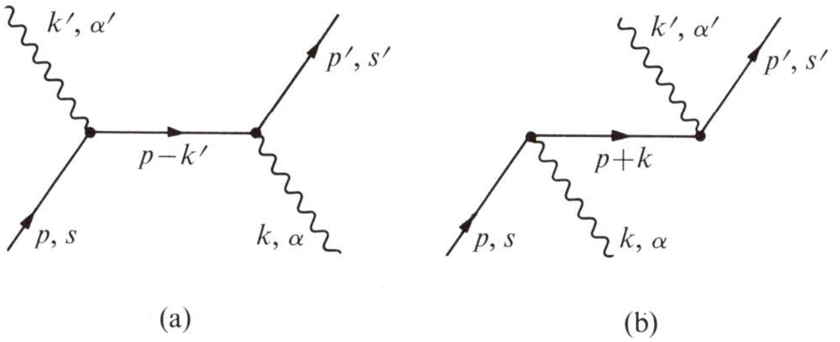

Fig. 4–7. Feynman diagrams for Compton scattering.

Feynman rules. If the reader has closely followed the derivations of (4.166) and (4.175) and, in addition, worked out a few similar problems, he will easily convince himself of the following set of prescriptions for writing $-i\mathscr{M}_{fi}$. First, draw all possible topologically inequivalent diagrams. So far as the external fermion lines go, for each real (as opposed to virtual) electron or positron annihilated or created, associate the following free-particle spinors:

Annihilated electron:	$u^{(s)}(\mathbf{p})$	
Annihilated positron:	$\bar{v}^{(s)}(\mathbf{p})$	
Created electron:	$\bar{u}^{(s)}(\mathbf{p})$	
Created positron:	$v^{(s)}(\mathbf{p})$	

For the photon use $\epsilon_\mu^{(\alpha)}$. Write $-e\gamma_\mu$ at each vertex. (Note that the factor i that appears in $-ie\bar{\psi}\gamma_\mu\psi A_\mu$ cancels with the $-i$ of $(-i)^n$ that appears in the nth-order

S-matrix element.) For a virtual electron (or a positron) represented by an internal line $\bullet\!\!\longrightarrow\!\!\bullet$ write

$$\frac{-i\gamma\cdot q + m}{i(q^2 + m^2 - i\epsilon)}, \tag{4.176}$$

where q is the four-momentum of the virtual particle when the internal line represents the propagation of an *electron going forward in time*. (The case when the virtual fermion is a positron is taken care of automatically.) Assume energy-momentum conservation at each vertex. If the energy-momentum of the virtual electron is already fixed by energy-momentum conservation (as in our examples of two-photon annihilation and Compton scattering), there is no q-space integration needed, nor is the factor $1/(2\pi)^4$ necessary. On the other hand, if the virtual four-momentum is not fixed, a "nontrivial integration" remains to be performed; we must integrate with $d^4q/(2\pi)^4$.‡ As for the ordering of the factors, just "follow the arrow." In addition, there are certain sign factors which will be explained later.

This would be a logical place to discuss the connection between the old-fashioned perturbation method based on energy denominators and the modern covariant perturbation method based on Lorentz-invariant denominators. However, since this connection can be seen in a simpler way in processes in which spin-zero virtual bosons propagate, and since spin is an inessential complication in the illustration of this point, we shall postpone this topic until Section 4–6.

Cross section for two-photon annihilation. The remaining part of this section is devoted to a detailed discussion of two-photon annihilation and other related processes. The relation between the $\mathscr{M}$-matrix element for two-photon annihilation (4.166) and the differential cross section can be obtained just as in the Λ-decay case treated in the previous section. The only new feature is the flux density, which is indicative of how fast the two beams hit each other; it is given by v_{rel}/V, where v_{rel} is the *relative* velocity of the electron and the positron as observed in the coordinate system in question. In analogy with (4.127) we have

$$d\sigma = (2\pi)^4 \frac{1}{v_{\text{rel}}} \frac{1}{2E_+} \frac{1}{2E_-} |\mathscr{M}_{fi}|^2 (2m)^2 \frac{d^3k_1}{(2\pi)^3 2\omega_1} \frac{d^3k_2}{(2\pi)^3 2\omega_2} \delta^{(4)}(p_+ + p_- - k_1 - k_2). \tag{4.177}$$

This formula is good in any coordinate system in which the two incident beams are collinear. In the center-of-mass (CM) system

$$v_{\text{rel}} = \frac{|\mathbf{p}_+|}{E_+} + \frac{|\mathbf{p}_-|}{E_-} = \frac{|\mathbf{p}|E_{\text{tot}}}{E_+E_-} \quad (\text{CM}) \tag{4.178}$$

with $|\mathbf{p}| = |\mathbf{p}_+| = |\mathbf{p}_-|$. It is worth keeping in mind that v_{rel} can exceed unity, that is, the relative velocity can be greater than the speed of light. In the so-called laboratory system, in which the moving positron collides with the electron at rest ($\mathbf{p}_- = 0$), we simply have

$$v_{\text{rel}} = |\mathbf{p}_+|/E_+ \quad (\text{lab}). \tag{4.179}$$

‡We shall encounter nontrivial integrations in Section 4–7.

It is interesting to note that $v_{\text{rel}}E_+E_-$ is the same in any Lorentz frame in which the two incident beams are collinear since we can readily prove

$$v_{\text{rel}}\,E_+E_- = \sqrt{(p_+\cdot p_-)^2 - m^4}. \tag{4.180}$$

Using (4.128) and (4.180), we see that (4.178) is Lorentz invariant. This is reasonable since the integrated cross section must be the same regardless of the coordinate system we use. Using (4.178) and the analog of (4.133), we can readily work out the differential cross section in the CM-system:

$$\left(\frac{d\sigma}{d\Omega}\right)_{\text{CM}} = \frac{1}{4}\frac{1}{(4\pi)^2}\frac{1}{E_{\text{tot}}^2}\frac{|\mathbf{k}|}{|\mathbf{p}|}|\mathscr{M}_{fi}|^2(2m)^2. \tag{4.181}$$

This expression is a special case of a very useful relation:

$$\left(\frac{d\sigma}{d\Omega}\right)_{\text{CM}} = \frac{1}{4}\frac{1}{(4\pi)^2}\frac{1}{E_{\text{tot}}^2}\frac{|\mathbf{p}_{\text{final}}|}{|\mathbf{p}_{\text{initial}}|}|\mathscr{M}_{fi}|^2\prod_{\substack{\text{external}\\ \text{fermion}}}(2m_{\text{fermi}}) \tag{4.182}$$

applicable to any reaction of the type $1 + 2 \to 3 + 4$. In the lab system we may be interested in the differential cross section for the emission of one of the photons, say photon 1. As usual, we can immediately integrate over k_2 because of the three-dimensional delta function that expresses momentum conservation. In evaluating $\delta(E_+ + E_- - \omega_1 - \omega_2)\,d^3k_1$, all we need to know is $\partial E_f/\partial|\mathbf{k}_1|$ with θ_1 and ϕ_1 fixed. Using the kinematical relations

$$\begin{aligned}|\mathbf{p}_+ - \mathbf{k}_1| &= \sqrt{|\mathbf{p}_+|^2 - 2|\mathbf{p}_+|\omega_1\cos\theta_1 + \omega_1^2} = |\mathbf{k}_2| = \omega_2,\\ m + E_+ &= \omega_1 + \omega_2,\end{aligned} \tag{4.183}$$

we obtain

$$\begin{aligned}\left(\frac{\partial E_f}{\partial|\mathbf{k}_1|}\right)_{\theta_1,\phi_1} &= \frac{\partial(\omega_1 + |\mathbf{p}_+ - \mathbf{k}_1|)}{\partial|\mathbf{k}_1|}\\ &= 1 + \frac{|\mathbf{k}_1| - |\mathbf{p}_+|\cos\theta_1}{\omega_2} = \frac{m(m+E_+)}{\omega_1\omega_2}.\end{aligned} \tag{4.184}$$

So the differential cross section in the lab system is

$$\begin{aligned}\left(\frac{d\sigma}{d\Omega}\right)_{\text{lab}} &= \frac{1}{4m|\mathbf{p}_+|}(2m)^2\frac{|\mathbf{k}_1|^2}{2\omega_1(2\pi)^3}\frac{2\pi}{2\omega_2}\frac{1}{(\partial E_f/\partial|\mathbf{k}_1|)_{\theta_1,\phi_1}}|\mathscr{M}_{fi}|^2\\ &= \frac{1}{4}\frac{1}{(4\pi)^2}\frac{1}{m^2|\mathbf{p}_+|}\frac{\omega_1^2}{(m+E_+)}|\mathscr{M}_{fi}|^2(2m)^2.\end{aligned} \tag{4.185}$$

It turns out that in this particular problem the evaluation of $|\mathscr{M}_{fi}|^2$ is simplest in the lab system. First

$$p_- = (0, 0, 0, im), \qquad \epsilon^{(\alpha)} = (\boldsymbol{\epsilon}^{(\alpha)}, 0). \tag{4.186}$$

This means that $\gamma\cdot p_-$ anticommutes with $\gamma\cdot\epsilon^{(\alpha)}$, hence

$$(-i\gamma\cdot p_-)(\gamma\cdot\epsilon^{(\alpha)})u^{(s)}(\mathbf{p}_-) = -(\gamma\cdot\epsilon^{(\alpha)})mu^{(s)}(\mathbf{p}_-), \tag{4.187}$$

where $\epsilon^{(\alpha)}$ may be $\epsilon^{(\alpha_1)}$ or $\epsilon^{(\alpha_2)}$. Moreover

$$\begin{aligned}(p_- - k_{1,2})^2 &= -m^2 - 2p_-\cdot k_{1,2}\\ &= -m^2 + 2m\omega_{1,2}.\end{aligned} \tag{4.188}$$

Thus (4.166) now becomes

$$\mathcal{M}_{fi} = ie^2 \bar{v}\left[\frac{\gamma\cdot\epsilon^{(\alpha_1)}\gamma\cdot k_2\gamma\cdot\epsilon^{(\alpha_2)}}{2m\omega_2} + \frac{\gamma\cdot\epsilon^{(\alpha_2)}\gamma\cdot k_1\gamma\cdot\epsilon^{(\alpha_1)}}{2m\omega_1}\right]u. \tag{4.189}$$

If the initial electron and positron are unpolarized, we must evaluate

$$\begin{aligned}\frac{1}{4}\sum_{s_+,s_-}|\mathcal{M}_{fi}|^2 &= \frac{e^4}{4}\sum_{s_+,s_-}\bar{u}^{(s_-)}\gamma_4 O^\dagger\gamma_4 v^{(s_+)}\bar{v}^{(s_+)}Ou^{(s_-)}\\ &= \frac{e^4}{4}\mathrm{Tr}\left[\gamma_4 O^\dagger\gamma_4\left(-\frac{i\gamma\cdot p_+ + m}{2m}\right)O\left(\frac{-i\gamma\cdot p_- + m}{2m}\right)\right],\end{aligned} \tag{4.190}$$

where we have denoted the expression inside the bracket of (4.189) by O. The factor $\frac{1}{4}$ appears because we must average over the four spin states of the e^-e^+ system. In evaluating $\gamma_4 O^\dagger\gamma_4$ it is important to remember that

$$\gamma_4(\gamma\cdot a)^\dagger\gamma_4 = \gamma_4[\boldsymbol{\gamma}\cdot\mathbf{a} + \gamma_4(ia_0)^*]\gamma_4 = -\gamma\cdot a \tag{4.191}$$

for any four-vector a_μ whose space components are purely real and whose fourth component is purely imaginary. The relation (4.191) can be generalized as follows:

$$\gamma_4[(\gamma\cdot a)(\gamma\cdot b)(\gamma\cdot c)\cdots(\gamma\cdot l)]^\dagger\gamma_4 = (-\gamma\cdot l)\cdots(-\gamma\cdot c)(-\gamma\cdot b)(-\gamma\cdot a). \tag{4.192}$$

For the trace arising from the square of the amplitude for Fig. 4–6(a) we must evaluate

$$\mathrm{Tr}[(\gamma\cdot\epsilon^{(\alpha_2)}\gamma\cdot k_2\gamma\cdot\epsilon^{(\alpha_1)})(i\gamma\cdot p_+ + m)(\gamma\cdot\epsilon^{(\alpha_1)}\gamma\cdot k_2\gamma\cdot\epsilon^{(\alpha_2)})(-i\gamma\cdot p_- + m)]. \tag{4.193}$$

First, terms linear in m are zero because the trace of the product of an odd number of the gamma matrices vanishes by (4.83). Terms proportional to m^2 also vanish because

$$\gamma\cdot k_2\gamma\cdot\epsilon^{(\alpha_1)}\gamma\cdot\epsilon^{(\alpha_1)}\gamma\cdot k_2 = k_2^2 = 0, \tag{4.194}$$

where we have used

$$(\gamma\cdot a)(\gamma\cdot a) = a^2, \tag{4.195}$$

which is a special case of (4.81). To evaluate the remaining terms we keep in mind that (4.195) can be used to "kill" the gamma matrices; so the trick is to cleverly rewrite the various factors using (4.81) until we reach the stage at which we can take advantage of (4.195). Specifically

$$\begin{aligned}&\mathrm{Tr}(\gamma\cdot\epsilon^{(\alpha_2)}\gamma\cdot k_2\gamma\cdot\epsilon^{(\alpha_1)}\gamma\cdot p_+\gamma\cdot\epsilon^{(\alpha_1)}\gamma\cdot k_2\gamma\cdot\epsilon^{(\alpha_2)}\gamma\cdot p_-)\\ &\quad= \mathrm{Tr}[\gamma\cdot k_2(2\epsilon^{(\alpha_1)}\cdot p_+ - \gamma\cdot p_+\gamma\cdot\epsilon^{(\alpha_1)})\gamma\cdot\epsilon^{(\alpha_1)}\gamma\cdot k_2(2\epsilon^{(\alpha_2)}\cdot p_- - \gamma\cdot p_-\gamma\cdot\epsilon^{(\alpha_2)})\gamma\cdot\epsilon^{(\alpha_2)}]\\ &\quad= -2\epsilon^{(\alpha_1)}\cdot p_+\,\mathrm{Tr}(\gamma\cdot k_2\gamma\cdot\epsilon^{(\alpha_1)}\gamma\cdot k_2\gamma\cdot p_-) + \mathrm{Tr}(\gamma\cdot k_2\gamma\cdot p_+\gamma\cdot k_2\gamma\cdot p_-)\\ &\quad= -2\epsilon^{(\alpha_1)}\cdot p_+\,\mathrm{Tr}[\gamma\cdot k_2(2\epsilon^{(\alpha_1)}\cdot k_2 - \gamma\cdot k_2\gamma\cdot\epsilon^{(\alpha_1)})\gamma\cdot p_-]\\ &\qquad+ \mathrm{Tr}[\gamma\cdot k_2(2p_+\cdot k_2 - \gamma\cdot k_2\gamma\cdot p_+)\gamma\cdot p_-]\\ &\quad= -16(\epsilon^{(\alpha_1)}p_+)(\epsilon^{(\alpha_1)}\cdot k_2)(k_2\cdot p_-) + 8(k_2\cdot p_+)(k_2\cdot p_-)\\ &\quad= -16(\epsilon^{(\alpha_1)}\cdot k_2)^2(k_2\cdot p_-) + 8(k_1\cdot p_-)(k_2\cdot p_-),\end{aligned} \tag{4.196}$$

where in the last step we have used $\epsilon^{(\alpha_1)}\cdot p_+ = \epsilon^{(\alpha_1)}\cdot k_2$ and $k_2\cdot p_+ = k_1\cdot p_-$ which follow from energy-momentum conservation and the transversality condition for the photon polarization vector. The trace that arises from the interference of Figs. 4–6(a) and 4–6(b) can be obtained in a similar manner using various tricks. The trace arising from the square of the amplitude for Fig. 4–6(b) can be written immediately just by interchanging the photon labels in (4.196). The net result can be shown to be

$$\frac{1}{4}\sum_{s_+s_-}|\mathscr{M}_{fi}|^2 = \frac{e^4}{2(2m)^2}\left[\frac{\omega_2}{\omega_1}+\frac{\omega_1}{\omega_2}+2-4(\boldsymbol{\epsilon}^{(\alpha_1)}\cdot\boldsymbol{\epsilon}^{(\alpha_2)})^2\right]. \tag{4.197}$$

This leads to the differential cross section for two-photon annihilation:

$$\left(\frac{d\sigma}{d\Omega}\right)_{\text{lab}} = \frac{\omega_1^2 r_0^2}{8\,|\mathbf{p}_+|\,(m+E_+)}\left[\frac{\omega_2}{\omega_1}+\frac{\omega_1}{\omega_2}+2-4(\boldsymbol{\epsilon}^{(\alpha_1)}\cdot\boldsymbol{\epsilon}^{(\alpha_2)})^2\right], \tag{4.198}$$

obtained by P. A. M. Dirac in 1930.

Let us now specialize to the low-energy limit of (4.198). With

$$\omega=\omega_1=\omega_2,\qquad \mathbf{k}=\mathbf{k}_1=-\mathbf{k}_2,\qquad |\mathbf{p}_+|=\beta_+ m \tag{4.199}$$

(valid when the incident positron is very slow), we get

$$\left(\frac{d\sigma}{d\Omega}\right)_{\text{lab}} = \frac{r_0^2}{4\beta_+}[1-(\boldsymbol{\epsilon}^{(\alpha_1)}\cdot\boldsymbol{\epsilon}^{(\alpha_2)})^2]. \tag{4.200}$$

If the photon polarization is not observed, we can sum over the photon polarization states. Choosing the z-axis long $\mathbf{k}=\mathbf{k}_1$, one can orient the polarization vector in the x-direction or in the y-direction. In this way we readily construct Table 4–1 for $1-(\boldsymbol{\epsilon}^{(\alpha_1)}\cdot\boldsymbol{\epsilon}^{(\alpha_2)})^2$. The polarization sum evidently gives rise to just 2. To obtain the total cross section for pair annihilation we must note that the final state involves two indistinguishable particles. Hence

$$\sigma^{\text{tot}} = \frac{1}{2}\int\left(\frac{d\sigma}{d\Omega}\right)d\Omega. \tag{4.201}$$

So we have

$$\sigma^{\text{tot}} = \frac{r_0^2}{4\beta_+}2\times 2\pi = \frac{\pi r_0^2}{\beta_+}\qquad (\beta_+\ll 1) \tag{4.202}$$

for the spin-average total cross section. The cross section varies inversely as the velocity of the incident positron. This is characteristic of an exothermic reaction. Although the cross section becomes infinite as $\beta_+\to 0$, the number of annihilations per unit time remains finite as the positron beam slows down in matter since the flux is proportional to β_+.

Coming back to (4.200) and Table 4–1, we note that the transition probability vanishes when $\boldsymbol{\epsilon}^{(1)}$ and $\boldsymbol{\epsilon}^{(2)}$ are parallel. This simple result, we argue, must have a simple explanation. To see this and other interesting features let us treat the problem of low-energy pair annihilation using the explicit forms of the free-particle

Table 4–1

THE VALUES OF $1-(\boldsymbol{\epsilon}^{(\alpha_1)}\cdot\boldsymbol{\epsilon}^{(\alpha_2)})^2$ IN (4.200)

	$\boldsymbol{\epsilon}^{(\alpha_2)}$ parallel to x-axis	$\boldsymbol{\epsilon}^{(\alpha_2)}$ parallel to y-axis
$\boldsymbol{\epsilon}^{(\alpha_1)}$ parallel to x-axis	0	1
$\boldsymbol{\epsilon}^{(\alpha_1)}$ parallel to y-axis	1	0

spinors. As the positron velocity goes to zero, the $\mathscr{M}$-matrix element becomes

$$\mathscr{M}_{fi} = (ie^2/2m\omega)\bar{v}^{(s_+)}(0)[\gamma\cdot\epsilon^{(\alpha_1)}\gamma\cdot k_2\gamma\cdot\epsilon^{(\alpha_2)} + \gamma\cdot\epsilon^{(\alpha_2)}\gamma\cdot k_1\gamma\cdot\epsilon^{(\alpha_1)}]u^{(s_-)}(0). \quad (4.203)$$

We can replace $\gamma\cdot\epsilon^{(\alpha_1)}\gamma\cdot k_2\gamma\cdot\epsilon^{(\alpha_2)}$ by $\boldsymbol{\gamma}\cdot\boldsymbol{\epsilon}^{(\alpha_1)}\boldsymbol{\gamma}\cdot\mathbf{k}_2\boldsymbol{\gamma}\cdot\boldsymbol{\epsilon}^{(\alpha_2)}$ since the photon polarization vectors have no time components and $\gamma_i\gamma_4\gamma_j$ vanishes when taken between $u^{(s_-)}(0)$ and $\bar{v}^{(s_+)}(0)$. The procedure is similar for the second term. Remembering (4.199), we obtain, for the expression inside the bracket of (4.203), the following:

$$\begin{aligned} -\boldsymbol{\gamma}\cdot\boldsymbol{\epsilon}^{(\alpha_1)}\boldsymbol{\gamma}\cdot\mathbf{k}\boldsymbol{\gamma}\cdot\boldsymbol{\epsilon}^{(\alpha_2)} &+ \boldsymbol{\gamma}\cdot\boldsymbol{\epsilon}^{(\alpha_2)}\boldsymbol{\gamma}\cdot\mathbf{k}\boldsymbol{\gamma}\cdot\boldsymbol{\epsilon}^{(\alpha_1)} \\ &= -[i\boldsymbol{\Sigma}\cdot(\boldsymbol{\epsilon}^{(\alpha_1)}\times\mathbf{k})]\boldsymbol{\gamma}\cdot\boldsymbol{\epsilon}^{(\alpha_2)} + \boldsymbol{\gamma}\cdot\boldsymbol{\epsilon}^{(\alpha_2)}[-i\boldsymbol{\Sigma}\cdot(\boldsymbol{\epsilon}^{(\alpha_1)}\times\mathbf{k})] \\ &= -2\mathbf{k}\cdot(\boldsymbol{\epsilon}^{(\alpha_1)}\times\boldsymbol{\epsilon}^{(\alpha_2)})\gamma_4\gamma_5, \end{aligned} \quad (4.204)$$

where we have used

$$(\boldsymbol{\gamma}\cdot\mathbf{a})(\boldsymbol{\gamma}\cdot\mathbf{b}) = \mathbf{a}\cdot\mathbf{b} + i\boldsymbol{\Sigma}\cdot(\mathbf{a}\times\mathbf{b}), \qquad \boldsymbol{\gamma}\cdot\mathbf{a} = i\gamma_4\gamma_5\boldsymbol{\Sigma}\cdot\mathbf{a}. \quad (4.205)$$

We now show that the e^-e^+ state with a zero relative momentum in the triplet spin state cannot annihilate into two photons. Because of rotational invariance it is sufficient to prove this statement for the $J_3 = 1$ state represented by

$$(\uparrow)_{e^-}(\uparrow)_{e^+} = b^{(1)\dagger}(0)d^{(1)\dagger}(0)\,|0\rangle. \quad (4.206)$$

The relevant free-particle spinors are $u^{(1)}(0)$ and $v^{(1)}(0)$. In accordance with (4.204) we evaluate

$$(\bar{v}\gamma_4\gamma_5 u)_{\text{triplet, } J_3=1} = v^{(1)\dagger}(0)\gamma_5 u^{(1)}(0) = (0,0,0,-1)\begin{pmatrix} 0 & -I \\ -I & 0 \end{pmatrix}\begin{pmatrix} 1 \\ 0 \\ 0 \\ 0 \end{pmatrix} = 0. \quad (4.207)$$

On the other hand, for the singlet state symbolically represented by

$$(1/\sqrt{2})[(\uparrow)_{e^-}(\downarrow)_{e^+} - (\downarrow)_{e^-}(\uparrow)_{e^+}] = (1/\sqrt{2})[b^{(1)\dagger}(0)d^{(2)\dagger}(0) - b^{(2)\dagger}(0)d^{(1)\dagger}(0)]\,|0\rangle, \quad (4.208)$$

we have‡

$$\begin{aligned} (v^\dagger\gamma_5 u)_{\text{singlet}} &= \frac{1}{\sqrt{2}}(0,0,1,0)\begin{pmatrix} 0 & -I \\ -I & 0 \end{pmatrix}\begin{pmatrix} 1 \\ 0 \\ 0 \\ 0 \end{pmatrix} \\ &\quad - \frac{1}{\sqrt{2}}(0,0,0,-1)\begin{pmatrix} 0 & -I \\ -I & 0 \end{pmatrix}\begin{pmatrix} 0 \\ 1 \\ 0 \\ 0 \end{pmatrix} \\ &= -\sqrt{2}. \end{aligned} \quad (4.209)$$

‡Note that this is one place where the minus sign in $v^{(1)}(\mathbf{p}) = -u^{(4)}(-\mathbf{p})$ is necessary and important.

To summarize, we get

$$\begin{aligned} |\mathcal{M}_{fi}|^2_{\text{triplet}} &= 0, \\ |\mathcal{M}_{fi}|^2_{\text{singlet}} &= (\sqrt{2}/2m\omega)^2 e^4 \,|2\mathbf{k}\cdot(\boldsymbol{\epsilon}^{(\alpha_1)} \times \boldsymbol{\epsilon}^{(\alpha_2)})|^2 \\ &= (2e^4/m^2)[1 - (\boldsymbol{\epsilon}^{(\alpha_1)}\cdot\boldsymbol{\epsilon}^{(\alpha_2)})^2]. \end{aligned} \tag{4.210}$$

These results are, of course, in agreement with the low-energy limit of (4.197) since

$$\tfrac{1}{4}\sum_{s_+ s_-} |\mathcal{M}_{fi}|^2 = \tfrac{3}{4}|\mathcal{M}_{fi}|^2_{\text{triplet}} + \tfrac{1}{4}|\mathcal{M}_{fi}|^2_{\text{singlet}}. \tag{4.211}$$

Symmetry considerations. We have shown that as $\mathbf{p}_+ \longrightarrow 0$, (a) the triplet state is forbidden to annihilate into two photons, and (b) the singlet state can annihilate into two photons provided $\boldsymbol{\epsilon}^{(\alpha_1)}\cdot\boldsymbol{\epsilon}^{(\alpha_2)} \neq \pm 1$. We shall now demonstrate that these features follow from very general symmetry considerations and are, in fact, independent of the perturbation approximation we have made. Let us first list possible two-photon states with zero total momentum. On the assumption that $\mathbf{k}\ (= \mathbf{k}_1 = -\mathbf{k}_2)$ is in the direction of the positive z-axis, $\boldsymbol{\epsilon}^{(\alpha_1)}$ or $\boldsymbol{\epsilon}^{(\alpha_2)}$ must lie in a plane parallel to the xy-plane. For photon 1 the two polarization vectors $\boldsymbol{\epsilon}^{(1)}$ and $\boldsymbol{\epsilon}^{(2)}$ can be chosen to be unit vectors in the positive x- and y-directions respectively. For photon 2 with momentum $-\mathbf{k}$, $\boldsymbol{\epsilon}^{(1)}$ can again be chosen to be in the positive x-direction, but then $\boldsymbol{\epsilon}^{(2)}$ must necessarily point in the *negative* y-direction, because the negative z-axis, the positive x-axis, and the negative y-axis taken in that order form a right-handed coordinate system. (If the reader is confused, he may look at Fig. 4–8.) In other words, we use the phase convention

$$\boldsymbol{\epsilon}^{(1)}(\mathbf{k}) = \boldsymbol{\epsilon}^{(1)}(-\mathbf{k}), \qquad \boldsymbol{\epsilon}^{(2)}(\mathbf{k}) = -\boldsymbol{\epsilon}^{(2)}(-\mathbf{k}), \tag{4.212}$$

where we have explicitly exhibited the dependence of $\boldsymbol{\epsilon}^{(\alpha)}$ on the photon momentum $\mathbf{k}$ or $-\mathbf{k}$. Two-photon states in which the polarization vectors are parallel must be linear combinations of the following two states:

$$a^\dagger_{\mathbf{k},1} a^\dagger_{-\mathbf{k},1}|0\rangle, \qquad a^\dagger_{\mathbf{k},2} a^\dagger_{-\mathbf{k},2}|0\rangle. \tag{4.213}$$

Likewise, states in which the two polarization vectors are perpendicular can be formed out of $a^\dagger_{\mathbf{k},1} a^\dagger_{-\mathbf{k},2}|0\rangle$ and $a^\dagger_{\mathbf{k},2} a^\dagger_{-\mathbf{k},1}|0\rangle$; it turns out, however, to be more

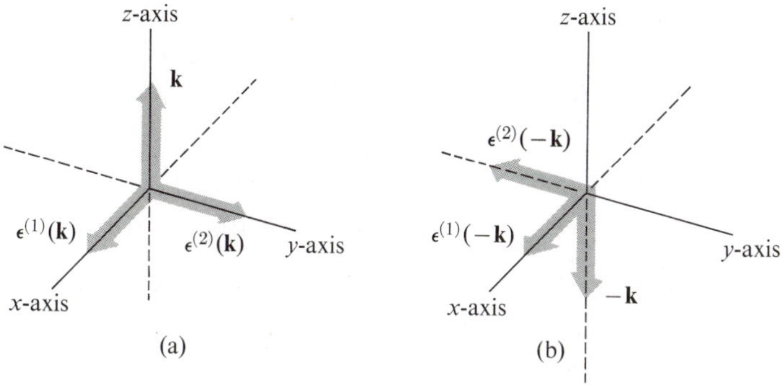

Fig. 4–8. Choice of polarization vectors: (a) photon 1, (b) photon 2.

convenient to consider the following orthogonal linear combinations:

$$\frac{1}{\sqrt{2}}(a^\dagger_{\mathbf{k},1}a^\dagger_{-\mathbf{k},2} + a^\dagger_{\mathbf{k},2}a^\dagger_{-\mathbf{k},1})|0\rangle, \tag{4.214a}$$

$$\frac{1}{\sqrt{2}}(a^\dagger_{\mathbf{k},1}a^\dagger_{-\mathbf{k},2} - a^\dagger_{\mathbf{k},2}a^\dagger_{-\mathbf{k},1})|0\rangle. \tag{4.214b}$$

The two-photon states (4.213), (4.214a), and (4.214b) all satisfy Bose-Einstein statistics automatically because the $a^\dagger$ commute.

The two-photon states we have constructed have defininte transformation properties under parity. First recall that the transverse electromagnetic field transforms as

$$\mathbf{A}(\mathbf{x}, t) \longrightarrow -\mathbf{A}(-\mathbf{x}, t) \tag{4.215}$$

under parity. If we go back to the plane-wave expansion of the field operator (2.60), we readily see that with the phase convention given by (4.212), the above transformation can be accomplished if the individual creation and annihilation operators transform under parity as follows:

$$a^\dagger_{\mathbf{k},\alpha} \rightleftarrows (-1)^\alpha a^\dagger_{-\mathbf{k},\alpha}, \qquad a_{\mathbf{k},\alpha} \rightleftarrows (-1)^\alpha a_{-\mathbf{k},\alpha} \qquad (\alpha = 1, 2). \tag{4.216}$$

An immediate consequence of this is that (4.213) and (4.214b) are even under parity but (4.214a) is odd under parity. Meanwhile, as shown in Section 3–11, the e^-e^+ system with $\mathbf{p}_+ = \mathbf{p}_- = 0$ is required by the Dirac theory to have a negative parity. It then follows that so long as parity conservation holds in the annihilation process, the two-photon system that results from the $\mathbf{p}_+ = \mathbf{p}_- = 0$ e^-e^+ system must necessarily be represented by a state vector of the type (4.214a) with *perpendicular* polarization vectors. This agrees with our earlier result based on perturbation theory.

Next we shall discuss the consequences of charge conjugation invariance. Neutral systems such as a single-photon state and an e^-e^+ state can be eigenstates of the charge conjugation operation C explained in Sections 3–10 and 3–11. By this we do not mean that all neutral systems are C-eigenstates; states such as a neutron and a hydrogen atom cannot be eigenstates of C, since under C they are transformed into an antineutron and an "antihydrogen atom" (a positron bound to an antiproton), etc. The eigenvalue of the charge conjugation operator is known as the charge conjugation parity or C-parity. We recall that charge conjugation interchanges $b^{(s)\dagger}(\mathbf{p})$ and $d^{(s)\dagger}(\mathbf{p})$ (cf. Eq. 3.404). Consider again a $\mathbf{p}_+ = \mathbf{p}_- = 0$ e^-e^+ system. The triplet state (4.206) transforms as

$$\begin{aligned} b^{(1)\dagger}(0)d^{(1)\dagger}(0)|0\rangle &\xrightarrow{\text{charge conj}} d^{(1)\dagger}(0)b^{(1)\dagger}(0)|0\rangle \\ &= -b^{(1)\dagger}(0)d^{(1)\dagger}(0)|0\rangle. \end{aligned} \tag{4.217}$$

Thus the C-parity of the triplet state is odd. Similarly, we readily see from the following that the C-parity of the singlet state (4.208) is even:

$$\begin{aligned} &(1/\sqrt{2})[b^{(1)\dagger}(0)d^{(2)\dagger}(0) - b^{(2)\dagger}(0)d^{(1)\dagger}(0)]|0\rangle \\ &\xrightarrow{\text{charge conj}} (1/\sqrt{2})[b^{(1)\dagger}(0)d^{(2)\dagger}(0) - b^{(2)\dagger}(0)d^{(1)\dagger}(0)]|0\rangle. \end{aligned} \tag{4.218}$$

(Note that the anticommutation relation between $b^\dagger$ and $d^\dagger$ plays a crucial role in obtaining these results.) We may mention, without going into a proof, that a consideration of this kind can be generalized to an e^-e^+ system of a definite orbital angular momentum (without the restriction $\mathbf{p}_+ = \mathbf{p}_- = 0$); the C-parity of an e^-e^+ system is given by $(-1)^{l+S}$ where l and S stand respectively for the relative orbital angular momentum and the total spin ($S = 1$ for the triplet, $S = 0$ for the singlet). Meanwhile the electromagnetic field A_μ is odd under charge conjugation since j_μ changes sign, but the interaction density $-j_\mu A_\mu$ must remain invariant. This means that the creation and annihilation operator for the transverse electromagnetic field must change sign since $\mathbf{A}$ contains $a^\dagger_{\mathbf{k},\alpha}$ and $a_{\mathbf{k},\alpha}$ linearly:

$$a^\dagger_{\mathbf{k},\alpha} \xrightarrow{\text{charge conj}} -a^\dagger_{\mathbf{k},\alpha}, \qquad a_{\mathbf{k},\alpha} \xrightarrow{\text{charge conj}} -a_{\mathbf{k},\alpha}. \tag{4.219}$$

An immediate consequence of this is that a state with an even (odd) number of photons is even (odd) under charge conjugation:

$$a^\dagger_{\mathbf{k}_1,\alpha_1} a^\dagger_{\mathbf{k}_2,\alpha_2} \cdots a^\dagger_{\mathbf{k}_n,\alpha_n} |0\rangle \xrightarrow{\text{charge conj}} (-1)^n a^\dagger_{\mathbf{k}_1,\alpha_1} a^\dagger_{\mathbf{k}_2,\alpha_2} \cdots a^\dagger_{\mathbf{k}_n,\alpha_n} |0\rangle, \tag{4.220}$$

no matter how the momentum and polarization of each photon are oriented. Returning now to the problem of two-photon annihilation of a $\mathbf{p}_+ = \mathbf{p}_- = 0$ e^-e^+ system, we see that

$$\begin{aligned} &\text{triplet } (\mathbf{p}_- = \mathbf{p}_+ = 0) \not\rightarrow \text{even number of photons,} \\ &\text{singlet } (\mathbf{p}_- = \mathbf{p}_+ = 0) \not\rightarrow \text{odd number of photons,} \end{aligned} \tag{4.221}$$

so long as the basic interaction is invariant under charge conjugation. Our earlier perturbation-theoretic result (4.210) is, of course, consistent with (4.221). As for three-photon annihilations, the perturbation-theoretic calculations were performed to order e^3 in the amplitude with the result that the $\mathbf{p}_+ = \mathbf{p}_- = 0$ singlet state cannot annihilate into three photons but the $\mathbf{p}_+ = \mathbf{p}_- = 0$ triplet state can. L. Michel, L. Wolfenstein, and D. Ravenhall showed that these features are a special case of (4.221), which follows rigorously from charge conjugation invariance regardless of the validity of perturbation theory.

It is also instructive to consider the selection rules imposed by angular momentum conservation. This time it is simpler to analyze the polarization state of a two-photon system ($\mathbf{k}_1 = -\mathbf{k}_2$) using the circular polarization language. Since the spin component of the photon along the propagation direction is either $+1$ or -1, and the orbital angular momentum cannot have a nonvanishing component in the propagation direction, we see that the total angular momentum component along $\mathbf{k} = \mathbf{k}_1 = -\mathbf{k}_2$ denoted by J_3 is either 0 or ± 2. Let us consider the $J_3 = 0$ case in detail. As is evident from Fig. 4–9, the two photons are both right-circularly polarized or both left-circularly polarized; the two possibilities are represented by the state vectors Ψ_{RR} and Ψ_{LL}. Consider now a 180° rotation about the x-axis. We denote this rotation opera-

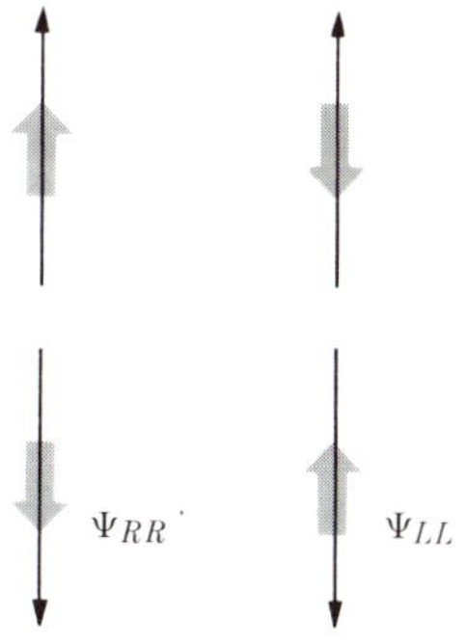

Fig. 4–9. The $J_3 = 0$ states of a two-photon system. The gray arrows indicate the spin direction.

tion by $\mathscr{R}_1(\pi)$. For the field operator we have

$$\begin{aligned} A_1(x_1, x_2, x_3, t) &\xrightarrow{\mathscr{R}_1(\pi)} A_1(x_1, -x_2, -x_3, t), \\ A_{2,3}(x_1, x_2, x_3, t) &\xrightarrow{\mathscr{R}_1(\pi)} -A_{2,3}(x_1, -x_2, -x_3, t). \end{aligned} \tag{4.222}$$

The individual creation operators of photons with momenta along the positive or the negative z-axis must then transform as‡

$$a^\dagger_{\mathbf{k},1} \xleftrightarrow{\mathscr{R}_1(\pi)} a^\dagger_{-\mathbf{k},1}, \qquad a^\dagger_{\mathbf{k},2} \xleftrightarrow{\mathscr{R}_1(\pi)} a^\dagger_{-\mathbf{k},2}. \tag{4.223}$$

Using this property and the commutation relations for $a^\dagger$, we get

$$\begin{aligned} \Psi_{RR} &= [(-1/\sqrt{2})(a^\dagger_{\mathbf{k},1} - ia^\dagger_{\mathbf{k},2})][(-1/\sqrt{2})(a^\dagger_{-\mathbf{k},1} - ia^\dagger_{-\mathbf{k},2})]\,|\,0\rangle \\ &\xrightarrow{\mathscr{R}_1(\pi)} [(-1/\sqrt{2})(a^\dagger_{\mathbf{k},1} - ia^\dagger_{\mathbf{k},2})][(-1/\sqrt{2})(a^\dagger_{-\mathbf{k},1} - ia^\dagger_{-\mathbf{k},2})]\,|\,0\rangle, \end{aligned} \tag{4.224}$$

that is, Ψ_{RR} is even under $\mathscr{R}_1(\pi)$. Furthermore, it is a well-known fact that any system with $J_3 = 0$ transforms under proper rotations like the spherical harmonics $Y^0_J(\theta, \phi)$ or the Legendre polynomial $P_J(\cos\theta)$. The transformation property of $P_J(\cos\theta)$ is particularly simple for $\mathscr{R}_1(\pi)$:

$$P_J(\cos\theta) \xrightarrow{\mathscr{R}_1(\pi)} P_J(-\cos\theta) = (-1)^J P_J(\cos\theta). \tag{4.225}$$

This means that ψ_{RR}, which is even under $\mathscr{R}_1(\pi)$, cannot have $J = 1, 3, 5, \ldots,$ where J is the *total* angular momentum of the two-photon system represented by ψ_{RR}. The case is similar for Ψ_{LL}.§ Since a $J = 1$ system cannot have $J_3 = \pm 2$ and the two-photon system whose relative momentum is along the z-axis has been shown to have either $J_3 = 0$ (Ψ_{RR} and/or Ψ_{LL}) or $J_3 = 2$ (Ψ_{RL} and/or Ψ_{LR}), we have proved that

$$J = 1 \text{ system} \not\rightarrow 2 \text{ photons}. \tag{4.226}$$

This far-reaching selection rule (based only on angular momentum conservation) was first deduced by C. N. Yang in 1950.|| From this general result we see that the "forbiddenness" of the two-photon annihilation of the triplet $\mathbf{p}_- = \mathbf{p}_+ = 0$ e^-e^+ system follows from angular momentum conservation as well as from charge conjugation invariance. Table 4–2 summarizes the requirements imposed by the various selection rules.# These selection rules, of course, apply equally well to systems having the same C, P, and J quantum numbers as to the e^-e^+ system.

---

‡The fact that $a^\dagger_{\pm\mathbf{k},2}$ does *not* go into $-a^\dagger_{\mp\mathbf{k},2}$ is due to the orientation of the linear polarization vectors for $a^\dagger_{\mathbf{k},2}|0\rangle$ and $a^\dagger_{-\mathbf{k},2}|0\rangle$ along the positive y- and the *negative* y-direction respectively [cf. our phase convention (4.212).]

§Actually, if the two-photon system is to be a parity eigenstate, we must consider $(1/\sqrt{2})$ $(\Psi_{RR} \pm \Psi_{LL})$. The parity considerations, however, are irrelevant in a discussion of the angular momentum selection rules.

||This selection rule was used to rule out the spin-one assignment for a neutral pion very early in the history of pion physics.

#Some of the results of Table 4–2 can be obtained just as simply without using the language of quantum field theory [cf. for example, Sakurai (1964), pp. 15–16, p. 45]. We have deliberately used field-theoretic language throughout mainly to illustrate how the transformation properties of the field operators are related to those of the state vectors.

Table 4–2

SELECTION RULES FOR THE ANNIHILATION OF AN ELECTRON-POSITRON SYSTEM WITH $\mathbf{p}_+ = \mathbf{p}_- = 0$

The statement "no by C(P, J)" means that the process is forbidden by charge conjugation invariance (parity, angular momentum conservation).

	2-photon annihilation ($\boldsymbol{\epsilon}^{(\alpha_1)} \parallel \boldsymbol{\epsilon}^{(\alpha_2)}$)	2-photon annihilation ($\boldsymbol{\epsilon}^{(\alpha_1)} \perp \boldsymbol{\epsilon}^{(\alpha_2)}$)	3-photon annihilation
Triplet ($C = -1, J = 1$)	No by C and J	No by C and J	Yes
Singlet ($C = +1, J = 0$)	No by P	Yes	No by C

Polarization measurement. Let us now look more closely at the two-photon annihilation of the singlet state. We have already shown that parity conservation requires that the two-photon system must be represented by the state vector (4.214a). Suppose there is a device that is sensitive to the linear polarization direction of one of the outgoing photons, say photon 1 with momentum $\mathbf{k}$. Equation (4.214a) tells us that there is equal probability for observing the photon polarization in the x-direction as in the y-direction. This is not at all surprising since there is nothing to disturb the cylindrical symmetry about the direction $\mathbf{k}$. We shall next consider a more sophisticated experimental arrangement in which photon polarizations are measured *in coincidence*. Imagine observer A and observer B who specialize in measuring the polarization of photon 1 (momentum $\mathbf{k}$) and that of photon 2 (momentum $-\mathbf{k}$) respectively. Suppose observer A finds that for a particular annihilation event the polarization direction of photon 1 is in the x-direction; then, according to (4.214a), observer B must find that the polarization vector of photon 2 has no component in the x-direction. This is quite a striking result if we realize that the two polarization measurements can be made at a widely separated distance long after the two photons have ceased to interact. In fact the alleged correlation is expected even if the two measurements are made at a spacelike distance so that there is no way of communicating to observer B the result of the measurement made by observer A before observer B performs his measurement. Since photon 2 appears unpolarized when the polarization of photon 1 is not measured, it appears as if photon 2 "gets to know" in which direction its polarization vector must point at the very instant the polarization of photon 1 is measured. In other words, having measured the polarization direction of photon 1, we can predict with certainty the result of measuring the polarization direction of photon 2 (which can be very far away from photon 1), despite the fact that both polarization directions are equally likely for photon 2 when the polarization measurement of photon 1 is not carried out.

If the reader finds all this disturbing, it is only natural. Some of the greatest minds in twentieth-century physics have worried about this problem. In the literature a peculiarity of this kind is known as the "Einstein-Podolsky-Rosen paradox."

In his famous autobiography, A. Einstein wrote on this subject as follows:

> But on one supposition we should, in my opinion, absolutely hold fast: the real factual situation of the system S_2 is independent of what is done with the system S_1 which is spatially separated from the former.

To analyze this paradox a little more closely, let us ask what constitutes a quantum-mechanical system in this particular problem. When we measure the polarization direction of photon 1, the quantum-mechnanical system on which we make a measurement is not the single-photon state $a^\dagger_{\mathbf{k},\alpha}|0\rangle$ but the *composite* two-photon system represented by the state vector (4.214a). In quantum mechanics it is wrong to regard a two-photon system as two separate single-photon systems even if the two photons are a million miles apart so that they cannot possibly interact. Any measurement we make on one of the photons is to be regarded as a measurement made on the entire two-photon system. When observer A finds with certainty that the polarization vector of photon 1 is in the x-direction, what he actually learns is that the two-photon system is in $a^\dagger_{\mathbf{k},1}a^\dagger_{-\mathbf{k},2}|0\rangle$ rather than in $a^\dagger_{\mathbf{k},2}a^\dagger_{-\mathbf{k},1}|0\rangle$. Any subsequent measurement observer B may perform must be consistent with the finding that the system at the time of the first measurement is represented by the state vector $a^\dagger_{\mathbf{k},1}a^\dagger_{-\mathbf{k},2}|0\rangle$; from this point of view it is perhaps not too peculiar that observer B necessarily finds that the polarization vector of photon 2 is in the y-direction. The Einstein-Podolsky-Rosen paradox ceases to be a paradox once we accept the view that the polarization measurement on photon 1 is actually a measurement to determine whether the *composite* two-photon system is in $a^\dagger_{\mathbf{k},1}a^\dagger_{-\mathbf{k},2}|0\rangle$ or in $a^\dagger_{\mathbf{k},2}a^\dagger_{-\mathbf{k},1}|0\rangle$.

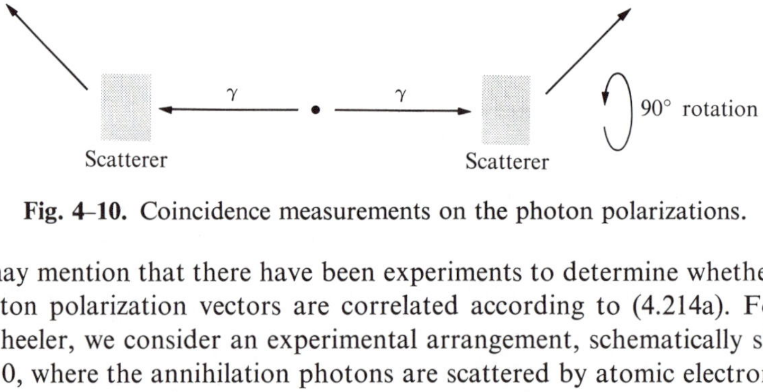

Fig. 4–10. Coincidence measurements on the photon polarizations.

We may mention that there have been experiments to determine whether or not the photon polarization vectors are correlated according to (4.214a). Following J. A. Wheeler, we consider an experimental arrangement, schematically shown in Fig. 4–10, where the annihilation photons are scattered by atomic electrons. As is evident from (2.170), Thomson scattering is an excellent means of analyzing the direction of the incident photon polarization; for instance, at $\theta = 90°$, the differential cross section varies as $\sin^2\phi$, where ϕ is the aximuthal angle of the scattered photon direction measured from the incident polarization direction. Since the energy of the γ ray is as large as the rest mass of the electron, it is better to use an analogous expression based on the Klein-Nishina formula (3.344); however, it is not difficult to show that the strong dependence on the polarization persists even if $E_\gamma \simeq m$. By measuring the coincidence counting rates using the arrangement shown in Fig. 4–10 and then repeating measurements with an arrangement where the right-hand half of Fig. 4–10 is rotated 90° around the propagation direction **k**, we can test whether there is, in fact, the perpendicular correlation indicated by (4.214a). The experimental results of C. S. Wu and I. Shaknov have fully confirmed that there indeed is such a correlation effect of the annihilation gamma rays as predicted by (4.214a).

Positronium lifetime. We shall now go back to the expression for the $\beta_+ \longrightarrow 0$ limit of the total cross section (4.202) to show how this quantity is simply related to the lifetime of an e^-e^+ bound state, known as *positronium*, whose existence was established experimentally in a series of elegant experiments performed by M. Deutsch and coworkers in the early 1950's. We first recall that $\sigma_{\text{tot}} v_+$, which has the dimension of volume divided by time, characterizes the reciprocal of the positron mean lifetime when there is one electron per unit volume. So

$$R = \sigma_{\text{tot}} v_+ \rho \tag{4.227}$$

is the reciprocal of the positron mean lifetime when the electron density is ρ. For instance, for positron annihilation in a medium in which there are N atoms of atomic number Z per unit volume, the ρ that appears in (4.227) is NZ. When we want to calculate the lifetime of a *bound* e^-e^+ system, the electron density (relative to the position of the positron) is just the square of the bound-state wave function evaluated at the origin; for instance, for the ground state of the positronium we get

$$\rho = |\psi_{1s}(\mathbf{x} = 0)|^2 = 1/[\pi(2a_0)^3], \tag{4.228}$$

where we have used the fact that the Bohr radius of the positronium is twice that of the hydrogen atom. In practice, since the electron (positron) velocity in the bound system is of the order of $\frac{1}{137}$ times the speed of light, we can use $\lim_{\beta_+ \to 0} \sigma_{\text{tot}}$ in (4.227) to compute the positronium lifetime. For the ground state we get

$$\begin{aligned} \Gamma(n = 1, {}^1S \longrightarrow 2\gamma) &= \lim_{v_+ \to 0} 4\, \sigma_{\text{tot}}^{(\text{unpol})} v_+ |\psi_{1s}(\mathbf{x} = 0)|^2 \\ &= 4\pi \left(\frac{\alpha}{m}\right)^2 \frac{1}{\pi(2/\alpha m)^3} \\ &= \tfrac{1}{2}\alpha^5 m, \end{aligned} \tag{4.229}$$

where the factor 4 is due to the fact that the singlet state does the whole job in (4.211). This gives for the mean lifetime of the $n = 1\ {}^1S$ state,

$$\tau_{\text{singlet}} = 2/(\alpha^5 m) \simeq 1.25 \times 10^{-10}\ \text{sec}, \tag{4.230}$$

as first calculated by J. Pirenne. The ${}^3S$ state which is forbidden to decay into two gamma rays is expected to be much more long-lived since the amplitude for three-photon annihilation must involve another power of e, as is evident from Fig. 4–11. Without evaluating the complicated diagrams shown in Fig. 4–11 we may guess that the lifetime of the ${}^3S$ state must be about 137 times longer. The actual answer (due to A. Ore and J. L. Powell) turns out to be

$$\frac{\tau_{\text{triplet}}}{\tau_{\text{singlet}}} = \frac{9\pi}{4(\pi^2 - 9)\alpha} \simeq 1115 \tag{4.231}$$

for the $n = 1$, s wave bound states.

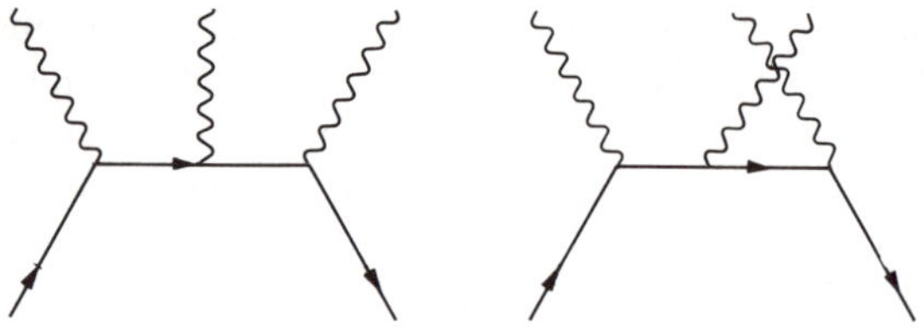

Fig. 4-11. Examples of Feynman diagrams for three-photon annihilation.

Cross section for Compton scattering. Earlier in this section we obtained the $\mathscr{M}$-matrix element for Compton scattering as well as that for two-photon annihilation (which we discussed in detail). It is interesting to note that the $\mathscr{M}$-matrix elements for two-photon annihilation and Compton scattering (Eqs. 4.166 and 4.175) are related by the following simple substitutions:

$$k_1, \epsilon^{(\alpha_1)} \longrightarrow -k, \epsilon^{(\alpha)}, \qquad k_2, \epsilon^{(\alpha_2)} \longrightarrow k', \epsilon^{(\alpha')},$$
$$p_- \longrightarrow p, \qquad p_+ \longrightarrow -p'. \tag{4.232}$$

As for the free-particle spinors we must interchange $\bar{v}^{(s_+)}(\mathbf{p}_+)$ and $\bar{u}^{(s')}(\mathbf{p}')$ [as well as $u^{(s_-)}(\mathbf{p}_-)$ and $u^{(s)}(\mathbf{p})$], but this change is automatically taken care of (apart from a minus sign) by (4.232) when we compute $|\mathscr{M}_{fi}|^2$ by the trace method; this is because

$$-\frac{i\gamma \cdot p_+ + m}{2m} \longrightarrow \frac{-i\gamma \cdot p' + m}{2m}. \tag{4.233}$$

Comparing Figs. 4-6 and 4-7, we see that this set of transformations simply amounts to changing the external lines corresponding to the (incoming) positron and one of the (outgoing) photons, photon 1, which appear in Fig. 4-6, into the external lines corresponding to the outgoing electron and the incoming photon which appear in Fig. 4-7. In other words, all we need to do is "bend" some of the external lines. Furthermore, simplification completely analogous to (4.187) and (4.188) takes place if we compute the $\mathscr{M}$-matrix element for Compton scattering in the lab system. For this reason $\sum_{s,s'} |\mathscr{M}_{fi}|^2$ for Compton scattering can be readily written once we have the expression (4.197) for two-photon annihilation. Apart from the sign change due to (4.233), the only other change we must note is a factor two coming from

$$\tfrac{1}{4} \sum_{s_+ s_-} \longrightarrow \tfrac{1}{2} \sum_{ss'} \tag{4.234}$$

since initially there are just two fermion spin states. So

$$\frac{1}{2} \sum_{ss'} |\mathscr{M}_{fi}|^2 = \frac{e^4}{(2m)^2} \left[\frac{\omega}{\omega'} + \frac{\omega'}{\omega} - 2 + 4(\boldsymbol{\epsilon}^{(\alpha)} \cdot \boldsymbol{\epsilon}^{(\alpha')})^2 \right]. \tag{4.235}$$

In the lab system the relative velocity is just unity (in natural units), and in computing the phase space factor, we must note that

$$\frac{\partial(E' + \omega')}{\partial(|\mathbf{k}'|)} = \frac{\omega' - \omega \cos \theta}{E'} + 1 = \frac{m\omega}{E'\omega'}, \tag{4.236}$$

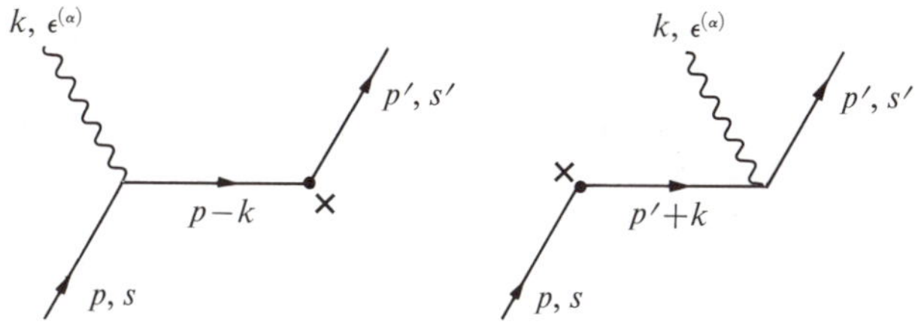

Fig. 4–12. *Bremsstrahlung.*

which is not difficult to obtain if we note the kinematical relations

$$E' = \sqrt{\omega^2 + \omega'^2 - 2\omega\omega' \cos\theta + m^2},$$
$$\omega' = \frac{m\omega}{m + \omega(1 - \cos\theta)}. \tag{4.237}$$

Putting the various factors together, we get the famous Klein-Nishina formula (3.344). If the incident photon is unpolarized, and if we do not observe the polarization of the outgoing photon, we can sum over the final polarization states and average over the incident polarization directions; this can be done using the techniques we discussed in connection with Thomson scattering (Section 2–5). The net result is

$$\frac{d\sigma}{d\Omega} = \frac{r_0^2}{2}\left(\frac{\omega'}{\omega}\right)^2\left(\frac{\omega}{\omega'} + \frac{\omega'}{\omega} - \sin^2\theta\right). \tag{4.238}$$

Note that because ω' is not equal to ω (unless $\omega \ll m$) but is dependent on the scattering angle via (4.237), the angular distribution is no longer forward-backward symmetric as it is in Thomson scattering. Formula (4.238) has been checked at a variety of energies. Already in the late 1920's W. Friedrich and M. Goldhaber had measured the angular distribution of γ-rays with $E_\gamma = 0.17$ m ($\lambda = 0.14$ Å) to show that the observed deviation from the Thomson formula is precisely what one expects from formula (4.238). The agreement achieved here is one of the earliest quantitative triumphs of the Dirac theory.

Bremsstrahlung and pair production. To finish this section we briefly mention two other processes which can be treated using the electron propagator: *bremsstrahlung* and *pair production* (in the Coulomb field of a nucleus). This time the A_μ that appears in the S-matrix expansion can be split into two parts,

$$A_\mu = A_\mu^{(e)} + A_\mu^{(q)}, \tag{4.239}$$

where $A_\mu^{(e)}$ is simply the Coulomb potential (4.56) considered in connection with Mott scattering and $A_\mu^{(q)}$ is the quantized transverse field mentioned in Chapter 2. For *bremsstrahlung* there are two topologically inequivalent graphs shown in Fig. 4–12, where the symbol × stands for a first-order interaction with the Coulomb potential. The essential point is that the annihilation of the incident photon in Compton scattering is now replaced by a first-order interaction with the Coulomb potential. As in the Mott scattering problem there is no over-all momentum conservation (as far as the electron-photon system goes) since the nucleus (assumed to be infinitely heavy) can take up any arbitrary amount of momentum. For this reason the S-matrix element cannot be written in the form of a four-dimensional delta function times a covariant matrix element $\mathscr{M}$. However, the S-matrix element is just as simple; the reader is urged to work out the details (Problem 4–10a) to obtain

$$S_{fi} = -i2\pi Ze^3\sqrt{\left(\frac{1}{2\omega V}\right)\left(\frac{m}{EV}\right)\left(\frac{m}{E'V}\right)}\delta(E' + \omega - E)\frac{1}{|\mathbf{p}' + \mathbf{k} - \mathbf{p}|^2}$$
$$\times \bar{u}^{(s')}(\mathbf{p}')\left[\gamma_4\frac{-i\gamma\cdot(p-k)+m}{i[(p-k)^2+m^2]}\gamma\cdot\epsilon^{(\alpha)} + \gamma\cdot\epsilon^{(\alpha)}\frac{-i\gamma\cdot(p'+k)+m}{i[(p'+k)^2+m^2]}\gamma_4\right]u^{(s)}(\mathbf{p}). \tag{4.240}$$

Once we have this expression for *bremsstrahlung* we can readily write the matrix element for pair production corresponding to Fig. 4–13 using substitutions analogous to (4.232). The formulas for the cross sections of *bremsstrahlung* and pair production were both derived in 1934 by H. A. Bethe and W. Heitler; they can be found in Heitler's book along with detailed comparison with experiments.‡ In Problem 4–10(b) and (c) the reader is asked to work out the cross section for *bremsstrahlung* when the incident electron is nonrelativistic and consequently the energy of the emitted photon is small compared to m.

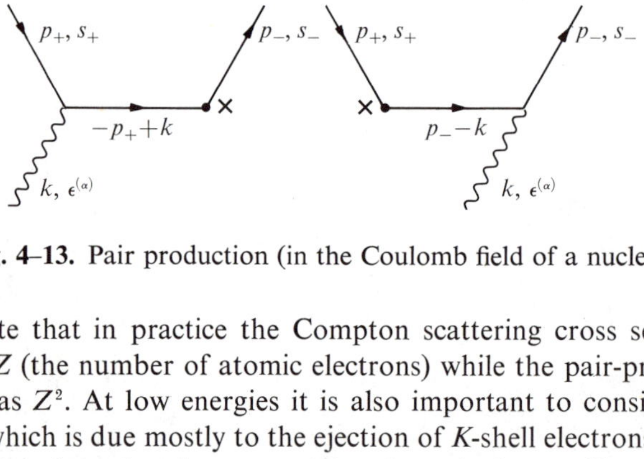

Fig. 4–13. Pair production (in the Coulomb field of a nucleus).

We may note that in practice the Compton scattering cross section must be multiplied by Z (the number of atomic electrons) while the pair-production cross section varies as Z^2. At low energies it is also important to consider the photoelectric effect which is due mostly to the ejection of K-shell electrons. According to Problem 2–4 the photoelectric cross section for a hydrogen-like atom varies inversely as the fifth power of the Bohr radius, hence the photoelectric cross section due to inner shell electrons is proportional to Z^5. As for the energy dependence, the photoelectric effect diminishes rapidly as $\omega^{-7/2}$ for $\omega \ll m$ (which can also be seen from Problem 2–4 if we note that $|\mathbf{k}_f|^2$ is proportional to ω) and as $1/\omega$ for $\omega \gg m$. The Compton cross section takes the constant Thomson value for $\omega \ll m$ (but ω is still large compared to the binding energy of an atom) and then drops rather rapidly as ω exceeds m; integrating (4.238) over angles, we obtain the total

‡Heitler (1954), pp. 242–268.

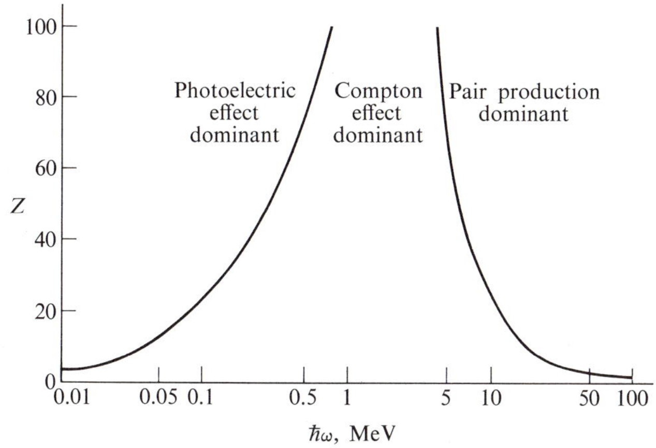

Fig. 4–14. Relative importance of photoelectric effect, Compton scattering, and pair production.

Compton cross section

$$\sigma_{\text{Comp}} = \sigma_{\text{Thom}} \frac{3}{8} \frac{m}{\omega} \left(\log \frac{2\omega}{m} + \frac{1}{2} \right), \qquad \omega \gg m. \tag{4.241}$$

It is evident that at high energies the pair-production process is the most important one. More quantitatively, Fig. 4–14 (taken from R. D. Evans's *Encyclopedia* article) illustrates which of the three processes is dominant in a particular energy region for a given value of Z.

4–5. FEYNMAN'S SPACE-TIME APPROACH TO THE ELECTRON PROPAGATOR

In the previous two sections we showed how to compute the matrix elements for some simple processes using the language of the second-quantized Dirac theory. In this section we shall demonstrate how the results we have rigorously derived from quantum field theory can be obtained more quickly by relying on a more *intuitive* method, due to R. P. Feynman, who based it directly on the particle wave equation of Dirac. We present this alternative approach in some detail since it provides an intuitive understanding of the essentials of the field-theoretic results derived in the previous two sections.

Green's function. Since Feynman's approach is based directly on the wave equation, we first review the way of solving an inhomogeneous differential equation in mathematical physics. Take the trivial example

$$\nabla^2 \phi(\mathbf{x}) = -\rho(\mathbf{x}). \tag{4.242}$$

As every "child" knows, the solution to Poisson's equation (4.242) with the boundary condition $\phi(\mathbf{x}) \longrightarrow 0$ as $|\mathbf{x}| \longrightarrow \infty$ is given by

$$\phi(\mathbf{x}) = \int G(\mathbf{x}, \mathbf{x}')\rho(\mathbf{x}')\, d^3x', \tag{4.243}$$

where
$$G(\mathbf{x}, \mathbf{x}') = \frac{1}{4\pi\,|\mathbf{x} - \mathbf{x}'|}. \tag{4.244}$$

The function $G(\mathbf{x}, \mathbf{x}')$ is called the *Green's function* for Poisson's equation and satisfies
$$\nabla^2 G(\mathbf{x}, \mathbf{x}') = -\delta^{(3)}(\mathbf{x} - \mathbf{x}'). \tag{4.245}$$

In other words, instead of solving (4.242) directly, we first solve the unit source problem (4.245); once we obtain the solution to this simpler problem, the solution to the more general problem (4.242) can be written immediately because of the superposition principle.

In relativistic electron theory we are interested in solving the Dirac equation
$$\left(\gamma_\mu \frac{\partial}{\partial x_\mu} + m\right)\psi = ie\gamma_\mu A_\mu \psi; \tag{4.246}$$

here (and throughout this section) ψ is a c-number wave function, not the quantized electron field. The right-hand side of (4.246) is very much like the source $\rho(\mathbf{x})$ in (4.242). In analogy with electrostatics we may argue that it is too difficult to tackle (4.246) directly; instead, let us first consider the simpler unit source problem;
$$\left(\gamma_\mu \frac{\partial}{\partial x_\mu} + m\right)K(x, x') = -i\delta^{(4)}(x - x'), \tag{4.247}$$

or, writing the Dirac indices explicitly, we have
$$[(\gamma_\mu)_{\alpha\beta}(\partial/\partial x_\mu) + m\delta_{\alpha\beta}]K_{\beta\gamma}(x, x') = -i\delta^{(4)}(x - x')\delta_{\alpha\gamma}. \tag{4.248}$$

The function $K(x, x')$ is the Green's function for the free-particle Dirac equation just as $G(\mathbf{x}, \mathbf{x}')$ is the Green's function for Poisson's equation. In complete analogy with (4.243) we expect that a solution ψ to the more complex differential equation (4.246) satisfies
$$\psi(x) = -\int K(x, x')e\gamma_\mu A_\mu(x')\psi(x')\, d^4x'. \tag{4.249}$$

That this is indeed the case can be proved by direct substitution:
$$\begin{aligned}\left(\gamma_\mu \frac{\partial}{\partial x_\mu} + m\right)&\left[-\int K(x, x')e\gamma_\mu A_\mu(x')\psi(x')\, d^4x'\right] \\ &= \int i\delta^{(4)}(x - x')e\gamma_\mu A_\mu(x')\psi(x')\, d^4x' \\ &= ie\gamma_\mu A_\mu \psi.\end{aligned} \tag{4.250}$$

Note, however, that even if we add to (4.249) any solution to the free-wave equation ($e = 0$), the differential equation (4.246) will still be satisfied. To be more explicit, let us consider
$$\psi(x) = \psi_0(x) - \int K(x, x')e\gamma_\mu A_\mu(x')\psi(x')\, d^4x', \tag{4.251}$$

where $\psi_0(x)$ is a solution to the Dirac equation when $e = 0$. We can easily verify that (4.251) satisfies (4.246) just as well. In a scattering problem, for instance, $\psi_0(x)$ may represent an incident plane wave which would be present even if there were no interaction.

There is an important difference between the differential equation in electrostatics (4.242) and the differential equation for the Dirac electron (4.246). The right-hand side of (4.242) does not contain ϕ; so we could write the solution (4.243) immediately in a closed form. In contrast, in the case of (4.246) the function ψ itself appears on the right-hand side; as a result, (4.251) is an integral equation. So, even if we obtain an explicit form of $K(x, x')$, as we shall in a moment, we cannot write the exact solution to (4.246) in a closed form. However, if a perturbation expansion in powers of e can be justified, we can obtain an approximate solution to (4.246) accurate to any desired power of e by the now familiar iteration method:

$$\begin{aligned}\psi(x) &= \psi_0(x) + \int d^4x'\, K(x, x')[-e\gamma_\mu A_\mu(x')]\psi_0(x') \\ &\quad + \int d^4x' \int d^4x''\, K(x, x')[-e\gamma_\mu A_\mu(x')]K(x', x'')[-e\gamma_\nu A_\nu(x'')]\psi_0(x'') \\ &\quad + \int d^4x' \int d^4x'' \int d^4x'''\, K(x, x')[-e\gamma_\mu A_\mu(x')]K(x', x'') \\ &\quad \times [-e\gamma_\nu A_\nu(x'')]K(x'', x''')[-e\gamma_\lambda A_\lambda(x''')]\psi_0(x''') + \cdots. \end{aligned} \tag{4.252}$$

Our next task is to obtain an explicit integral representation of $K(x, x')$. First, because of translational invariance (in both space and time) it is clear that $K(x, x')$ is a function of $x - x'$ only. So without loss of generality we can set $x' = 0$. We solve (4.247) using the four-dimensional Fourier-transform method just as we solve (1.36) using the three-dimensional Fourier-transform method. Let us define $\tilde{K}(p)$ such that

$$K(x, 0) = [1/(2\pi)^4] \int \tilde{K}(p) e^{ip\cdot x}\, d^4p. \tag{4.253}$$

If we substitute this expression for $K(x, 0)$ in (4.247) with $x' = 0$, we get

$$\begin{aligned}[1/(2\pi)^4] \int (i\gamma\cdot p + m)\tilde{K}(p) e^{ip\cdot x}\, d^4p &= -i\delta^{(4)}(x) \\ &= -[i/(2\pi)^4] \int e^{ip\cdot x}\, d^4p. \end{aligned} \tag{4.254}$$

So the differential equation (4.247) is now reduced to a simple algebraic equation in momentum-space,

$$(i\gamma\cdot p + m)\tilde{K}(p) = -i, \tag{4.255}$$

or

$$\tilde{K}(p) = -i/(i\gamma\cdot p + m), \tag{4.256}$$

where $1/(i\gamma\cdot p + m)$ is understood to be a 4×4 matrix such that if we multiply with $(i\gamma\cdot p + m)$ from the right or from the left, we obtain the 4×4 identity matrix. Clearly we can write (4.256) as

$$\tilde{K}(p) = -\frac{i(-i\gamma\cdot p + m)}{(i\gamma\cdot p + m)(-i\gamma\cdot p + m)} = \frac{-i\gamma\cdot p + m}{i(p^2 + m^2)}, \tag{4.257}$$

since $(i\gamma\cdot p)^2 = -p^2$. Going back to coordinate (x) space, we get, using (4.253),

$$K(x, 0) = -\frac{i}{(2\pi)^4} \int d^4p \frac{(-i\gamma\cdot p + m)}{p^2 + m^2} e^{ip\cdot x}, \tag{4.258}$$

or, more generally,

$$K(x, x') = -\frac{i}{(2\pi)^4}\int d^4p \frac{(-i\gamma\cdot p + m)}{p^2 + m^2} e^{ip\cdot(x-x')}. \tag{4.259}$$

The integrand of (4.259) is already familiar from the previous section. When we integrate along the real p_0-axis, there are poles at $p_0 = \pm E = \pm\sqrt{|\mathbf{p}|^2 + m^2}$:

$$K(x, x') = -\frac{i}{(2\pi)^4}\int d^3p\, e^{i\mathbf{p}\cdot(\mathbf{x}-\mathbf{x}')} \int_{-\infty}^{\infty} dp_0 \frac{(-i\gamma\cdot p + m)e^{-ip_0(x_0-x_0')}}{(-p_0 + E)(p_0 + E)}. \tag{4.260}$$

The particular form of $K(x, x')$ depends on the particular manner in which we go around the poles in the complex p_0-plane. That this kind of ambiguity exists is to be expected on physical grounds, since we have not yet specified the boundary conditions to be used in connection with the differential equation (4.247).

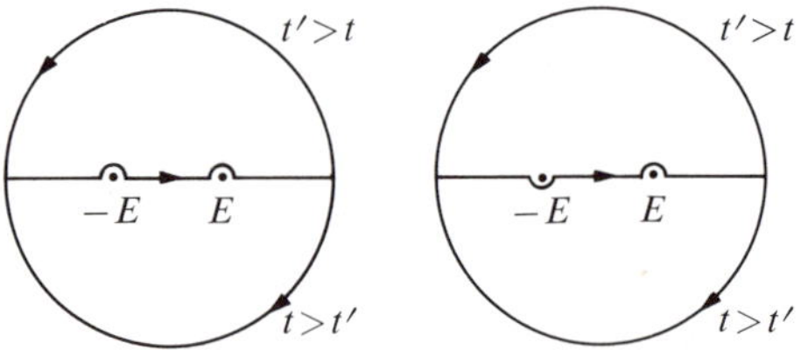

(a) $K_{\text{ret}}(x, x')$ (b) $K_F(x, x')$

Fig. 4–15. Prescriptions for the integration (4.260) in the complex p_0-plane.

K_F and K_{ret}. We shall now consider two types of $K(x, x')$ corresponding to different boundary conditions. First, we shall look at what is known as the *retarded* Green's function with the property

$$K_{\text{ret}}(x, x')\begin{cases} \neq 0, & t > t', \\ = 0, & t < t'. \end{cases} \tag{4.261}$$

In the complex p_0-plane, K_{ret} corresponds to the choice of contour indicated in Fig. 4–15(a). When $t > t'$, we must close the contour in the lower half-plane; hence we pick up the *contributions due to both poles*. In contrast, when $t < t'$, we receive *no contribution from either pole*. Alternatively we may use the contour indicated in Fig. 4–15(b). It is evident from Fig. 4–5 and Eq. (4.159) that $K(x, x')$ then coincides with the expression $\langle 0|T(\psi(x)\bar\psi(x'))|0\rangle = -(\frac{1}{2})S_F(x - x')$, discussed extensively in the previous section. The Green's function obtained by following the prescription shown in Fig. 4–15(b) is denoted by $K_F(x, x')$. In other words

$$\begin{aligned} K_F(x, x') &= -\frac{i}{(2\pi)^4}\int d^4p \frac{(-i\gamma\cdot p + m)e^{ip\cdot(x-x')}}{p^2 + m^2 - i\epsilon} \\ &= \langle 0|T(\psi(x)\bar\psi(x'))|0\rangle = -\tfrac{1}{2}S_F(x - x'). \end{aligned} \tag{4.262}$$

We already proved in the previous section that if we integrate in the *clockwise* direction around $p_0 = +E$, we obtain (4.154), while if we integrate in the *counterclockwise* direction around $p_0 = -E$, we obtain (4.157). For K_{ret} with $t > t'$

we must go around *both* E and $-E$ in the *clockwise* direction. We therefore obtain

$$K_{\text{ret}}(x, x') = \begin{cases} \lim\limits_{V\to\infty} \sum\limits_{\mathbf{p},s} (m/EV)[u^{(s)}(\mathbf{p})\bar{u}^{(s)}(\mathbf{p})e^{ip\cdot(x-x')} + v^{(s)}(\mathbf{p})\bar{v}^{(s)}(\mathbf{p})e^{-ip\cdot(x-x')}] & \text{for} \quad t > t', \\ 0 \quad \text{for} \quad t' > t. \end{cases} \tag{4.263}$$

In contrast, for K_F we have the old result

$$K_F(x, x') = \lim_{V\to\infty} \sum_{\mathbf{p},s} \frac{m}{EV} \begin{cases} u^{(s)}(\mathbf{p})\bar{u}^{(s)}(\mathbf{p})e^{ip\cdot(x-x')} & \text{for} \quad t > t', \qquad (4.264\text{a}) \\ [-v^{(s)}(\mathbf{p})\bar{v}^{(s)}(\mathbf{p})e^{-ip\cdot(x-x')}] & \text{for} \quad t' > t. \qquad (4.264\text{b}) \end{cases}$$

Note again that p_0 in (4.259) through (4.262) is, in general, not equal to $E = \sqrt{|\mathbf{p}|^2 + m^2}$, whereas p_0 in (4.263) and (4.264) is identical with E.

The burning question is: Which of the Green's functions, K_{ret} or K_F, is the appropriate one to use in connection with the perturbation expansion (4.252)? In the Schrödinger theory we can show that the correct Green's function is the one that vanishes for $t' > t$; this is evident intuitively because the wave function at t can be influenced by the presence of a potential acting at t' if t is later than t', but *not* if t is earlier. (This point will be clearer when the reader works out Problem 4-11.) So we may be tempted to use K_{ret} in the relativistic electron theory as well. We shall now demonstrate, however, that the use of K_{ret} is incompatible with the hole-theoretic interpretation of the Dirac theory.

Let us consider a potential problem assuming the validity of first-order perturbation theory. If K in the perturbation expansion (4.252) is taken to be K_{ret}, we have, with the aid of (4.263),

$$\begin{aligned} \psi(x) &\simeq \psi_0(x) + \int d^4x'\, K_{\text{ret}}(x, x')[-e\gamma_\mu A_\mu^{(e)}(x')]\psi_0(x') \\ &= \psi_0(x) + \sum_{\mathbf{p}'} \sum_{s'=1}^{2} c_{\mathbf{p}',s'}^{(+)}(t)[\sqrt{m/E'V}\, u^{(s')}(\mathbf{p}')e^{ip'\cdot x}] \\ &\quad + \sum_{\mathbf{p}'} \sum_{s'=1}^{2} c_{\mathbf{p}',s'}^{(-)}(t)[\sqrt{m/E'V}\, v^{(s')}(\mathbf{p}')e^{-ip'\cdot x}], \end{aligned} \tag{4.265}$$

where

$$\begin{aligned} c_{\mathbf{p}',s'}^{(+)}(t) &= -e \int d^3x' \int_{-\infty}^{t} dt' \sqrt{m/E'V}\, e^{-ip'\cdot x'} \bar{u}^{(s')}(\mathbf{p}')\gamma_\mu \psi_0(x') A_\mu^{(e)}(x'), \\ c_{\mathbf{p}',s'}^{(-)}(t) &= -e \int d^3x' \int_{-\infty}^{t} dt' \sqrt{m/E'V}\, e^{ip'\cdot x'} \bar{v}^{(s')}(\mathbf{p}')\gamma_\mu \psi_0(x') A_\mu^{(e)}(x'). \end{aligned} \tag{4.266}$$

Let us assume, for definiteness, that $\psi_0(x)$ is the normalized wave function for a positive-energy plane wave characterized by $\mathbf{p}$ and s. According to the usual interpretation of wave mechanics, $|c_{\mathbf{p}',s'}^{(+)}(t)|^2$ gives the probability for finding the electron at t in a positive-energy state characterized by $\mathbf{p}', s'$ (assumed to be different from $\mathbf{p}, s$) when the electron is known with certainty to be in state $\mathbf{p}, s$ in the remote past. In fact, from Section 4-3, we see that $c_{\mathbf{p}',s'}^{(+)}(\infty)$ is precisely the first-order S-matrix for the transition of an electron characterized by $(\mathbf{p}, s)$ into another state characterized by $(\mathbf{p}', s')$ (cf. Eq. 4.64). This is reasonable because a positive-energy electron state is known to be able to make a transition into some other positive-energy state in the presence of an external potential. On the other hand, $c_{\mathbf{p}',s'}^{(-)}(t)$ is much more disturbing. Recalling that $v^{(s)}(\mathbf{p})e^{-ip\cdot x}$ is a negative-energy wave function because of the relation (cf. Eq. 3.384),

$$v^{(1,2)}(\mathbf{p})e^{-ip\cdot x} = \mp u^{(4,3)}(-\mathbf{p})e^{i(-\mathbf{p})\cdot\mathbf{x}-i(-E)t} \tag{4.267}$$

(where E is *positive*), we see that for a large value of t there is a finite probability for finding the electron (initially in a positive-energy state) in a negative-energy state. (See Problem 3–12 for a specific example of this kind.) Thus the use of K_{ret} leads to the catastrophic result that in the presence of an external potential a positive-energy electron can make a transition into a negative energy state, in violent contradiction to the hole theory.

Let us now use K_F in the perturbation expansion (4.252). To first order we get, with the help of (4.264),

$$\begin{aligned}\psi(x) &\simeq \psi_0(x) + \int d^4x'\, K_F(x,x')[-e\gamma_\mu A_\mu^{(e)}(x')]\psi_0(x')\\ &= \psi_0(x) + \sum_{\mathbf{p}'}\sum_{s'=1}^{2} c_{\mathbf{p}',s'}^{(+)}(t)[\sqrt{m/E'V}\, u^{(s')}(\mathbf{p}')e^{ip'\cdot x}]\\ &\quad + \sum_{\mathbf{p}'}\sum_{s'=1}^{2} c_{\mathbf{p}',s'}^{(-)}(t)[\sqrt{m/E'V}\, v^{(s')}(\mathbf{p}')e^{-ip'\cdot x}],\end{aligned} \tag{4.268}$$

where

$$\begin{aligned}c_{\mathbf{p}',s'}^{(+)}(t) &= -e\int d^3x' \int_{-\infty}^{t} dt'\, \sqrt{m/E'V}\, e^{-ip'\cdot x'}\bar{u}^{(s')}(\mathbf{p}')\gamma_\mu\psi_0(x')A_\mu^{(e)}(x'),\\ c_{\mathbf{p}',s'}^{(-)}(t) &= +e\int d^3x' \int_{t}^{\infty} dt'\, \sqrt{m/E'V}\, e^{ip'\cdot x'}\bar{v}^{(s')}(\mathbf{p}')\gamma_\mu\psi_0(x')A_\mu^{(e)}(x').\end{aligned} \tag{4.269}$$

Note that this time the limits of the t' integrations are different for $c^{(+)}$ and $c^{(-)}$. Everything is satisfactory as $t \longrightarrow \infty$. Letting ψ_0 stand again for a normalized plane-wave solution characterized by $\mathbf{p}$ and s, we see that $c_{\mathbf{p}',s'}^{(+)}(\infty)$ is identical to the $S^{(1)}$-matrix element for the scattering of an electron initially in $(\mathbf{p}, s)$ into another electron state characterized by $(\mathbf{p}', s')$, just as in the case of (4.266). However, the undesirable feature of (4.266) is now absent; $c_{\mathbf{p}',s'}^{(-)}$ does go to zero as $t \longrightarrow \infty$ when K_F is used. Physically this means that the positive-energy electron cannot make an energy-conserving ("real" as opposed to "virtual") transition into a negative-energy state even in the presence of a time-dependent external potential. Thus the catastrophic transition into a negative-energy state is now forbidden.

On the other hand, the behavior of (4.269) as $t \longrightarrow -\infty$ is somewhat puzzling. Although $c_{\mathbf{p}',s'}^{(+)}$ goes to zero as $t \longrightarrow -\infty$, $c_{\mathbf{p}',s'}^{(-)}(-\infty)$ does not, in general, vanish. As a result, $\psi(t = -\infty)$ is *not* equal to ψ_0. This is quite strange; in fact, as we shall show in a moment, its ultimate physical interpretation requires a new postulate concerning the meaning of the Dirac wave function. Fortunately some of the results obtained in Section 4–3 help us to understand the physical meaning of $c_{\mathbf{p}',s'}^{(-)}(-\infty)$; we note that, apart from a minus sign (which is not significant‡), $c_{\mathbf{p}',s'}^{(-)}$

‡Had we used $d^\dagger b^\dagger|0\rangle$ rather than $b^\dagger d^\dagger|0\rangle$ as the initial state in discussing $e^- + e^+ \longrightarrow$ vacuum (Section 4–3), the S-matrix element (4.114) would have been opposite in sign.

$(-\infty)$ is nothing more than the first-order S-matrix element (4.114) for $e^- + e^+ \longrightarrow$ vacuum in the presence of $A_\mu^{(e)}$ when ψ_0 is set equal to a normalized (positive-energy electron) plane wave and $(\mathbf{p}', s')$ is identified with the momentum-spin index of the annihilated positron. This encourages us to identify $c_{\mathbf{p}',s'}^{(-)}(t)\sqrt{m/EV}\, v^{(s')}(\mathbf{p}')e^{-ip'\cdot x}$ as the wave function for an available *empty* negative-energy state (with $-\mathbf{p}'$ and $r' = 5 - s'$) into which the incident electron is capable of making a transition in the presence of $A_\mu^{(e)}$; as the time goes on, such a state becomes rarer and rarer because of pair annihilation, and, as $t \longrightarrow \infty$, $c_{\mathbf{p}',s'}^{(-)}$, in fact, goes to *zero*. Alternatively we may identify $|c_{\mathbf{p}',s'}^{(-)}(t)|^2$ as the probability for finding a *positron* in state $(\mathbf{p}', s')$ which will certainly be annihilated by the time $t = \infty$. To compute the S-matrix for pair annihilation, we simply go backward in time and ask, what is the probability amplitude for finding, at $t = -\infty$, the positron which is doomed to be annihilated at later times. The nonvanishing of $c_{\mathbf{p}',s'}^{(-)}$ for large negative values of t, which may, at first sight, seem rather puzzling, is seen to be quite welcome if we realize that the external potential can not only scatter electrons but also annihilate electron-positron pairs.

In order to fully appreciate the physical significance of each term in (4.268) and (4.269) it is instructive to examine the properties of the integral operator

$$\int d^3x'\, K_F(x, x')\gamma_4 \tag{4.270}$$

acting on a Dirac wave function whose space-time coordinate is x'. Suppose the time ordering is "ordinary" in the sense that $t > t'$. When the operator (4.270) acts on a free-particle positive-energy wave function, we simply get the wave function for the same momentum-spin state evaluated at a later time:

$$\begin{aligned}\int d^3x'\, K(x, x')\gamma_4\sqrt{m/EV}\, u^{(s)}(\mathbf{p})e^{ip\cdot x'}\\ &= \sum_{\mathbf{p}'}\sum_{s'} (m/E'V)u^{(s')}(\mathbf{p}')u^{(s')\dagger}(\mathbf{p}')\sqrt{m/EV}\, u^{(s)}(\mathbf{p})e^{ip'\cdot x}\int_V d^3x'\, e^{-ip'\cdot x'+ip\cdot x'}\\ &= \sqrt{m/EV}\, u^{(s)}(\mathbf{p})e^{ip\cdot x},\end{aligned} \tag{4.271}$$

where we have made use of the orthogonality property of the free-particle spinor (3.106). On the other hand, when the integral operator (4.270) acts on a negative-energy wave function (4.267), it obviously gives zero so long as $t > t'$ [recall $u^{(s')\dagger}(-\mathbf{p})v^{(s)}(\mathbf{p}) = \pm u^{(s')\dagger}(-\mathbf{p})u^{(3\text{ or }4)}(-\mathbf{p}) = 0$]. Similarly, when the time ordering is "opposite," that is, $t' > t$, the negative of the integral operator (4.270) acting on a negative-energy wave function gives the wave function for the same negative-energy state evaluated at an *earlier* time; when (4.270) with $t' > t$ acts on a positive-energy wave function it gives zero. To summarize:

$$\int d^3x'\, K_F(x, x')\gamma_4 \begin{Bmatrix}\sqrt{m/EV}\, u^{(s)}(\mathbf{p})e^{ip\cdot x'}\\ \sqrt{m/EV}\, v^{(s)}(\mathbf{p})e^{-ip\cdot x'}\end{Bmatrix} = \begin{cases}\sqrt{m/EV}\, u^{(s)}(\mathbf{p})e^{ip\cdot x}\theta(t - t')\\ -\sqrt{m/EV}\, v^{(s)}(\mathbf{p})e^{-ip\cdot x}\theta(t' - t).\end{cases} \tag{4.272}$$

In other words, the integral operator (4.270) carries positive-energy solutions forward in time and negative-energy ones backward in time.

Returning now to the first line of (4.268), we first note that $-e\gamma_\mu\psi_0(x')A_\mu^{(e)}(x')$ represents the interaction of the incident electron with the external potential at x'. As a result of this interaction, we expect both positive- and negative-energy waves originating at x'. Looking into the future, the Green's function $K_F(x, x')$ with $t > t'$ carries positive-energy electron states forward in time. Looking into the past, the Green's function $K_F(x, x')$ with $t < t'$ carries negative-energy electron states backward in time. But we argued earlier that the probability amplitude $c_{\mathbf{p},s}^{(-)}$ for such a negative-energy state represents the probability amplitude for the corresponding positron state which is doomed to be annihilated at later times. We can therefore visualize pair annihilation as the scattering of a positive-energy electron into a negative-energy state which propagates backward in time (Fig. 4–16b) in complete analogy with ordinary scattering (Fig. 4–16a). The positron which is to be annihilated at x' is simply an electron "scattered backward to earlier times." From this point of view a negative-energy electron going backward in time is equivalent to a positron going forward in time. This equivalence which was suggested in the previous section by examining the propagation of a virtual electron (positron) is now seen to make sense also for a *real* electron (positron).

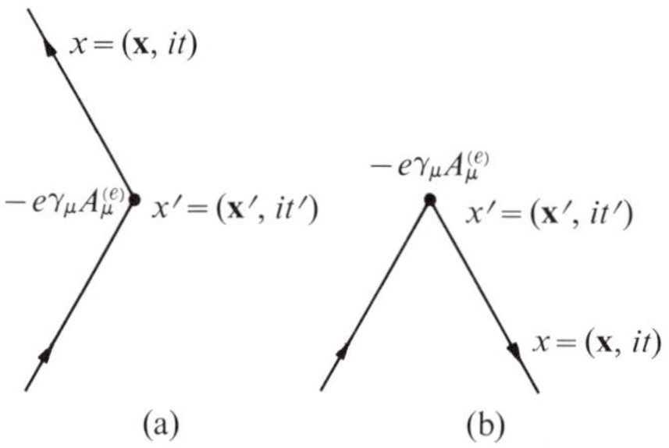

Fig. 4–16. Graphical interpretation of (4.268): (a) ordinary scattering ($t > t'$), (b) pair annihilation ($t < t'$).

So far we have treated electron scattering and pair annihilation in detail. Considering $\psi_0(x)$ in (4.268) to be a negative-energy free-particle wave function, we can formulate positron scattering and pair production to first order in the external potential in an entirely analogous manner. In connection with electron scattering and pair annihilation we have seen that, with $\psi_0(x)$ taken to be a positive-energy wave function, the use of K_F (rather than K_{ret}) ensures the boundary condition that in the remote future ψ has no component of the type $v^{(s)}e^{-ip\cdot x}$. In the formulation of problems of positron scattering and pair production, the correct boundary condition which we must use is that in the remote *past* ψ has no component of the type $u^{(s)}e^{ip\cdot x}$; this is because neither in positron scattering nor in pair production should there be a positive-energy electron going forward in time at $t = -\infty$. With ψ_0 taken to be a negative-energy wave function, it is easy to see that the use of K_F automatically guarantees the above boundary condition.

More generally, to determine the wave function at all times in the Feynman theory, it is necessary to specify the positive-energy components in the remote past and the negative-energy components in the remote future. This is because the positive-energy electron propagates forward in time while the negative-energy electron propagates backward in time. In contrast, we may recall that the Schrödinger wave function can be completely determined at all times once we know the wave function in the remote past.

Feynman's approach to Compton scattering. The utility of Feynman's view that the electron paths can "zigzag" forward or backward in time can be fully appreciated when we consider more complex problems. For this reason we now examine the second-order term in the perturbation expansion (4.252). For definiteness consider the case of Compton scattering. The $A_\mu(x)$ now represents the equivalent time-dependent potential, discussed in Chapter 2, corresponding to the emission and absorption of a photon: $(1/\sqrt{2\omega V})\epsilon_\mu^{(\alpha)}e^{-ik\cdot x}$ for emission and $(1/\sqrt{2\omega V})\epsilon_\mu^{(\alpha)}e^{ik\cdot x}$ for absorption. We look at (4.252) to second order with ψ_0 taken to be the wave function that represents the incident positive-energy plane wave characterized by $(\mathbf{p}, s)$. With K set to K_F we get

$$\begin{aligned}\psi(x) \simeq \sqrt{m/EV}\, u^{(s)}(\mathbf{p})e^{ip\cdot x} &+ \int d^4x' \int d^4x''\, K_F(x, x')[-e\gamma_\mu(1/\sqrt{2\omega V})\epsilon_\mu^{(\alpha)}e^{ik\cdot x}] \\ &\times K_F(x', x'')[-e\gamma_\nu(1/\sqrt{2\omega' V})\epsilon_\nu^{(\alpha_1)}e^{-ik'\cdot x''}][\sqrt{m/EV}\, u^{(s)}(\mathbf{p})e^{ip\cdot x''}] \\ &+ \begin{Bmatrix} k \longleftrightarrow k' \\ \alpha \longleftrightarrow \alpha' \end{Bmatrix},\end{aligned} \tag{4.273}$$

where we have omitted a term linear in e which cannot contribute to scattering. As $t \longrightarrow \infty$, we can replace the first of the two Green's functions, $K_F(x, x')$, by (4.264a) since t is necessarily later than t'. Thus

$$\psi(x) \xrightarrow{t\to\infty} \sqrt{m/EV}\, u^{(s)}(\mathbf{p})e^{ip\cdot x} + \sum_{\mathbf{p}'}\sum_{s'} c^{(+)}_{\mathbf{p}',s'}(\infty)[\sqrt{m/E'V}\, u^{(s')}(\mathbf{p}')e^{ip'\cdot x}], \tag{4.274}$$

where

$$\begin{aligned}c^{(+)}_{\mathbf{p}',s'}(\infty) = \int d^4x' \int d^4x''\, &(\sqrt{m/E'V}\, \bar{u}^{(s')}(\mathbf{p}')e^{-ip'\cdot x'})[-e\gamma_\mu(1/\sqrt{2\omega V})\epsilon_\mu^{(\alpha)}e^{ik\cdot x}] \\ &\times K_F(x', x'')[-e\gamma_\nu(1/\sqrt{2\omega' V})\, \epsilon_\nu^{(\alpha')}e^{-ik'\cdot x''}](\sqrt{m/EV}\, u^{(s)}(\mathbf{p})e^{ip\cdot x''}) \\ &+ \begin{Bmatrix} k \longleftrightarrow k' \\ \alpha \longleftrightarrow \alpha' \end{Bmatrix}.\end{aligned} \tag{4.275}$$

When we recall that $K_F(x', x'')$ is equal to $\langle 0 | T(\psi(x')\bar{\psi}(x'')) | 0 \rangle$, the coefficient $c^{(+)}_{\mathbf{p}',s'}(\infty)$ is seen to be completely identical to the second-order S-matrix (4.174) for Compton scattering which we derived in the previous section, using the language of quantum field theory. That we can derive the correct matrix element in such a simple manner is the main advantage of Feynman's space-time approach.

In Feynman's language we can read the various factors in the integral (4.275) from the right to the left as follows. First, the incident electron characterized by $(\mathbf{p}, s)$ emits a photon $(\mathbf{k}', \alpha')$ at x''. Subsequently the electron goes from x'' to x'; if the time ordering is "ordinary," that is, $t' > t''$, then the first form of K_F given by

(4.264a) ensures that a positive-energy virtual electron goes forward in time, as shown in Fig. 4–17(a). On the other hand, if the time ordering is opposite, that is, $t'' > t'$, then $K_F(x, x')$ carries a negative-energy virtual electron originating at x'' backward in time to x'; this is shown in Fig. 4–17(b). Meanwhile, using the field-theoretic formalism, we have already shown that $\langle 0 | T(\psi(x')\bar{\psi}(x'')) | 0 \rangle$ represents the propagation of a positive-energy electron from x'' to x' when the time ordering is "ordinary" and the propagation of a positive-energy positron from x' to x'' when the time ordering is "opposite." Comparing the two points of view and recalling that $K_F(x', x'')$ is identical with $\langle 0 | T(\psi(x')\bar{\psi}(x'')) | 0 \rangle$ we see once again that negative-energy electrons going backward in time represent positrons going forward in time. Finally at x', the virtual electron (positive energy or negative energy depending on whether $t' > t''$ or $t' < t''$) absorbs the incident photon $(\mathbf{k}, \alpha)$ and goes on as a positive-energy electron $(\mathbf{p}', s')$ propagating forward in time. As we emphasized earlier, a great virtue of covariant perturbation theory is that we can combine the terms corresponding to two different time-ordered graphs, Fig. 4–17(a) and (b), into a single simple expression.

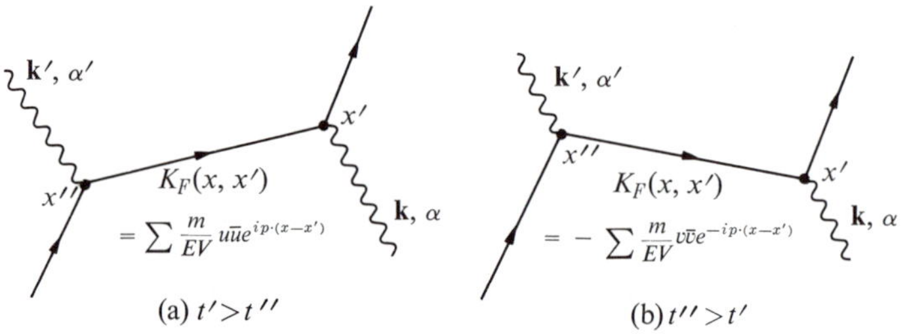

Fig. 4–17. Examples of time-ordered graphs for Compton scattering. Note that (a) and (b) *together* correspond to a *single* Feynman graph (Fig. 4–7a).

We have seen that the results which we rigorously derived in the previous two sections from quantum field theory can be written rather quickly when we use Feynman's space-time approach. There is no doubt that the calculational steps become shorter and simpler when done in Feynman's way. This is perhaps not so surprising since Feynman's method which is based directly on the covariant wave equation of Dirac does not conceal the relativistic invariance of the theory at each state of the calculations. In contrast, since the customary Hamiltonian formalism, which was used to derive the calculational methods of the previous two sections, is based on the Schrödinger-like equation (4.24) which singles out the time coordinate in the very beginning, we must work somewhat harder to arrive at the same covariant matrix element (for example, by combining separately noncovariant pieces in a clever way).

In this text we first demonstrated how to derive the matrix elements for some simple processes, using quantum field theory, and then showed how the same matrix elements can be obtained much more quickly when we use the space-time approach of Feynman. It is true that we can hardly overestimate the simplicity and the intuitive appeal of Feynman's approach. We must, at the same time, admit, however, that the logical structure of Feynman's method is somewhat more involved. This is because a formalism based on the *single*-particle wave equation is used to describe processes such as pair creation, pair annihilation, and even vacuum polarization, which involve *more than one* particle at a given time. In fact, the careful reader might have already noted that the physical interpretations of (4.268) given go beyond the usual interpretation of wave mechanics in which the square of the Fourier coefficient represents the probability for finding a *single* particle in the state in question at a given time. For this reason we prefer to regard Feynman's graphical method as a convenient pictorial device that enables us to keep track of the various terms in the matrix elements which we derive rigorously from quantum field theory. Our philosophy closely parallels that of F. J. Dyson who in his 1951 lectures at Cornell University made the following remarks:

> In this course we follow the pedestrian route of logical development, starting from the general principles of quantization applied to covariant field equations, and deriving from these principles first the existence of particles and later the results of the Feynman theory. Feynman, by the use of imagination and intuition, was able to build a correct theory and get the right answers to problems much quicker than we can. It is safer and better for us to use the Feynman space-time pictures not as the basis for our calculation but only as a help in visualizing the formulae which we derive rigorously from field theory. In this way we have the advantages of the Feynman theory, its correctness and its simplification of calculations, without its logical disadvantages.

Historically it was S. Tomonaga and J. Schwinger who first recognized the importance of the interaction representation for performing covariant calculations in quantum field theory. Using this representation, they were able to isolate in an unambiguous way the finite (observable) part of the radiative corrections to the basic γ_μ vertex (a topic to be discussed in Section 4–7). Subsequently, in two famous 1949 papers, "Theory of Positrons" and "Space-time Approach to Quantum Electrodynamics," R. P. Feynman developed a set of "rules" which are based directly on the Dirac and Maxwell equations but not (explicitly) on quantum field theory; using his "rules" he could obtain the correct results much more quickly and efficiently than Tomonaga and Schwinger. Meanwhile, starting with the S-matrix expansion of the field operators (as we have done in this book), F. J. Dyson demonstrated that the Feynman rules are direct and rigorous consequences of quantum field theory and that the approach of Tomonaga and Schwinger is, in fact, mathematically equivalent to the simpler approach of Feynman.

Whether the reader prefers to derive the Feynman rules rigorously from quantum field theory or more intuitively from the wave equations à la Feynman, there is no doubt that the graphs and rules which Feynman taught us to use have had profound influences on a number of areas of physics—not only quantum electrodynamics and high-energy (elementary particle) physics but also nuclear many-body problems, superconductivity, hard-sphere Bose gases, polaron problems etc. In less than two decades the graphical techniques of Feynman have become one of the most indispensable and powerful tools in attacking a broad class of problems in theoretical physics.

4–6. MØLLER SCATTERING AND THE PHOTON PROPAGATOR; ONE-MESON EXCHANGE INTERACTIONS

Scalar meson exchange. In the previous two sections we learned how to compute second-order processes which involve virtual electrons and positrons or, more generally, virtual spin-$\frac{1}{2}$ particles. In this section we shall discuss processes which involve virtual photons or, more generally, virtual bosons. (Processes that involve the propagation of both a virtual fermion and a virtual boson at the same time will be treated in the next section.) Our main discussion centers on the electron-electron interaction brought about by the exchange of a single "covariant photon." Let us, however, first treat the simpler problem of the exchange of a spinless meson.

We shall consider a spin-$\frac{1}{2}$ particle of mass m which we call a "nucleon"; the corresponding field, denoted by $\psi(x)$, is assumed to satisfy the Dirac equation and to be quantized according to the rules of Jordan and Wigner. In addition, we shall consider a spin-zero neutral particle of mass μ which we call a "meson." The field associated with the meson, denoted by $\phi(x)$, is assumed to satisfy the Klein-Gordon equation and to be subject to the usual quantization rules (cf. Problem 2–3). The "nucleon" and the "meson" interact via Yukawa coupling:

$$\mathscr{H}_{\text{int}} = G\bar{\psi}\psi\phi. \tag{4.276}$$

We do not claim that our "nucleon" and "meson" bear any similarity to the nucleon and the meson observed in nature; as we emphasized in Section 3–10, the observed low-mass mesons are pseudoscalar mesons which do not admit a coupling of the type (4.276) because of parity conservation in strong interactions.

Let us consider the elastic scattering of two (identical) nucleons to order G^2 in the amplitude. Initially we have two nucleons characterized by $(\mathbf{p}_1, s_1)$ and $(\mathbf{p}_2, s_2)$; the two final nucleons are characterized by $(\mathbf{p}_1', s_1')$ and $(\mathbf{p}_2', s_2')$. Since there is no meson in the initial state or in the final state, the initial and the final state are the vacuum state so far as the meson operators are concerned. The second-order matrix element ("form 1") reads

$$\begin{aligned} S_{fi}^{(2)} = (-iG)^2 \int d^4x_1 \int_{t_2<t_1} d^4x_2 \langle N_1', N_2' \,|\, \bar{\psi}_\alpha(x_1)\psi_\beta(x_1)\bar{\psi}_\gamma(x_2)\psi_\delta(x_2) \,|\, N_1, N_2\rangle \\ \times (1)_{\alpha\beta}(1)_{\gamma\delta}\langle 0 \,|\, \phi(x_1)\phi(x_2) \,|\, 0\rangle, \end{aligned} \tag{4.277}$$

where

$$|N_1, N_2\rangle = b_2^\dagger b_1^\dagger |0\rangle, \qquad |N_1', N_2'\rangle = b_2'^\dagger b_1'^\dagger |0\rangle \tag{4.278}$$

with the obvious abbreviation $b_1^\dagger = b^{(s_1)\dagger}(\mathbf{p}_1)$, etc. Although the ordering of $b_1'^\dagger$ and $b_2'^\dagger$ is somewhat arbitrary, with the choice (4.278) we get $S^{(0)} = 1$ when state 1 and state 2 coincide with state 1′ and state 2′ respectively. In (4.277), $\bar{\psi}\psi$ is written as $\bar{\psi}_\alpha\psi_\beta(1)_{\alpha\beta}$ so that our method here can be immediately generalized to the γ_μ(or the γ_5) coupling case. This time, since there are two nucleons in the initial state, both $\psi(x_1)$ and $\psi(x_2)$ must be used to annihilate the incident nucleons; similarly both $\bar{\psi}(x_1)$ and $\bar{\psi}(x_2)$ must be used to create the outgoing nucleons. The terms containing $\bar{\psi}^{(+)}$ and $\psi^{(-)}$ make no contributions since there are no antinucleons to be annihilated or created. Hence we can replace the product of the fermion operators in (4.277) by

$$\bar{\psi}_\alpha(x_1)\psi_\beta(x_1)\bar{\psi}_\gamma(x_2)\psi_\delta(x_2) \longrightarrow \bar{\psi}_\alpha^{(-)}(x_1)\psi_\beta^{(+)}(x_1)\bar{\psi}_\gamma^{(-)}(x_2)\psi_\delta^{(+)}(x_2). \tag{4.279}$$

Note that if we are computing the S-matrix to order G^2, there are no virtual nucleons; every one of the fermion operators in (4.277) is needed to annihilate and create the *real* nucleons in the initial and the final states. As for the meson operators, we can obviously replace $\langle 0|\phi(x_1)\phi(x_2)|0\rangle$ by $\langle 0|\phi^{(+)}(x_1)\phi^{(-)}(x_2)|0\rangle$, where $\phi^{(+)}(\phi^{(-)})$ is the annihilation part (the creation part) of the meson field:

$$\begin{aligned} \phi^{(+)} &= \sum_{\mathbf{k}} \frac{1}{\sqrt{2\omega V}} a_{\mathbf{k}} e^{ik\cdot x}, \\ \phi^{(-)} &= \sum_{\mathbf{k}} \frac{1}{\sqrt{2\omega V}} a_{\mathbf{k}}^\dagger e^{-ik\cdot x} \end{aligned} \tag{4.280}$$

with $\omega = \sqrt{|\mathbf{k}|^2 + \mu^2}$. Physically $\langle 0|\phi^{(+)}(x_1)\phi^{(-)}(x_2)|0\rangle$ means that a virtual meson is created at x_2 and subsequently (note $t_1 > t_2$) annihilated at x_1, as is clear from our experience with the fermion case in Section 4–4. We shall say more about this vacuum expectation value later.

Coming back to the fermion operators, we now move $\psi^{(+)}$ to the right and $\psi^{(-)}$ to the left so that the product of the annihilation operators may immediately annihilate the nucleons in the initial state:

$$\begin{aligned} &\bar{\psi}_\alpha^{(-)}(x_1)\psi_\beta^{(+)}(x_1)\bar{\psi}_\gamma^{(-)}(x_2)\psi_\delta^{(+)}(x_2) \\ &\quad = -\bar{\psi}_\alpha^{(-)}(x_1)\bar{\psi}_\gamma^{(-)}(x_2)\psi_\beta^{(+)}(x_1)\psi_\delta^{(+)}(x_1) + \bar{\psi}_\alpha^{(-)}(x_1)\{\psi_\beta^{(+)}(x_1), \bar{\psi}_\gamma^{(-)}(x_2)\}\psi_\delta^{(+)}(x_2). \end{aligned} \tag{4.281}$$

It is not difficult to convince oneself that the last term of (4.281), which is just $\bar{\psi}_\alpha^{(-)}(x_1)\psi_\delta^{(+)}(x_2)$ times a c-number function of x_1 and x_2, does not contribute to the scattering problem in which we are interested; it just gives rise to the amplitude for a process in which nucleon 1 and nucleon 2 proceed in space-time independently of each other even though one of them suffers a self-energy interaction.‡ Our next task is to evaluate $\psi_\beta^{(+)}(x_1)\psi_\delta^{(+)}(x_2)$ acting on $|N_1, N_2\rangle$ from the left and $-\bar{\psi}_\alpha^{(-)}(x_1)\bar{\psi}_\gamma^{(-)}(x_2)$ acting on $\langle N_1', N_2'|$ from the right. Let $b_{(1)}$ and $b_{(2)}$ stand for typical annihilation operators contained in the fields $\psi^{(+)}(x_1)$ and $\psi^{(+)}(x_2)$ respectively. So far as the creation and annihilation operators are concerned, a typical term arising from $\psi_\beta^{(+)}(x_1)\psi_\delta^{(+)}(x_2)|N_1, N_2\rangle$ looks like

$$\begin{aligned} b_{(1)}b_{(2)}b_2^\dagger b_1^\dagger|0\rangle &= [b_{(1)}b_{(2)}, b_2^\dagger b_1^\dagger]|0\rangle \\ &= [-b_{(1)}b_2^\dagger\{b_1^\dagger, b_{(2)}\} + b_{(1)}\{b_2^\dagger, b_{(2)}\}b_1^\dagger]|0\rangle \\ &= [-\{b_{(1)}, b_2^\dagger\}\delta_{1(2)} + \{b_{(1)}, b_1^\dagger\}\delta_{2(2)}]|0\rangle \\ &= (-\delta_{(1)2}\delta_{1(2)} + \delta_{(1)1}\delta_{2(2)})|0\rangle, \end{aligned} \tag{4.282}$$

where we have used formula (3.414). Similarly we can compute the Hermitian

‡The diagram corresponding to this is an example of what is known as a *disconnected graph.*

conjugate of $-\bar{\psi}_\alpha^{(-)}(x_1)\bar{\psi}_\gamma^{(-)}(x_2)$ acting on $\langle N_1', N_2'|$ from the right as follows:

$$[-b'_{(2)}b'_{(1)}b_2'^{\dagger}b_1'^{\dagger}|0\rangle]^\dagger = [(\delta_{2'(2')}\delta_{1'(1')} - \delta_{(1')2'}\delta_{1'(2')})|0\rangle]^\dagger, \tag{4.283}$$

where $b'_{(1)}$ and $b'_{(2)}$ stand for typical annihilation operators contained in $[\bar{\psi}_\alpha^{(-)}(x_1)]^\dagger$ and $[\bar{\psi}_\gamma^{(-)}(x_2)]^\dagger$ respectively. The physical meaning of (4.282) can be seen as follows. The operator $\psi^{(+)}(x_2)$ can be used to annihilate either nucleon 2 or nucleon 1; if $\psi^{(+)}(x_2)$ annihilates nucleon 2, then $\psi^{(+)}(x_1)$ must necessarily annihilate nucleon 1, and vice versa.

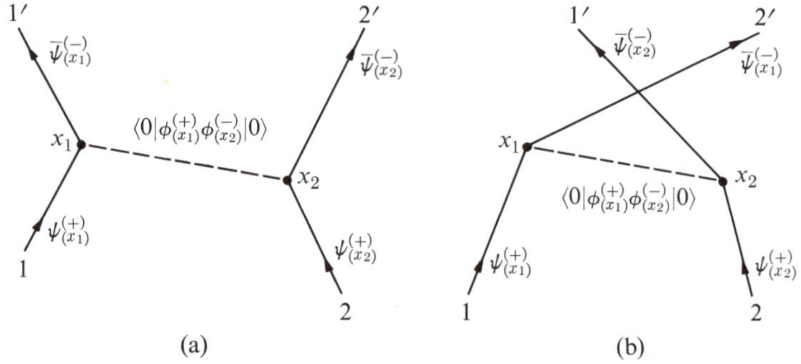

Fig. 4–18. Examples of time-ordered graphs in nucleon-nucleon scattering ($t_1 > t_2$).

Let us focus on the case where $\psi^{(+)}(x_2)$ and $\psi^{(+)}(x_1)$ annihilate nucleon 2 and nucleon 1 respectively; this corresponds to the second term of the last line of (4.282). Looking at (4.283) we see that there are still two possibilities: (a) $\bar{\psi}^{(-)}(x_2)$ creates nucleon 2', and $\bar{\psi}^{(-)}(x_1)$ creates nucleon 1', or (b) $\bar{\psi}^{(-)}(x_2)$ creates nucleon 1', and $\bar{\psi}^{(-)}(x_1)$ creates nucleon 2'. These two possibilities can be visualized by drawing time-ordered graphs as in Fig. 4–18, where we have recalled that $\langle 0|\phi^{(+)}(x_1)\phi^{(-)}(x_2)|0\rangle$ gives the propagation of a virtual meson from x_2 to x_1. It is very important to note that (4.283) tells us that we associate a minus sign with Fig. 4–18(b). From our experience with the scattering of two identical particles in nonrelativistic wave mechanics we recognize that this is precisely the minus sign associated with an antisymmetrized final state as demanded by the Pauli exclusion principle. Our formalism based on the second-quantized Dirac theory *automatically* gives the full force of Fermi-Dirac statistics. We also note that no other factors such as $\sqrt{2}$, etc., are needed; the only other thing we must keep in mind is that in obtaining the total cross section we integrate over only half of the solid angles, just as in the case of two-photon annihilation (cf. Eq. 4.201). We must still consider the case where $\psi_\delta^{(+)}(x_2)$ annihilates nucleon 1, and $\psi_\beta^{(+)}(x_1)$ annihilates nucleon 2 [cf. the first term of (4.282)]. However, since we are integrating over both x_1 and x_2, we may as well make the interchanges $x_1 \leftrightarrow x_2$, $\alpha\beta \leftrightarrow \gamma\delta$ so that $\psi_\delta^{(+)}(x_2)$ annihilates nucleon 2, and $\psi_\beta^{(+)}(x_1)$ annihilates nucleon 1, just as before. We must, of course, keep in mind that t_1 is now *earlier* than t_2 ("form 2"). Noting that the product of the first term of (4.282) with $(1) \leftrightarrow (2)$ and (4.283) with $(1') \leftrightarrow (2')$ is equal to the product of the second term of (4.282) and (4.283), we see that the S-matrix (4.277) becomes

$$S_{fi}^{(2)} = (-iG)^2 \int d^4x_1 \int d^4x_2 \sqrt{\left(\frac{m}{E_1'V}\right)\left(\frac{m}{E_2'V}\right)\left(\frac{m}{E_1V}\right)\left(\frac{m}{E_2V}\right)}$$
$$\times [(\bar{u}_1' e^{-ip_1'\cdot x_1} u_1 e^{ip_1\cdot x_1})(\bar{u}_2' e^{-ip_2'\cdot x_2} u_2 e^{ip_2\cdot x_2}) - (\bar{u}_2' e^{-ip_2'\cdot x_1} u_1 e^{ip_1\cdot x_1})(\bar{u}_1' e^{-ip_1'\cdot x_2} u_2 e^{ip_2\cdot x_2})]$$
$$\times [\langle 0 | \phi^{(+)}(x_1)\phi^{(-)}(x_2) | 0\rangle \theta(t_1 - t_2) + \langle 0 | \phi^{(+)}(x_2)\phi^{(-)}(x_1) | 0\rangle \theta(t_2 - t_1)]. \tag{4.284}$$

From (4.284) we observe that the problem is now reduced to that of evaluating the vacuum expectation value of a time-ordered product which we *newly* define as follows:

$$\langle 0 | T(\phi(x)\phi(x')) | 0\rangle = \begin{cases} \langle 0 | \phi(x)\phi(x') | 0\rangle & \text{for} \quad t > t', \\ \langle 0 | \phi(x')\phi(x) | 0\rangle & \text{for} \quad t' > t. \end{cases} \tag{4.285}$$

Unlike (4.152), the T-product of two boson field operators is defined without any minus sign even when the time ordering is "opposite" ($t' > t$); in the S-matrix occurring in practical problems the various field operators not used in annihilating or creating real particles always arrange themselves in such a way that we have the combination (4.153) for fermions and the combination (4.285) (or its analogs) for bosons. Clearly (4.285) is equal to

$$\langle 0 | \phi(x)\phi(x') | 0\rangle = \langle 0 | \phi^{(+)}(x)\phi^{(-)}(x') | 0\rangle$$
$$= \lim_{V\to\infty} \frac{1}{V} \sum_{\mathbf{k}} \sum_{\mathbf{k}'} \frac{1}{2\sqrt{\omega\omega'}} \exp[i(k\cdot x - k'\cdot x')]\langle 0 | a_{\mathbf{k}} a_{\mathbf{k}'}^{\dagger} | 0\rangle$$
$$= \frac{1}{(2\pi)^3} \int \frac{d^3k}{2\omega} \exp[i\mathbf{k}\cdot(\mathbf{x} - \mathbf{x}') - i\omega(t - t')] \tag{4.286}$$

for $t > t'$. Similarly for $t' > t$, (4.285) reduces to

$$\langle 0 | \phi(x')\phi(x) | 0\rangle = \langle 0 | \phi^{(+)}(x')\phi^{(-)}(x) | 0\rangle$$
$$= \frac{1}{(2\pi)^3} \int \frac{d^3k}{2\omega} \exp[i\mathbf{k}\cdot(\mathbf{x} - \mathbf{x}') - i(-\omega)(t - t')]. \tag{4.287}$$

Again, just as in the fermion case [cf. (4.154) through (4.162)], we get, for both $t > t'$ and $t' > t$,

$$\langle 0 | T(\phi(x)\phi(x')) | 0\rangle = -\frac{i}{(2\pi)^4} \int d^4k \frac{e^{ik\cdot(x-x')}}{k^2 + \mu^2 - i\epsilon} \tag{4.288}$$

with

$$k^2 = |\mathbf{k}|^2 - |k_0|^2, \tag{4.289}$$

where k_0 is a variable, not, in general, equal to $\omega = \sqrt{|\mathbf{k}|^2 + \mu^2}$. Having studied the fermion case in detail, we can easily understand the physical meaning of (4.288) (known as the meson propagator). When t is later than t' a virtual meson created at x' propagates from x' to x and then gets annihilated at x; when t' is later than t, a virtual meson propagates from x to x'. Thus the combination of the vacuum expectation values indicated by (4.285) automatically keeps track of the correct "causal sequence" of a virtual process in which a meson is emitted by one of the nucleons and is subsequently absorbed by the other nucleon, a point emphasized by E. C. G. Stueckelberg prior to the advent of Feynman's theory. We again mention that the notion of t being later or earlier than t' is not a Lorentz invariant concept when x and x' are spacelike and that the meson propagator does not vanish outside the light cone $[(x-x')^2 > 0]$; yet the combination (4.285) is Lorentz invariant as is evident from its explicit Fourier integral form (4.288). In the literature one often defines $\Delta_F(x-x')$ so that

$$\frac{1}{2}\Delta_F(x-x') = -\frac{i}{(2\pi)^4}\int d^4k\,\frac{e^{ik\cdot(x-x')}}{k^2+\mu^2-i\epsilon}. \tag{4.290}$$

The previously defined S_F is related to Δ_F via the expression

$$[S_F(x-x')]_{\alpha\beta} = \left(\gamma_\mu\frac{\partial}{\partial x_\mu} - m\right)_{\alpha\beta}\Delta_F(x-x') \tag{4.291}$$

with m replacing μ in the expression for Δ_F.

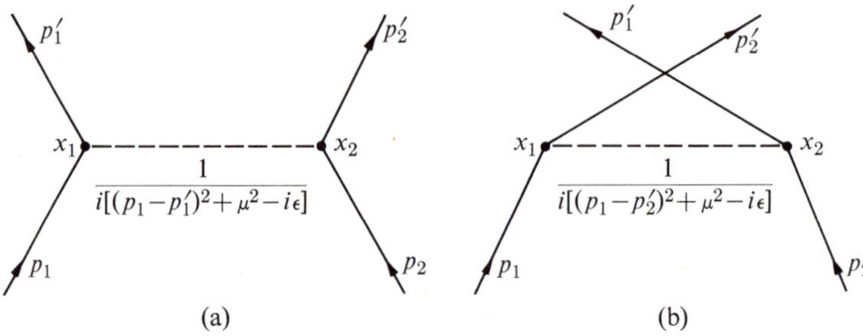

Fig. 4–19. Feynman graphs for nucleon-nucleon scattering: (a) direct scattering, (b) exchange scattering.

Substituting (4.288) in (4.284) and integrating over x_1 and x_2, we easily obtain the explicit form of the covariant $\mathscr{M}$-matrix defined, as usual, by

$$S_{fi} = \delta_{fi} - i(2\pi)^4\delta^{(4)}(p_1+p_2-p_1'-p_2')\sqrt{\left(\frac{m}{E_1'V}\right)\left(\frac{m}{E_2'V}\right)\left(\frac{m}{E_1V}\right)\left(\frac{m}{E_2V}\right)}\,\mathscr{M}_{fi}. \tag{4.292}$$

To order G^2 we get from (4.284)

$$-i\mathscr{M}_{fi} = (-iG)^2\left[\frac{(\bar{u}_1'u_1)(\bar{u}_2'u_2)}{i[(p_1-p_1')^2+\mu^2-i\epsilon]} - \frac{(\bar{u}_2'u_1)(\bar{u}_1'u_2)}{i[(p_1-p_2')^2+\mu^2-i\epsilon]}\right], \tag{4.293}$$

just as we obtained (4.166) from (4.163) in the problem of two-photon annihilation. The various factors in (4.293) can be read immediately from the Feynman graphs Fig. 4–19 where, in contrast to the time-ordered graphs Fig. 4–18, the virtual meson is emitted either at x_2 or at x_1 (and subsequently absorbed at x_1 or at x_2), depending on whether $t_1 > t_2$ or $t_2 > t_1$. Again the "rule" for writing $-i\mathscr{M}$ is remarkably simple. Just as we associate $(-i\gamma\cdot q + m)/[i(q^2+m^2-i\epsilon)]$ with each virtual spin-$\frac{1}{2}$ fermion of four-momentum q, we associate $1/[i(q^2+\mu^2-i\epsilon)]$ with each virtual spinless meson of four-momentum q. Recall that in computing the energy-momentum of the virtual particle we assume energy-momentum conservation at each vertex. (One may argue that the four-momentum of the virtual meson in Fig. 4–19(a) is $p_1 - p_1'$ $(= p_2' - p_2)$ or $p_1' - p_1$ depending on whether the meson is emitted by nucleon 1 or nucleon 2; but in the boson case this is of no importance since the meson propagator is a function of q^2 only.) Because of the scalar coupling (4.276), we have $(-iG)$ at each vertex in place of the $(-e\gamma_\mu)$ appropriate for the electromagnetic interaction. In addition, we multiply the result by $+1$ for Fig. 4–19(a) ("direct" fermion-fermion scattering) and -1 for Fig. 4–19(b) ("exchange" fermion-fermion scattering).

Before we discuss the more complicated problem of photon exchange, we shall briefly outline two alternative methods for deriving the $\mathscr{M}$-matrix element (4.293). First, we show how we may have guessed it from semiclassical considerations with a nonquantized meson field. According to Yukawa's idea, the presence of nucleon 1 generates a ("classical") meson field which we denote by $\phi^{(1)}$. When the interaction Lagrangian is given by the negative of (4.276), the field $\phi^{(1)}$ satisfies the inhomogeneous Klein-Gordon equation

$$\Box\phi^{(1)}(x) - \mu^2\phi^{(1)}(x) = Gs^{(1)}(x), \tag{4.294}$$

where $s^{(1)}(x)$ is the scalar invariant constructed out of the initial and final wave functions for nucleon 1, viz.

$$s^{(1)} = \sqrt{(m/E_1'V)(m/E_1V)}\,\bar{u}_1'\,u_1\,e^{i(p_1-p_1')\cdot x}. \tag{4.295}$$

We solve the differential equation (4.294) assuming that the space-time dependence of $\phi^{(1)}$ is also given by $e^{i(p_1-p_1')\cdot x}$. We then get

$$\phi^{(1)}(x) = -\frac{G}{(p_1-p_1')^2+\mu^2}\sqrt{\left(\frac{m}{E'V}\right)\left(\frac{m}{E_1V}\right)}\,\bar{u}_1'\,u_1\,e^{i(p_1-p_1')\cdot x}. \tag{4.296}$$

Since we are interested in the interaction between nucleon 1 and nucleon 2, we compute the effect of nucleon 1 on nucleon 2. The Hamiltonian for the interaction of nucleon 2 with the field $\phi^{(1)}$ (generated by nucleon 1) to lowest-order is

$$\begin{aligned} H_{\text{int}} &= G\int \phi^{(1)}(x)s^{(2)}(x)\,d^3x \\ &= G\int \phi^{(1)}(x)\sqrt{\left(\frac{m}{E_2'V}\right)\left(\frac{m}{E_2V}\right)}\,\bar{u}_2'u_2e^{i(p_2-p_2')\cdot x}. \end{aligned} \tag{4.297}$$

But from the S-matrix expansion (4.36) we see that $(-i)$ times the time integral of H_{int} is to be identified with the $(S-1)$ matrix element to lowest order. Therefore we get, to order G^2,

$$\begin{aligned} S_{fi}^{(2)} &= -(-i)G^2\sqrt{\left(\frac{m}{E_1'V}\right)\left(\frac{m}{E_1V}\right)\left(\frac{m}{E_2'V}\right)\left(\frac{m}{E_2V}\right)} \\ &\quad\times\int d^4x\,\frac{(\bar{u}_1'u_1)(\bar{u}_2'u_2)}{(p_1-p_1')^2+\mu^2}\exp\left[i(p_1+p_2-p_1'-p_2')\cdot x\right] \\ &= (2\pi)^4\delta^{(4)}(p_1+p_2-p_1'-p_2') \\ &\quad\times\sqrt{\left(\frac{m}{E_1'V}\right)\left(\frac{m}{E_1V}\right)\left(\frac{m}{E_2'V}\right)\left(\frac{m}{E_2V}\right)}\frac{(-G^2)(\bar{u}_1'u_1)(\bar{u}_2'u_2)}{i[(p_1-p_1')^2+\mu^2]}, \end{aligned} \tag{4.298}$$

in complete agreement with the first term of (4.293) which represents "direct scattering." We can obtain the second term of (4.293) by taking into account the antisymmetrization rule for the final-state two-fermion wave function.

There is yet a third method for obtaining (4.293) which goes as follows. As in the first method, we use this time a formalism in which the meson field is quantized, in the sense that we talk about the emission and absorption of a virtual meson; however, the S-matrix is computed according to the rules of the old-fashioned second-order perturbation theory based on energy-difference denominators. The scattering process is visualized as taking place in two steps. First, the initial state $|N_1, N_2\rangle$ makes a transition into either of the following two types of intermediate states: (a) an intermediate state made up of a virtual meson, nucleon 1, and nucleon 2′ and (b) an intermediate state made up of a virtual meson, nucleon 1′, and nucleon 2. The intermediate state of either type then makes a transition into the final state $|N_1', N_2'\rangle$. The existence of two types of intermediate states, of course, arises from the fact that there are two time-ordered graphs corresponding to the single graph in Fig. 4–19(a); type (a) intermediate states arise when nucleon 2 first emits a virtual meson which is subsequently absorbed by nucleon 1, as represented by the time-ordered graph Fig. 4–18(a), whereas type (b) intermediate states arise when the virtual meson is emitted by nucleon 1 and subsequently absorbed by nucleon 2. According to the rules of Chapter 2 the emission and absorption of a meson is equivalent to an interaction of the nucleon with the time-dependent (scalar) potentials $\sqrt{1/2\omega V}\, e^{-ik\cdot x}$ and $\sqrt{1/2\omega V}\, e^{ik\cdot x}$. When nucleon 2 emits a meson of momentum $\mathbf{k}$, energy ω, to become nucleon 2′ (without any accompanying change in nucleon 1), we associate with this process a time-*independent* matrix element

$$
\begin{aligned}
H_{2'2}^{(\mathrm{emis})} &= G\sqrt{(m/E_2'V)(m/E_2V)(1/2\omega V)} \int_V d^3x \bar{u}_2' u_2\, e^{i(\mathbf{p}_2-\mathbf{p}_2'-\mathbf{k})\cdot x} \\
&= G\sqrt{(m/E_2'V)(m/E_2V)(1/2\omega V)}\,\bar{u}_2' u_2 V\, \delta_{\mathbf{p}_2,\mathbf{p}_2'+\mathbf{k}},
\end{aligned}
\tag{4.299}
$$

where ω satisfies the energy-momentum relation

$$
\omega = \sqrt{|\mathbf{k}|^2 + \mu^2}.
$$

Similarly, when nucleon 1 absorbs the virtual meson to become nucleon 1′, we associate with this process

$$
H_{1'1}^{(\mathrm{abs})} = G\sqrt{(m/E_1'V)(m/E_1V)(1/2\omega V)}\,\bar{u}_1' u_1 V\, \delta_{\mathbf{p}_1+\mathbf{k},\mathbf{p}_1'}. \tag{4.300}
$$

In addition, the rules of the old-fashioned perturbation theory tell us that we must divide the product of the perturbation matrix elements by the "energy denominator," which is

$$
(E_1 + E_2) - (E_1 + E_2') - \omega
$$

for a virtual process involving a type (a) intermediate state. In this way we obtain the effective Hamiltonian matrix element between the two nucleons in the form

$$\sum_{\mathbf{k}} \left(\frac{H^{(\text{abs})}_{1'1} H^{(\text{emis})}_{2'2}}{(E_1 + E_2) - (E_1 + E_2') - \omega} + \frac{H^{(\text{abs})}_{2'2} H^{(\text{emis})}_{1'1}}{(E_1 + E_2) - (E_1' + E_2) - \omega} \right)$$

$$= \sqrt{\left(\frac{m}{E_1 V}\right)\left(\frac{m}{E_2 V}\right)\left(\frac{m}{E_1' V}\right)\left(\frac{u}{E_2' V}\right)} \frac{G^2 V}{2\omega} \delta_{\mathbf{p}_1+\mathbf{p}_2,\,\mathbf{p}_1'+\mathbf{p}_2'} \bar{u}_1' u_1 \bar{u}_2' u_2$$

$$\times \left[\frac{1}{E_2 - E_2' - \omega} + \frac{1}{E_1 - E_1' - \omega} \right]$$

$$= \sqrt{\left(\frac{m}{E_1 V}\right)\left(\frac{m}{E_2 V}\right)\left(\frac{m}{E_1' V}\right)\left(\frac{m}{E_2' V}\right)} (2\pi)^3 \delta^{(3)}(\mathbf{p}_1 + \mathbf{p}_2 - \mathbf{p}_1' - \mathbf{p}_2') G^2 \bar{u}_1' u_1 \bar{u}_2' u_2$$

$$\times \left[\frac{1}{2\omega(E_2 - E_2' - \omega)} + \frac{1}{2\omega(E_1 - E_1' - \omega)} \right], \tag{4.301}$$

where we have let $V \to \infty$. The $S^{(2)}$ matrix element differs from the above expression simply by a factor $(-2\pi i)$ times $\delta(E_1 + E_2 - E_1' - E_2')$. We now observe that the bracket in the last line of (4.301) can be reduced to a remarkably simple form. Taking advantage of over-all energy conservation, we can replace $E_2 - E_2'$ by $-(E_1 - E_1')$. Hence

$$\frac{1}{2\omega(E_2 - E_2' - \omega)} + \frac{1}{2\omega(E_1 - E_1' - \omega)} = -\left(\frac{1}{2\omega}\right)\left[\frac{\omega - (E_1 - E_1') + \omega + (E_1 - E_1')}{\omega^2 - (E_1 - E_1')^2}\right]$$

$$= -\frac{1}{|\mathbf{k}|^2 + \mu^2 - (E_1 - E_1')^2}$$

$$= -\frac{1}{(p_1 - p_1')^2 + \mu^2}. \tag{4.302}$$

Apart from a factor i this is precisely the propagator for a spin-zero meson in momentum space. We thus see that (4.301) times $-2\pi i\ \delta(E_1 + E_2 - E_1' - E_2')$ is equal to the direct-scattering term of the $S^{(2)}$-matrix element computed earlier using the covariant method (cf. Eqs. 4.292 and 4.293). Note once again that by combining two noncovariant expressions we have succeeded in obtaining a single, more concise, covariant expression.

The above calculation illustrates the basic differences between the (prewar) old-fashioned method based on energy denominators such as $E_2 - E_2' - \omega$ and the (postwar) covariant method based on relativistically invariant denominators such as $(p_1 - p_1')^2 + \mu^2$. In the old-fashioned method momentum is conserved at each vertex, as indicated by the Kronecker deltas in (4.299) and (4.300). However, the intermediate state is not assumed to be an energy-conserving one; indeed, it is precisely the energy difference between the initial and the intermediate state that appears as the energy denominator. On the other hand, in the *covariant method*, where energy and momentum are treated on the same footing, we get automatically *both* energy conservation and momentum conservation at each vertex because of the appearance of a four-dimensional integral of the type $\int d^4x \exp[i(p_2 - p_2' - k)\cdot x]$. The exchanged virtual meson, however, is *not* visualized (in the sense of a Feynman diagram) as satisfying the energy-momentum relation $\omega = \sqrt{|\mathbf{k}|^2 + \mu^2}$; instead, we have, in general,

$$k^2 + \mu^2 = |\mathbf{k}|^2 - |k_0|^2 + \mu^2 \neq 0. \tag{4.303}$$

Indeed, it is precisely the difference between $k^2 = (p_1 - p_1')^2$ and $-\mu^2$ that appears as the denominator of the covariant meson propagator in momentum space. From this example we can appreciate the real advantage of covariant perturbation theory in which we insist that the space and the time integrations be performed *simultaneously*.

Matrix element for Møller scattering. We are now prepared to make an attack on our main problem—the scattering of two relativistic electrons, commonly known as Møller scattering after C. Møller, who first treated this problem in 1931. First, recall that the basic Lagrangian density of quantum electrodynamics is

$$\mathscr{L} = -\frac{1}{4}F_{\mu\nu}F_{\mu\nu} - \bar{\psi}\left(\gamma_\mu \frac{\partial}{\partial x_\mu} + m\right)\psi + ie\bar{\psi}\gamma_\mu\psi A_\mu. \tag{4.304}$$

Using the techniques of Appendix A applied to the Dirac electron case, we can deduce from this Lagrangian density the total Hamiltonian as follows:

$$H = H_0 + H_I, \tag{4.305}$$

where

$$H_0 = \frac{1}{2}\int (|\mathbf{E}^{(\perp)}|^2 + |\mathbf{B}^{(\perp)}|^2)\, d^3x + \int \bar{\psi}(\boldsymbol{\gamma}\cdot\boldsymbol{\nabla} + m)\psi\, d^3x, \tag{4.306}$$

and

$$H_I = -ie\int \bar{\psi}\boldsymbol{\gamma}\psi\cdot\mathbf{A}^{(\perp)}\, d^3x + \frac{e^2}{2}\int d^3x_1 \int d^3x_2 \frac{(\bar{\psi}\gamma_4\psi)_{\mathbf{x}_1, t}(\bar{\psi}\gamma_4\psi)_{\mathbf{x}_2, t}}{4\pi|\mathbf{x}_1 - \mathbf{x}_2|}. \tag{4.307}$$

By $(\bar{\psi}\gamma_4\psi)_{\mathbf{x}_1, t}$ we mean $\bar{\psi}(\mathbf{x}_1, t)\gamma_4\psi(\mathbf{x}_1, t)$; the superscript $(\perp)$ stands for "transverse." In this method only the transverse part of A_μ is treated dynamically, and the basic interaction (4.307) is seen to consist of the interaction of the transverse electromagnetic field with $\mathbf{j} = ie\bar{\psi}\boldsymbol{\gamma}\psi$ and the *instantaneous* Coulomb interaction between charge densities. We quantize only the transverse part of A_μ according to the method of Chapter 2 so that

$$A_j^{(\perp)}(x) = \frac{1}{\sqrt{V}}\sum_{\mathbf{k}}\sum_{\alpha=1}^{2}\frac{1}{\sqrt{2\omega}}(\epsilon_j^{(\alpha)} a_{\mathbf{k}\alpha} e^{ik\cdot x} + \epsilon_j^{(\alpha)} a_{\mathbf{k}\alpha}^\dagger e^{-ik\cdot x}). \tag{4.308}$$

When the Dirac field is also quantized, both $\bar{\psi}\boldsymbol{\gamma}\psi$ and $\bar{\psi}\gamma_4\psi$ are to be understood as operators properly antisymmetrized so that $\langle 0|\bar{\psi}\gamma_\mu\psi|0\rangle$ vanishes (cf. Eq. 4.59). The S-matrix expansion of Section 4–2 can now be carried out with H_I taken to be (4.307).

Because the interaction Hamiltonian consists of two terms, the problem of the interaction between two electrons is a little more involved than that of the nucleon-nucleon interaction brought about via the exchange of a scalar meson.

Fortunately the first term of (4.307) is seen to be completely analogous to the space integral of the scalar coupling density (4.276). Therefore, so far as the exchange of transverse photons is concerned, everything we have done with the nucleon-nucleon interaction via one-meson exchange can be carried out just as well for the Møller interaction provided that we replace the identity matrix by γ_i and the coupling constant G by $-ie$. Since the steps analogous to (4.277) through (4.284) are essentially the same, we can immediately write the analog of (4.284)

$$S_{fi}^{(\text{trans})} = (-e)^2\sqrt{(m/E_1'V)(m/E_2'V)(m/E_1V)(m/E_2V)}\int d^4x_1\int d^4x_2$$
$$\times\,[(\bar{u}_1'e^{-ip_1'\cdot x_1}\gamma_i u_1 e^{ip_1\cdot x_1})(\bar{u}_2'e^{-ip_2'\cdot x_2}\gamma_j u_2 e^{ip_2\cdot x_2})$$
$$-(\bar{u}_2'e^{-ip_2'\cdot x_1}\gamma_i u_1 e^{ip_1\cdot x_1})(\bar{u}_1'e^{-ip_1'\cdot x_2}\gamma_j u_2 e^{ip_2\cdot x_2})]\,\langle 0\,|\,T(A_i^{(\perp)}(x_1)A_j^{(\perp)}(x_2))\,|\,0\rangle \tag{4.309}$$

where the T product of transverse electromagnetic fields is defined as in (4.285). Let us now evaluate the vacuum expectation value of the T-product. First, if $t > t'$, we get

$$\langle 0\,|\,T(A_i^{(\perp)}(x)A_j^{(\perp)}(x'))\,|\,0\rangle \overset{t>t'}{=} \langle 0\,|\,A_i^{(\perp)}(x)A_j^{(\perp)}(x')\,|\,0\rangle$$
$$= \frac{1}{V}\sum_{\mathbf{k}}\sum_{\mathbf{k}'}\sum_{\alpha}\sum_{\alpha'}\frac{1}{2\sqrt{\omega\omega'}}\epsilon_i^{(\alpha)}\epsilon_j^{(\alpha')}e^{i(k\cdot x-k'\cdot x')}\langle 0\,|\,a_{\mathbf{k}\alpha}a_{\mathbf{k}'\alpha'}^{\dagger}\,|\,0\rangle$$
$$= \frac{1}{V}\sum_{\mathbf{k}}\sum_{\alpha}\frac{1}{2\omega}\epsilon_i^{(\alpha)}\epsilon_j^{(\alpha)}e^{ik\cdot(x-x')}. \tag{4.310}$$

Comparing this with (4.286), we note that the only new feature is the polarization sum $\sum_{\alpha=1}^{2}\epsilon_i^{(\alpha)}\epsilon_j^{(\alpha)}$. This summation can be readily performed by recalling that $\boldsymbol{\epsilon}^{(1)}, \boldsymbol{\epsilon}^{(2)}$, and $\mathbf{k}/|\mathbf{k}|$ form an orthonormal set of unit vectors. Suppose we *define*

$$\boldsymbol{\epsilon}^{(3)} = \mathbf{k}/|\mathbf{k}|. \tag{4.311}$$

Clearly $\epsilon_i^{(\alpha)}$ with $\alpha = 1, 2, 3$ can be regarded as an element of the orthogonal matrix that transforms three unit vectors along the usual x-, y-, and z-axes into the new set $\boldsymbol{\epsilon}^{(1)}$, $\boldsymbol{\epsilon}^{(2)}$, and $\boldsymbol{\epsilon}^{(3)} = \mathbf{k}/|\mathbf{k}|$; hence we expect to obtain the orthogonality condition‡

$$\sum_{\alpha=1}^{3}\epsilon_i^{(\alpha)}\epsilon_j^{(\alpha)} = \delta_{ij} \tag{4.312}$$

or, equivalently,

$$\sum_{\alpha=1}^{2}\epsilon_i^{(\alpha)}\epsilon_j^{(\alpha)} = \delta_{ij} - k_ik_j/|\mathbf{k}|^2. \tag{4.313}$$

The remaining steps in the evaluation of the vacuum expectation value are completely analogous to (4.286) through (4.288). We get

$$\langle 0\,|\,T(A_i^{(\perp)}(x)A_j^{(\perp)}(x'))\,|\,0\rangle = \frac{1}{(2\pi)^4}\int d^4k\left(\delta_{ij} - \frac{k_ik_j}{|\mathbf{k}|^2}\right)\frac{e^{ik\cdot(x-x')}}{i(k^2-i\epsilon)}. \tag{4.314}$$

‡See, for example, Goldstein (1951), pp. 97–98.

Inserting (4.314) into (4.309) and integrating over d^4x_1 and d^4x_2, we obtain (to order e^2) that part of the covariant $\mathscr{M}$-matrix which is due to the exchange of transverse photons as follows:

$$-i\mathscr{M}_{fi}^{(\text{trans})} = \frac{(-e)^2}{i(q^2 - i\epsilon)}[(\bar{u}_1'\boldsymbol{\gamma}u_1)(\bar{u}_2'\boldsymbol{\gamma}u_2) - (\bar{u}_1'\boldsymbol{\gamma}\cdot\hat{\mathbf{q}}u_1)(\bar{u}_2'\boldsymbol{\gamma}\cdot\hat{\mathbf{q}}u_2)]$$
$$- \frac{(-e)^2}{i(q'^2 - i\epsilon)}[(\bar{u}_2'\boldsymbol{\gamma}u_1)\cdot(\bar{u}_1'\boldsymbol{\gamma}u_2) - (\bar{u}_2'\boldsymbol{\gamma}\cdot\hat{\mathbf{q}}'u_1)(\bar{u}_1'\boldsymbol{\gamma}\cdot\hat{\mathbf{q}}'u_2)], \quad (4.315)$$

where

$$q = p_1 - p_1' = p_2' - p_2, \qquad q' = p_1 - p_2' = p_1' - p_2, \quad (4.316)$$

and $\hat{\mathbf{q}}$ and $\hat{\mathbf{q}}'$ are unit vectors along $\mathbf{q}$ and $\mathbf{q}'$ respectively. This is as far as we shall go with the exchange of transverse photons for the time being.

Let us now look at the second term of (4.307), which represents the instantaneous Coulomb interaction. If we are computing the electron-electron interaction to order e^2 in the amplitude, we need to consider this term only to *first* order. Using the rule for writing $S^{(1)}$, we get

$$S_{fi}^{(\text{Coul})} = -\frac{ie^2}{2}\int dt \int d^3x_1 \int d^3x_2 \frac{\langle e_1', e_2' | (\bar{\psi}\gamma_4\psi)_{\mathbf{x}_1,t}(\bar{\psi}\gamma_4\psi)_{\mathbf{x}_2,t} | e_1, e_2\rangle}{4\pi|\mathbf{x}_1 - \mathbf{x}_2|}$$
$$= -\frac{ie^2}{2}\int d^4x_1 \int d^4x_2 \delta(t_2 - t_1)\frac{\langle e_1', e_2' | \bar{\psi}(x_1)\gamma_4\psi(x_1)\bar{\psi}(x_2)\gamma_4\psi(x_2) | e_1, e_2\rangle}{4\pi|\mathbf{x}_1 - \mathbf{x}_2|} \quad (4.317)$$

for the Coulomb contribution. The manner in which the fermion fields operate on the initial and final states is the same as before except that this time we need not worry about whether t_1 is earlier or later than t_2 since the whole interaction is instantaneous. The factor $\frac{1}{2}$ just cancels with a factor 2 arising from the fact that $\psi(x_2)$ can annihilate electron 1 as well as electron 2 (cf. Eq. 4.282 and 4.283). Hence (4.317) becomes

$$S_{fi}^{(\text{Coul})} = -ie^2\sqrt{\left(\frac{m}{E_1'V}\right)\left(\frac{m}{E_2'V}\right)\left(\frac{m}{E_1V}\right)\left(\frac{m}{E_2V}\right)}\int d^4x_1 \int d^4x_2 \frac{\delta(t_2 - t_1)}{4\pi|\mathbf{x}_2 - \mathbf{x}_1|}$$
$$\times [(\bar{u}_1' e^{-ip_1'\cdot x_1}\gamma_4 u_1 e^{ip_1\cdot x_1}) \times (\bar{u}_2' e^{-ip_2'\cdot x_2}\gamma_4 u_2 e^{ip_2\cdot x_2})$$
$$- (\bar{u}_2' e^{-ip_2'\cdot x_1}\gamma_4 u_1 e^{ip_1\cdot x_1})(\bar{u}_1' e^{-ip_1'\cdot x_2}\gamma_4 u_2 e^{ip_2\cdot x_2})]. \quad (4.318)$$

Let us look just at the direct term since, apart from a minus sign, the exchange term is obtained from the direct term by interchanging the final-state labels $p_1' \longleftrightarrow p_2'$, $s_1' \longleftrightarrow s_2'$. For the space-time dependence, it is convenient to rewrite the product of the plane waves as follows:

$$\exp(-ip_1'\cdot x_1 + ip_1\cdot x_1 - ip_2'\cdot x_2 + ip_2\cdot x_2)$$
$$= \exp[-i(p_2' - p_2)\cdot(x_2 - x_1)] \exp[i(p_1 + p_2 - p_1' - p_2')\cdot x_1]. \quad (4.319)$$

Since the d^4x_2 integration is over all space-time, instead of integrating over d^4x_2, we can integrate over d^4x_3, where $x_3 = x_2 - x_1$. The t_3-integration just kills the δ-function, whereas the d^3x_3-integration gives the three-dimensional Fourier transform of the Coulomb potential, $1/|\mathbf{p}_2' - \mathbf{p}_2|^2$. The d^4x_1-integration gives $(2\pi)^4$ times the four-dimensional δ-function that appears in the definition of the $\mathscr{M}$-

matrix element. Thus, we obtain

$$
\begin{aligned}
-i\mathscr{M}_{fi}^{(\text{Coul})} &= e^2\left[\frac{(\bar{u}_1'\gamma_4 u_1)(\bar{u}_2'\gamma_4 u_2)}{i\,|\mathbf{p}_2' - \mathbf{p}_2|^2} - \frac{(\bar{u}_2'\gamma_4 u_1)(\bar{u}_1'\gamma_4 u_2)}{i\,|\mathbf{p}_1' - \mathbf{p}_2|^2}\right] \\
&= (-e)^2\left[\frac{(\bar{u}_1'\gamma_4 u_1)(\bar{u}_2'\gamma_4 u_2)}{i\,|\mathbf{q}|^2} - \frac{(\bar{u}_2'\gamma_4 u_1)(\bar{u}_1'\gamma_4 u_2)}{i\,|\mathbf{q}'|^2}\right]. \qquad (4.320)
\end{aligned}
$$

We should emphasize at this stage that neither $\mathscr{M}_{fi}^{(\text{trans})}$ nor $\mathscr{M}_{fi}^{(\text{Coul})}$, taken *separately*, is covariant. We shall now show, however, that when we sum the two, we end up with a compact expression which is not only covariant but also much simpler. To see this explicitly, we first rewrite the direct-scattering term of (4.320) as follows:

$$
\begin{aligned}
(-e)^2\frac{(\bar{u}_1'\gamma_4 u_1)(\bar{u}_2'\gamma_4 u_2)}{i\,|\mathbf{q}|^2} &= (-e)^2(\bar{u}_1'\gamma_4 u_1)(\bar{u}_2'\gamma_4 u_2)\left[\frac{|\mathbf{q}|^2 - |q_0|^2 - i\epsilon}{|\mathbf{q}|^2}\right]\frac{1}{i(q^2 - i\epsilon)} \\
&= (-e)^2\left[\frac{(\bar{u}_1'\gamma_4 u_1)(\bar{u}_2'\gamma_4 u_2)}{i(q^2 - i\epsilon)} + \frac{(iq_0)^2}{|\mathbf{q}|^2}\frac{(\bar{u}_1'\gamma_4 u_1)(\bar{u}_2'\gamma_4 u_2)}{i(q^2 - i\epsilon)}\right]. \qquad (4.321)
\end{aligned}
$$

Meanwhile, because of the properties of the free-particle spinor (3.105), we have

$$\bar{u}_1'\gamma\cdot q u_1 = -i\bar{u}_1'[i\gamma\cdot(p_1 - p_1')]u_1 = 0 \qquad (4.322)$$

or

$$iq_0\bar{u}_1'\gamma_4 u_1 = -\bar{u}_1'\boldsymbol{\gamma}\cdot\mathbf{q}u_1. \qquad (4.323)$$

Similarly, remembering (4.316), we get

$$iq_0\bar{u}_2'\gamma_4 u_2 = -\bar{u}_2'\boldsymbol{\gamma}\cdot\mathbf{q}u_2. \qquad (4.324)$$

We therefore find

$$(iq_0)^2(\bar{u}_1'\gamma_4 u_1)(\bar{u}_2'\gamma_4 u_2) = (\bar{u}_1'\boldsymbol{\gamma}\cdot\mathbf{q}u_1)(\bar{u}_2'\boldsymbol{\gamma}\cdot\mathbf{q}u_2), \qquad (4.325)$$

and (4.321) now becomes

$$(-e)^2\frac{(\bar{u}_1'\gamma_4 u_1)(\bar{u}_2'\gamma_4 u_2)}{i\,|\mathbf{q}|^2} = (-e)^2\frac{(\bar{u}_1'\gamma_4 u_1)(\bar{u}_2'\gamma_4 u_2)}{i(q^2 - i\epsilon)} + (-e)^2\frac{(\bar{u}_1'\boldsymbol{\gamma}\cdot\hat{\mathbf{q}}u_1)(\bar{u}_2'\boldsymbol{\gamma}\cdot\hat{\mathbf{q}}u_2)}{i(q^2 - i\epsilon)}. \qquad (4.326)$$

This leads to a remarkable result; the second term of (4.326) cancels exactly the second term in the first bracket of (4.315). A similar cancellation also takes place in the exchange-scattering amplitude. Hence

$$
\begin{aligned}
-i\mathscr{M}_{fi} &= -i(\mathscr{M}_{fi}^{(\text{trans})} + \mathscr{M}_{fi}^{(\text{Coul})}) \\
&= (-e)^2\left[\frac{(\bar{u}_1'\gamma_\mu u_1)(\bar{u}_2'\gamma_\mu u_2)}{i[(p_1 - p_1')^2 - i\epsilon]} - \frac{(\bar{u}_2'\gamma_\mu u_1)(\bar{u}_1'\gamma_\mu u_2)}{i[(p_1 - p_2')^2 - i\epsilon]}\right]. \qquad (4.327)
\end{aligned}
$$

The final result (4.327) is simple and covariant; so is the basic Lagrangian (4.304) of quantum electrodynamics with which we started. This suggests that a somewhat awkward formalism has been used to derive the simple matrix element (4.327). Indeed we could actually have guessed (4.327) much more quickly from semi-classical considerations analogous to (4.294) through (4.298), that is, by arguing that the presence of electron 1 induces a ("classical") four-vector potential $A_\mu^{(1)}$

satisfying (in the Lorentz gauge)

$$\Box A_\mu^{(1)} = -ie\sqrt{(m/E_1'V)(m/E_1V)}\bar{u}_1'\gamma_\mu u_1 e^{i(p_1 - p_1')\cdot x} \tag{4.328}$$

which interacts with electron 2. It was precisely this kind of reasoning that led C. Møller, in 1931, to write the matrix element for electron-electron interaction. The quantum-electrodynamical derivation given here is essentially that of a 1932 paper of H. Bethe and E. Fermi except that we have used the perturbation-theoretic formalism based on covariant denominators rather than that based on energy-difference denominators.

Covariant photon propagator. It is also possible to derive (4.327) more elegantly by constructing a manifestly covariant quantum theory of the Maxwell field. The result (4.327), when compared to (4.293), strongly suggests that this can be accomplished if we quantize not just the transverse parts of A_μ but also its longitudinal and fourth components. Instead of starting with the formalism of Appendix A in which only the transverse parts of A_μ are treated as dynamical fields, let us suppose that we can take all four components of A_μ to be four *independent Hermitian* Klein-Gordon fields of zero mass subject to the usual rules of field quantization. We expand A_μ as follows:

$$A_\mu = (1/\sqrt{V}) \sum_{\mathbf{k}} \sum_{\alpha=1}^{4} (1/\sqrt{2\omega})(a_{\mathbf{k}\alpha}\epsilon_\mu^{(\alpha)} e^{ik\cdot x} + a_{\mathbf{k}\alpha}^\dagger \epsilon_\mu^{(\alpha)*} e^{-ik\cdot x}). \tag{4.329}$$

Note that the α indices now run from 1 to 4. The polarization vectors are so chosen that

$$\begin{aligned} \epsilon^{(1,2)} &= \epsilon^{(1,2)*} = (\hat{\mathbf{n}}^{(1,2)}, 0) \qquad \text{with} \quad \hat{\mathbf{n}}^{(1,2)} \perp \mathbf{k}, \hat{\mathbf{n}}^{(1)} \perp \hat{\mathbf{n}}^{(2)}; \\ \epsilon^{(3)} &= \epsilon^{(3)*} = (\mathbf{k}/|\mathbf{k}|, 0); \qquad \epsilon^{(4)} = -\epsilon^{(4)*} = (0, 0, 0, i) \end{aligned} \tag{4.330}$$

Note the orthogonality relation

$$\sum_{\alpha=1}^{4} \epsilon_\mu^{(\alpha)} \epsilon_\nu^{(\alpha)*} = \delta_{\mu\nu}. \tag{4.331}$$

The creation and annihilation operators are assumed to satisfy the following:

$$[a_{\mathbf{k},\alpha}, a_{\mathbf{k}',\alpha'}^\dagger] = \delta_{\mathbf{k}\mathbf{k}'}\delta_{\alpha\alpha'}, \qquad [a_{\mathbf{k},\alpha}, a_{\mathbf{k}',\alpha'}] = [a_{\mathbf{k}',\alpha}^\dagger, a_{\mathbf{k}',\alpha'}^\dagger] = 0, \tag{4.332}$$

just as in the transverse case. We may call $a_{\mathbf{k},3}^\dagger$ and $a_{\mathbf{k},4}^\dagger$ the creation operators for a longitudinal and a timelike photon respectively. In this formalism we forget about the instantaneous Coulomb interaction; the electron-electron interaction is visualized as taking place via the exchange of *four* types of photons—two transverse, one longitudinal, and one timelike. To compute the $\mathscr{M}$-matrix element for Møller scattering all we need to do is evaluate $\langle 0 | T(A_\mu(x)A_\nu(x')) | 0 \rangle$. This can be easily accomplished because of (4.331) and (4.332); we obtain‡

$$\langle 0 | T(A_\mu(x)A_\nu(x')) | 0 \rangle = \delta_{\mu\nu} \int \frac{d^4k}{(2\pi)^4} \frac{e^{ik\cdot(x-x')}}{i(k^2 - i\epsilon)}, \tag{4.333}$$

‡In the literature one often defines $D_F(x - x')$ by

$$\frac{1}{2} D_F(x - x') = -\frac{i}{(2\pi)^4} \int \frac{d^4k\, e^{ik\cdot(x-x')}}{(k^2 - i\epsilon)} = \lim_{\mu\to 0} \frac{1}{2}\Delta_F(x - x').$$

which is called the *covariant photon propagator.* Using this, we can immediately write the covariant matrix element (4.327). Thus, to obtain the covariant result, it is simpler to use a formalism in which all four components of A_μ are quantized.

There are, however, two important features which we overlooked in the previous paragraph. First, in treating all four components of A_μ as kinematically independent, we have completely disregarded the Lorentz condition; actually this turns out to be not too serious because the source of the electromagnetic field is conserved. Second, we have assumed that $A_4(x)$ is a Hermitian field operator, whereas we know from the correspondence with classical electrodynamics that the expectation value of $A_4(x)$ must be purely imaginary. To overcome these difficulties a rather elaborate formalism has been developed by K. Bleuler and S. Gupta to justify what we have done in the preceding paragraph. The proposal of Bleuler and Gupta is, among other things, to modify the definition of a scalar product in Hilbert space by introducing the notion of indefinite metric (negative probabilities); as a result, $\langle A_4 \rangle$ becomes imaginary despite the hermiticity of the field operator A_4. We shall not discuss this elaborate method in any further detail since we have already learned how to arrive at the correct covariant matrix element (4.327) by using the alternative noncovariant method. A good discussion of the Bleuler-Gupta formalism can be found in many books.‡

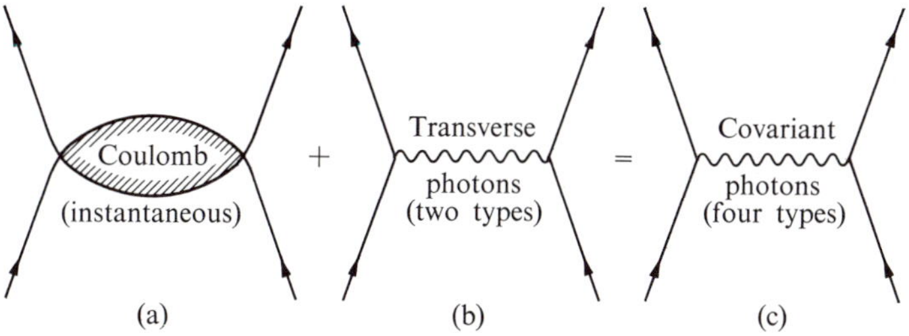

Fig. 4–20. Exchange of "covariant photons."

In any case it is convenient to visualize the covariant matrix element (4.327) as representing the exchange of four kinds of "covariant photons." Looking back at the Coulomb contribution (4.326), we can say that the so-called instantaneous Coulomb interaction is actually equivalent to an interaction due to the emission and subsequent absorption of a time-like photon and a longitudinal photon, considered *together*. Taking into account the possibility that the *four* types of "covariant photons" can be exchanged between two electrons, we can completely forget about the existence of the instantaneous Coulomb interaction. This is indicated pictorially in Fig. 4–20. From now on by a photon we will mean a *covariant photon.*

It can be shown that this equivalence of the instantaneous Coulomb interaction and the exchange of a timelike and a longitudinal photon (taken together) holds

‡See, for example, Mandl (1959), pp. 57–71; Källén (1958), pp. 181–204.

not just for this particular problem of electron-electron interaction treated to order e^2 in the amplitude but also for any problem in quantum electrodynamics so long as the source of the electromagnetic field satisfies the continuity equation. Thus we can add to our previous list of "rules" for writing $-i\mathscr{M}$ the following prescription:

In a process which involves the propagation of a covariant photon, just insert

$$\delta_{\mu\nu}/[i(q^2 - i\epsilon)] \tag{4.334}$$

in the same way as we insert $1/[i(q^2 + \mu^2 - i\epsilon)]$ for a spinless meson and $(-i\gamma\cdot q + m)/[i(q^2 + m^2 - i\epsilon)]$ for a spin-$\frac{1}{2}$ fermion.

We have argued that a *virtual* photon can be visualized as having four states of polarization. On the other hand, we know that a *real* photon, or a free photon, has *only two states* of polarization. At first sight this may appear disturbing; after all the so-called real photon should be the limit of a virtual photon where $|\mathbf{q}|$ approaches q_0. As a matter of fact most real photons of physical interest are, strictly speaking, virtual in the sense that they are emitted at some place and absorbed at some other place. To analyze this feature for the particular problem of electron-electron interaction, we simply note that as $|\mathbf{q}|$ approaches $|q_0|$, the first term of (4.321), which represents the exchange of a timelike photon, tends to cancel the second term of (4.321), which, because of (4.325), represents the exchange of a longitudinal photon; this cancellation becomes complete when the photon becomes real, that is, when $|\mathbf{q}| = |q_0|$. If one wishes, one may say that the photon *always* has four states of polarization regardless of whether it is virtual or real, but that whenever it is real, the timelike photon and the longitudinal photon always give rise to contributions which are equal in magnitude but opposite in sign.

Cross section for Møller scattering. Let us now compute the cross section for Møller scattering. The connection between the differential cross section and the $\mathscr{M}$-matrix element can readily be obtained as in the case of two-photon annihilation. Quite generally, for any elastic fermion-fermion scattering we have in the CM system

$$\left(\frac{d\sigma}{d\Omega}\right)_{\text{CM}} = \frac{1}{4}\frac{1}{(4\pi)^2}\frac{(4m_1m_2)^2}{E_{\text{tot}}^2}|\mathscr{M}_{fi}|^2, \tag{4.335}$$

which is a special case of (4.182). If the initial electron beam is unpolarized, and if we are not interested in observing the final spin states, we must sum over the initial and final spin states and divide the result by four. Our task is to compute $a, b,$ and c in

$$\frac{1}{4}\sum_{s_1}\sum_{s_1'}\sum_{s_2}\sum_{s_2'}|\mathscr{M}_{fi}|^2 = \frac{ae^4}{(p_1 - p_1')^4} + \frac{be^4}{(p_1 - p_2')^4} - \frac{ce^4}{(p_1 - p_1')^2(p_1 - p_2')^2}. \tag{4.336}$$

For the coefficient a we have

$$\begin{aligned} a &= \frac{1}{4}\sum_{s_1}\sum_{s_1'}\sum_{s_2}\sum_{s_2'}(\bar{u}_1\gamma_4\gamma_\nu\gamma_4u_1'\bar{u}_1'\gamma_\mu u_1)(\bar{u}_2\gamma_4\gamma_\nu\gamma_4u_2'\bar{u}_2'\gamma_\mu u_2) \\ &= \frac{1}{4}\mathrm{Tr}[\Lambda^{(e^-)}(\mathbf{p}_1)\gamma_\nu\Lambda^{(e^-)}(\mathbf{p}_1')\gamma_\mu]\,\mathrm{Tr}\,[\Lambda^{(e^-)}(\mathbf{p}_2)\gamma_\nu\Lambda^{(e^-)}(\mathbf{p}_2')\gamma_\mu], \end{aligned} \tag{4.337}$$

where we have used the abbreviation

$$\Lambda^{(e^-)}(\mathbf{p}_1) = (-i\gamma\cdot p_1 + m)/2m, \tag{4.338}$$

etc. The coefficient b can be obtained from (4.337) by the substitution $p_1' \longleftrightarrow p_2'$. Finally for the coefficient c we must evaluate the long trace

$$\begin{aligned} c &= \frac{2}{4} \sum_{s_1} \sum_{s_1'} \sum_{s_2} \sum_{s_2'} (\bar{u}_1'\gamma_\mu u_1)(\bar{u}_2'\gamma_\mu u_2)(\bar{u}_2\gamma_\nu u_1')(\bar{u}_1\gamma_\nu u_2') \\ &= \frac{1}{2} \sum_{s_1} \sum_{s_1'} \sum_{s_2} \sum_{s_2'} (\bar{u}_2'\gamma_\mu u_2)(\bar{u}_2\gamma_\nu u_1')(\bar{u}_1'\gamma_\mu u_1)(\bar{u}_1\gamma_\nu u_2') \\ &= \frac{1}{2} \mathrm{Tr}[\Lambda^{(e^-)}(\mathbf{p}_2')\gamma_\mu \Lambda^{(e^-)}(\mathbf{p}_2)\gamma_\nu \Lambda^{(e^-)}(\mathbf{p}_1')\gamma_\mu \Lambda^{(e^-)}(\mathbf{p}_1)\gamma_\nu]. \end{aligned} \tag{4.339}$$

The evaluation of (4.337) is facilitated if we use (4.80) and its generalization

$$\mathrm{Tr}\,[(\gamma\cdot a)(\gamma\cdot b)(\gamma\cdot c)(\gamma\cdot d)] = 4[(a\cdot b)(c\cdot d) - (a\cdot c)(b\cdot d) + (a\cdot d)(b\cdot c)], \tag{4.340}$$

which can be readily proved. The evaluation of (4.339) may look more formidable, but actually it is not so bad if we use

$$\begin{aligned} \gamma_\mu(\gamma\cdot a)\gamma_\mu &= -2\gamma\cdot a, \\ \gamma_\mu(\gamma\cdot a)(\gamma\cdot b)\gamma_\mu &= 2[(\gamma\cdot a)(\gamma\cdot b) + (\gamma\cdot b)(\gamma\cdot a)] = 4a\cdot b, \\ \gamma_\mu(\gamma\cdot a)(\gamma\cdot b)(\gamma\cdot c)\gamma_\mu &= -2(\gamma\cdot c)(\gamma\cdot b)(\gamma\cdot a), \end{aligned} \tag{4.341}$$

where we are to sum over the μ-indices .These formulas are not difficult to prove using the basic anticommutation relation for the gamma matrices; for instance, the proof of the second formula of (4.341) goes as follows:

$$\begin{aligned} \gamma_\mu(\gamma_\nu a_\nu)(\gamma_\lambda b_\lambda)\gamma_\mu &= [-(\gamma_\nu a_\nu)\gamma_\mu + 2a_\mu]\gamma_\lambda b_\lambda \gamma_\mu \\ &= -(\gamma\cdot a)(-\gamma_\lambda b_\lambda \gamma_\mu + 2b_\mu)\gamma_\mu + 2(\gamma\cdot b)(\gamma\cdot a) \\ &= 2[(\gamma\cdot a)(\gamma\cdot b) + (\gamma\cdot b)(\gamma\cdot a)] \\ &= 4a\cdot b, \end{aligned} \tag{4.342}$$

where we have used $\gamma_\mu\gamma_\mu = 4$ (μ-indices summed). We shall not evaluate these traces in detail since in Problem 4–13 the reader is asked to work out similar expressions for electron-proton scattering. The net result is that we can express a, b, and c in terms of three Lorentz-invariant scalar products

$$\begin{aligned} p_1\cdot p_2 &= p_1'\cdot p_2', \\ p_1\cdot p_1' &= p_2\cdot p_2', \\ p_1\cdot p_2' &= p_2\cdot p_1'. \end{aligned} \tag{4.343}$$

Actually among the three scalar products only two are independent since the three quantities,

$$\begin{aligned} s &= -(p_1 + p_2)^2 = -(p_1' + p_2')^2 = 2m^2 - 2p_1\cdot p_2, \\ t &= -(p_1 - p_1')^2 = -(p_2' - p_2)^2 = 2m^2 + 2p_1\cdot p_1', \\ u &= -(p_1 - p_2')^2 = -(p_1' - p_2)^2 = 2m^2 + 2p_1\cdot p_2', \end{aligned} \tag{4.344}$$

turn out to satisfy one relation‡

$$s + t + u = 4m^2. \tag{4.345}$$

To prove (4.345) we first verify it in the CM-system, where $E = E_1 = E_1' = E_2 = E_2'$, $\mathbf{p} = \mathbf{p}_2 = -\mathbf{p}_1$, $\mathbf{p}' = \mathbf{p}_2' = -\mathbf{p}_1'$, $|\mathbf{p}| = |\mathbf{p}'|$:

$$\begin{aligned} (p_1 + p_2)^2 &= -E_{\text{tot}}^2 = -(2E)^2 = -4(|\mathbf{p}|^2 + m^2), \\ (p_1 - p_1')^2 &= |\mathbf{p} - \mathbf{p}'|^2 = 2|\mathbf{p}|^2(1 - \cos\theta), \\ (p_1 - p_2') &= |\mathbf{p} + \mathbf{p}'|^2 = 2|\mathbf{p}|^2(1 + \cos\theta). \end{aligned} \tag{4.346}$$

Since s, t, and u are invariant scalars, the relation (4.345) which has been verified in the CM-system is valid in *any* coordinate system.

Even if we do not evaluate the traces (4.337) and (4.339), we can immediately write the nonrelativistic limit of the differential cross section for Møller scattering, because $\bar{u}_1'\gamma_k u_1 \approx 0$ and $\bar{u}_1'\gamma_4 u_1 \approx 1$ for $s_1 = s_1'$ and 0 for $s_1 \neq s_1'$ in the nonrelativistic limit ($\mathbf{p} = \mathbf{p}' = 0$). Using (4.327), (4.335), and (4.346), we readily get the spin-average cross section

$$\left(\frac{d\bar{\sigma}}{d\Omega}\right)_{\text{CM}} \overset{\text{NR}}{=} \frac{r_0^2}{16\beta^4}\left[\frac{1}{\sin^4(\theta/2)} + \frac{1}{\cos^4(\theta/2)} - \frac{1}{\sin^2(\theta/2)\cos^2(\theta/2)}\right], \tag{4.347}$$

where β is the velocity of one of the electrons in units of c (not the relative velocity of the two incoming or outgoing electrons). Clearly the first term of (4.347) is just the Rutherford term; the second term is necessary because the whole expression must be symmetric under $\theta \to \pi - \theta$ for the elastic scattering of two identical particles. The third term, which is due to the interference of direct scattering and exchange scattering, clearly shows the effect of the Pauli principle, which tends to reduce the cross section at $\theta \approx \pi/2$. We may mention that the exact expression for the scattering of two nonrelativistic electrons differs from the above perturbation-theoretic result in that the third term of (4.347) is multiplied by cos $\{(\alpha/2\beta)\log[\tan^2(\theta/2)]\}$. The extreme relativistic limit of the Møller scattering cross section is also relatively simple to obtain. In evaluating the indicated traces in (4.337) and (4.339), $\Lambda^{(e-)}(\mathbf{p}_1)$, etc. can be replaced by $-i\gamma\cdot p_1/2m$, etc.; we then get

$$\left(\frac{d\bar{\sigma}}{d\Omega}\right)_{\text{CM}} \overset{\text{ER}}{=} \frac{r_0^2}{8}\left(\frac{m}{E}\right)^2\left[\frac{1+\cos^4(\theta/2)}{\sin^4(\theta/2)} + \frac{1+\sin^4(\theta/2)}{\cos^4(\theta/2)} + \frac{2}{\sin^2(\theta/2)\cos^2(\theta/2)}\right]. \tag{4.348}$$

Effective potential; the Breit interaction. In nonrelativistic wave mechanics (as well as in classical mechanics) we introduce the notion of a potential to be used in conjunction with the Schrödinger equation. In contrast, in relativistic field theory we visualize the interactions as arising from the *exchange of quanta*. We may naturally ask how the two concepts are related. It is not difficult to establish the desired connection for slowly moving particles if we are concerned just with lowest-order

‡The variables s, t, and u are sometimes called the Mandelstam variables after S. Mandelstam, who studied the analytic properties of meson-nucleon scattering amplitudes as functions of $-(p_\pi + p_N)^2$, $-(p_\pi - p_\pi')^2$, and $-(p_\pi - p_N')^2$.

processes. According to the Born approximation in the Schrödinger theory the differential cross section for the elastic scattering of particle 1 (mass m_1) and particle 2 (mass m_2) is given in natural units by‡

$$\left(\frac{d\sigma}{d\Omega}\right)_{\text{CM}} = \left|\frac{m_{\text{red}}}{2\pi}\int e^{-i\mathbf{p}'\cdot\mathbf{x}}V(\mathbf{x})e^{i\mathbf{p}\cdot\mathbf{x}}\,d^3x\right|^2, \tag{4.349}$$

where m_{red} stands for the reduced mass,

$$m_{\text{red}} = \frac{m_1 m_2}{m_1 + m_2}, \tag{4.350}$$

and $\mathbf{p}$ and $\mathbf{p}'$ are the initial and finial momenta of one of the particles (say, particle 2) in the CM-system. This formula can be easily generalized to cases where the particles have spins and the potential V depends on momentum $(-i\boldsymbol{\nabla})$ and spin. Coming back to (4.335), we see that for slowly moving fermions the covariant matrix element $\mathscr{M}_{fi}$ is related to the differential cross section as follows:

$$\left(\frac{d\sigma}{d\Omega}\right)_{\text{CM}} \overset{\text{NR}}{=} \left|\left(\frac{1}{4\pi}\right)\left(\frac{2m_1 m_2}{m_1 + m_2}\right)\mathscr{M}_{fi}\right|^2. \tag{4.351}$$

Comparing (4.349) and (4.351) we obtain

$$\mathscr{M}_{fi} = \int Ve^{-i\mathbf{q}\cdot\mathbf{x}}\,d^3x. \tag{4.352}$$

with

$$\mathbf{q} = \mathbf{p}_1' - \mathbf{p}, \tag{4.353}$$

where we have chosen the phase of (4.352) in such a way that a repulsive force (positive potential) corresponds to a positive $\mathscr{M}_{fi}$ in conformity with the definition of the covariant matrix element (4.292). See also (4.38) and (4.39). Fourier-transforming (4.352) we get

$$V = \frac{1}{(2\pi)^3}\int \mathscr{M}_{fi}e^{i\mathbf{q}\cdot\mathbf{x}}\,d^3q. \tag{4.354}$$

This formula enables us to construct an effective three-dimensional potential (to be used in connection with the Schrödinger equation) once we know the nonrelativistic limit of the covariant matrix element. The essential point is that the so-called potential is nothing more than the three-dimensional Fourier transform of the lowest-order $\mathscr{M}$-matrix element.§ In the case of scattering of two identical fermions the $\mathscr{M}$-matrix element that appears in (4.354) is just that part of the covariant matrix element which corresponds to direct scattering since the use of antisymmetric wave functions in nonrelativistic wave mechanics automatically takes care of the contributions due to exchange scattering.

---

‡See, for example, Dicke and Wittke (1960), pp. 293–295; Merzbacher (1961), pp. 226–227.

§In higher orders, however, a more careful consideration becomes necessary since the iteration of lower-order potentials also gives higher-order matrix elements.

Let us see how the rule expressed in (4.354) works in the case of Møller interaction. In the CM-system the direct term of (4.327) is reduced to

$$\mathscr{M}_{fi}^{(\text{direct})} = \frac{e^2(\bar{u}_1'\gamma_4 u_1)(\bar{u}_2'\gamma_4 u_2)}{|\mathbf{q}|^2} + \frac{e^2(\bar{u}_1'\boldsymbol{\gamma} u_1)(\bar{u}_2'\boldsymbol{\gamma} u_2)}{|\mathbf{q}|^2}. \tag{4.355}$$

At very low energies $\bar{u}_1'\gamma_4 u_1$ can be replaced by $\delta_{s_1 s_1'}$, $\bar{u}_1'\boldsymbol{\gamma} u_2$ by zero; similarly for $\bar{u}_2'\gamma_\mu u_2$. Hence the insertion of (4.355) into (4.354) immediately gives the repulsive Coulomb potential $e^2/(4\pi r)$, as expected from our earlier considerations. To see the physical meaning of the second term of (4.355), which begins to be important when β is not too small compared to unity, we first recall (cf. Eq. 3.215) that

$$\bar{u}_1'\boldsymbol{\gamma} u_1 = \chi^{(s_1')\dagger}\left[-\frac{i(\mathbf{p}_1 + \mathbf{p}_1')}{2m} + \frac{\boldsymbol{\sigma}^{(1)} \times (\mathbf{p}_1' - \mathbf{p}_1)}{2m}\right]\chi^{(s_1)}. \tag{4.356}$$

Note that this vector is "transverse" in the sense that it has no component in the direction $\mathbf{p}_1 - \mathbf{p}_1'$. Calling the first and second terms of (4.356) respectively "current" and "dipole," we see that the second term of (4.355) gives rise to a current-current interaction, a dipole-dipole interaction, and a current-dipole interaction, As an example, we shall consider the dipole-dipole interaction in detail. Remembering $\mathbf{q} = \mathbf{p}_2' - \mathbf{p}_2 = -\mathbf{p}_1' + \mathbf{p}_1$, we can readily take the Fourier transform of the dipole-dipole matrix as follows:

$$\begin{aligned} -\frac{e^2}{(2\pi)^3}\int \frac{(\boldsymbol{\sigma}^{(1)} \times \mathbf{q})\cdot(\boldsymbol{\sigma}^{(2)} \times \mathbf{q})e^{i\mathbf{q}\cdot\mathbf{x}}}{(2m)^2|\mathbf{q}|^2} d^3q & \\ = \left(\frac{e}{2m}\right)^2(\boldsymbol{\sigma}^{(1)} \times \boldsymbol{\nabla})\cdot(\boldsymbol{\sigma}^{(2)} \times \boldsymbol{\nabla})\left(\frac{1}{4\pi r}\right) & \\ = -\frac{e\boldsymbol{\sigma}^{(1)}}{2m}\cdot\boldsymbol{\nabla} \times \left[\frac{e\boldsymbol{\sigma}^{(2)}}{2m} \times \boldsymbol{\nabla}\left(\frac{1}{4\pi r}\right)\right]. & \end{aligned} \tag{4.357}$$

We recognize that $[(e\boldsymbol{\sigma}^{(2)}/2m) \times \boldsymbol{\nabla}(1/4\pi r)]$ is precisely the vector potential generated by the magnetic moment of electron 2; hence (4.357) is just the interaction energy between the magnetic moments of the two electrons expected from classical considerations.‡ As for the current-dipole interaction, which goes like $[\boldsymbol{\sigma}^{(1)} \times (\mathbf{p}_1' - \mathbf{p}_1)]\cdot(\mathbf{p}_2 + \mathbf{p}_2')$, we can use the identity

$$\begin{aligned} \int e^{-i\mathbf{p}'\cdot\mathbf{x}}\frac{1}{r}\frac{dV}{dr}\boldsymbol{\sigma}\cdot[\mathbf{x} \times (-i\boldsymbol{\nabla})]e^{i\mathbf{p}\cdot\mathbf{x}}d^3x & \\ = -i\boldsymbol{\sigma}\cdot(\mathbf{p} \times \mathbf{p}')\int Ve^{-i\mathbf{q}\cdot\mathbf{x}}d^3x & \end{aligned} \tag{4.358}$$

(valid for any spherically symmetric V) to prove that it gives rise to a spin-orbit force (cf. Problem 4–14). The spin-orbit potential that arises in this manner should not be confused with the spin-orbit potential in the Thomas sense, discussed in Section 3–3. The Thomas interaction is actually buried in the *first* term of (4.355)

‡See our earlier discussion on the hyperfine structure of the hydrogen atom given in Section 3–8 (especially Eq. 3.327).

as we may readily verify using

$$(\bar{u}_1'\gamma_4 u_1) \approx \chi^{(s_1')\dagger}\left[1 + \frac{\mathbf{p}_1\cdot\mathbf{p}_1'}{(2m)^2} + \frac{i\boldsymbol{\sigma}^{(1)}\times(\mathbf{p}_1'\times\mathbf{p}_1)}{(2m)^2}\right]\chi^{(s_1)}. \tag{4.359}$$

As a result, the spin-orbit interaction between two electrons is *three* times as strong as the Thomas interaction:

$$V^{(L-S)} = -\frac{3}{2m^2}\frac{1}{r}\frac{d}{dr}\left(\frac{e^2}{4\pi r}\right)\mathbf{L}\cdot\left(\frac{\boldsymbol{\sigma}^{(1)}+\boldsymbol{\sigma}^{(2)}}{2}\right). \tag{4.360}$$

Prior to Møller's work, G. Breit worked out all these correction terms to the Coulomb potential using essentially classical arguments and applied them to the He atom; hence they are collectively known as the Breit interaction between two electrons.

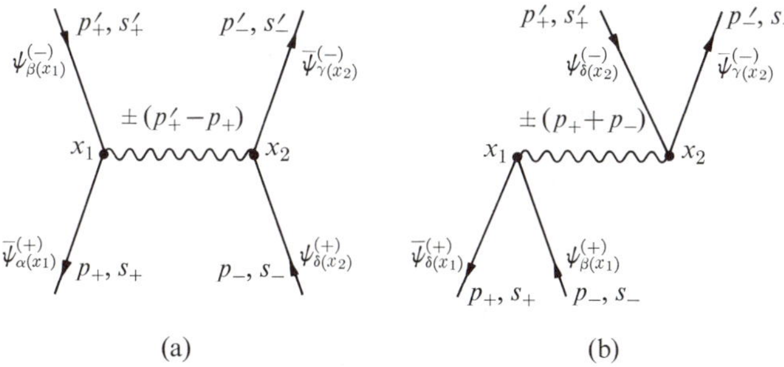

Fig. 4–21. Bhabha scattering: (a) direct diagram, (b) annihilation diagram.

Bhabha scattering; positronium. Having studied the Møller amplitude in detail, we can readily obtain the amplitude for electron-positron scattering, or Bhabha scattering (so called after H. J. Bhabha who treated this problem in 1935). To order e^2 the products of the electron-positron operators that give nonvanishing contributions are

$$\bar{\psi}_\alpha^{(+)}(x_1)\psi_\beta^{(-)}(x_1)\bar{\psi}_\gamma^{(-)}(x_2)\psi_\delta^{(+)}(x_2),$$

$$\bar{\psi}_\alpha^{(-)}(x_1)\psi_\beta^{(+)}(x_1)\bar{\psi}_\gamma^{(+)}(x_2)\psi_\delta^{(-)}(x_2) \underset{x_1\longleftrightarrow x_2}{\overset{\alpha\beta\longleftrightarrow\gamma\delta}{\longleftrightarrow}} \bar{\psi}_\gamma^{(-)}(x_2)\psi_\delta^{(+)}(x_2)\bar{\psi}_\alpha^{(+)}(x_1)\psi_\beta^{(-)}(x_1),$$

$$\bar{\psi}_\alpha^{(+)}(x_1)\psi_\beta^{(+)}(x_1)\bar{\psi}_\gamma^{(-)}(x_2)\psi_\delta^{(-)}(x_2),$$

$$\bar{\psi}_\alpha^{(-)}(x_1)\psi_\beta^{(-)}(x_1)\bar{\psi}_\gamma^{(+)}(x_2)\psi_\delta^{(+)}(x_2) \underset{x_1\longleftrightarrow x_2}{\overset{\alpha\beta\longleftrightarrow\gamma\delta}{\longleftrightarrow}} \bar{\psi}_\gamma^{(-)}(x_2)\psi_\delta^{(-)}(x_2)\bar{\psi}_\alpha^{(+)}(x_1)\psi_\beta^{(+)}(x_1). \tag{4.361}$$

We can see that the first two terms give rise to the fermion lines in Fig. 4–21(a). (Note again that for the second term we interchange $x_1 \longleftrightarrow x_2$ and redefine the Dirac indices so that the second term looks similar to the first one.) The third and fourth terms are somewhat novel. Take, for instance, $\bar{\psi}_\gamma^{(-)}(x_2)\psi_\delta^{(-)}(x_2)$ in the third term; this pair of operators can create an electron-positron pair at x_2. Similarly $\bar{\psi}_\alpha^{(+)}(x_1)\psi_\beta^{(+)}(x_1)$ can annihilate an electron-positron pair at x_1. Thus the third and the fourth terms give rise to the fermion lines in Fig. 4–21(b), known as the *annihi-*

lation diagram. The manner in which the covariant photon propagator joints the products of the Dirac spinors is the same as in the Møller case for both Fig. 4–21(a) and (b). The only treacherous point is the sign of each diagram. To begin let us arrange the first term so that the annihilation (creation) operators stand on the right (left). Apart from an uninteresting term that does not contribute to the scattering process, the first term is just

$$\psi_\beta^{(-)}(x_1)\bar{\psi}_\gamma^{(-)}(x_2)\bar{\psi}_\alpha^{(+)}(x_1)\psi_\delta^{(+)}(x_2). \tag{4.362}$$

This is to be sandwiched between the initial state $b^{(s_-)\dagger}(\mathbf{p}_-)d^{(s_+)\dagger}(\mathbf{p}_+)|0\rangle$ and the final state $b^{(s'_-)\dagger}(\mathbf{p}'_-)d^{(s'_+)\dagger}(\mathbf{p}'_+)|0\rangle$. But

$$\begin{aligned}\bar{\psi}_\alpha^{(+)}(x_1)\psi_\delta^{(+)}(x_2)&b^{(s_-)\dagger}(\mathbf{p}_-)d^{(s_+)\dagger}(\mathbf{p}_+)|0\rangle\\ &=\sqrt{(m/E_+V)(m/E_-V)}\bar{v}_\alpha u_\delta \exp(ip_-\cdot x_2+ip_+\cdot x_1)|0\rangle,\\ \langle 0|d^{(s'_+)}(\mathbf{p}'_+)&b^{(s'_-)}\psi_\beta^{(-)}(x_1)\bar{\psi}_\gamma^{(-)}(x_2)\\ &=-\sqrt{(m/E'_+V)(m/E'_-V)}v'_\beta \bar{u}'_\gamma \exp(-ip'_-\cdot x_2-ip'_+\cdot x_1)\langle 0|,\end{aligned} \tag{4.363}$$

where u_δ, $\bar{u}'_\gamma$, etc. stand for $u_\delta^{(s_-)}(\mathbf{p}_-)$, $\bar{u}_\gamma^{(s'_-)}(\mathbf{p}'_-)$, etc.; we therefore associate a *minus* sign with Fig. 4–21(a). Since the ordering of the electron and positron operators in the third term of (4.361) differs from that of the first term by an odd number of permutations, we must associate a *plus* sign with the annihilation diagram, Fig. 4–21(b). As a result the $\mathscr{M}$-matrix element for Bhabha scattering is written as follows:

$$-i\mathscr{M}_{fi}=(-e)^2\left[-\frac{(\bar{v}'\gamma_\mu v)(\bar{u}'\gamma_\mu u)}{i[(p_+-p'_+)^2-i\epsilon]}+\frac{(\bar{u}'\gamma_\mu v)(\bar{v}'\gamma_\mu u)}{i[(p_++p_-)^2-i\epsilon]}\right]. \tag{4.364}$$

Comparing Fig. 4–21 with the analogous Feynman diagram for Møller scattering [or comparing (4.364) with (4.327)], we see that Bhabha scattering and Møller scattering are related by the simple set of substitutions

$$\begin{aligned} p_1&\longrightarrow -p'_+, & p'_1&\longrightarrow -p_+,\\ p_2&\longrightarrow p_-, & p'_2&\longrightarrow p'_- \end{aligned} \tag{4.365}$$

(apart from an overall minus sign). From this point of view the annihilation diagram of Bhabha scattering is related to the exchange diagram of Møller scattering. Indeed, in the hole-theoretic description the annihilation mechanism of Fig. 4–21(b) corresponds to a process in which the incident electron undergoes exchange scattering with one of the negative-energy electrons in the Dirac sea; hence we expect a relative minus sign between the direct and the annihilation diagrams. We again see that such sign factors are automatically taken care of if we use the second-quantized Dirac theory.

Using (4.354), we can derive the potential corresponding to each of the two terms in (4.364). In the nonrelativistic limit the first term obviously gives a Coulomb potential since $\bar{v}'\gamma_4 v=\delta_{s_+s'_+}$, $\bar{u}'\gamma_4 u=\delta_{s_-s'_-}$, and $\bar{v}'\boldsymbol{\gamma}v=\bar{u}'\boldsymbol{\gamma}u=0$ for $\mathbf{p}_+=\mathbf{p}_-=\mathbf{p}'_+=\mathbf{p}'_-=0$. The sign of the Coulomb potential corresponds to an attraction since the covariant $\mathscr{M}$-matrix element is negative. This, of course, is just what we expect from elementary considerations; e^- and e^+ are known to attract each other. The potential corresponding to the second term is more interesting.

The reader may show in Problem 4–15 that this annihilation term gives rise to a repulsive short-ranged [$\delta^{(3)}(x)$ type] potential that affects just the ^3S-state.

At this point we shall briefly comment on how the perturbation-theoretic matrix elements between initial and final plane-wave states can be used in bound-state problems. Obviously we cannot directly apply perturbation theory to compute bound-state energies. However, we can first obtain approximate bound-state energies by solving exactly the Schrödinger (or the Dirac) equation with a potential that is relatively simple to handle (for example, the Coulomb potential); we then use the usual perturbation-theoretic method to estimate the effect of the difference between the true (total) interaction and the potential interaction already considered.‡ Let us see how this method works in the case of a positronium. We first obtain approximate energy levels of the positronium using the Schrödinger (or the single-particle Dirac) theory with the usual Coulomb potential. Having obtained the Coulomb wave functions and energy levels we can estimate the effects of the dipole-dipole interaction, the annihilation interaction, etc., not previously considered in solving the Schrödinger equation with the $-e^2/(4\pi r)$ potential. Using this method we readily obtain the hyperfine splitting of the ground state of positronium as follows (Problem 4–15):§

$$E(n=1, {}^3S) - E(n=1, {}^1S) = (\tfrac{2}{3} + \tfrac{1}{2})\alpha^2 Ry_\infty, \tag{4.366}$$

which is numerically equal to 8.2×10^{-4} eV. In this formula, Ry_∞ represents $(\alpha^2/2)m$ as usual; the $\frac{2}{3}$ is due to the dipole-dipole interaction contained in Fig. 4–21(a), whereas the $\frac{1}{2}$ arises from the repulsion (in the ^3S state only) caused by the annihilation force represented by Fig. 4–21(b). We may mention parenthetically that higher-order corrections due to the anomalous magnetic moment, the annihilation force involving two virtual photons, etc., slightly modify this formula so that the final result is

$$E(n=1, {}^3S) - E(n=1, {}^1S) = \alpha^2 Ry_\infty\left[\frac{2}{3} + \frac{1}{2} - \frac{\alpha}{\pi}\left(\log 2 + \frac{16}{9}\right)\right], \tag{4.367}$$

as shown by R. Karplus and A. Klein. Numerically formula (4.367) gives 2.0337 $\times\ 10^5$ Mc for the microwave frequency corresponding to the ^3S–^1S energy difference. This theoretical value is in an exceedingly good agreement with the experimental value, $(2.0338 \pm 0.0004) \times 10^5$ Mc, obtained by M. Deutsch and collaborators. In any case, even setting aside the small corrections of order $\alpha^3 Ry_\infty$, it is abundantly clear that we must take into account the annihilation force of

‡An alternative, more formal, procedure that leads to the same results makes use of what is known as the *bound interaction representation.* See, for example, Jauch and Rohrlich (1955) pp. 306–321.

§Although the hyperfine splitting of the hydrogen atom is smaller than the fine structure splitting by a factor of about (m_e/m_p), (see Eq. 3.328), the hyperfine splitting and fine-structure splittings are of the same order for positronium. For this reason the hyperfine splitting of positronium is often referred to simply as a fine-structure splitting.

Fig. 4-21(b) (which has no analog in the hydrogen atom) if we are to obtain the correct splitting between the two hyperfine states of positronium. The necessity of the annihilation force has also been established experimentally in Bhabha scattering at $p^{(\text{lab})} \sim 3m$.

One-pion exchange potential. The relation (4.354) can also be applied to meson theory to derive so-called one-meson exchange potentials. In the case of scalar coupling the first term of the Fourier transform of the covariant matrix element (4.293) immediately leads to the familiar Yukawa potential

$$V^{(S)} = -\frac{G^2}{4\pi}\frac{e^{-\mu r}}{r}, \tag{4.368}$$

obtained in Section 1-3 from classical considerations. Note that the sign of the force between two nucleons due to the exchange of a scalar meson is *attractive*, in contrast to the Coulomb case.

It is known empirically that the lightest meson that can be exchanged between two nucleons is a *pseudoscalar* π meson. If we use the pseudoscalar coupling

$$\mathscr{H}_{\text{int}} = iG\bar{\psi}\gamma_5\psi\phi \tag{4.369}$$

in place of the scalar coupling, we obtain for the direct scattering part of the covariant matrix element the following:

$$-i\mathscr{M}_{fi}^{(\text{dir})} = \frac{G^2(\bar{u}_1'\gamma_5 u_1)(\bar{u}_2'\gamma_5 u_2)}{i[(p_1 - p_1')^2 + \mu^2]}. \tag{4.370}$$

Comparing this with the analogous expression in the Møller case, we see that, apart from the finite mass of the exchanged boson, the only change necessary is $-e\gamma_\mu \to G\gamma_5$. In the nonrelativistic limit we just have

$$\bar{u}_1'\gamma_5 u_1 = \left(\chi^{(s_1')\dagger}, -\chi^{(s_1')\dagger}\frac{\boldsymbol{\sigma}^{(1)}\cdot\mathbf{p}_1'}{2m}\right)\begin{pmatrix} 0 & -I \\ -I & 0\end{pmatrix}\begin{pmatrix}\chi^{(s_1)} \\ \dfrac{\boldsymbol{\sigma}^{(1)}\cdot\mathbf{p}_1}{2m}\chi^{(s_1)}\end{pmatrix} = -\chi^{(s_1')\dagger}\frac{\boldsymbol{\sigma}^{(1)}\cdot\mathbf{q}}{2m}\chi^{(s_1)}, \tag{4.371}$$

and

$$\bar{u}_2'\gamma_5 u_2 = \chi^{(s_2')\dagger}\frac{\boldsymbol{\sigma}^{(2)}\cdot\mathbf{q}}{2m}\chi^{(s_2)}. \tag{4.372}$$

So (4.370) leads to

$$\mathscr{M}_{fi}^{(\text{dir})} \xrightarrow{NR} -\frac{G^2(\boldsymbol{\sigma}^{(1)}\cdot\mathbf{q})(\boldsymbol{\sigma}^{(2)}\cdot\mathbf{q})}{(2m)^2(|\mathbf{q}|^2 + \mu^2)}, \tag{4.373}$$

where m and μ are respectively the masses of the nucleon and the meson.‡ This

---

‡In place of the pseudoscalar coupling we may use a pseudovector coupling

$$\mathscr{H}_{\text{int}} = \frac{iF}{\mu}\bar{\psi}\gamma_5\gamma_\mu\psi\frac{\partial}{\partial x_\mu}\phi.$$

So far as the one-meson exchange interaction is concerned, the result is identical to (4.373), provided we make the replacement $(G^2/4\pi) \to (F^2/4\pi)(2m/\mu)^2$.

gives rise to the potential

$$V^{(PS)} = \frac{G^2}{4\pi}\frac{1}{(2m)^2}(\boldsymbol{\sigma}^{(1)}\cdot\nabla)(\boldsymbol{\sigma}^{(2)}\cdot\nabla)\frac{e^{-\mu r}}{r}. \tag{4.374}$$

It is not difficult to show by direct differentiation that for $r \neq 0$, (4.374) can also be written as

$$V^{(PS)} = \frac{G^2}{4\pi}\left(\frac{\mu}{2m}\right)^2\left\{(\boldsymbol{\sigma}^{(1)}\cdot\boldsymbol{\sigma}^{(2)}) + \left[\frac{1}{(\mu r)^2} + \frac{1}{(\mu r)} + \frac{1}{3}\right]S_{12}\right\}\frac{e^{-\mu r}}{r}, \tag{4.375}$$

where S_{12} is the well-known tensor operator‡

$$S_{12} = [3(\boldsymbol{\sigma}^{(1)}\cdot\mathbf{x})(\boldsymbol{\sigma}^{(2)}\cdot\mathbf{x})/r^2] - \boldsymbol{\sigma}^{(1)}\cdot\boldsymbol{\sigma}^{(2)}. \tag{4.376}$$

The potential (4.375) is appropriate for describing the exchange of a neutral pseudoscalar meson, for example, the p-p (or n-n) interaction due to π^0 exchange. In discussing the p-n case, however, we must take into account the fact that the neutron (proton) can emit a $\pi^-(\pi^+)$ meson to become a proton (neutron), and that the proton (neutron) can absorb a $\pi^-(\pi^+)$ meson to become a neutron (proton). It turns out that the principle of charge independence in nuclear physics requires that the coupling constant for $\mathrm{n} \longleftrightarrow \mathrm{p} + \pi^-$, for instance, be related to that for $\mathrm{p} \longleftrightarrow \mathrm{p} + \pi^0$ by a simple numerical factor. As discussed in almost any textbook of high-energy nuclear physics, the pion-nucleon interaction consistent with charge independence may be taken to be

$$\mathscr{H}_{\text{int}} = iG[\sqrt{2}\,\phi_{\text{ch}}\bar{\psi}_{\text{n}}\gamma_5\psi_{\text{p}} + \sqrt{2}\,\phi_{\text{ch}}^{\dagger}\bar{\psi}_{\text{p}}\gamma_5\psi_{\text{n}} + \phi_3(\bar{\psi}_{\text{p}}\gamma_5\psi_{\text{p}} - \bar{\psi}_{\text{n}}\gamma_5\psi_{\text{n}})], \tag{4.377}$$

where ϕ_3, ϕ_{ch}, and $\phi_{\text{ch}}^{\dagger}$ are the field operators respectively for π^0, π^-, and π^+ (for example, ϕ_{ch} annihilates π^- and creates π^+).§ If we compute the one-pion-exchange potential using (4.377), we obtain the potential (4.375) multiplied by the isospin factor (first derived by N. Kemmer in 1938):

$$\boldsymbol{\tau}^{(1)}\cdot\boldsymbol{\tau}^{(2)} = \begin{cases} 1 & \text{for } T = 1, \\ -3 & \text{for } T = 0, \end{cases} \tag{4.378}$$

where T stands for the total isospin of the two-nucleon system; $T = 1$ for ^1S, ^3P, ^1D, etc., and $T = 0$ (not accessible to p-p, n-n) for ^3S, ^1P, ^3D, etc.

As we shall see in a moment, the interaction constant G that appears in (4.377) is rather large; therefore the use of perturbation-theoretic arguments is, in general, not very fruitful in meson theory. The one-pion exchange potential which we have derived, however, turns out to be of some use when we discuss the collision of two nucleons with large impact parameters or, equivalently, when we consider the higher l partial-wave amplitudes of the nucleon-nucleon scattering matrix. The

‡From the transformation properties of S_{12} under various symmetry operations we infer that this operator can change the relative orbital-angular momentum of the two-nucleon system without changing the total angular momentum, parity, and the symmetry of the spin-wave function; hence it connects, for instance, the $^3S_1$ state with the $^3D_1$ state. See, for example, Blatt and Weisskopf (1952), pp. 96–108; Preston (1952), pp. 77–83.

§See, for example, Nishijima (1964), pp. 75–89; Sakurai (1964), pp. 197–222.

reason is that the exchange of many pions is unimportant at distances considerably larger than $1/\mu \approx 1.4 \times 10^{-13}$ cm. Crudely speaking, this is a direct consequence of the uncertainty principle which requires that the energy-conservation violating emission and reabsorption of an object with mass M can take place with an appreciable probability only for a time interval comparable to or shorter than $1/M$, giving rise to a potential whose range is of the order $1/M$. (A more quantitative argument shows that the two-pion exchange and the three-pion exchange potential indeed fall off as $e^{-2\mu r}/r$ and $e^{-3\mu r}/r$ at large distances.) In other words, we expect that the tail end of the nucleon-nucleon potential is correctly given by the one-pion exchange potential, a point emphasized by M. Taketani and others. Meanwhile because of the centrifugal barrier, the higher l-waves are sensitive only to the tail end of the nucleon-nucleon potential. Therefore from the higher l phase shifts of the nucleon-nucleon scattering amplitude we can estimate the pion-nucleon coupling constant G, as has been done by M. J. Moravcsik and coworkers. The coupling constant determined in this way turns out to be

$$G^2/4\pi \approx 14, \tag{4.379}$$

which agrees rather well with the value determined in entirely different ways, for example, from the dispersion relations for pion-nucleon scattering and low-energy photoproduction of charged and neutral pions.

Another feature of the one-pion exchange interaction worth taking seriously is its *tensor force*. It turns out that the exchanges of two pions and other heavy mesons do not contribute much to the tensor force; hence one-pion exchange is the main contributor to the tensor force known to exist between two nucleons. In the case of ${}^3S(T=0)$, the sign of the tensor force predicted by the one-pion exchange potential is attractive since $\boldsymbol{\tau}^{(1)}\cdot\boldsymbol{\tau}^{(2)} = -3$. The sign obtained in this way correctly accounts for the sign of the quadrupole moment of the deuteron (which is "cigar-like" rather than "tangerine-like"). For this reason, among the various meson theories proposed in the 1930's, the theory based on pseudoscalar particles of isospin unity (that is, with electric charges 0, $\pm e$) was preferred by J. Schwinger and others long before the discovery of the pion.

Vector meson exchange. In this section we discussed in detail the propagator rules for a spin-zero meson and a photon. In nature there also exist *spin-one* (vector) mesons with nonvanishing rest masses (ρ, ω, ϕ, K^*); it is therefore not entirely academic to obtain the propagator rule for a spin-one meson. The analog of (4.334) for a finite-rest-mass, spin-one meson turns out to be

$$\frac{\delta_{\mu\nu} + (q_\mu q_\nu/\mu^2)}{i(q^2 + \mu^2 - i\epsilon)}. \tag{4.380}$$

The derivation of this rule from quantum field theory is somewhat involved. We can, however, make (4.380) plausible by looking at the wave equation for a spin-one particle coupled to a source j_μ (cf. Problem 1–4):

$$\Box\phi_\mu - \frac{\partial}{\partial x_\nu}\left(\frac{\partial\phi_\nu}{\partial x_\mu}\right) - \mu^2\phi_\mu = j_\mu. \tag{4.381}$$

Taking the divergence of both sides, we get

$$-\mu^2\frac{\partial\phi_\mu}{\partial x_\mu}=\frac{\partial}{\partial x_\mu}j_\mu. \tag{4.382}$$

The wave equation (4.381) therefore becomes

$$\begin{aligned}\Box\phi_\mu-\mu^2\phi_\mu&=j_\mu+\frac{\partial}{\partial x_\mu}\left(\frac{\partial\phi_\nu}{\partial x_\nu}\right)\\&=j_\mu-\frac{1}{\mu^2}\frac{\partial}{\partial x_\mu}\frac{\partial}{\partial x_\nu}j_\nu\\&=\left(\delta_{\mu\nu}-\frac{1}{\mu^2}\frac{\partial^2}{\partial x_\mu\partial x_\nu}\right)j_\nu.\end{aligned} \tag{4.383}$$

We may now regard the last line of (4.383) as a "new source" and proceed as in (4.294) through (4.298). In this way we obtain the spin-one meson propagator (4.380) in momentum space. If the vector meson is coupled to a conserved source, we can, of course, forget about the $q_\mu q_\nu/\mu^2$ term in (4.380) since $(\partial/\partial x_\nu)j_\nu$ vanishes.

It is evident that if the source of the vector meson is proportional to $i\bar{\psi}\gamma_\mu\psi$, then the potential due to the exchange of a vector meson is similar to the photon exchange potential discussed in connection with the Møller interaction except for the occurrence of $e^{-\mu r}/r$ in place of $1/r$; in particular it contains a *repulsive* central potential and an associated spin-orbit potential of the same sign as the Thomas potential. Proton-proton scattering experiments performed by 0. Chamberlain, E. Segrè, T. Ypsilantis, and others in 1956 revealed the presence of a short-ranged repulsion and a spin-orbit force which are somewhat difficult to explain using a meson theory based on the exchange of uncorrelated pseudoscalar pions. It was then conjectured by G. Breit and J. J. Sakurai that a unified explanation of the short-ranged repulsion and the spin-orbit force would be possible if there existed vector mesons of masses several times the pion mass. (The posssible existence of such vector mesons was also conjectured by Y. Nambu and by W. R. Frazer and J. Fulco to explain the electromagnetic structure of the nucleon.) Subsequent experiments by B. C. Maglić, W. D. Walker, and others have confirmed that there do indeed exist heavy vector mesons, known as ω^0 and $\rho^{\pm,0}$, that can be exchanged between two nucleons.

4–7. MASS AND CHARGE RENORMALIZATION; RADIATIVE CORRECTIONS

The last section of this book will be devoted to higher-order processes in quantum electrodynamics. We tacitly assume that the Feynman rules which we have derived from field theory for certain second-order processes can be generalized to more complicated cases. The field-theoretic proof of the Feynman rules for the most general nth order S-matrix is not entirely trivial. Just as there are two (2!) ways of writing $S^{(2)}$ [as indicated by (4.48)], there are $n!$ alternative ways of writing $S^{(n)}$. In any given problem we can cleverly manipulate the time orderings of the various field operators in the S-matrix element so that the end result can be expressed simply in terms of the electron and/or the photon propagators and the wave functions of annihilated and created particles. To achieve this in a systematic way it is convenient to use a formalism (based on chronological and normal products) developed by F. J. Dyson and G. C. Wick (discussed in almost any standard text on quantum field theory).‡

Electron self-energy. Before we discuss high-order processes such as radiative corrections to the external potential problem, it is advantageous to look at the self-energy of the relativistic electron using the rules of covariant perturbation theory. The relevant matrix element (to order e^2) is represented by Fig. 4–22 in which the internal wavy line represents the propagation of a covariant photon. This single diagram represents in old-fashioned language both the instantaneous interaction of the electron with the Coulomb field created by the electron itself and the self-interaction due to the emission and reabsorption of a virtual transverse photon. (We recall that for a *nonrelativistic* electron a self-energy effect of the latter type was already considered in Section 2–8.) Note also that when $t_2 > t_1$, Fig. 4–22 represents a virtual process in which there is an electron-positron pair in the intermediate state. (Pair creation at x_1 is followed by pair annihilation at x_2.)

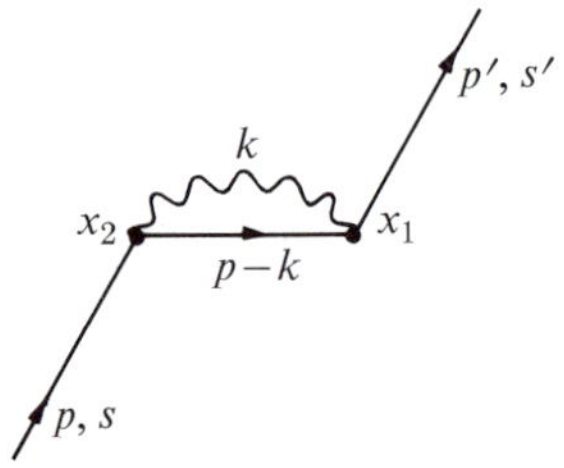

Fig. 4–22
Electron self-energy.

Using the Feynman rules we can write the second-order matrix element for an electron in state $(\mathbf{p}, s)$ to make a transition to state $(\mathbf{p}', s')$ in the absence of any external field as follows:

$$S_{fi} = -i(2\pi)^4\delta^4(p - p')\sqrt{(m/EV)(m/E'V)}\bar{u}^{(s')}(\mathbf{p}') \Sigma\,(p)u^{(s)}(\mathbf{p}), \quad (4.384)$$

where $\Sigma\,(p)$ is a 4×4 matrix given by

$$-\Sigma\,(p) = (-e)^2 \int \frac{d^4k}{(2\pi)^4}\frac{1}{i(k^2 - i\epsilon)}\gamma_\mu \frac{-i\gamma\cdot(p-k)+m}{i[(p-k)^2 + m^2 - i\epsilon]}\gamma_\mu. \quad (4.385)$$

Note that there is one nontrivial integration to be performed in four-dimensional momentum space. This integration is expected since the virtual photon can have an arbitrary four-momentum even if we assume energy-momentum conservation at the two vertices.

‡A particularly lucid discussion of the Dyson-Wick formalism can be found in Mandl (1959), pp. 87–104. Akhietzer and Berestetskii (1965), pp. 309–316, discuss in detail the reduction of the most general third-order S-matrix element in quantum electrodynamics.

It can be seen, on invariance grounds, that $\sum(p)$ can be expressed as a linear combination of the identity matrix and the 4×4 matrix $\gamma \cdot p$. It is more convenient, however, to consider an expansion of $\sum(p)$ about the point $-i\gamma \cdot p = m$ of the form

$$\Sigma(p) = A + (i\gamma \cdot p + m)B + (i\gamma \cdot p + m)^2 \Sigma^f(p), \tag{4.386}$$

where A, B, and $\Sigma^f(p)$ do not contain gamma matrices. Clearly, when $\sum(p)$ is sandwiched between free-particle spinors as in (4.384), only the A term contributes. In order to understand the physical meaning of A let us consider a new "fictitious interaction"

$$\mathscr{H}_{\text{int}} = \delta m \bar{\psi} \psi. \tag{4.387}$$

Note that this can also be regarded as an additive contribution to the mass term of the free-field Dirac Hamiltonian. The first-order S-matrix element induced by this fictitious interaction is just

$$S_{fi} = -i(2\pi)^4 \delta^4(p - p')\sqrt{(m/EV)(m/E'V)}\delta m \bar{u}^{(s')}(\mathbf{p}')u^{(s)}(\mathbf{p}). \tag{4.388}$$

If we replace $\Sigma(p)$ by A in (4.384) and then compare it with (4.388), we readily see that A can be understood as a change in the electron mass due to the interaction mechanism represented by Fig. 4–22 according to the relation

$$A = \delta m. \tag{4.389}$$

So far as the free electron is concerned, a theory in which we start with $m = m_{\text{bare}}$ (the mass the electron would have if e were zero) and take into account the self-energy diagram Fig. 4–22, is equivalent to a theory in which we start with $m = m_{\text{bare}} + \delta m$ and omit the diagram in Fig. 4–22. When the electron is not free, however, the self-energy diagram has a further effect, as we shall see later. Therefore, in more general cases, it is better to start with

$$m = m_{\text{bare}} + \delta m, \tag{4.390}$$

take into account the self-energy diagram, and *subtract* the effect due to the fictitious interaction (4.387). To put it another way, when we use the *observed* mass, $m_{\text{bare}} + \delta m$, in the free-field equation, the full interaction density is given by the usual interaction density *minus* $\delta m \bar{\psi} \psi$ so that we must consider *both* Figs. 4–23(a)

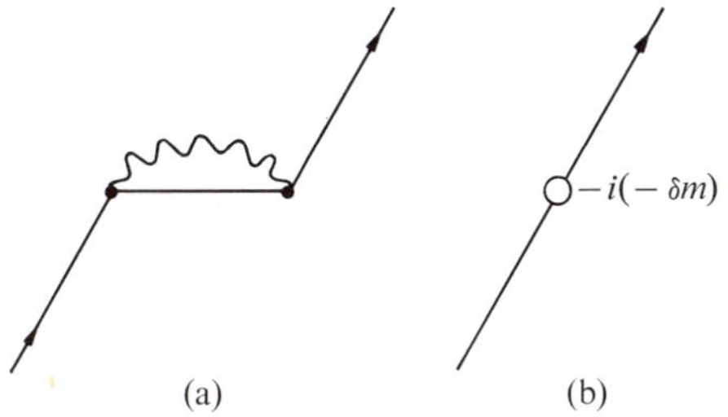

Fig. 4–23. Diagrams to be considered when we use the *observed* mass in the free-field equation.

and (b), where the latter represents the *negative* of the first-order effect of the interaction density (4.387). Comparing this with what we did in Section 2–8, we see that this subtraction procedure (first formulated by Z. Koba and S. Tomonaga in 1947) is a covariant generalization of H. A. Kramer's idea of mass renormalization.

As we shall show in Appendix E, the constants A and B in (4.386) are both infinite although $\sum^f(p)$ can be shown to be finite. These infinities arise from the fact that the energy-momentum of virtual photons, represented in Fig. 4–22, can take an arbitrarily high value. These divergence difficulties are, in fact, one of the most unsatisfactory features of present-day theoretical physics. We are forced to argue that at short distances the emission and absorption of virtual photons of very high four-momenta are not correctly accounted for by the usual propagator rule. On the other hand, we do know that the propagator rule works well when the magnitude of k^2 is small. Therefore, to obtain a finite result for $\sum$, we may proceed by artificially modifying the Fourier transform of the covariant photon propagator according to the prescription

$$\frac{\delta_{\mu\nu}}{i(k^2 - i\epsilon)} \longrightarrow \frac{\delta_{\mu\nu}}{i(k^2 - i\epsilon)} C(k^2), \tag{4.391}$$

where the function $C(k)^2$ (known as the cut-off function) satisfies

$$\begin{aligned} C(k^2) &\approx 1 \quad \text{for} \quad |k^2| \ll \Lambda^2, \\ C(k^2) &\approx 0 \quad \text{for} \quad |k^2| \gg \Lambda^2. \end{aligned} \tag{4.392}$$

In (4.392) the (positive and real) constant Λ is assumed to be much greater than the electron mass (and, as we shall discuss later, probably larger than the muon mass by at least an order of magnitude). There are, of course, many functions which we can use to represent $C(k^2)$; a particularly simple one (which is actually used in Appendix E to compute A) is

$$C(k^2) = \frac{\Lambda^2}{\Lambda^2 + k^2}. \tag{4.393}$$

Inserting the modified photon propagator (4.391) into (4.385), we can express the (formally divergent) constant A in terms of Λ. This calculation is done in Appendix E. The result is

$$A = \frac{3\alpha m}{2\pi} \log\left(\frac{\Lambda}{m}\right). \tag{4.394}$$

That the self-energy of the free Dirac electron is positive and logarithmically dependent on the cut-off with coefficient $3\alpha m/2\pi$ was first demonstrated by V. F. Weisskopf in 1939.

It is instructive to contrast this result with that for the self-energy effects of a nonrelativistic electron. In the nonrelativistic (Schrödinger) theory the Coulomb part of the electron self-energy is just the expectation value of

$$(e^2/2)[1/(4\pi|\mathbf{x} - \mathbf{x}'|)]$$

as $\mathbf{x}'$ approaches $\mathbf{x}$:

$$
\begin{aligned}
E_{\text{self}}^{\text{Coul}} &= \frac{e^2}{2}\int \psi_{\text{NR}}^*(\mathbf{x})\left[\int \frac{d^3x'}{4\pi\,|\mathbf{x}-\mathbf{x}'|}\delta^{(3)}(\mathbf{x}-\mathbf{x}')\right]\psi_{\text{NR}}(\mathbf{x})\,d^3x \\
&= \frac{e^2}{2}\int \psi_{\text{NR}}^*(\mathbf{x})\left[\int d^3p\int \frac{d^3x'}{(2\pi)^3}\frac{e^{i\mathbf{p}\cdot(\mathbf{x}-\mathbf{x}')}}{4\pi\,|\mathbf{x}-\mathbf{x}'|}\right]\psi_{\text{NR}}(\mathbf{x})\,d^3x \\
&= \frac{e^2}{2(2\pi)^3}\int \frac{d^3p}{|\mathbf{p}|^2} \\
&= \frac{\alpha}{\pi}\int d|\mathbf{p}|. \qquad (4.395)
\end{aligned}
$$

This diverges *linearly* with $|\mathbf{p}|$. We saw in Section 2–8 that the self-energy of a nonrelativistic electron due to the emission and absorption of transverse photons also diverges linearly with the virtual photon momentum (cf. Eq. 2.248). A qualitative argument as to why the electron self-energy is much less divergent in the relativistic case goes as follows: In the Dirac electron theory we must take into account the creation of a virtual electron-positron pair in the vicinity of the electron; because of the Pauli exclusion principle, the original electron gets effectively repelled by the virtual electron. This gives rise to a spread-out charge distribution for the physical electron, hence a smaller self-energy.‡ Note that the characteristic distance at which this pair effect takes place is of the order of $1/m$, which is 137 times larger than the classical radius of the electron, a distance in the vicinity of which the classical theory (and also the nonrelativistic Schrödinger theory) of the electron runs into difficulty. For this reason it is completely unrealistic to discuss the electron self-energy within the framework of classical electrodynamics.

Suppose we ask what value of Λ we must have in order that the self-energy (4.394) represents a substantial part of the observed electron mass. Clearly we get

$$\Lambda \sim me^{2\pi/3\alpha} \approx 10^{100}m, \qquad (4.396)$$

which corresponds to a distance of 10^{-111} cm. This value for the cut-off Λ is *ridiculously* high; in this connection we may mention that the mass of the entire universe is estimated to be about $10^{80}m$. There will certainly be a lot of "new physics" before the energy-momentum of a virtual photon reaches such a high value. Thus the idea of attributing most of the observed electron mass to the interaction mechanism represented by Fig. 4–22 is a very dubious proposition. In what follows we assume that even though $A = \delta m$ is formally divergent, it is still "small" in the sense that it is of the order of $\frac{1}{137}$ times the electron mass. Note in this connection that even if we set $\Lambda = 10$ BeV, A is only $0.015\,m$.

To understand the physical meaning of the coefficient B in the expansion (4.386) it is profitable to consider a diagram in which the self-energy interaction of Fig.

‡It is amusing to remark in this connection that the electromagnetic self-energy of a charged Klein-Gordon particle can be shown to diverge *quadratically* with Λ.

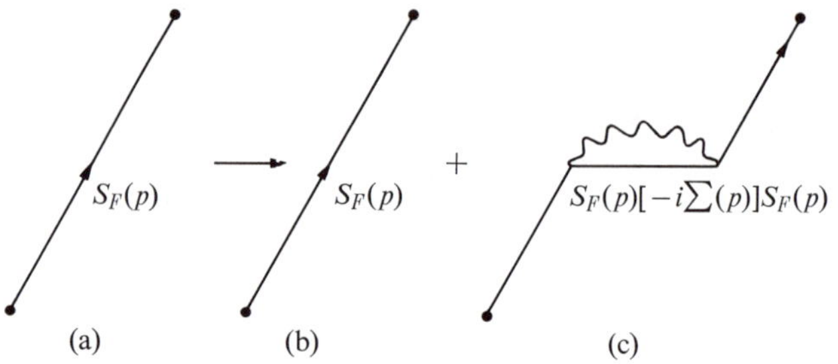

Fig. 4–24. Modification of the electron propagator.

4–22 now appears as *part* of a more complicated diagram. To this end let us first look at an internal electron line represented by Fig. 4–24(a). To simplify the notation we define

$$S_F(p) = \frac{-i\gamma\cdot p + m}{i(p^2 + m^2)} = \frac{1}{i(i\gamma\cdot p + m)}, \tag{4.397}$$

where the $-i\epsilon$ prescription is implicit. The question which we now ask ourselves is how $S_F(p)$ changes in the presence of a self-energy interaction. Clearly, for every internal electron line of the kind shown in Fig. 4–24(a) we can consider a higher-order diagram represented by Fig. 4–24(c). This means that the electron propagator (in momentum space) is modified as follows:

$$S_F(p) \longrightarrow S_F(p) + S_F(p)[-i\Sigma(p)]S_F(p), \tag{4.398}$$

where the m that appears in $S_F(p)$ is understood to be the *bare mass*. Before we evaluate this, we note that for any operators X and Y (which need not commute), we have the useful identity

$$\frac{1}{X+Y} = \frac{1}{X} - \frac{1}{X}Y\frac{1}{X} + \frac{1}{X}Y\frac{1}{X}Y\frac{1}{X} - \cdots, \tag{4.399}$$

as can be verified by multiplying by $X + Y$ from the right. Let us insert in (4.398) the expression for $\Sigma(p)$ given by (4.386), ignoring the finite $\Sigma^f(p)$ term.‡ We see that the modification (4.398) amounts to the substitution

$$\frac{1}{i(i\gamma\cdot p + m)} \longrightarrow \frac{1}{i(i\gamma\cdot p + m)} + \frac{1}{i(i\gamma\cdot p + m)}(-iA)\frac{1}{i(i\gamma\cdot p + m)} - \frac{B}{i(i\gamma\cdot p + m)}$$

$$= \frac{(1-B)}{i[i\gamma\cdot p + (m + A)]} + 0(\alpha^2), \tag{4.400}$$

where we have taken advantage of (4.399). As stated earlier, both A and B are dependent on a cut-off parameter and become infinite as the cut-off is made to go to infinity. Nevertheless the modified propagator [the last line of (4.400)] is strikingly

‡It can be shown that the $\Sigma^f(p)$ term gives a momentum-transfer-dependent structure to the electron propagator when p^2 is very far from $-m^2$; because of its finiteness, $\Sigma^f(p)$ is not relevant for the purpose of circumventing the divergence difficulties of quantum electrodynamics.

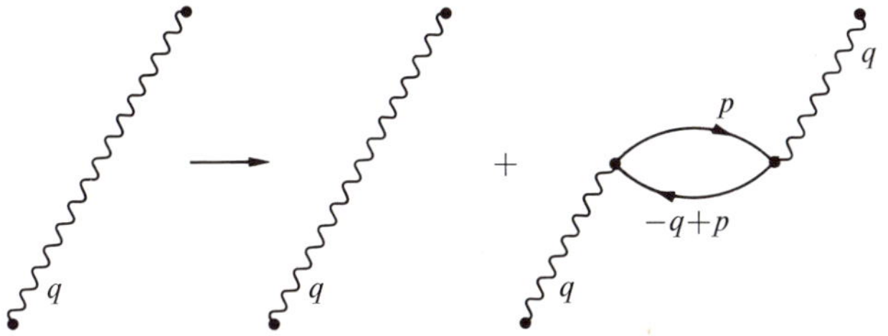

Fig. 4–25. Modification of the photon propagator.

similar in form to the original (bare particle) propagator. First, we see that the new mass of the modified propagator is $m + A$, in agreement with our earlier interpretation of the coefficient A. Second, apart from the above-mentioned change in the mass, the modified propagator and the original propagator differ by an overall multiplicative constant $1 - B$. If we call the modified propagator in momentum space $S'_F(p)$ and ignore the change in the electron mass, we have

$$S'_F(p) = (1 - B)S_F(p) \tag{4.401}$$

in the vicinity of $-i\gamma \cdot p = m$. We shall comment on the physical implications of the constant B when we discuss vertex corrections. At this stage we simply remark that the factor $1 - B$ can be shown to be regarded as the probability of finding a *bare* electron in a *physical* ("dressed") electron state. The constant $1 - B$ is known as the wave-function renormalization constant for the electron (to order α).‡

Photon self-energy; vacuum polarization. We are now in a position to discuss the modification of the covariant photon propagator in a similar fashion. The photon propagator is modified because the covariant photon can virtually disintegrate into an electron-positron pair for a certain fraction of the time, as indicated in Fig. 4–25.§ The last diagram of Fig. 4–25 with an electron "closed loop" is known as a "photon self-energy diagram" in analogy with Fig. 4–22. Calling

$$D_F(q) = -i/q^2, \tag{4.402}$$

(where the $-i\epsilon$ prescription is again implicit), we can write the modification represented by Fig. 4–25 as follows:

$$\begin{aligned}\delta_{\mu\nu} D_F(q) \longrightarrow \delta_{\mu\nu} D_F(q) &+ \delta_{\mu\lambda} D_F(q)[-i\Pi_{\lambda\sigma}(q)]\delta_{\sigma\nu} D_F(q)\\ &= \{\delta_{\mu\nu} + D_F(q)[-i\Pi_{\mu\nu}(q)]\}D_F(q),\end{aligned} \tag{4.403}$$

‡Quite generally, the proportionality constant between $S'_F(p)$ (in the vicinity of $-i\gamma \cdot p = m$) and $S_F(p)$ is denoted by Z_2 so that $S'_F(p) = Z_2 S_F(p)$ at $-i\gamma \cdot p \approx m$. To order α we have $Z_2 = 1 - B$.

§This modification should not be confused with the modification (4.391) which is an *artificial device* to make the electron self-energy finite.

which can be recognized as the analog of (4.398) for the photon. We must now compute the quantity $\Pi_{\mu\nu}$, which is known as the *polarization tensor*. We first write the result and then explain where it comes from:

$$-i\Pi_{\mu\nu}(q) = -(-e)^2 \int \frac{d^4p}{(2\pi)^4} \mathrm{Tr}\left\{\frac{-i\gamma\cdot(-q+p)+m}{i[(-q+p)^2+m^2-i\epsilon]}\gamma_\mu \frac{-i\gamma\cdot p+m}{i(p^2+m^2-i\epsilon)}\gamma_\nu\right\}. \tag{4.404}$$

The appearance of the two electron propagators is easy to understand. As in the electron self-energy case, there is one nontrivial momentum space integration to be performed since the virtual energy-momenta of the internal lines are not completely fixed even though we assume energy-momentum conservation at each of the two vertices. The only features of (4.404) that are not completely obvious from the set of rules given earlier are the appearance of the minus sign and the trace symbol. To understand these new features we may go back to the S-matrix expansion to order e^2. As far as the electron operators are concerned, both the initial and the final state are the vacuum state. So we must consider, for $t_1 > t_2$,

$$\begin{aligned}
&(\gamma_\mu)_{\alpha\beta}(\gamma_\nu)_{\gamma\delta}\langle 0|\bar\psi_\alpha^{(+)}(x_1)\psi_\beta^{(+)}(x_1)\bar\psi_\gamma^{(-)}(x_2)\psi_\delta^{(-)}(x_2)|0\rangle \\
&\quad = (\gamma_\mu)_{\alpha\beta}(\gamma_\nu)_{\gamma\delta}\langle 0|\psi_\beta^{(+)}(x_1)\bar\psi_\gamma^{(-)}(x_2)\bar\psi_\alpha^{(+)}(x_1)\psi_\delta^{(-)}(x_2)|0\rangle \\
&\quad = (\gamma_\mu)_{\alpha\beta}(\gamma_\nu)_{\gamma\delta}\langle 0|\psi_\beta^{(+)}(x_1)\bar\psi_\gamma^{(-)}(x_2)|0\rangle\langle 0|\bar\psi_\alpha^{(+)}(x_1)\psi_\delta^{(-)}(x_2)|0\rangle \\
&\quad = -\langle 0|T(\psi_\delta(x_2)\bar\psi_\alpha(x_1))|0\rangle(\gamma_\mu)_{\alpha\beta}\langle 0|T(\psi_\beta(x_1)\bar\psi_\gamma(x_2))|0\rangle(\gamma_\nu)_{\gamma\delta}.
\end{aligned} \tag{4.405}$$

From this we can understand why we must associate a minus sign with the closed loop;‡ the appearance of the trace in (4.404) is also obvious from the manner in which the Dirac indices appear in the last line of (4.405).

It is important to remark that, unlike $\Sigma(p)$, $\Pi_{\mu\nu}(q)$ is *not* a 4×4 matrix. The general structure of $\Pi_{\mu\nu}(q)$ can be inferred from the requirement of Lorentz covariance. There is only one four-vector available, namely the four-momentum of the covariant photon q. So we can write $\Pi_{\mu\nu}(q)$ as the sum of an invariant scalar function of q^2 times $\delta_{\mu\nu}$ and another invariant scalar function times $q_\mu q_\nu$. For $q=0$, however, the polarization tensor $\Pi_{\mu\nu}$ can be proportional to $\delta_{\mu\nu}$ only; so it is more illuminating to write it as

$$\Pi_{\mu\nu}(q) = D\delta_{\mu\nu} + \delta_{\mu\nu}q^2\Pi^{(1)}(q^2) + q_\mu q_\nu\Pi^{(2)}(q^2), \tag{4.406}$$

where D is a constant [not to be confused with $D_F(q)$] defined by§

$$\Pi_{\mu\nu}(0) = D\delta_{\mu\nu}. \tag{4.407}$$

Let us now look at the modification of the covariant photon propagator when all four-components of q are small; as a concrete example, we may consider the elastic scattering of two very slow electrons due to one-photon exchange. Keeping

‡This can be seen to be a general rule applicable to *any* closed fermion loop.
§It is obvious that the relation $q=0$ necessarily implies $q^2=0$; the converse, however, is not true.

only the leading term $D\delta_{\mu\nu}$ of (4.406), we see that (4.403) now becomes (cf. Eq. 4.399)

$$\frac{\delta_{\mu\nu}}{iq^2} \longrightarrow \frac{\delta_{\mu\nu}}{iq^2} + \frac{1}{iq^2}(-iD\delta_{\mu\nu})\frac{1}{iq^2}$$

$$= \frac{\delta_{\mu\nu}}{i(q^2 + D)} + 0(\alpha^2). \tag{4.408}$$

But this is nothing more than the propagator for a neutral-vector meson of mass $\sqrt{D}$ coupled to a conserved source (cf. Eq. 4.380). In the static limit it gives rise to a potential between two electrons that falls off like a repulsive Yukawa potential $\exp(-\sqrt{D}r)/r$. In reality, however, we know that the *observed photon* mass is zero and that the interaction between two slow electrons goes like $1/r$ at large distances; therefore, if the theory is correct, it must give $D = 0$. Yet, according to (4.404) and (4.406), D is given by

$$D = \frac{ie^2}{4}\int\frac{d^4p}{(2\pi)^4}\mathrm{Tr}\left[\frac{(-i\gamma\cdot p + m)\gamma_\mu(-i\gamma\cdot p + m)\gamma_\mu}{(p^2 + m^2 - i\epsilon)^2}\right]$$

$$= ie^2\int\frac{d^4p}{(2\pi)^4}\frac{(2p^2 + 4m^2)}{(p^2 + m^2 - i\epsilon)}, \tag{4.409}$$

where we have used

$$\sum_{\mu=1}^{4}\Pi_{\mu\mu}(0) = D\sum_{\mu=1}^{4}\delta_{\mu\mu} = 4D \tag{4.410}$$

and (4.341). It is not difficult to convince oneself that almost any "honest" calculation gives $D \neq 0$. In fact, using the techniques of Appendix E with a convergence factor $[\Lambda^2/(p^2 + \Lambda^2)]^2$, one can readily show that D is a positive, real constant that depends quadratically on Λ. In other words a straightforward calculation based on (4.409) gives the following result: If the bare mass of the photon is zero, the square of the physical photon mass goes like the square of the cut-off.

There have been numerous sophisticated papers written to explain why this quadratically divergent constant D must be discarded. For instance, W. Pauli and F. Villars noted that if we subtract from the integrand of (4.404) a similar expression with m replaced by a very large mass and then integrate with a suitable convergence factor, we get a much more reasonable result (even though such a procedure has no meaning in terms of physically realizable particles). Others prefer to start with a theory of a neutral vector boson of a finite bare mass μ_{bare} such that the observed mass becomes *exactly* zero *after* the boson mass is renormalized, that is, $\delta\mu^2 = -\mu^2_{\text{bare}}$ (even though this sounds like a very artificial explanation for the simple observed fact, $\mu^2_{\text{physical}} = 0$).‡ We shall not present these arguments in detail since they essentially say that the formalism must be set up in such a way that the observed photon mass is strictly zero. Instead, we ask: If the quadratically divergent constant D has been removed, does $\Pi_{\mu\nu}(q)$ contain any interesting physics?

---

‡This method is discussed in some detail in Jauch and Rohrlich (1955), pp. 191–192.

Before we discuss the $\Pi^{(1)}$ and $\Pi^{(2)}$ that appear in (4.406), we wish to remark that there is one property of $\Pi_{\mu\nu}$ which we have not yet fully exploited. Suppose we consider the influence of an externally applied potential on an electron as in Section 4–3. In addition to Fig. 4–2(a), which is of order e times the parameter that characterizes the strength of the external potential (for example, Ze in the Coulomb potential case), there is an additional higher order contribution, represented by Fig. 4–26, known as a "vacuum polarization diagram," where $\times$ stands for a first-order interaction with the external potential. This diagram gives an amplitude $\sim$ 137 times smaller than that of Fig. 4–2(a). The corresponding S-matrix element is proportional to

$$\frac{e^3 \bar{u}' \gamma_\mu u}{q^2} \Pi_{\mu\nu}(q) \tilde{A}_\nu^{(e)}(q), \tag{4.411}$$

where $\tilde{A}_\nu^{(e)}(q)$ is the four-dimensional Fourier transform of the external potential. The appearance of $\Pi_{\mu\nu}$ can be understood by noting that the virtual electron-positron pair in Fig. 4–26 gives rise to an integral identical with the polarization tensor considered in connection with the modification of the photon propagator (Fig. 4–25). Comparing (4.411) with the direct scattering term of the Møller amplitude, we can visualize Fig. 4–26 as the interaction between $e^2 \Pi_{\mu\nu} \tilde{A}_\nu^{(e)}$ and the charge-current of the electron via one-photon exchange exactly in the same way as we visualize Fig. 4-20(c) to represent the interaction of the charge-current of electron 1 with that of electron 2 via one-photon exchange. From this point of view $e^2 \Pi_{\mu\nu} \tilde{A}_\nu^{(e)}$ is nothing more than the Fourier transform of the charge-current induced in the vacuum due to the presence of the external potential. Note that $\Pi_{\mu\nu}$ characterizes the proportionality between the induced current and the external potential; that is why $\Pi_{\mu\nu}$ is called the *polarization tensor*. Since this induced current is expected to satisfy the continuity equation, we must have

$$q_\mu \Pi_{\mu\nu}(q) \tilde{A}_\nu^{(e)}(q) = 0 \tag{4.412}$$

in momentum space. This relation must hold for any external potential so that we have the important result

$$q_\mu \Pi_{\mu\nu}(q) = 0. \tag{4.413}$$

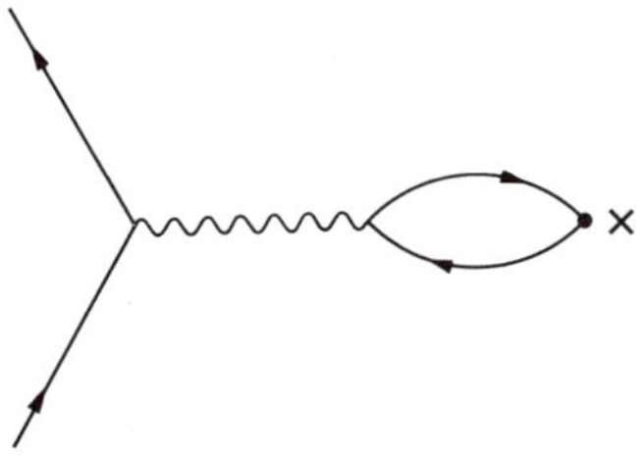

Fig. 4–26. Vacuum polarization.

From the requirement that the induced current be unchanged under gauge transformations on the external potential, we can also infer that

$$\Pi_{\mu\nu}(q) q_\nu = 0. \tag{4.414}$$

But, since $\Pi_{\mu\nu}$ is symmetric in μ and ν (cf. Eq. 4.406), we do not learn anything new from (4.414). Applying the requirement (4.413) to $\Pi_{\mu\nu}$ as given by (4.406), we obtain

$$q_\nu[D + q^2 \Pi^{(1)}(q^2) + q^2 \Pi^{(2)}(q^2)] = 0, \tag{4.415}$$

from which follows‡

$$\Pi^{(2)} = (-D/q^2) - \Pi^{(1)}. \tag{4.416}$$

In other words,

$$\Pi_{\mu\nu}(q) = (\delta_{\mu\nu}q^2 - q_\mu q_\nu)\Pi(q^2), \tag{4.417}$$

where we have defined $\Pi(q^2)$ to be $-\Pi^{(2)}(q^2)$.

In order to actually calculate $\Pi(q^2)$ it is convenient to set $\mu = \nu$ in (4.417) and sum over the μ indices

$$\sum_{\mu=1}^{4} \Pi_{\mu\mu}(q) = \left(q^2 \sum_{\mu=1}^{4} \delta_{\mu\mu} - q^2\right)\Pi(q^2) = 3q^2\Pi(q^2). \tag{4.418}$$

Now, if D (which is equal to $\sum_{\mu=1}^{4} \Pi_{\mu\mu}(0)/4$) were zero, we could as well write

$$\sum_{\mu=1}^{4} \Pi_{\mu\mu}(q) = \sum_{\mu=1}^{4} [\Pi_{\mu\mu}(q) - \Pi_{\mu\mu}(0)]. \tag{4.419}$$

Even though almost any "honest" calculation gives $D \neq 0$, the way we compute $\Pi(q^2)$ must be consistent with assigning a null value to the integral that appears in (4.409); otherwise the observed photon mass would not vanish. This means that, when we actually compute $\Pi(q^2)$, we must use the expression

$$\Pi(q^2) = (1/3q^2) \sum_{\mu=1}^{4} [\Pi_{\mu\mu}(q) - \Pi_{\mu\mu}(0)] \tag{4.420}$$

[with $\Pi_{\mu\mu}(q)$ given by (4.404)] rather than (4.418) because that part of $\Pi_{\mu\mu}(q)$ which leads to a nonvanishing photon mass must be discarded. Both $\sum_\mu \Pi_{\mu\mu}(q)$ and $\sum_\mu \Pi_{\mu\mu}(0)$ are quadratically divergent; however, the difìerence that appears in (4.420) turns out to be only logarithmically divergent. Using (4.404), (4.420), and the techniques of Appendix E, we can evaluate $\Pi(q^2)$ in a straightforward manner. We shall not present this calculation in detail because we later present an alternative (somewhat simpler) method for calculating $\Pi(q^2)$. The result is

$$\Pi(q^2) = C + q^2\Pi^f(q^2), \tag{4.421}$$

where

$$C = \frac{\alpha}{3\pi} \log\left(\frac{\Lambda^2}{m^2}\right) \tag{4.422}$$

and

$$q^2\Pi^f(q^2) = -\frac{2\alpha}{\pi}\int_0^1 dz\, z(1-z) \log\left[1 + \frac{q^2 z(1-z)}{m^2}\right]. \tag{4.423}$$

Note that the constant C is logarithmically divergent with the cut-off Λ, whereas $\Pi^f(q^2)$ is completely finite. When q^2 is small, (4.423) gives

$$\Pi^f(q^2) = -\frac{\alpha}{\pi m^2}\left[\frac{1}{15} - \frac{1}{40}\left(\frac{q^2}{m^2}\right) + \cdots\right]. \tag{4.424}$$

---

‡Many textbooks state that D must vanish as a result of the requirement (4.413) or (4.414). This argument, however, is not completely satisfactory since on the basis of Lorentz invariance and charge conservation (or gauge invariance) *alone*, we cannot prove that $\Pi^{(2)}$ has no pole at $q^2 = 0$.

Historically speaking, the fact that D and C are respectively quadratically and logarithmically divergent was already recognized in 1934 by P. A. M. Dirac, W. Heisenberg, and R. E. Peierls. The structure of $\Pi^f(q^2)$ was also discussed in the 1930's by R. Serber, M. E. Rose, and W. Pauli. We have rederived these old results using the language of covariant perturbation theory.

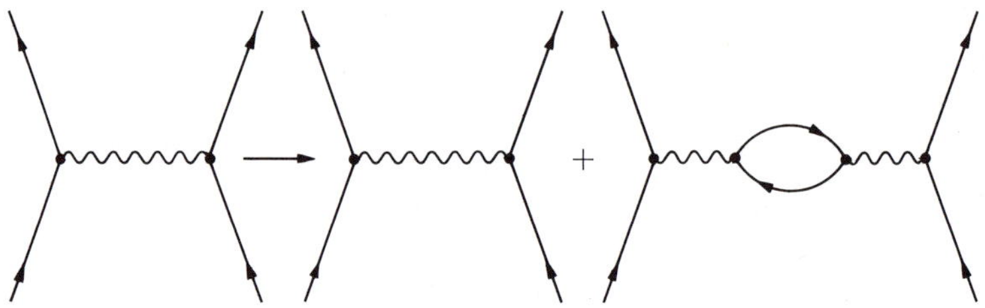

Fig. 4–27. Change in the Møller scattering amplitude.

We are now in a position to examine the physical meaning of C and Π^f. Let us recall that we have been discussing the modification of the photon propagator due to an electron closed loop. The effect of this modification on the Møller scattering amplitude is represented by Fig. 4–27. Clearly this change amounts to

$$\begin{aligned}\frac{e^2(\bar{u}_1'\gamma_\mu u_1)(\bar{u}_2'\gamma_\mu u_2)}{iq^2} &\longrightarrow e^2(\bar{u}_1'\gamma_\mu u_1)\frac{1}{iq^2}\left[\delta_{\mu\nu} - \frac{i\Pi_{\mu\nu}(q^2)}{iq^2}\right](\bar{u}_2'\gamma_\nu u_2)\\ &= e^2(\bar{u}_1'\gamma_\mu u_1)\frac{1}{iq^2}\left\{\delta_{\mu\nu} - \frac{(\delta_{\mu\nu}q^2 - q_\mu q_\nu)[C + q^2\Pi^f(q^2)]}{q^2}\right\}(\bar{u}_2'\gamma_\nu u_2)\\ &= e^2(1-C)\frac{(\bar{u}_1'\gamma_\mu u_1)(\bar{u}_2'\gamma_\mu u_2)}{iq^2} + ie^2\Pi^f(q^2)(\bar{u}_1'\gamma_\mu u_1)(\bar{u}_2'\gamma_\mu u_2),\end{aligned} \tag{4.425}$$

where in the last step we have taken advantage of the fact that the $q_\mu q_\nu$ term does not contribute because of (4.322). Let us treat the case where q^2 is small. Evidently only the first term of the last line of (4.425) is important. It can be seen that the constant C is completely analogous to the constant B discussed in connection with the electron self-energy diagram. The modified photon propagator at small q^2 differs from the original propagator by a factor $(1 - C)$; calling the modified photon propagator in momentum space $\delta_{\mu\nu}D_F'(q)$, we have

$$D_F'(q) = (1 - C)D_F(q) \tag{4.426}$$

in the vicinity of $q^2 = 0$. This is quite analogous to (4.401). The constant $1 - C$ is known as the "wave-function renormalization constant" for the photon (to order α).‡

Instead of regarding $1 - C$ as a correction factor to be applied to the photon propagator, we can alternatively regard it as a correction factor to be applied to the

‡In the literature the proportionality factor between $D_F'(q)$ (in the vicinity of $q^2 = 0$) and $D_F(q)$ is quite generally denoted by Z_3 so that $D_F'(q) = Z_3 D_F(q)$ at $q^2 \approx 0$. To order α we have $Z_3 = 1 - C$.

coupling constant. From (4.425) we notice that at very small momentum transfers the modified Møller amplitude is the same as the lowest-order Møller amplitude except that the electric charge is decreased (note $C > 0$) as follows:

$$e^2 \longrightarrow e^2(1 - C). \tag{4.427}$$

The charge of the electron which we physically measure may be deduced from the strength of the interaction between two electrons at small momentum transfers. This means that the *observable* physical charge e_{obs} is to be identified with $\sqrt{1 - C}$ times the *bare* charge e_{bare} (the electric charge which we would measure if the closed-loop effect were absent);

$$e_{\text{obs}} = \sqrt{1 - C}\, e_{\text{bare}}. \tag{4.428}$$

Thus, as far as small q^2 effects are concerned, the modification represented by Fig. 4–27 simply *decreases* the effective coupling constant. What is observed to be 1/137.04 is not $e_{\text{bare}}^2/4\pi$ but rather $(1 - C)$ times that quantity. Just as in the case of the electron self-energy nobody knows what e_{bare} is since nobody knows how quantum electrodynamics must be modified at small distances. If we employ the usual cut-off procedure, we see that even with a cut-off of order 10 BeV, the decrease in the value of $e^2/4\pi$ can be computed from (4.422) to be only $\sim 3\%$.

The basic idea which we have discussed in the preceeding paragraph is known as *charge renormalization.* The observable physical charge denoted by e_{obs} in (4.428) is often called the *renormalized charge* just as the observed mass given by $m_{\text{bare}} + \delta m$ is called the renormalized mass. Historically speaking, at the first Solvay Conference convened after the empirical discovery of the positron, P. A. M. Dirac presented an argument to show that the observed electronic charge must be smaller than the bare charge by an amount $\sim 1/137$ times the bare charge because of the creation and annihilation of virtual electron-positron pairs.

At this stage one may argue that the electronic charge can be measured not only by the interaction strength between two electrons but also by the response of the electron to an *externally applied field.* It is easy to see that in this case, just as before, the appropriate correction factor to be applied to the bare electric charge is $\sqrt{1 - C}$. To see this for the particular case of the Coulomb potential we notice that the S matrix element for Mott scattering (4.65) must be modified because of the vacuum polarization graph as follows (cf. Fig. 4–26):

$$\begin{aligned} &-\frac{Ze^2\bar{u}'\gamma_4 u}{|\mathbf{q}|^2} \\ &\longrightarrow -\frac{Ze^2\bar{u}'\gamma_4 u}{|\mathbf{q}|^2} + \frac{(-e)(\bar{u}'\gamma_\mu u)(-iC - iq^2\Pi^f)(q^2\delta_{\mu 4} - q_\mu q_4)}{iq^2}\left(\frac{Ze}{|\mathbf{q}|^2}\right) \\ &= -\frac{Ze^2(1 - C)\bar{u}'\gamma_4 u}{|\mathbf{q}|^2} - q^2\Pi^f(q^2)\left(-\frac{Ze^2\bar{u}'\gamma_4 u}{|\mathbf{q}|^2}\right), \end{aligned} \tag{4.429}$$

where we have used $q^2 = |\mathbf{q}|^2$ and omitted the factor $-2\pi i\,\delta(E - E')$ $\sqrt{(m/EV)(m/E'V)}$ which is common to every term in (4.429). We see from (4.429) that *both* $(-Ze)$ and e get multiplied by $\sqrt{1 - C}$, which, of course, is equivalent to saying that we multiply the electron charge by $\sqrt{1-C}$ and the external Coulomb potential by $\sqrt{1-C}$. It is amusing to note that this decrease in the effective charge of the nucleus is in agreement with the qualitative discussion of the vacuum polarization, given in Section 3–9, according to which the virtual electron-positron cloud tends to neutralize the positive electric charge of the nucleus.

To understand the physical meaning of Π^f let us turn our attention to the second term of the last line of (4.429), which, according to (4.424), is just $-Ze^2\alpha/(15\pi m^2)$ when $|\mathbf{q}|^2$ is not too large. Since the effective interaction potential between the nucleus and the electron (assumed to be nonrelativistic) is just the three-dimensional Fourier transform of (4.429), we see that the Coulomb interaction has been modified as follows:

$$-\frac{Ze^2_{\text{bare}}}{4\pi r} \longrightarrow -\frac{Ze^2_{\text{bare}}(1-C)}{4\pi r} - \frac{Ze^2_{\text{bare}}\alpha_{\text{bare}}}{15\pi m^2}\delta^{(3)}(\mathbf{x})$$

$$= -\frac{Ze^2_{\text{obs}}}{4\pi r} - \frac{Ze^2_{\text{obs}}\alpha_{\text{obs}}}{15\pi m^2}\delta^{(3)}(\mathbf{x}) + 0(\alpha^2_{\text{obs}}Ze^2_{\text{obs}}). \tag{4.430}$$

When the electron is scattered with a very small momentum transfer, the form of the potential "experienced" by the electron is just that of the usual $1/r$ potential, since scattering with a small momentum transfer is sensitive only to the outer region of the potential; only the magnitude of the charge is decreased by a universal factor, a change taken care of in any case by charge renormalization (cf. Eq. 4.428). On the other hand, according to the intuitive treatment of vacuum polarization given in Section 3–9, when the momentum transfer becomes somewhat larger, or, equivalently, when the impact parameter becomes smaller, the electron is expected to start penetrating the polarization cloud with a resulting increase in the effective interaction strength. The δ function term in (4.430) is seen to represent precisely this effect. Note that the sign of this δ function interaction corresponds to an attraction (as it should from the simple argument of Section 3–9). We may mention that this term gives rise to an energy difference between the $2s\frac{1}{2}$ and the $2p\frac{1}{2}$ states of the hydrogen atom since the expectation value of the δ function potential is zero for $l \neq 0$ states and finite for $l = 0$ states. Qualitatively we get

$$\Delta E_{nl} = -\frac{\alpha e^2}{15\pi m^2}|\psi(0)|^2 = -\frac{8\alpha^3}{15\pi n^3}Ry_\infty\delta_{l0}, \tag{4.431}$$

as first shown by E. A. Uehling in 1935. The energy shift (4.431) contributes -27 Mc to the total Lamb shift of $+1058$ Mc. To the extent that the experimental and the theoretical value of the Lamb shift agree to an accuracy of a few tenths of 1 Mc, we see that the *vacuum polarization effect* is a real physical phenomenon which must be taken seriously. We wish to emphasize that the formally divergent constant C is not directly observable (because it is absorbed in the definition of the observed electric charge), whereas we do get finite numbers for physically observable phenomena such as the Uehling effect.

We should, of course, consider also the vacuum polarization effects due to virtual $\mu^+\mu^-$ pairs, $\pi^+\pi^-$ pairs, $\bar{p}p$ pairs, etc., but these are much less important because $\Pi^f(q^2)$ at small q^2 is inversely proportional to the mass squared of the charged particle; for example, the vacuum polarization effect due to $\mu^+\mu^-$ pairs is $(205)^2$ times *less* important than that due to e^+e^- pairs. This means, in particular, that the Uehling effect in a *muonic atom* (an atom in which one of the orbital electrons is replaced by a negative muon) is due mainly to virtual e^+e^- pairs, not $\mu^+\mu^-$ pairs. We might mention that measurements of the X-ray energies emitted by high Z muonic atoms (performed by V. Fitch, J. Rainwater, and others) indeed confirm the existence of the vacuum polarization term. The vacuum polarization effect also plays a significant role in the electromagnetic corrections to low-energy proton-proton scattering.

Let us now go back to expression (4.423). Following the $m^2 \to m^2 - i\epsilon$ prescription, we can readily show that $\Pi(q^2)$ possesses an imaginary part given by

$$\begin{aligned}\operatorname{Im}\Pi(q^2) &= q^2 \operatorname{Im}\Pi^f(q^2)\\ &= (2\alpha/\pi)\int_0^1 dz\, z(1-z)\pi\theta[z(1-z)+(m^2/q^2)]\\ &= (\alpha/3)[1-(2m^2/q^2)]\sqrt{1+(4m^2/q^2)}\,\theta(-q^2-4m^2). \qquad (4.432)\end{aligned}$$

In other words, the polarization tensor becomes complex when $q^2 < -4m^2$. The reader who has worked out Problem 4–5 will note that, apart from $2|a/2|^2\omega^2$, (4.432) is identical to the probability that the external three-vector potential $\mathbf{A} = (0, 0, a\cos\omega t)$, $\omega > 2m$, creates an electron-positron pair per unit volume, per unit time, where ω^2 corresponds to $-q^2$. This similarity is not accidental. To understand this point we go back to the unitarity equation (4.45c) of Section 4–2. Let the initial and the final state be the vacuum state in the presence of the external potential $\mathbf{A} = (0, 0, a\cos\omega t)$. The unitarity relation (4.45c) now becomes

$$S_{ii}^{(2)*} + S_{ii}^{(2)} = -\sum_n |S_{ni}^{(1)}|^2, \qquad (4.433)$$

where i stands for the vacuum state. $S_{ii}^{(2)}$ is then the amplitude for the vacuum to remain the vacuum with two electromagnetic vertices. But, apart from certain kinematical factors which the reader may work out in Problem 4–16, $S_{ii}^{(2)}$ is essentially $-i\Pi(q^2)$ evaluated at $q^2 = -\omega^2$, since the only way to have two electromagnetic interactions and still end up with the vacuum state is to have an electron-positron pair in the *intermediate* state. This means that the left-hand side of (4.433) is essentially the negative of the imaginary part of $\Pi(q^2)$. We now argue that the $S_{ni}^{(1)}$ that appears on the right-hand side of (4.433) must represent the amplitude for the external potential to create an electron-positron pair and that state n can only stand for an electron-positron pair state; this is because there are no states other than e^+e^- states which are connected to the vacuum state via the *first-order* S matrix. The elements $S_{ii}^{(2)}$ and $S_{ni}^{(1)}$ are graphically represented in Fig. 4–28. Thus the unitarity equation (4.433) essentially says that the imaginary part of $\Pi(q^2)$ is proportional to the probability for pair creation; this is precisely the relation which we inferred by comparing (4.432) to Problem 4–5.

We argued earlier that the polarization tensor $\Pi_{\mu\nu}$ is essentially a proportionality constant that relates the induced current to the external field. From this point of

$S_{ii}^{(2)}$(Vacuum → Vacuum)
(a)

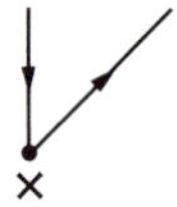

$S_{ni}^{(1)}$(Vacuum → e⁻e⁺)
(b)

Fig. 4–28. The S-matrix elements in (4.433).

view $\Pi(q^2)$ can be regarded as the dielectric constant of the vacuum. That this dielectric constant becomes complex for $q^2 < -4m^2$ is to be expected on physical grounds, since the Maxwell field loses energy because of the creation of real pairs.

We are now in a position to discuss an important relation between the real and imaginary parts of $\Pi(q^2)$. Let us first recall that $e^2\Pi_{\mu\nu}(q)\tilde{A}_\nu^{(e)}(q)$ is the Fourier transform of the induced charge-current due to an externally applied potential whose Fourier transform is $\tilde{A}_\nu^{(e)}(q)$. In other words, $\Pi_{\mu\nu}(q)\tilde{A}_\nu^{(e)}(q)$ characterizes the response of the vacuum to the applied field. Since this response must be causally related to the applied field, an argument completely analogous to the one used to derive the Kramers-Kronig (dispersion) relation for the scattering of light may now be repeated (cf. Eqs. 2.203 through 2.209). In particular, considering a three-vector potential which takes the form of a $\delta(t)$ function pulse, we see that the "causality principle"‡ requires that

$$\int \Pi(q^2)\big|_{q^2=-\omega^2} e^{-i\omega t} d\omega = 0 \quad \text{for} \quad t<0. \tag{4.434}$$

This in turn implies that we can write a dispersion relation for $\Pi(q^2)$ with respect to the variable $\omega = \sqrt{-q^2}$:

$$\text{Re}\,\Pi(q^2)\big|_{q^2=-\omega^2} = \frac{1}{\pi}\,\text{Pr}\int_{-\infty}^{\infty} \frac{d\omega'}{(\omega'-\omega)}\,\text{Im}\,\Pi(q'^2)\big|_{q'^2=-\omega'^2}. \tag{4.435}$$

We can rewrite this using the variable ω^2:

$$\begin{aligned}\text{Re}\,\Pi(q^2)\big|_{q^2=-\omega^2} &= \frac{1}{\pi}\,\text{Pr}\int_0^{\infty} d\omega'\left(\frac{1}{\omega'-\omega}+\frac{1}{\omega'+\omega}\right)\text{Im}\,\Pi(q'^2)\big|_{q'^2=-\omega^2}\\ &= \frac{1}{\pi}\,\text{Pr}\int_0^{\infty}\frac{d(\omega'^2)}{\omega'^2-\omega^2}\,\text{Im}\,\Pi(q'^2)\big|_{q'^2=-\omega'^2}.\end{aligned} \tag{4.436}$$

Since the imaginary part of Π vanishes for $\omega'^2 < 4m^2$, we have the final result

$$\text{Re}\,\Pi(q^2) = \frac{1}{\pi}\,\text{Pr}\int_{4m^2}^{\infty}\frac{\text{Im}\,\Pi(-s)}{s+q^2}\,ds. \tag{4.437}$$

We remark once again that Im Π $(-s)$ is essentially the probability for pair creation in the vacuum by the external potential of Problem 4–5 with $\omega^2 = s$. The function

‡The connection between the vanishing of the Fourier transform of $\Pi(q^2)$ and the causality principle can be discussed in a more rigorous manner when one uses the Heisenberg representation of the electron field operator. See Källén (1958), pp. 279–285.

$\Pi(q^2)$ that appears on the left-hand side of (4.437), however, is supposed to characterize more complicated phenomena such as charge renormalization and the Uehling effect. In order to calculate these (less elementary) vacuum polarization effects, it is sufficient to solve Problem 4–5, which is one of the most elementary problems in relativistic quantum mechanics. We can then use the dispersion relation (4.437) to evaluate Re $\Pi(q^2)$. This powerful method based on unitarity and causality has been advocated particularly by G. Källén, and by R. N. Euwema and J. A. Wheeler.

To see how this dispersion theoretic approach works in detail, we compute the coefficient C responsible for charge renormalization. In this case we can set $q^2 = 0$ in (4.437) to obtain a sum rule for the renormalization constant as follows:

$$C = \Pi(0) = \frac{1}{\pi} \,\mathrm{Pr} \int_{4m^2}^{\infty} ds \frac{\mathrm{Im}\,\Pi(-s)}{s}. \tag{4.438}$$

According to (4.432), which the reader is urged to prove in Problem 4–16 from the unitary relation (4.432) and Problem 4–5, we have

$$\lim_{s\to\infty} \mathrm{Im}\,\Pi(-s) = \lim_{s\to\infty} [-s\,\mathrm{Im}\,\Pi^f(-s)] = \alpha/3. \tag{4.439}$$

Substituting this asymptotic value in (4.438), we immediately see that C is a logarithmically divergent constant that depends on $s_{\max}$, the square of a cut-off energy, as follows:

$$C = \frac{\alpha}{3\pi} \log\left(\frac{s_{\max}}{m^2}\right) + \text{a finite constant}. \tag{4.440}$$

Since a finite constant is of no physical significance when it appears together with a divergent expression, we see that (4.440) is in exact agreement with (4.422), which was obtained by evaluating a much more complicated integral. Note also that this approach offers a direct (and rigorous) explanation of the sign of the charge renormalization. The constant C must be *positive* because the probability for pair creation is positive definite; as a result the correction factor $\sqrt{1 - C}$ to be applied to the bare charge (or to the electromagnetic potential) is less than unity. Instead of (4.437) it is also instructive to write a once-subtracted dispersion relation for $\Pi(q^2)$ by applying the now-familiar analyticity argument to $\Pi(q^2)/\omega$ rather than to $\Pi(q^2)$ itself. This time the dispersion integral is completely convergent, and we can readily obtain the finite part of $\Pi(q^2)$, denoted earlier by $q^2\Pi^f$, responsible for the Uehling effect, etc. This is left as an exercise (Problem 4–16).

Before we leave the subject of vacuum polarization, it will be worth while to make some remarks on the effect of a free radiation field on the vacuum. When $A_\mu^{(\mathrm{rad})}$ represents a classical free radiation field satisfying the D'Alembertian equation (with no source term) and the Lorentz condition, there is no induced current $e^2\Pi_{\mu\nu}\tilde{A}_\nu^{(\mathrm{rad})}$ because

$$(q^2\delta_{\mu\nu} - q_\mu q_\nu)\tilde{A}_\nu^{(\mathrm{rad})} = 0. \tag{4.441}$$

Thus for free photons or for traveling electromagnetic waves in empty space, there is no vacuum polarization effect that depends linearly on the electromagnetic field. This agrees with our common-sense idea of the vacuum. When nonlinear effects are considered, however, this is no longer true. Two photon beams can scatter each other because of the polarization of the vacuum brought about by a "square diagram," as represented in Fig. 4–29, a possibility first discussed by H. Euler.‡ The cross section for this process, known as the scattering of light by light, can be computed to be about 4×10^{-31} cm^2 at $\omega \sim m$, which is too small to be observed experimentally. When two of the photon interaction vertices in Fig. 4–29 are replaced by interactions with the Coulomb potential of a nucleus, we get a diagram for a process known as Delbrück scattering (after M. Delbrück who first argued that such a process is to be expected on theoretical grounds). This is shown in Fig. 4–30. Delbrück scattering, whose angular distribution is characterized by a very sharp forward peak, has been observed experimentally by R. R. Wilson.

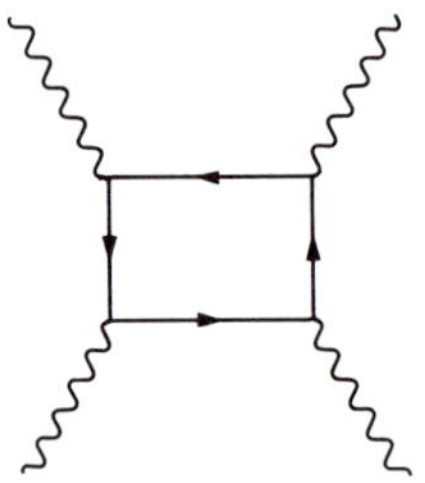

Fig. 4–29. Scattering of light by light.

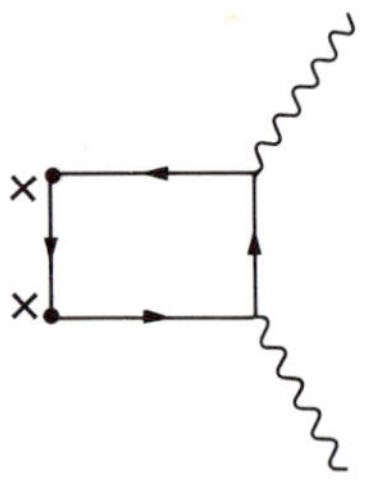

Fig. 4–30. Delbrück scattering.

Radiative corrections. We argued earlier that the multiplicative factor $1 - C$ can be regarded either as the correction factor to be applied to the covariant photon propagator or as the correction factor to be applied to $\alpha = e^2/4\pi$. Since the correction factor $1 - B$ for the electron propagator is quite analogous to $1 - C$, we might ask whether this factor also contributes to the renormalization of the electric charge. To answer this question, for instance, for the particular case of an external potential problem, it is necessary to consider systematically all possible diagrams that contribute corrections to the first-order scattering amplitude of Fig. 4–2(a). There are altogether *six* such diagrams, as shown in Fig. 4–31, to order e^2 times the strength of the original first-order amplitude of Section 4–2. They are collectively known as *radiative corrections* to electron scattering by an external field. The first one, Fig. 4–31(a), is the vacuum polarization diagram already discussed. Figure 4–31(b) is referred to as "vertex correction"; it will be discussed in detail in a moment. Figures 4–31(c) and (d) are the self-energy corrections to the initial and final electron lines. Figures 4–31(e) and (f) are necessary whenever we use the observed (already renormalized) mass for the electron (recall the arguments given in connection with Fig. 4–23); clearly (e), or (f), cancels with that part of (c), or (d), which goes as the coefficient A.

‡In addition to Fig. 4–29 there are five other diagrams which can be obtained by permuting the orders of the photon interactions. Each diagram turns out to be separately divergent logarithmically, but the sum can be shown to be finite.

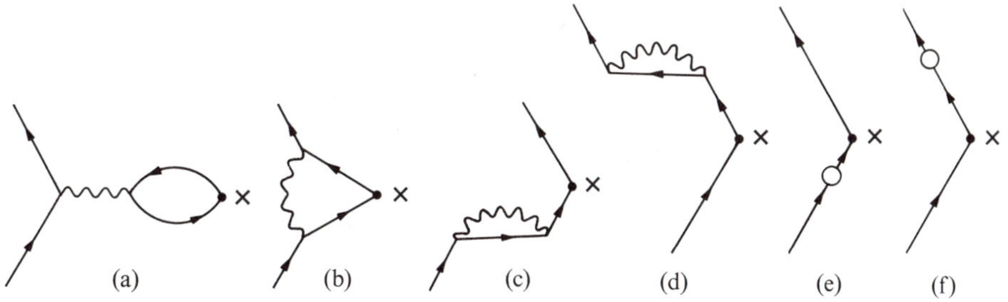

Fig. 4–31. Radiative corrections to the scattering by an external field.

We shall now look at Fig. 4–31(c) more carefully. Apart from the mass correction, which is taken care of by the counter graph (e), (c) is seen to give rise to a modification of the free-particle spinor for the incident electron as follows (cf. Eq. 4.386):

$$u^{(s)}(\mathbf{p}) \longrightarrow u^{(s)}(\mathbf{p}) + \frac{1}{i(i\gamma\cdot p + m)}(-i)\Big[(i\gamma\cdot p + m)B + (i\gamma\cdot p + m)^2 \textstyle\sum^f(p)\Big]u^{(s)}(\mathbf{p}). \tag{4.442}$$

We can safely drop the last $\sum^f$ term since $u^{(s)}(\mathbf{p})$ satisfies the free-particle spinor equation (3.105). We may next be tempted to cancel $(i\gamma\cdot p + m)$ with $1/(i\gamma\cdot p + m)$ to obtain $1 - B$ as the correction factor to be applied to the incident spinor. However, this procedure cannot be justified when we realize that $(i\gamma\cdot p + m)$ acting on $u^{(s)}(\mathbf{p})$ is zero, which means that the correction factor is indeterminate. A careful treatment (due to F. J. Dyson who used an argument based on the adiabatic switchings of the interaction in the remote past and the remote future) reveals that the right correction factor due to Fig. 4–31(c) is not $1 - B$ but $1 - (B/2)$. We can see this in a nonrigorous way as follows. Recall first that when the electron appears in an internal line, as in the case of $S_F(p)$, we get $1 - B$ as the correction factor (cf. Eq. 4.401). Now the electron propagator [whose Fourier transform is $S_F(p)$] is *bilinear* in the electron spinor; in contrast, in an external electron line the electron spinor appears *linearly*. Hence the appropriate correction factor due to the self-energy interaction mechanism of Fig. 4–31(c) may be inferred to be just

$$\sqrt{1 - B} = 1 - \tfrac{1}{2}B + 0(\alpha^2) \tag{4.443}$$

A similar argument holds for Fig. 4–31(d).‡

As in the case of the photon propagator, we can regard these multiplicative factors as corrections to be applied to the electric charge. We then get

$$e \longrightarrow [1 - \tfrac{1}{2}B - \tfrac{1}{2}B + 0(\alpha^2)]e \approx (1 - B)e, \tag{4.444}$$

due to Fig. 4–31(c) and (d).

We must now turn our attention to the vertex correction represented by Fig. 4–31(b). Using the Feynman rules, we see that this diagram modifies the γ_μ vertex

‡The whole problem of wave-function renormalization can be most unambiguously handled using what is known as the "reduction technique," which is beyond the scope of the present book. See, for instance, Chapter 13 of Gasiorowicz (1966).

that represents the first-order interaction of the electron with the external potential as follows:

$$-e\gamma_\mu \longrightarrow -e[\gamma_\mu + \Lambda_\mu(p', p)], \tag{4.445}$$

where $\Lambda_\mu(p', p)$ is a 4×4 matrix given by

$$\Lambda_\mu(p', p) = (-e)^2 \int \frac{d^4k}{(2\pi)^4} \frac{1}{i(k^2 - i\epsilon)} \gamma_\nu \left\{ \frac{-i\gamma\cdot(p' - k) + m}{i[(p' - k)^2 + m^2 - i\epsilon]} \right\} \gamma_\mu$$
$$\times \left\{ \frac{-i\gamma\cdot(p - k) + m}{i[(p - k)^2 + m^2 - i\epsilon]} \right\} \gamma_\nu. \tag{4.446}$$

When $\Lambda_\mu(p', p)$ is sandwiched between the free-particle spinors $\bar{u}'$ (from the left) and u (from the right), it is possible to write it as

$$\Lambda_\mu(p', p)\,|_{i\gamma\cdot p' = -m,\, i\gamma\cdot p = -m} = L\gamma_\mu + \Lambda_\mu^f(p', p), \tag{4.447}$$

where L and $\Lambda_\mu^f(p', p)$ are *uniquely* defined by

$$\lim_{p' \to p} \bar{u}'\Lambda_\mu(p', p)u = L\bar{u}'\gamma_\mu u, \tag{4.448a}$$

$$\Lambda_\mu^f(p', p) = [\Lambda_\mu(p', p) - \Lambda_\mu(p, p)]\,|_{i\gamma\cdot p' = -m,\, i\gamma\cdot p = -m}. \tag{4.448b}$$

We might wonder why $\lim_{p' \to p} \Lambda_\mu(p', p)$ taken between the free-particle spinors depends just on γ_μ and not on p_μ; the reason is that we can always re-express any dependence on p_μ by taking advantage of the analog of the Gordon decomposition (3.203) for the free-particle spinors:

$$(p' + p)_\mu \bar{u}'u = 2im\bar{u}'\gamma_\mu u + i(p' - p)_\nu \bar{u}'\sigma_{\mu\nu}u. \tag{4.449}$$

As we explained in connection with the vacuum polarization, the electric charge that we physically observe may be determined from the scattering of an electron with a very small momentum transfer ($p' \to p$). Therefore the correction factor to be applied to the electric charge due to Fig. 4–31(b) is just $1 + L$.‡

To summarize the various correction factors to be applied to the electric charge, we have

$$e \longrightarrow \sqrt{1 - C}(1 - B)(1 + L)e, \tag{4.450}$$

where $\sqrt{1 - C}$ comes from the vacuum polarization graph, Fig. 4–31(a), as discussed earlier, $1 - B$ from 4–31(c) and (d), and $1 + L$ from 4–31(b); these factors are due respectively to the photon field renormalization, the electron wave function renormalization, and the vertex correction. In quantum electrodynamics, however, something rather unexpected and remarkable takes place: the coefficients B and L become equal. To prove this important equality we first rewrite (4.385) and (4.446) as

$$\Sigma(p) = ie^2 \int \frac{d^4k}{(2\pi)^4} D_F(k)\gamma_\nu S_F(p - k)\gamma_\nu, \tag{4.451}$$

---

‡Quite generally the *ratio* of the electric charge corrected by the vertex correction to the bare charge is known as $1/Z_1$. To order e^2 we have $L = (1/Z_1) - 1$.

and

$$\Lambda_\mu(p', p) = e^2 \int \frac{d^4k}{(2\pi)^4} D_F(k)\gamma_\nu S_F(p' - k)\gamma_\mu S_F(p - k)\gamma_\nu. \tag{4.452}$$

But quite generally we have

$$\partial S_F/\partial p_\mu = S_F(p)\gamma_\mu S_F(p), \tag{4.453}$$

which can be proved by first differentiating both sides of

$$S_F(p)\, S_F^{-1}(p) = 1 \tag{4.454}$$

with respect to p_μ,

$$\frac{\partial S_F}{\partial p_\mu} S_F^{-1}(p) + S_F(p)\frac{\partial}{\partial p_\mu}[i(i\gamma \cdot p + m)] = 0, \tag{4.455}$$

and then multiplying by $S_F(p)$ from the right. Note that (4.453) essentially says that the insertion of the γ_μ vertex in an internal electron line without any energy-momentum transfer is equivalent to the differentiation of the electron propagator with respect to p_μ. Using this important identity, we can rewrite $\lim_{p' \to p}\Lambda_\mu(p', p)$ as follows:

$$\begin{aligned} L\gamma_\mu &= \lim_{p' \to p} \Lambda_\mu(p', p)\,|_{i\gamma \cdot p' = -m,\, i\gamma \cdot p = -m} \\ &= e^2 \int \frac{d^4k}{(2\pi)^4} D_F(k)\gamma_\nu \frac{\partial S_F(p-k)}{\partial (p-k)_\mu}\gamma_\nu \\ &= e^2 \int \frac{d^4k}{(2\pi)^4} D_F(k)\gamma_\nu \frac{\partial S_F(p-k)}{\partial p_\mu}\gamma_\nu. \end{aligned} \tag{4.456}$$

Meanwhile, from the definition of B given by (4.386) and the expression (4.451) for $\Sigma(p)$, we get

$$\begin{aligned} \bar{u}^{(s')}(\mathbf{p}) B\gamma_\mu u^{(s)}(\mathbf{p}) &= \bar{u}^{(s')}(\mathbf{p})(-i\partial\Sigma/\partial p_\mu) u^{(s)}(\mathbf{p}) \\ &= \bar{u}^{(s')}(\mathbf{p})\left[e^2 \int \frac{d^4k}{(2\pi)^4} D_F(k)\gamma_\nu \frac{\partial S_F(p-k)}{\partial p_\mu}\gamma_\nu\right] u^{(s)}(\mathbf{p}) \\ &= \bar{u}^{(s')}(\mathbf{p}) L\gamma_\mu u^{(s)}(\mathbf{p}), \end{aligned} \tag{4.457}$$

from which follows the equality

$$B = L. \tag{4.458}$$

This then implies that

$$(1 - B)(1 + L) = 1 + 0(\alpha^2). \tag{4.459}$$

The modification (4.450) now reads

$$e \longrightarrow \sqrt{1 - C}\, e, \tag{4.460}$$

to order α, which means that the charge renormalization is due to the vacuum polarization *only*. This is an extremely gratifying result; it shows that the net correction factor to be applied to the electric charge is independent of the *type* of charged Dirac particles. Without such a relation there would be no guarantee that the renormalized charge of the electron is the same as that of the muon since B and L are explicitly dependent on the fermion mass (as well as on the cut-off Λ).

The relation (4.460), which is a consequence of the equality (4.458), tells us that if the bare charges of the electron and the muon are equal, then their renormalized charges must also be equal since the same coefficient C appears in both the electron and the muon case. As shown by J. C. Ward, the statement that the charge renormalization is due only to the photon field renormalization can be generalized to *all orders* in perturbation theory. For this reason, the equality in (4.458) or its higher-order analog is in known the literature as the Ward identity.‡

Presumably a cancellation mechanism of the kind (4.459) is also at work for the case of the electric charge of the proton which possesses mesonic interactions. Here the major part of the nucleon wave-function renormalization arises from the fact that the proton can emit or absorb virtual pions (and other heavy mesons as well). In this case one must prove that a diagram such as Fig. 4–32(a) does not alter the charge of the proton when considered together with Fig. 4–32(b), etc. We may mention that the experimentally observed proton charge is equal in magnitude to the electron charge to an accuracy of one part in 10^{22}.

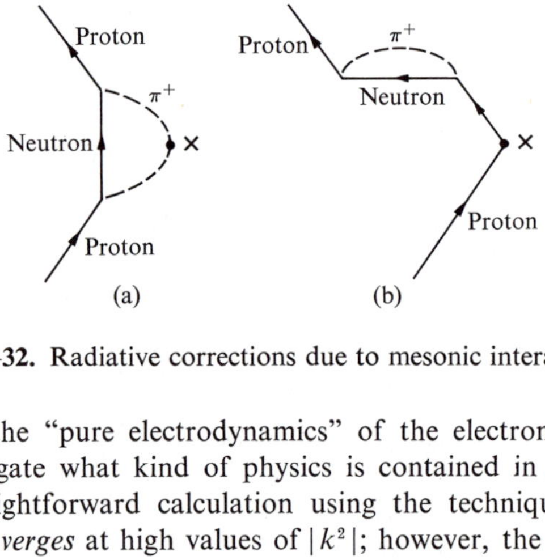

Fig. 4–32. Radiative corrections due to mesonic interactions.

Coming back to the "pure electrodynamics" of the electron and the muon, we must still investigate what kind of physics is contained in $\Lambda_\mu^f(p', p)$ defined by (4.448b). A straightforward calculation using the techniques of Appendix E shows that Λ_μ^f *converges* at high values of $|k^2|$; however, the integral diverges logarithmically at *low* values of $|k|^2$ as $|k|^2 \longrightarrow 0$. This is an example of what is known as the "infrared catastrophe." A possible way to dispose of this difficulty in a covariant manner (due to R. P. Feynman) is to temporarily assign a very small but finite mass to the photon; this can be accomplished by modifying the photon propagator that appears in (4.446) according to the prescription

$$\frac{1}{i(k^2 - i\epsilon)} \longrightarrow \frac{1}{i(k^2 + \lambda_{\min}^2 - i\epsilon)}, \tag{4.461}$$

where $\lambda_{\min}$ is the "mass" of the photon. With this device virtual photons of very low energies ($\omega \lesssim \lambda_{\min}$) do not get emitted or absorbed. We shall comment on the physical meaning of this artificial procedure in a later part of this section. The

‡In terms of Z_1, Z_2, and Z_3 the Ward identity reads $Z_1 = Z_2$ so that $e_{\text{obs}} = \sqrt{Z_3}\, e_{\text{bare}}$.

net result is that, for small values of $q = p' - p$, $\Lambda_\mu^f(p', p)$ can be written as follows:

$$\Lambda_\mu^f(p', p) = -\left[\frac{\alpha}{3\pi}\frac{q^2}{m^2}\left(\log\frac{m}{\lambda_{\min}} - \frac{3}{8}\right)\right]\gamma_\mu + \left(\frac{\alpha}{2\pi}\right)\left(\frac{1}{2m}\right)q_\nu\sigma_{\nu\mu}. \tag{4.462}$$

In Appendix E we shall discuss how to derive this formula and show in detail that the coefficient of $q_\nu\sigma_{\nu\mu}$ is indeed $(\alpha/2\pi)(1/2m)$. The important point to be noted at this stage is that once the constant L is cancelled by the constant B, the observable effects are completely independent of the divergence of Λ_μ at *high* values of $|k^2|$.

To summarize, when q is small, apart from the charge renormalization, the total radiative corrections due to Fig. 4–31 (a) through (f) modify the first-order S-matrix (4.64) as follows:

$$\begin{aligned} &-e\sqrt{\left(\frac{m}{EV}\right)\left(\frac{m}{E'V}\right)}\,\bar{u}'\gamma_\mu u\tilde{A}_\mu^{(e)}(q) \\ &\longrightarrow -e\sqrt{\left(\frac{m}{EV}\right)\left(\frac{m}{E'V}\right)}\,\bar{u}'\left\{\gamma_\mu\left[1 - \frac{\alpha}{3\pi}\frac{q^2}{m^2}\left(\log\frac{m}{\lambda_{\min}} - \frac{3}{8} - \frac{1}{5}\right)\right]\right. \\ &\qquad\qquad \left. + \left(\frac{\alpha}{2\pi}\right)\left(\frac{1}{2m}\right)\sigma_{\nu\mu}q_\nu\right\} u\tilde{A}_\mu^{(e)}(q), \end{aligned} \tag{4.463}$$

where $q = p' - p$ and

$$\tilde{A}_\mu^{(e)}(q) = \int d^4x e^{-i(p'-p)\cdot x} A_\mu^{(e)}(x). \tag{4.464}$$

Note that the factor $-\frac{1}{5}$ that multiplies $-(\alpha/3\pi)(q^2/m^2)$ arises from the vacuum polarization graph (cf. Eqs. 4.424 and 4.429).

We shall first comment on the physical significance of the last term of (4.463). Suppose we consider an interaction Hamiltonian density (cf. Eq. 3.211)

$$\begin{aligned} \mathscr{H}_{\text{int}} &= -\frac{e\kappa}{2m}\left[\frac{1}{2}F_{\nu\mu}^{(e)}\bar{\psi}\sigma_{\nu\mu}\psi\right] \\ &= -\frac{e\kappa}{2m}\frac{\partial A_\mu^{(e)}}{\partial x_\nu}\bar{\psi}\sigma_{\nu\mu}\psi. \end{aligned} \tag{4.465}$$

We can readily compute the first-order S-matrix element for the scattering of an electron due to the interaction (4.465). The result is

$$\begin{aligned} S_{fi} &= \frac{ie\kappa}{2m}\sqrt{\left(\frac{m}{EV}\right)\left(\frac{m}{E'V}\right)}\,\bar{u}'\sigma_{\nu\mu}u\int d^4x e^{-i(p'-p)\cdot x}\frac{\partial A_\mu^{(e)}}{\partial x_\nu} \\ &= \frac{ie\kappa}{2m}\sqrt{\left(\frac{m}{EV}\right)\left(\frac{m}{E'V}\right)}\,\bar{u}'\sigma_{\nu\mu}u(+iq_\nu)\int d^4x A_\mu^{(e)}e^{-iq\cdot x} \\ &= -\frac{e\kappa}{2m}\sqrt{\left(\frac{m}{EV}\right)\left(\frac{m}{E'V}\right)}\,\bar{u}'\sigma_{\nu\mu}uq_\nu\tilde{A}_\mu^{(e)}(q), \end{aligned} \tag{4.466}$$

which is identical to the last term of (4.463) with κ set equal to $(\alpha/2\pi)$. Meanwhile we have already shown in Section 3–5 that the interaction (4.465) together with the usual $ie\bar{\psi}\gamma_\mu\psi A_\mu^{(e)}$ interaction results in a total magnetic moment of the electron given by $(1 + \kappa)(e/2m)$ (see Eq. 3.212). This means that, as a result of the vertex correction Fig. 4–31(b), the electron appears to have a total magnetic moment given by‡

$$\mu_{\text{el}} = \left(1 + \frac{\alpha}{2\pi}\right)\left(\frac{e}{2m}\right), \tag{4.467}$$

as first shown by J. Schwinger in 1948 prior to the precise confirming measurement of the electron magnetic moment by P. Kusch. Subsequently the fourth-order (α^2) corrections to the electron magnetic moment were computed by C. Sommerfield and A. Petermann. The theoretical value for the magnetic moment to this order is

$$\begin{aligned} \mu_{\text{el}} &= \left[1 + \frac{\alpha}{2\pi} - 0.328\left(\frac{\alpha}{2\pi}\right)^2\right]\frac{e}{2m} \\ &= 1.0011596\left(\frac{e}{2m}\right). \end{aligned} \tag{4.468}$$

This value is in excellent agreement with the latest experimental value $[1 + (\alpha/2\pi) - (0.327 \pm 0.005)(\alpha/\pi)^2](e/2m)$ obtained by H. R. Crane and co-workers.

We shall now comment on Feynman's method for treating the infrared problem based on the fictitious photon mass $\lambda_{\min}$ introduced in (4.461). As an illustrative example, we shall show how the dependence on $\lambda_{\min}$ disappears for the particular problem of small-angle Mott scattering at nonrelativistic energies provided that we ask "physically sensible questions." When we compute the elastic scattering cross section to order $(Z\alpha)^2\alpha$ using the S-matrix (4.463) with $\tilde{A}_\mu^{(e)}(q)$ given by (4.66), we clearly get

$$\frac{d\sigma}{d\Omega} = \left(\frac{d\sigma}{d\Omega}\right)^{(0)}\left[1 - \frac{2\alpha}{3\pi m^2}|\mathbf{q}|^2\log\left(\frac{m}{\lambda_{\min}}\right) + \text{terms independent of } \lambda_{\min}\right], \tag{4.469}$$

where $(d\sigma/d\Omega)^{(0)}$ is the (uncorrected) Mott cross section calculated to lowest order as in Section 4–3. Now, according to Problem 4–10, the *bremsstrahlung* cross section per unit energy for the emission of a very soft photon by a nonrelativistic electron can be written as

$$\frac{d^2\sigma}{d\Omega_{\mathbf{p}'}\,d\omega} = \left(\frac{d\sigma}{d\Omega}\right)^{(0)}\frac{2\alpha|\mathbf{q}|^2}{3\pi m^2\omega}. \tag{4.470}$$

As a consequence of the $1/\omega$ behavior, the *bremsstrahlung* cross section integrated over the photon energy diverges logarithmically at low values of ω; this divergence would be absent if the photon had a finite mass. The cross section for the emission of a finite mass photon whose energy is between $\omega = \lambda_{\min}$ and $\omega = \omega_0$ can be shown to be

$$\int_{\lambda_{\min}}^{\omega_0}\left(\frac{d^2\sigma}{d\Omega_{\mathbf{p}'}\,d\omega}\right)_{\text{finite mass photon}} d\omega = \left(\frac{d\sigma}{d\Omega}\right)^{(0)}\frac{2\alpha}{3\pi}\frac{|\mathbf{q}|^2}{m^2}\left[\log\left(\frac{2\omega_0}{\lambda_{\min}}\right) - \frac{5}{6}\right], \tag{4.471}$$

where ω_0 is assumed to be some photon energy at which the low-energy approxima-

‡We emphasize that (4.465) with $\kappa = \alpha/2\pi$ is not to be regarded as a "fundamental interaction." What we have shown is that the fundamental $ie\bar{\psi}\gamma_\mu\psi A_\mu$ interaction gives rise to a higher-order effect which simulates a first-order effect of (4.465).

tion we have been making is still valid.‡ We now make the crucial observation that in any Mott scattering experiment there is always the possibility of the emission of photons which are *too soft* to be detected. In reality it is the energy resolution of the experimental apparatus that decides whether a particular event with the emission of a very soft photon should be counted as a scattering event or thrown away as a *bremsstrahlung* event; the worse the energy resolution, the more the events which are counted as "scattering events." In other words, what we physically measure in any realistic experiment is the probability for elastic scattering *plus* the probability for "slightly inelastic" scattering with the emission of a photon with energy less than ΔE where ΔE is the minimum photon energy detectable by the experimental arrangement in question. Thus the *observable* scattering cross section is actually the *sum* of (4.469) and (4.471) with ω_0 set to ΔE. But this sum is seen to be completely independent of the fictitious photon mass λ_{min} so that λ_{min} can now be made to go to zero.§ Thus if we ask the "right" question, the so-called infrared catastrophe is *not* a catastrophe at all. (In contrast, the divergences which we have encountered in computing the electron self-energy and the charge renormalization constant are real "catastrophes"; to solve these problems we would have to drastically modify the present form of quantum field theory.)

We have shown how the elimination of the infrared divergence comes about only for the particular problem of low-energy Mott scattering to order $(Z\alpha)^2\alpha$ in the cross section. In a good-resolution experiment the emission of many very soft photons can become important, which means that higher-order effects cannot be ignored. In fact, the whole perturbation argument breaks down when $\alpha \log (m/\lambda_{min})$ becomes comparable to unity, which occurs at $\lambda_{min} \approx e^{-137}\, m \approx 10^{-60}\, m$. Fortunately the elimination of the infrared catastrophe can be achieved to all orders in perturbation theory, as shown by F. Bloch and A. Nordsieck back in 1937.‖

To summarize, once we renormalize the mass and the charge, the radiative corrections to electron scattering by an external Coulomb field are completely free of divergences of any kind. This important point was first demonstrated in 1948 by D. Itô, Z. Koba, and S. Tomonaga and by J. Schwinger.

---

‡Note that the factor that multiplies $\log \lambda_{min}$ is expected from (4.470); when we integrate (4.470) with respect to ω from ω_{min} to some energy ω_0 (greater than ω_{min}), we obtain exactly the same factor that multiplies $\log \omega_{min}$. The additive constant $\log 2 - \frac{5}{6}$ in (4.471) arises because

a) a finite mass photon can be longitudinally polarized (spin perpendicular to the propagation direction) and
b) the relation $|\mathbf{k}| = \omega$ no longer holds.

Since these points are rather technical, we shall not discuss them in detail. The reader who is interested in the derivation of (4.471) may consult Jauch and Rohrlich (1951), pp. 336–338.

§One might legitimately ask whether the $\lambda_{min} \to 0$ limit is a smooth one since we know that the number of the photon spin states must decrease discontinuously from 3 to 2. Fortunately, in any problem involving fictitious finite-mass photons, the probability for the emission of a longitudinally polarized photon turns out to be $\lambda_{min}^2/(|\mathbf{k}|^2 + \lambda_{min}^2)$ times the probability for the emission of transversely polarized photons, a point demonstrated by F. Coester and J. M. Jauch.

‖For the Bloch-Nordsieck method consult Akhietzer and Berestetzkii (1965) pp. 413–422.

Lamb shift. As a final application of (4.463) we shall show how this expression may be used to obtain a finite answer to the historic problem of the Lamb shift. As usual, it is convenient to first construct an effective potential that corresponds to the S matrix element (4.463), where $\tilde{A}_\mu(q)$ is now the Fourier transform of the Coulomb potential (cf. Eq. 4.66). As is familiar from our earlier discussion on the vacuum polarization, the q^2-dependent term of (4.463) just gives rise to a $\delta^{(3)}(\mathbf{x})$ function potential. To find the effective potential corresponding to the anomalous moment term we may use the results of Problem 3–6 in which the reader is asked to work out the lowest-order matrix element of the $(\kappa e/2m)\,\bar{\psi}\,(\frac{1}{2}\sigma_{k4}F_{k4})\psi$ interaction between free-particle states. Taking the Fourier transform of the matrix element obtained there and taking advantage of (4.358), we get for the effective potential corresponding to the anomalous moment interaction

$$V^{(\text{anom mom})} = \left(\frac{\alpha}{2\pi}\right)\left(\frac{e}{2m}\right)\left[\frac{e}{2m}\,\delta^{(3)}(\mathbf{x}) + \frac{e}{4\pi m r^3}\,\boldsymbol{\sigma}\cdot\mathbf{L}\right]. \tag{4.472}$$

Note that the first term (the analog of the Foldy term in the electron-neutron interaction) affects just s-states while the second "spin-orbit" term affects all but s-states. To summarize, apart from the leading term $-e^2/(4\pi r)$, the effective potential due to (4.463) is

$$V^{(\text{eff})} = \frac{4\alpha^2}{3m^2}\left(\log\frac{m}{\lambda_{\min}} - \frac{3}{8} - \frac{1}{5} + \frac{3}{8}\right)\delta^{(3)}(\mathbf{x}) + \frac{\alpha^2}{4\pi m^2 r^3}(\boldsymbol{\sigma}\cdot\mathbf{L}). \tag{4.473}$$

If the photon mass were finite, we could immediately make use of the effective potential (4.473) to compute the Lamb shift. Unfortunately, as it stands, the expression diverges as $\lambda_{\min}\to 0$; this divergence, of course, stems from the fact that the original integral (4.446) is divergent at low values of $|k^2|$. At this stage it is fruitful to recall that Bethe's nonrelativistic treatment of the Lamb shift, discussed in Section 2–8, is completely free of divergence difficulties of this kind even though it is in difficulty at *high values* of the virtual-photon energy. According to (2.253), (2.255), and (2.259) Bethe's expression for the energy shift of an atomic level A corresponds to the effective potential

$$V^{\text{eff}}_{\text{Bethe}} = \frac{4\alpha^2}{3m^2}\log\frac{E_\gamma^{(\max)}}{\langle E_I - E_A\rangle_{\text{av}}}\,\delta^{(3)}(\mathbf{x}), \tag{4.474}$$

where E_I is the energy of an atomic level that can be reached from state A by an electric-dipole transition. It is now clear how we must proceed to obtain a completely finite expression for the Lamb shift. We use Bethe's formula (4.474) to estimate the contributions from very soft virtual photons; the nonrelativistic approximation used by Bethe should be a very good one if the photon wavelength is of the order of or larger than the atomic radius, that is, for $E_\gamma \lesssim \alpha m$. For the contribution from virtual photons of energies higher than αm we should use formula (4.473), which is divergence-free as far as the high-energy contributions are concerned. However, it is not so straightforward to "join" the two expressions since the low-energy cut-off that appears in (4.473) is expressed in terms of the mass of a covariant photon with four polarization states whereas the cut-off that appears in Bethe's calculation is the maximum energy of a zero-mass virtual photon with only two (transverse) polarization states.‡ A careful treatment (first done correctly by J. B. French) shows that the right correspondence between Feynman's $\lambda_{\min}$ and Bethe's $E_\gamma^{(\max)}$ is

$$\log \lambda_{\min} \longrightarrow \log 2E_\gamma^{(\max)} - \tfrac{5}{6}. \tag{4.475}$$

Since the derivation of this result is somewhat technical, we do not wish to perform it in this book; we shall simply mention that the difference between $\log \lambda_{\min}$ and $\log E_\gamma^{(\max)}$ is precisely the additive constant that appears in the *bremsstrahlung* cross section for a finite-mass photon (4.471).§ Using this result and the effective interactions (4.473) and (4.474), the atomic level shift is seen to be completely independent of $E_\gamma^{(\max)}$ or $\lambda_{\min}$; for a hydrogen-like atom characterized by n and l we get

$$\Delta E_{nl} = \frac{4\alpha^2}{3m^2}\left[\log\frac{m}{2\langle E_I - E_{nl}\rangle_{\text{av}}} + \frac{11}{24} - \frac{1}{5} + \frac{3}{8}\right] |\psi_{nl}(0)|^2 + \frac{\alpha^2}{4\pi m^2}\begin{Bmatrix} l \\ (-l-1)\end{Bmatrix}\int \frac{1}{r^3}|\psi_{nl}|^2\, d^3x \quad \text{for} \quad \begin{cases} j = l + \frac{1}{2} \\ j = l - \frac{1}{2}\end{cases} \tag{4.476}$$

where we have replaced $\boldsymbol{\sigma}\cdot\mathbf{L}$ in (4.473) by its eigenvalue. For s-states of the hydrogen atom the energy shift (4.476) reduces to

$$\Delta E_{l=0} = \frac{8\alpha^3}{3\pi}\frac{1}{n^3}Ry_\infty\left[\log\frac{m}{2\langle E_I - E_{n,\,l=0}\rangle_{\text{av}}} + \frac{11}{24} - \frac{1}{5} + \frac{3}{8}\right], \tag{4.477}$$

since the last term of (4.476), due to the "spin-orbit" interaction, is zero. For $l \neq 0$ states of the hydrogen atom we can use

$$\int \frac{1}{r^3}|\psi_{nl}|^2\, d^3x = \frac{1}{l(l+\frac{1}{2})(l+1)n^3a^3} \tag{4.478}$$

to obtain

$$\Delta E_{l\neq 0} = \frac{\alpha^3}{2\pi}\frac{1}{n^3}Ry_\infty \begin{cases} \dfrac{1}{(l+\frac{1}{2})(l+1)} & \text{for} \quad j = l + \frac{1}{2}, \\ \left[-\dfrac{1}{l(l+\frac{1}{2})}\right] & \text{for} \quad j = l - \frac{1}{2}. \end{cases} \tag{4.479}$$

Finally we get for the energy difference between the $2s\frac{1}{2}$ and the $2p\frac{1}{2}$ levels of the hydrogen atom

$$E(2s\tfrac{1}{2}) - E(2p\tfrac{1}{2}) = \frac{\alpha^3}{3\pi}Ry_\infty\left[\log\frac{m}{2\langle E_I - E_{n=2,\,l=0}\rangle_{\text{av}}} + \frac{11}{24} - \frac{1}{5} + \frac{3}{8} + \frac{1}{8}\right], \tag{4.480}$$

---

‡To show that the joining of Feynman's result to Bethe's result is indeed nontrivial, we may mention that this treacherous point caused a considerable amount of confusion (even among theoretical physicists of Nobel prize caliber) in the first attempts to obtain a finite result for the Lamb shift.

§The reader who is interested in the detailed derivation of (4.475) may consult Feynman (1961a), pp. 152–157; Bjorken and Drell (1964), pp. 173–176; Akhietzer and Berestetzkii (1965), pp. 423–429.

which corresponds to 1051 Mc. Various small corrections to formula (4.480) have been estimated; they include the use of the Dirac wave function in place of the Schrödinger wave function, the finite mass of the proton, and the fourth order corrections (terms of order $\alpha^4 Ry_\infty$). The *theoretical* value then becomes (1057.70 ± 0.15) Mc, to be compared with the *experimental* value (1057.77 ± 0.10) Mc.‡

Historically formula (4.480) was first derived in 1949 independently by N. M. Kroll and W. E. Lamb, by J. B. French and V. F. Weisskopf, and by H. Fukuda, Y. Miyamoto, and S. Tomonaga, who all used the old-fashioned perturbation method based on energy difference denominators. The derivation which we have sketched is that of Feynman, which is considerably simpler and shorter. What is more important, the separation of formally infinite (but not directly observable) quantities from finite observable quantities is much less ambiguous when we calculate the various terms in Feynman's way, which makes relativistic covariance apparent at each stage of the derivation.

Outlook. In this section we have shown how, despite the divergence difficulties inherent in the present form of the theory, we can extract finite numbers that can be compared to experimentally measured quantities for certain simple higher-order processes. One may naturally ask whether this procedure based on the renormalization of the mass and charge can be applied to other processes in quantum electrodynamics and to amplitudes involving even higher powers of α. The answer given by F. J. Dyson, A. Salam, and others is affirmative; once the mass and charge are redefined there are no further infinities for any purely quantum-electrodynamical processes to all finite orders in perturbation theory. In fact we have an unambiguous and workable set of prescriptions that enables us to calculate any purely quantum-electrodynamical effect to an arbitrary degree of accuracy in terms of just two parameters, the observed electron (or muon) mass and the observed charge.§

On the other hand, the very existence of divergences inherent in the theory makes us suspect that quantum electrodynamics must be modified at *short distances* or, equivalently, at *high energies*. We may recall that, in order to obtain finite values for the mass and the charge of the electron, we had to introduce cut-offs. We now wish to demonstrate that such a cut-off procedure is basically unsatisfactory. Take, for instance, the prescription for modifying the photon propagator,

$$\frac{\delta_{\mu\nu}}{ik^2} \longrightarrow \frac{\delta_{\mu\nu}}{ik^2}\left(\frac{\Lambda^2}{\Lambda^2 + k^2}\right), \tag{4.481}$$

‡More recent measurements of the Lamb shift by R. T. Robiscoe give (1058.05 ± 0.10) Mc. This value is somewhat different from the current theoretical estimate (1057.50 ± 0.11) Mc. At this writing it is not known whether this small discrepancy can be taken as evidence for breakdown of quantum electrodynamics.

§The reader who is interested in seeing how the renormalization program can be carried out in more general cases to all orders in perturbation theory may consult Chapter 19 of Bjorken and Drell (1965).

which was used earlier to obtain a finite value for the electron self-energy. Writing it as

$$\frac{\delta_{\mu\nu}}{ik^2}\left(\frac{\Lambda^2}{\Lambda^2 + k^2}\right) = \frac{\delta_{\mu\nu}}{ik^2} - \frac{\delta_{\mu\nu}}{i(k^2 + \Lambda^2)}, \tag{4.482}$$

we see that except for a minus sign the second part of (4.482) is the propagator for a neutral-vector boson of mass Λ. But because of the minus sign the hypothetical vector boson is coupled with $-(e^2/4\pi)$ in place of $(e^2/4\pi)$; this means that the coupling constant is purely imaginary. An interaction density that gives rise to (4.482) must therefore contain a term which is not Hermitian. Now we have seen in Section 4–2 that a non-Hermitian interaction results in a nonunitary S-matrix element. In other words, the modification (4.482) violates probability conservation.

Up to now, despite many heroic attempts, nobody has succeeded in satisfactorily modifying the theory without abandoning some of the cherished principles of twentieth-century physics—Lorentz invariance, the probabilistic interpretation of state vectors, the local nature of the interaction between $j_\mu(x)$ and $A_\mu(x)$ (that is, the field operators interact at the same space-time point), etc. It appears likely that to overcome the divergence difficulties in quantum field theory we really need new physical principles. P. A. M. Dirac views the present situation as follows:

> It would seem that we have followed as far as possible the path of logical development of the ideas of quantum mechanics as they are at present understood. The difficulties, being of a profound character, can be removed only by some drastic change in the foundations of the theory, probably a change as drastic as the passage from Bohr's orbit theory to the present quantum mechanics.

Just as the modification of classical physics was stimulated by a series of historic experiments that demonstrated the quantum nature of light and the wave nature of the electron, it is not inconceivable that some future experiments in quantum electrodynamics may throw light on the question of how the present theory must be modified. We may recall in this connection that in calculating finite (cut-off independent) quantities in quantum electrodynamics we have tacitly assumed that the cut-off Λ (which characterizes in a phenomenological way a possible modification of quantum electrodynamics at distances of order $1/\Lambda$) can be made to go to infinity; otherwise the calculated values of the Lamb shift, the anomalous moment of the electron, etc., would be dependent on Λ (cf. Problem 4–17). To turn the argument around, if the quantities we calculate disagreed with observation, we might gain some hints as to how the theory must be modified. The fact that the results of observations made so far agree extremely well with the theoretically predicted values (calculated under the assumption $\Lambda \longrightarrow \infty$) can be used to set a lower limit on Λ or, more generally, limits on the validity of the present quantum electrodynamics at short distances. For instance, from the fact that the observed anomalous part of the muon magnetic moment (measured by a CERN group) agrees with the theoretically predicted value to an accuracy of 0.4%, we may conclude (using Problem 4–17) that the cut-off Λ must be greater than 1 BeV (or $1/\Lambda < 2 \times 10^{-14}$ cm).

Quantum electrodynamics is not a finished subject. It is true that, thanks to the success of the renormalization procedure, quantum electrodynamics has come to enjoy a "status of peaceful coexistence with its divergences" (as S. D. Drell puts it); yet we are still very far from constructing a complete and finite theory. The advance in quantum electrodynamics made in the period 1947–1950 is likely to be viewed by future historians of physics as the last successful attempt to extract physically sensible results from the relativistic quantum theory whose inherent difficulties at a fundamantal level were already recognized in the early 1930's.

PROBLEMS

4–1. (a) Derive (4.38) and (4.39). Note how the "$i\varepsilon$" arises as we attempt to give a meaning to the integral

$$\int_{-\infty}^{t_1} dt_2\, e^{i(E_n - E_i)t_2}.$$

How would (4.39) be modified if $H_I^{(S)}$ had a harmonic time dependence $e^{\pm i\omega t}$?

(b) Show that the unitarity relation (4.54) implies for a decaying state $|i\rangle$ the analog of the optical theorem

$$\frac{1}{\tau_i} = -2\,\mathrm{Im}\, T_{ii}.$$

(c) In Section 2–8 it was shown that the imaginary part of the energy shift of an atomic level computed to second order is related via (2.237) to the mean lifetime computed using first-order perturbation theory. Discuss this connection using the relation derived in (b).

4–2. Suppose the Coulomb potential transformed relativistically like a scalar field (rather than like the fourth-component of a vector field) so that the interaction of the electron with the Coulomb potential would read

$$\mathscr{H}_{\text{int}} = e\bar{\psi}\psi\phi^{(e)}, \qquad \phi^{(e)} = -Ze/(4\pi r).$$

Show that both the shape and the energy dependence of the differential cross section would be completely different from those given by (4.85) at high energies even though the two differential cross sections are identical at nonrelativistic energies.

4–3. It is intuitively obvious that the amplitude for the scattering of a positron by a Coulomb field computed to first order is equal in magnitude but opposite in sign to the electron scattering amplitude for the same momentum-spin states. Prove this statement rigorously. Show also that the interference terms between the first-order and the second-order amplitude are opposite in sign for electron and positron scattering. (*Note:* In reality the electron scattering cross section and the positron scattering cross section differ significantly for scattering from high Z nuclei, for example, by as much as a factor of 6 for $Z = 80$, $\theta = 90°$, $E/m = 2$.)

4–4. (a) Prove (4.76).

(b) Show that the spin-projection operator (4.101) can be used without any modification for processes involving positrons.

4–5. Prove that the probability per unit volume per unit time that the external potential (4.57) [$\mathbf{A} = (0, 0, a \cos \omega t)$, $A_4 = 0$] creates an electron-positron pair in the vacuum is given by

$$R = \frac{2}{3}\left(\frac{e^2}{4\pi}\right)\left(\frac{|a|}{2}\right)^2 \omega^2\left(1 + \frac{2m^2}{\omega^2}\right)\sqrt{1 - \frac{4m^2}{\omega^2}}.$$

What is the angular distribution of electron-positron pairs?

4–6. The ω-meson (whose rest energy is 783 MeV) is an electrically neutral $J = 1^-$ particle that decays via strong interactions into $\pi^+ + \pi^- + \pi^0$ most of the time. Since a virtual process

$$\omega \longrightarrow \text{pair of strongly interacting particles} \longrightarrow \gamma$$

is allowed, the ω-meson is expected to decay into $e^+ + e^-$ some of the time. This decay interaction may be represented phenomenologically by

$$\mathscr{H}_{\text{int}} = ig\phi_\mu \bar{\psi}\gamma_\mu\psi$$

where ϕ_μ is the neutral vector field corresponding to the ω meson. Experiments have indicated that

$$\Gamma(\omega \longrightarrow \mathrm{e}^+ + \mathrm{e}^-)/\Gamma_{\text{tot}} \approx 10^{-4},$$

where Γ_{tot} is known to be about 10 MeV. Estimate $g^2/4\pi$. Express your final answer in the form

$$g^2/4\pi = (\text{dimensionless number}) \times (\tfrac{1}{137})^2.$$

(*Note:* In the rest system of the ω meson the fourth-component of the polarization vector vanishes, cf. Problem 1–4.)

4–7. Repeat Problem 2–5 by considering the interaction density

$$\mathscr{H}_{\text{int}} = \frac{\kappa e}{(m_\Lambda + m_\Sigma)}(\bar{\psi}_\Lambda \sigma_{\mu\nu}\psi_\Sigma)\left(\frac{1}{2}F_{\mu\nu}\right) + \text{Hc}$$

and compare the results. Do not make any nonrelativistic approximation. [*Hint:* Simplify the $\mathscr{M}$-matrix as much as you can before you square it and take the trace.]

4–8. Discuss the results of correlated polarization measurements on the two-photon state (4.214a) when (a) *both* observers have detection devices sensitive to the circular polarization, (b) observer A and observer B have detection devices sensitive to the linear and the circular polarization, respectively.

4–9. (a) Compute the differential cross section for the scattering of a π^+ meson by an unpolarized proton in the CM-system to lowest nonvanishing order using the interaction density (4.377).

(b) The coupling constant G is estimated from nuclear forces to be $G^2/4\pi \approx 14$. Show that if this value is used, the computed $d\sigma/d\Omega$ at very low kinetic energies ($\lesssim 20$ MeV) is in serious disagreement with observation:

$$(d\sigma/d\Omega)_{\text{observed}} \approx (0.1 - 0.2)\ \text{mb/ster}$$

independent of θ after Coulomb corrections are made. (*Note:* This shows that the use of perturbation theory is unjustified, and/or that the pseudoscalar coupling theory is wrong.)

4–10. (a) Derive the S-matrix element for *bremsstrahlung* (4.240). Simplify it as much as you can.

(b) Show that the *bremsstrahlung* cross section for the emission of a very soft photon of momentum **k**, polarization vector $\boldsymbol{\epsilon}^{(\alpha)}$, by a nonrelativistic electron is given by

$$\frac{d^3\sigma}{d\Omega_{\mathbf{k}}\, d\Omega_{\mathbf{p}'}\, d\omega} = \frac{\alpha}{4\omega m^2}\left(\frac{d\sigma}{d\Omega}\right)_{\text{Mott}} |\boldsymbol{\epsilon}^{(\alpha)}\cdot(\mathbf{p}' - \mathbf{p})|^2,$$

where $(d\sigma/d\Omega)_{\text{Mott}}$ stands for (4.85). Sum over the photon polarizations and integrate over the photon direction to obtain

$$\frac{d^2\sigma}{d\Omega_{\mathbf{p}'}\, d\omega} = \frac{2\alpha}{3\pi}\frac{|\mathbf{p}' - \mathbf{p}|^2}{\omega m^2}\left(\frac{d\sigma}{d\Omega}\right)_{\text{Mott}}.$$

[*Hint:* The polarization sum is facilitated if you take advantage of (4.313).]

4–11. (a) Show that in the nonrelativistic Schrödinger theory the Green function satisfying

$$\left(i\frac{\partial}{\partial t} + \frac{1}{2m}\nabla^2\right)G(x, x') = i\delta^{(4)}(x - x')$$

is given by

$$\begin{aligned} G(x, x') &= \frac{1}{(2\pi)^3}\int d^3p \exp\,[i\mathbf{p}\cdot(\mathbf{x} - \mathbf{x}') - i(|\mathbf{p}|^2/2m)(t - t')] \\ &= i\int \frac{d^3p}{(2\pi)^3}\exp\,[i\mathbf{p}\cdot(\mathbf{x} - \mathbf{x}')\int_{-\infty}^{\infty}\frac{d\omega}{2\pi}\frac{\exp\,[i\omega(t - t')]}{[\omega - (|\mathbf{p}|^2/2m)]}. \end{aligned}$$

How must we perform the integration along the real ω-axis if $G(x, x')$ is to vanish for $t' > t$?

(b) Using the Green's function with the above boundary condition, compute the transition amplitude for the elastic scattering of a Schrödinger particle by a time-dependent potential

$$\mathscr{V}(\mathbf{x}, t) = V(\mathbf{x})\cos\omega t$$

to lowest nonvanishing order. Show that the transition amplitude we obtain in this way is identical to the one obtained using the conventional second-order time-dependent perturbation theory.

4–12. (a) In Section 3–9 the amplitude for Thomson scattering was computed using the hole theory and the old-fashioned perturbation method. Setting $p = (0, im)$, $k = k' = 0$ in (4.275) and using the explicit forms of $K_F(x, x')$ given by (4.264), demonstrate the equivalence of the old-fashioned method to the calculational method of Section 4–5 for this particular process.

(b) When O. Klein and Y. Nishina derived formula (3.344) in 1929, the hole-theoretic interpretation of the Dirac theory had not yet been formulated. They therefore assumed that the negative-energy states are completely empty and a positive-energy electron can make a virtual transition into a negative-energy state. Why were they able to obtain the correct formula?

(c) Using the example of Compton scattering, convince yourself that the minus sign in (4.264b) can be visualized as arising from the Pauli exclusion principle.

4–13. It is evident that if the proton behaved as if it were just a heavy muon, the $\mathscr{M}$-matrix element for electron-proton scattering would be given by

$$\mathscr{M}_{fi} = -\frac{e^2(\bar{u}_{\mathrm{p}}'\gamma_\mu u_{\mathrm{p}})(\bar{u}_{\mathrm{e}}'\gamma_\mu u_{\mathrm{e}})}{q^2},$$

with $q = p^{(\mathrm{p})'} - p^{(\mathrm{p})} = p^{(\mathrm{e})} - p^{(\mathrm{e})'}$. Actually the strong interactions modify the γ_μ-vertex of the proton current in such a way that $\bar{u}'_{\mathrm{p}}\gamma_\mu u_{\mathrm{p}}$ must be replaced by

$$\bar{u}'_{\mathrm{p}}[\gamma_\mu F_1(q^2) + (\kappa/2m_{\mathrm{p}})\sigma_{\nu\mu} q_\nu F_2(q^2)]u_{\mathrm{p}},$$

where κ is the anomalous proton magnetic moment in units of $e/(2m_{\mathrm{p}})$ and $F_1(q^2)$ and $F_2(q^2)$ (which can be shown to be real for $q^2 \geqslant 0$) are the Dirac and the Pauli form factors of the proton normalized so that $F_1(0) = F_2(0) = 1$. Derive the famous Rosenbluth formula for electron-proton scattering in the laboratory system (the proton initially at rest):

$$\left(\frac{d\sigma}{d\Omega}\right)_{\mathrm{lab}} = \frac{\alpha^2 \cos^2(\theta/2)}{4E^2 \sin^4(\theta/2)} \frac{1}{\{1 + [2E \sin^2(\theta/2)/m_{\mathrm{p}}]\}}$$
$$\times \left\{|F_1(q^2)|^2 + \frac{q^2}{4m_{\mathrm{p}}^2}\left[2|F_1(q^2) + \kappa F_2(q^2)|^2 \tan^2\left(\frac{\theta}{2}\right) + \kappa^2 |F_2(q^2)|^2\right]\right\}$$

with

$$q^2 = \frac{4E^2 \sin^2(\theta/2)}{1 + (2E/m_{\mathrm{p}}) \sin^2(\theta/2)},$$

where the incident laboratory energy of the electron denoted by E is assumed to be much greater then m_{e}. [*Hint:* Simplify the $\mathcal{M}$-matrix element as much as you can before you square it and take the trace.]

4–14. (a) Prove (4.358).
(b) Show that the spin-orbit potential between two electrons is given by (4.360).

4–15. (a) Show that the annihilation diagram of Fig. 4–21(b) has no effect on the singlet state of an e^+e^- system with $\mathbf{p}_+ = \mathbf{p}_- = 0$, first, using the explicit forms of the free-particle (at-rest) wave functions, next, using only symmetry arguments.
(b) Construct an effective potential that corresponds to Fig. 4–21(b). Show that this "annihilation potential" affects only the ^3S state.
(c) Show that the energy difference (hyperfine splitting) between the ^3S and ^1S states of positronium with $n = 1$ is given by (4.366).

4–16. (a) Express in terms of $\Pi(q^2)$ the S-matrix element (up to second order in e) for the vacuum to remain the vacuum in the presence of the external potential considered in Problem 4–5. Using this result and Problem 4–5, show that the unitary equation (4.433) indeed gives (4.432).
(b) Applying the usual analyticity argument to $\Pi(q^2)/\omega$ with $q^2 = -\omega^2$, write a once-subtracted dispersion relation for $\Pi(q^2)$. Evaluate the dispersion integral to obtain expressions for $\Pi^f(q^2)$ valid for $q^2 \approx 0$ (cf. Eq. 4.424), and for $q^2 \gg m^2$.

4–17. Apply the techniques of Appendix E to evaluate the anomalous magnetic moment of the muon with a *finite* Λ. Discuss how the predicted (Schwinger) value of the magnetic moment would be altered if Λ were 500 MeV, 1 GeV, or 10 GeV.

APPENDIX A

ELECTRODYNAMICS IN THE RADIATION (COULOMB) GAUGE

The purpose of this appendix is to derive the basic Hamiltonian of electrodynamics subject to the requirement that the three-vector potential $\mathbf{A}$ satisfies the transversality condition. We shall first show that it is always possible to consider a gauge transformation of the type (1.82) such that in the new gauge the transversality condition

$$\nabla \cdot \mathbf{A}(\mathbf{x}, t) = 0 \tag{A–1}$$

is satisfied. To see this explicitly we just set

$$\begin{aligned} \mathbf{A}^{(\text{new})}(\mathbf{x}, t) &= \mathbf{A}^{(\text{old})}(\mathbf{x}, t) + \nabla\chi(\mathbf{x}, t), \\ A_0^{(\text{new})}(\mathbf{x}, t) &= A_0^{(\text{old})}(\mathbf{x}, t) + (1/c)[\partial\chi(\mathbf{x}, t)/\partial t], \end{aligned} \tag{A–2}$$

where

$$\chi(\mathbf{x}, t) = \frac{1}{4\pi}\int \frac{d^3x'}{|\mathbf{x} - \mathbf{x}'|}\nabla' \cdot \mathbf{A}^{(\text{old})}(\mathbf{x}', t). \tag{A–3}$$

We then have

$$\nabla \cdot \mathbf{A}^{(\text{new})} = \nabla \cdot (\mathbf{A}^{(\text{old})} + \nabla\chi) = 0, \tag{A–4}$$

which shows that the transversality condition is satisfied for the new vector potential. On the other hand, from (1.75), which holds quite generally, we get

$$\nabla^2 A_0 - \frac{1}{c^2}\frac{\partial^2 A_0}{\partial t^2} + \frac{1}{c}\frac{\partial}{\partial t}\left(\nabla \cdot \mathbf{A} + \frac{1}{c}\frac{\partial A_0}{\partial t}\right) = -\rho \tag{A–5}$$

for the zeroth component of A_μ. When the three-vector potential satisfies the transversality condition, it is seen that A_0 is determined by Poisson's equation

$$\nabla^2 A_0(\mathbf{x}, t) = -\rho(\mathbf{x}, t), \tag{A–6}$$

which does not contain any time derivative. The solution to this differential equation satisfying the boundary condition $A_0 \longrightarrow 0$ at infinity is

$$A_0(\mathbf{x}, t) = \frac{1}{4\pi}\int \frac{\rho(\mathbf{x}', t)d^3x'}{|\mathbf{x} - \mathbf{x}'|}. \tag{A–7}$$

We wish to emphasize that A_0 in (A–7) is an *instantaneous* Coulomb potential in the sense that the value of A_0 at t is determined by the charge distribution at the *same* instant of time; in particular, it should not be confused with the *retarded* Coulomb potential

$$A_0(\mathbf{x}, t) = \frac{1}{4\pi}\int d^3x' \left.\frac{\rho(\mathbf{x}', t')}{|\mathbf{x} - \mathbf{x}'|}\right|_{t' = t - \frac{|\mathbf{x}-\mathbf{x}'|}{c}} \tag{A–8}$$

which we obtain by solving the D'Alembertian equation for A_0.‡ To summarize what we have done so far:

a) we can always choose the vector potential so that the transversality condition (A–1) is satisfied;

b) if A satisfies the transversality condition, then A_0 is given by (A–7).

The gauge in which A satisfies (A–1) and A_0 is given by (A–7) is called the radiation (or Coulomb) gauge.

The total Hamiltonian for the charged particles and the electromagnetic field is given by the space integral of

$$\mathscr{H} = \mathscr{H}_{\text{em}} + \mathscr{H}_{\text{matter}} + \mathscr{H}_{\text{int}}. \tag{A–9}$$

For the Hamiltonian density of the electromagnetic field we have, from (1.73),

$$\begin{aligned}\mathscr{H}_{\text{em}} &= \frac{\partial \mathscr{L}_{\text{em}}}{\partial(\partial A_\mu/\partial x_4)}\frac{\partial A_\mu}{\partial x_4} - \mathscr{L}_{\text{em}} \\ &= \frac{1}{2}(|\mathbf{B}|^2 + |\mathbf{E}|^2) - i\mathbf{E}\cdot\boldsymbol{\nabla}A_4 \\ &= \frac{1}{2}(|\mathbf{B}|^2 + |\mathbf{E}|^2) - \rho A_0 + \boldsymbol{\nabla}\cdot(A_0\mathbf{E}),\end{aligned} \tag{A–10}$$

where the last term of the last line is of no significance since it vanishes when we evaluate $\int \mathscr{H}_{\text{em}} d^3x$. As for the interaction part of the Hamiltonian density, we write it as

$$\mathscr{H}_{\text{int}} = -j_\mu A_\mu/c, \tag{A–11}$$

where j_μ is the charge-current (four-vector) density. Combining (A–10) and (A–11), we get

$$\begin{aligned}H_{\text{em}} + H_{\text{int}} &= \int (\mathscr{H}_{\text{em}} + \mathscr{H}_{\text{int}})d^3x \\ &= \frac{1}{2}\int (|\mathbf{B}|^2 + |\mathbf{E}|^2)d^3x - \int[(\mathbf{j}\cdot\mathbf{A})/c]d^3x.\end{aligned} \tag{A–12}$$

At first sight this appears to be a very strange result because the ρA_0 term has completely disappeared. We shall come back to this "puzzle" in a moment.

Our next task is to show that the Hamiltonian (A–12) can be written in a somewhat different form when we work in the radiation gauge. First, we recall that, quite generally, the electric field $\mathbf{E}$ is derivable from

$$\mathbf{E} = -\boldsymbol{\nabla}A_0 - \frac{1}{c}\frac{\partial \mathbf{A}}{\partial t}. \tag{A–13}$$

We claim that in the radiation gauge the irrotational and the solenoidal part of $\mathbf{E}$ defined by

$$\begin{gathered}\mathbf{E} = \mathbf{E}_{||} + \mathbf{E}_{\perp}, \\ \boldsymbol{\nabla}\times\mathbf{E}_{||} = 0, \qquad \boldsymbol{\nabla}\cdot\mathbf{E}_{\perp} = 0.\end{gathered} \tag{A–14}$$

‡Note that the D'Alembertian equation for A_0 follows from (A–5) when the vector potential satisfies the Lorentz condition, not the transversality condition. For the derivation of (A–8) see, for example, Panofsky and Phillips (1958), pp. 212–214.

are uniquely given by

$$\mathbf{E}_{||} = -\nabla A_0, \qquad \mathbf{E}_{\perp} = -\frac{1}{c}\frac{\partial \mathbf{A}^{(\perp)}}{\partial t}, \tag{A–15}$$

where the superscript $(\perp)$ is to remind us of the transversality condition (A–1). This can be proved by noting: (a) $\nabla \times (\nabla A_0)$ vanishes, (b) the divergence of $\partial \mathbf{A}^{(\perp)}/\partial t$ vanishes because of the transversality condition (A–1), and (c) the decomposition of $\mathbf{E}$ into $\mathbf{E}_{||}$ and $\mathbf{E}_{\perp}$ is unique.

Meanwhile we can show that the space integral of $|\mathbf{E}|^2$ is actually given by

$$\int |\mathbf{E}|^2 d^3x = \int |\mathbf{E}_{\perp}|^2 d^3x + \int \rho A_0 d^3x \tag{A–16}$$

in the radiation gauge. To prove this we first note that

$$\begin{aligned}\int |\mathbf{E}|^2 d^3x &= \int [|\mathbf{E}_{\perp}|^2 - 2(\nabla A_0)\cdot\mathbf{E}_{\perp} + |\mathbf{E}_{||}|^2] d^3x \\ &= \int |\mathbf{E}_{\perp}|^2 d^3x + 2\int A_0(\nabla\cdot\mathbf{E}_{\perp}) d^3x + \int |\mathbf{E}_{||}|^2 d^3x \\ &= \int (|\mathbf{E}_{\perp}|^2 + |\mathbf{E}_{||}|^2) d^3x. \end{aligned} \tag{A–17}$$

Furthermore,

$$\begin{aligned}\int |\mathbf{E}_{||}|^2 d^3x &= \int (\nabla A_0)\cdot(\nabla A_0) d^3x \\ &= \int \nabla\cdot(A_0 \nabla A_0) d^3x - \int A_0 \nabla^2 A_0 d^3x \\ &= \int \rho A_0 d^3x. \end{aligned} \tag{A–18}$$

Combining (A–17) and (A–18), we see that the space integral of $|\mathbf{E}|^2$ is indeed given by (A–16). Coming back to the expression for the Hamiltonian given by (A–12), we note that the ρA_0-term which, we thought, had disappeared is actually buried in the space integral of $|\mathbf{E}|^2$:

$$H_{\text{em}} + H_{\text{int}} = \tfrac{1}{2}\int (|\mathbf{B}|^2 + |\mathbf{E}_{\perp}|^2) d^3x + \int \left(\tfrac{1}{2}\rho A_0 - \frac{1}{c}\mathbf{j}\cdot\mathbf{A}^{(\perp)}\right) d^3x. \tag{A–19}$$

Note, however, that the ρA_0-term now appears with a factor $\frac{1}{2}$.

We are finally in a position to eliminate A_0 altogether (or equivalently $\mathbf{E}_{||}$) from the Hamiltonian. Using (A–7), we can write the ρA_0-term of (A–19) as

$$\frac{1}{2}\int \rho A_0 d^3x = \frac{1}{2}\int d^3x \int d^3x' \frac{\rho(\mathbf{x}, t)\rho(\mathbf{x}', t)}{4\pi|\mathbf{x} - \mathbf{x}'|}. \tag{A–20}$$

Hence

$$\begin{aligned} H_{\text{em}} + H_{\text{int}} &= \frac{1}{2}\int (|\mathbf{B}|^2 + |\mathbf{E}_{\perp}|^2) d^3x \\ &\quad + \frac{1}{2}\int d^3x \int d^3x' \frac{\rho(\mathbf{x}, t)\rho(\mathbf{x}', t)}{4\pi|\mathbf{x} - \mathbf{x}'|} - \frac{1}{c}\int \mathbf{j}\cdot\mathbf{A}^{(\perp)} d^3x \end{aligned} \tag{A–21}$$

or, in terms of $\mathbf{A}^{(\perp)}$, $\mathbf{j}$, and ρ only,

$$H_{\text{em}} + H_{\text{int}} = \frac{1}{2}\int\left(|\nabla \times \mathbf{A}^{(\perp)}|^2 + \left|\frac{1}{c}\frac{\partial \mathbf{A}^{(\perp)}}{\partial t}\right|^2\right)d^3x$$
$$- \frac{1}{c}\int \mathbf{j}\cdot\mathbf{A}^{(\perp)}d^3x + \frac{1}{2}\int d^3x \int d^3x' \frac{\rho(\mathbf{x}, t)\rho(\mathbf{x}', t)}{4\pi|\mathbf{x} - \mathbf{x}'|}. \quad \text{(A–22)}$$

In this form the electrostatic potential A_0 (or the irrotational part of **E**) has been completely eliminated in favor of the instantaneous Coulomb interaction. This means that we can forget about the dynamical degrees of freedom associated with A_0 or $\mathbf{E}_\perp$. In quantizing the Maxwell field it is sufficient to quantize just the transverse vector potential $\mathbf{A}^{(\perp)}$ provided we work in the radiation gauge. This procedure was first advocated by E. Fermi in 1932.

From (A–21) and (A–22) we see that the only term in the interaction Hamiltonian which involves the electromagnetic potential explicitly is the space integral of $-(1/c)\,\mathbf{j}\cdot\mathbf{A}^{(\perp)}$. In nonrelativistic quantum mechanics this is given just by the difference

$$\sum_i \frac{1}{2m_i}\left(\mathbf{p}^{(i)} - \frac{e_i\mathbf{A}^{(\perp)}}{c}\right)^2 - \sum_i \frac{(\mathbf{p}^{(i)})^2}{2m_i}. \quad \text{(A–23)}$$

As for the Coulomb part, we can use the familiar form

$$\rho = \sum_i e_i\delta^{(3)}(\mathbf{x} - \mathbf{x}^{(i)}) \quad \text{(A–24)}$$

to obtain

$$\frac{1}{2}\int d^3x \int d^3x' \frac{\rho(\mathbf{x}, t)\rho(\mathbf{x}', t)}{4\pi|\mathbf{x} - \mathbf{x}'|} = \sum_{i>j}\frac{e_ie_j}{4\pi|\mathbf{x}^{(i)} - \mathbf{x}^{(j)}|} + \frac{1}{2}\sum_i \frac{e_i^2}{4\pi|\mathbf{x}^{(i)} - \mathbf{x}^{(i)}|}. \quad \text{(A–25)}$$

The last term which represents the interaction of a charged particle with the Coulomb field due to the same charged particle is infinite; hence it is usually dropped. However, in discussing the self-energy of the Dirac electron, it is necessary to include the relativistic analog of this term, as shown in Section 4–7.

APPENDIX B

GAMMA MATRICES

I. Our notation (also used by Pauli, Källén, Rose, Mandl, Akhietzer and Berestetzkii, etc.) is given below:

$$\{\gamma_\mu, \gamma_\nu\} = 2\delta_{\mu\nu},$$

where $\mu, \nu = 1, 2, \cdots 4, \qquad \gamma_\mu^\dagger = \gamma_\mu;$

$$\gamma_5 = \gamma_1\gamma_2\gamma_3\gamma_4 = \frac{1}{4!}\epsilon_{\mu\nu\lambda\sigma}\gamma_\mu\gamma_\nu\gamma_\lambda\gamma_\sigma;$$

$$\{\gamma_5, \gamma_\mu\} = 0,$$

where $\mu = 1, \cdots 4, \qquad \gamma_5^\dagger = \gamma_5, \qquad \gamma_5^2 = 1;$

$$a\cdot b = \mathbf{a}\cdot\mathbf{b} + a_4 b_4 = \mathbf{a}\cdot\mathbf{b} - a_0 b_0, \qquad a_\mu = (\mathbf{a}, a_4) = (\mathbf{a}, ia_0);$$

$$\gamma_k = -i\beta\alpha_k \overset{\text{DP}}{=} \begin{pmatrix} 0 & -i\sigma_k \\ i\sigma_k & 0 \end{pmatrix}, \qquad k = 1, 2, 3,$$

$$\gamma_4 = \beta \overset{\text{DP}}{=} \begin{pmatrix} I & 0 \\ 0 & -I \end{pmatrix},$$

$$\gamma_5 \overset{\text{DP}}{=} -\begin{pmatrix} 0 & I \\ I & 0 \end{pmatrix},$$

$$\sigma_{\mu\nu} = \frac{[\gamma_\mu, \gamma_\nu]}{2i} = -i\gamma_\mu\gamma_\nu \qquad (\mu \neq \nu),$$

$$\sigma_{ij} = -\gamma_5\alpha_k = \Sigma_k \overset{\text{DP}}{=} \begin{pmatrix} \sigma_k & 0 \\ 0 & \sigma_k \end{pmatrix} \qquad (ijk \text{ cyclic}),$$

$$\sigma_{k4} = -\sigma_{4k} = -\gamma_5\sigma_{ij} = \alpha_k \overset{\text{DP}}{=} \begin{pmatrix} 0 & \sigma_k \\ \sigma_k & 0 \end{pmatrix} \qquad (ijk \text{ cyclic}),$$

$$i\gamma_5\gamma_k \overset{\text{DP}}{=} \begin{pmatrix} \sigma_k & 0 \\ 0 & -\sigma_k \end{pmatrix},$$

$$i\gamma_5\gamma_4 \overset{\text{DP}}{=} i\begin{pmatrix} 0 & I \\ -I & 0 \end{pmatrix}.$$

For free-particle wave functions

$$\left(\gamma_\mu \frac{\partial}{\partial x_\mu} + \frac{mc}{\hbar}\right)\psi = 0, \qquad (i\gamma\cdot p + mc)u^{(s)}(\mathbf{p}) = 0.$$

II. Notation used by Messiah, Bjorken and Drell, Schweber, etc.:

$$\{\gamma^\mu, \gamma^\nu\} = 2g^{\mu\nu}, \qquad \mu, \nu = 0, 1, 2, 3;$$

$$g_{\mu\nu} = g^{\mu\nu}, \qquad g_{00} = 1, \qquad g_{kk} = -1, \qquad g_{\mu\nu} = 0 \quad \text{if} \quad \mu \neq \nu;$$

$$\gamma^{0\dagger} = \gamma^0, \qquad \gamma^{k\dagger} = -\gamma^k;$$

$$a\cdot b = a^\mu b_\mu = a_0 b_0 - \mathbf{a}\cdot\mathbf{b}, \qquad a^\mu = (a_0, \mathbf{a}),$$

$$a_\mu = g_{\mu\nu}a^\nu = (a_0, -\mathbf{a});$$

$$\gamma^0 = \beta \overset{\text{DP}}{=} \begin{pmatrix} I & 0 \\ 0 & -I \end{pmatrix}, \qquad \gamma^k = \beta\alpha_k \overset{\text{DP}}{=} \begin{pmatrix} 0 & \sigma_k \\ -\sigma_k & 0 \end{pmatrix}.$$

For free-particle wave functions,

$$\left(-i\gamma^\mu \frac{\partial}{\partial x^\mu} + \frac{mc}{\hbar}\right)\psi = 0, \qquad (\gamma\cdot p - mc)u^{(s)}(\mathbf{p}) = 0.$$

III. Useful relations (*our notation only*):

$$(\gamma\cdot a)(\gamma\cdot b) = 2a\cdot b - (\gamma\cdot b)(\gamma\cdot a),$$

$$(\boldsymbol{\gamma}\cdot\mathbf{a})(\boldsymbol{\gamma}\cdot\mathbf{b}) = (\boldsymbol{\alpha}\cdot\mathbf{a})(\boldsymbol{\alpha}\cdot\mathbf{b}) = \mathbf{a}\cdot\mathbf{b} + i\boldsymbol{\Sigma}\cdot(\mathbf{a}\times\mathbf{b}),$$

$$\gamma_4[(\gamma\cdot a)(\gamma\cdot b)\cdots(\gamma\cdot l)]^\dagger\gamma_4 = (-\gamma\cdot l)\cdots(-\gamma\cdot b)(-\gamma\cdot a),$$

if $a = (\mathbf{a}, ia_0)$ with a_k and a_0 real, etc.;

$$\left.\begin{aligned} &\gamma_\mu\gamma_\mu = 4 \\ &\gamma_\mu(\gamma\cdot a)\gamma_\mu = -2\gamma\cdot a \\ &\gamma_\mu(\gamma\cdot a)(\gamma\cdot b)\gamma_\mu = 2(\gamma\cdot a)(\gamma\cdot b) + 2(\gamma\cdot b)(\gamma\cdot a) \\ &\qquad\qquad\qquad\quad = 4a\cdot b \\ &\gamma_\mu(\gamma\cdot a)(\gamma\cdot b)(\gamma\cdot c)\gamma_\mu = -2(\gamma\cdot c)(\gamma\cdot b)(\gamma\cdot a) \end{aligned}\right\}\mu\text{-indices summed};$$

$$\mathrm{Tr}(\underbrace{\gamma_\mu\gamma_\nu\cdots\gamma_\rho}_{\text{odd number}}) = 0,$$

$$\mathrm{Tr}(\underbrace{\gamma_5\gamma_\mu\cdots}_{\text{less than }4\gamma\text{ (for }4\gamma\text{ nonvanishing only if all four are different)}}) = 0,$$

$$\mathrm{Tr}(\Gamma_A) = 0,$$

where

$$\Gamma_A = \gamma_\mu, \sigma_{\mu\nu}, i\gamma_5\gamma_\mu, \gamma_5,$$

$$\mathrm{Tr}[(\gamma\cdot a)(\gamma\cdot b)] = 4a\cdot b,$$

$$\mathrm{Tr}[(\gamma\cdot a)(\gamma\cdot b)(\gamma\cdot c)(\gamma\cdot d)] = 4[(a\cdot b)(c\cdot d) - (a\cdot c)(b\cdot d) + (a\cdot d)(b\cdot c)].$$

IV. Free-particle wave functions in the Dirac-Pauli representation:

a) For $E = \sqrt{|\mathbf{p}|^2c^2 + m^2c^4} > 0$,

$$\psi^{(1,2)} = \sqrt{\frac{mc^2}{EV}}\, u^{(1,2)}(\mathbf{p}) \exp[i(\mathbf{p}\cdot\mathbf{x}/\hbar) - (iEt/\hbar)],$$

where

$$u^{(1)}(\mathbf{p}) = \sqrt{\frac{E + mc^2}{2mc^2}} \begin{pmatrix} 1 \\ 0 \\ p_3c/(E + mc^2) \\ (p_1 + ip_2)/(E + mc^2) \end{pmatrix},$$

$$u^{(2)}(\mathbf{p}) = \sqrt{\frac{E + mc^2}{2mc^2}} \begin{pmatrix} 0 \\ 1 \\ (p_1 - ip_2)c/(E + mc^2) \\ -p_3c/(E + mc^2) \end{pmatrix}.$$

b) For $E = -\sqrt{|\mathbf{p}|^2 c^2 + m^2c^4} < 0$,

$$\psi^{(3,4)} = \sqrt{\frac{mc^2}{|E|V}} u^{(3,4)}(\mathbf{p}) \exp [i(\mathbf{p}\cdot\mathbf{x}/\hbar) + (i|E|t/\hbar)],$$

where

$$u^{(3)}(\mathbf{p}) = \sqrt{\frac{|E| + mc^2}{2mc^2}} \begin{pmatrix} -p_3c/(|E| + mc^2) \\ -(p_1 + ip_2)c/(|E| + mc^2) \\ 1 \\ 0 \end{pmatrix},$$

$$u^{(4)}(\mathbf{p}) = \sqrt{\frac{|E| + mc^2}{2mc^2}} \begin{pmatrix} -(p_1 - ip_2)/(|E| + mc^2) \\ p_3c/(|E| + mc^2) \\ 0 \\ 1 \end{pmatrix}.$$

c) For "positron spinors"

$$v^{(1)}(\mathbf{p}) = -u^{(4)}(-\mathbf{p}), \qquad v^{(2)}(\mathbf{p}) = u^{(3)}(-\mathbf{p}).$$

APPENDIX C

PAULI'S FUNDAMENTAL THEOREM

Before we prove Pauli's fundamental theorem that guarantees the representation independence of the gamma matrices, we shall summarize the properties of the gamma matrices obtained in the main text. In Section 3–5 we showed that we can construct sixteen 4 $\times$ 4 matrices denoted by Γ_A with $A = 1, \ldots, 16$. They are

$$\Gamma_A = 1, \quad \gamma_\mu, \quad \sigma_{\mu\nu}, \quad i\gamma_5\gamma_\mu, \quad \text{and} \quad \gamma_5, \tag{C–1}$$

each of which satisfies

$$\Gamma_A^2 = 1. \tag{C–2}$$

Using the anticommutation relation between γ_μ and γ_ν alone (that is, without recourse to any particular representation), we can show by direct inspection the following two additional properties:

a) Define a constant η_{AB} by

$$\Gamma_A\Gamma_B = \eta_{AB}\Gamma_C. \tag{C–3}$$

Then

$$\eta_{AB} = +1, -1, +i \quad \text{or} \quad -i. \tag{C–4}$$

b) Matrices Γ_A and Γ_B either commute or anticommute (that is, $\Gamma_A\Gamma_B = \pm\Gamma_B\Gamma_A$). Moreover, for every Γ_A ($\neq 1$), there exists at least one Γ_B with the property

$$\Gamma_B\Gamma_A\Gamma_B = -\Gamma_A. \tag{C–5}$$

As examples of property (b), we may mention that (C–5) can be checked for $\Gamma_A = i\gamma_5\gamma_3$ and σ_{12} as follows:

$$\gamma_5(i\gamma_5\gamma_3)\gamma_5 = -i\gamma_5\gamma_3, \qquad \gamma_2\sigma_{12}\gamma_2 = -\sigma_{12}. \tag{C–6}$$

As we already mentioned in the main text (see footnote on p. 106), using property (b) we can readily prove the following which we call property (c).

c) Any 4 $\times$ 4 matrix X can be expanded in a unique linear combination

$$X = \sum_{A=1}^{16} x_A\Gamma_A. \tag{C–7}$$

To prove Pauli's fundamental theorem we must first demonstrate what is known as Schur's lemma which is stated as follows:
Suppose that there is a 4 $\times$ 4 matrix with the property

$$[X, \gamma_\mu] = 0 \quad \text{for every} \quad \gamma_\mu \quad (\mu = 1, 2, 3, 4); \tag{C–8}$$

then X is a multiple of the identity matrix.

The proof of this lemma goes as follows. Using property (c) we can expand X

$$X = x_A\Gamma_A + \sum_{C\neq A} x_C\Gamma_C, \tag{C–9}$$

where Γ_A ($\neq 1$) is any one of the 15 Γ_A matrices in (C–1) (excluding the identity matrix). If X commutes with every γ_μ, then it must necessarily commute with Γ_B with $B = 1, \ldots, 16$:

$$\Gamma_B X = X\Gamma_B, \qquad \Gamma_B X\Gamma_B = X. \tag{C–10}$$

Because of property (b), for the particular Γ_A appearing in (C–9) we can find (at least one) Γ_B such that (C–5) is satisfied. With such Γ_B we get

$$\begin{aligned} X &= \Gamma_B X\Gamma_B \\ &= x_A\Gamma_B\Gamma_A\Gamma_B + \sum_{C\neq A} x_C\Gamma_B\Gamma_C\Gamma_B \\ &= -x_A\Gamma_A + \sum_{C\neq A} (\pm 1)x_C\Gamma_C, \end{aligned} \tag{C–11}$$

where we have used (C–9), (C–10), and properties (b) and (c). Comparing (C–11) with (C–9) and recalling property (c), we conclude that

$$x_A = -x_A = 0. \tag{C–12}$$

Now Γ_A is arbitrary except that it cannot be the identity matrix. The relation (C–12) then means that, when we expand X as in (C–7) all the expansion coefficients are zero except for the coefficient of the identity matrix. It therefore follows that X is a multiple of the identity matrix. (Of course, X can be identically zero.)

We are now in a position to prove Pauli's fundamental theorem which can be stated as follows:

Given a set of 4×4 matrices $\{\gamma_\mu\}$ and another set of 4×4 matrices $\{\gamma'_\mu\}$ with $\mu = 1, \ldots, 4$ such that

$$\{\gamma_\mu, \gamma_\nu\} = 2\delta_{\mu\nu}, \tag{C–13}$$

and

$$\{\gamma'_\mu, \gamma'_\nu\} = 2\delta_{\mu\nu}, \tag{C–14}$$

there exists a nonsingular 4×4 matrix S with the property

$$\gamma'_\mu = S\gamma_\mu S^{-1}. \tag{C–15}$$

Moreover, S is unique up to a multiplicative constant.

To prove this we first define Γ'_A in the same way we defined Γ_A, for example, if $\Gamma_A = i\gamma_5\gamma_4 = i\gamma_1\gamma_2\gamma_3$, then $\Gamma'_A = i\gamma'_5\gamma'_4 = i\gamma'_1\gamma'_2\gamma'_3$, etc. It is evident that

$$\Gamma'_A\Gamma'_B = \eta_{AB}\Gamma'_C \tag{C–16}$$

with the same η_{AB} as in (C–3). We consider S of the form

$$S = \sum_A \Gamma'_A F\Gamma_A, \tag{C–17}$$

where F is some (arbitrary) 4×4 matrix. Multiplying both sides of (C–17) by Γ'_B from the left and Γ_B from the right, we get

$$\Gamma'_B S\Gamma_B = \sum_A \Gamma'_B\Gamma'_A F\Gamma_A\Gamma_B. \tag{C–18}$$

Because of (C–2), which also holds for the primed gammas,

$$\Gamma'_B\Gamma'_A = \Gamma'^{-1}_B\Gamma'^{-1}_A = (\Gamma'_A\Gamma'_B)^{-1} = \eta^{-1}_{AB}\Gamma'^{-1}_C = \eta^{-1}_{AB}\Gamma'_C. \tag{C–19}$$

We can therefore rewrite (C–18) as follows:

$$\Gamma'_B S\Gamma_B = \sum_A \eta^{-1}_{AB}\Gamma'_C F\eta_{AB}\Gamma_C = \sum_C \Gamma'_C F\Gamma_C. \tag{C–20}$$

But, because of (C–17), we get

$$\Gamma'_B S\Gamma_B = S. \tag{C–21}$$

Note that this is true for any Γ_B.

We claim that F can be chosen so that $S \neq 0$. To prove this, suppose S were zero for every F. Choose F so that

$$F_{\alpha\beta} = \delta_{\alpha\alpha'}\delta_{\beta\beta'} \tag{C–22}$$

for example, for $\alpha' = 1$, $\beta' = 2$;

$$F = \begin{pmatrix} 0 & 1 & 0 & 0 \\ 0 & 0 & 0 & 0 \\ 0 & 0 & 0 & 0 \\ 0 & 0 & 0 & 0 \end{pmatrix}. \tag{C–23}$$

Clearly the assumption $S = 0$, together with the choice (C–22), enables us to rewrite (C–17) as

$$\sum_A (\Gamma'_A)_{\delta\alpha'}(\Gamma_A)_{\beta'\epsilon} = 0 \tag{C–24}$$

for the $\delta\epsilon$ element of the matrix S. Since S is identically zero for every F by assumption, (C–24) must hold for all β' and ϵ, which leads to the matrix equation

$$\sum_A (\Gamma'_A)_{\delta\alpha'}\Gamma_A = 0. \tag{C–25}$$

This clearly contradicts the linear independence of the Γ-matrices. So there must exist F for which $S \neq 0$.

Our next task is to prove that S is nonsingular. We construct

$$S' = \sum_C \Gamma_C F'\Gamma'_C, \tag{C–26}$$

where F' is some 4×4 matrix. Repeating steps analogous to (C–18) through (C–21), we get

$$\Gamma_B S'\Gamma'_B = S'. \tag{C–27}$$

Moreover, F' can be so chosen that S' is not zero. Consider

$$S'S = \Gamma_B S'\Gamma'_B\Gamma'_B S\Gamma_B = \Gamma_B S'S\Gamma_B. \tag{C–28}$$

This is true for every Γ_B. Hence

$$[S'S, \gamma_\mu] = 0. \tag{C–29}$$

By Schur's lemma $S'S$ is a multiple of the identity matrix

$$S'S = c, \tag{C–30}$$

where c is a constant. Thus

$$S^{-1} = S'/c. \tag{C–31}$$

Since S' is not zero, we have shown that S^{-1} indeed exists. Coming back to (C–21), we can multiply both sides of (C–21) by S^{-1} from the right:

$$\Gamma'_B S \Gamma_B S^{-1} = 1, \tag{C–32}$$

or

$$S\Gamma_B S^{-1} = \Gamma'_B, \tag{C–33}$$

which is of the form (C–15). We have thus exhibited the existence of S with the property (C–15).

To prove the uniqueness of S, let us assume that there exist S_1 and S_2 both satisfying (C–33). We then have

$$S_1\gamma_\mu S_1^{-1} = S_2\gamma_\mu S_2^{-1} \tag{C–34}$$

or

$$S_2^{-1}S_1\gamma_\mu = \gamma_\mu S_2^{-1}S_1. \tag{C–35}$$

Because of Schur's lemma, $S_2^{-1}S_1$ is seen to be a multiple of the identity matrix. Thus

$$S_1 = aS_2,$$

where a is a multiplicative constant. This concludes the proof of Pauli's fundamental theorem.

APPENDIX D

FORMULAS AND RULES IN COVARIANT PERTURBATION THEORY

I. Definition of the $\mathcal{M}$-matrix:

$$S_{fi} = \delta_{fi} - i(2\pi)^4\delta^{(4)}(P_f^{(\text{tot})} - P_i^{(\text{tot})})\sqrt{\prod_{\text{ext}}(n_j/V)}\mathcal{M}_{fi},$$

where

$$n_j = \begin{cases} m_j/E_j & \text{for fermion} \\ 1/2E_j & \text{for boson.} \end{cases}$$

II. Relation of $\mathcal{M}_{fi}$ to transition probabilities and cross sections ("covariant Golden Rule"):

a) Decay $1 \longrightarrow 2 + 3 + \ldots + n$. The differential decay rate dw is

$$dw = \frac{1}{2E_1}|\mathcal{M}_{fi}|^2[\prod_{\substack{\text{ext}\\\text{fermions}}}(2m_{\text{fermi}})]\frac{d^3p_2}{(2\pi)^3 2E_2}\cdots\frac{d^3p_n}{(2\pi)^3 2E_n}(2\pi)^4\delta^{(4)}\left(p_1 - \sum_{j=2}^{n} p_j\right).$$

To eliminate the $\delta^{(4)}$ function first integrate over the (three-) momentum of one of the final-state particles, and then use

$$d^3p_k\delta\left(E_1 - \sum_{j=2}^{n} E_j\right) = \frac{|\mathbf{p}_k|^2 d\Omega_k}{\left(\partial\left(\sum_{j=2}^{n} E_j\right)\Big/\partial|\mathbf{p}_k|\right)_{\theta_k,\phi_k}} = \frac{|\mathbf{p}_k|^2 d|\mathbf{p}_k| d\phi_k}{\left(\partial\left(\sum_{j=2}^{n} E_j\right)\Big/\partial(\cos\theta_k)\right)_{|\mathbf{p}_k|,\phi_k}}.$$

For $1 \longrightarrow 2 + 3$ the (partial) decay rate is given in the rest frame of particle 1 by

$$\Gamma(1 \longrightarrow 2 + 3) = \frac{1}{2}\frac{1}{(4\pi)^2}\frac{|\mathbf{p}_{\text{fin}}|}{m_1^2}\int d\Omega \sum_{\substack{\text{final}\\\text{spin}}}|\mathcal{M}_{fi}|^2 \prod_{\substack{\text{ext}\\\text{fermions}}}(2m_{\text{fermi}}),$$

where $|\mathbf{p}_{\text{fin}}| = |\mathbf{p}_2| = |\mathbf{p}_3|$.

b) Differential cross section for $1 + 2 \longrightarrow 3 + 4 + \cdots + n$ when the momenta of particles 1 and 2 are collinear:

$$d\sigma = \frac{1}{v_{\text{rel}}}\frac{1}{2E_1}\frac{1}{2E_2}|\mathcal{M}_{fi}|^2[\prod_{\substack{\text{ext}\\\text{fermions}}}(2m_{\text{fermi}})]\frac{d^3p_3}{(2\pi)^3 2E_3}\cdots\frac{d^3p_n}{(2\pi)^3 2E_n}$$

$$\times (2\pi)^4\delta^{(4)}\left(p_1 + p_2 - \sum_{j=3}^{n} p_j\right);$$

$$v_{\text{rel}}E_1E_2 = \begin{cases} E_{\text{tot}}|\mathbf{p}_{\text{in}}| & \text{in the CM system} \quad (|\mathbf{p}_{\text{in}}| = |\mathbf{p}_1| = |\mathbf{p}_2|) \\ m_2|\mathbf{p}_1| & \text{in the rest system of particle 2 (lab system)} \\ \sqrt{(p_1\cdot p_2)^2 - (m_1m_2)^2} & \text{in general.} \end{cases}$$

For $1 + 2 \longrightarrow 3 + 4$, the differential cross section in the CM-system is given by

$$\left(\frac{d\sigma}{d\Omega}\right)_{\text{CM}} = \frac{1}{4}\frac{1}{(4\pi)^2}\frac{1}{E_{\text{tot}}^2}\frac{|\mathbf{p}_{\text{fin}}|}{|\mathbf{p}_{\text{in}}|}|\mathscr{M}_{fi}|^2 \prod_{\substack{\text{ext}\\ \text{fermions}}} (2m_{\text{fermi}}).$$

One may sum over the final spin (polarization) states. If the initial particles are unpolarized, one may sum over the initial spin states and divide by the number of the spin states of the initial system.

III. Rules for writing $-i\mathscr{M}_{fi}$:

a) *External lines*

$\epsilon_\mu^{(\alpha)}$ for each absorbed photon (or spin-one boson) 

$\epsilon_\mu^{(\alpha)}$ for each emitted photon (or spin-one boson) 

(If $\epsilon_\mu^{(\alpha)}$ represents circular polarization, it should be complex-conjugated when it appears in the final state.)

1 for each spinless boson absorbed or emitted 

$u^{(s)}(\mathbf{p})$ for each absorbed spin-$\frac{1}{2}$ fermion 

$\bar{v}^{(s)}(\mathbf{p}) = (-1)^s\bar{u}^{(5-s)}(-\mathbf{p})$ for each absorbed spin-$\frac{1}{2}$ antifermion 

$\bar{u}^{(s)}(\mathbf{p})$ for each emitted spin-$\frac{1}{2}$ fermion 

$v^{(s)}(\mathbf{p}) = (-1)^s u^{(5-s)}(-\mathbf{p})$ for each emitted spin-$\frac{1}{2}$ antifermion 

b) *Vertex factors*

$$-e\gamma_\mu \text{ if } \mathscr{H}_{\text{int}} = -ie\,\bar{\psi}\gamma_\mu\psi A_\mu$$
$$-iG \text{ if } \mathscr{H}_{\text{int}} = G\,\bar{\psi}\psi\phi$$
$$G\gamma_5 \text{ if } \mathscr{H}_{\text{int}} = iG\,\bar{\psi}\gamma_5\psi\phi, \text{ etc.}$$

In general, take $-i\mathscr{H}_{\text{int}}$ (more precisely $i\mathscr{L}_{\text{int}}$) and replace the field operators by the appropriate free-particle wave functions. Omit $e^{\pm ip\cdot x}$ and the factors already taken care of by external lines (for example, $\epsilon_\mu^{(\alpha)}$ and $u^{(s)}(\mathbf{p})$) and normalization constants ($\sqrt{m/EV}$). The remainder is the *vertex factor*.‡

c) *Internal lines*

Photon ("covariant photon"): $\dfrac{\delta_{\mu\nu}}{i(q^2 - i\epsilon)}$ 

Spin-one boson: $\dfrac{1}{i(q^2 + \mu^2 - i\epsilon)}\left(\delta_{\mu\nu} + \dfrac{q_\mu q_\nu}{\mu^2}\right)$ 

Spinless boson: $\dfrac{1}{i(q^2 + \mu^2 - i\epsilon)}$ 

Spin-$\frac{1}{2}$ fermion: $\dfrac{-i\gamma\cdot q + m}{i(q^2 + m^2 - i\epsilon)}$ 

---

‡If the interaction density involves derivatives of field operators, $\mathscr{H}_{\text{int}}$ differs from $-\mathscr{L}_{\text{int}}$. It can be shown (using an argument originally given by P. T. Matthews for the case of the pseudovector coupling of a pseudoscalar field) that the vertex factor in the Feynman diagram should be read directly from $i\mathscr{L}_{\text{int}}$ rather than from $-i\mathscr{H}_{\text{int}}$. In this connection we may remark that it is possible to formulate an S matrix expansion using $\mathscr{L}_{\text{int}}$ (without recourse to $\mathscr{H}_{\text{int}}$). See Bogoliubov and Shirkov (1959), pp. 206–226.

d) *Caution*

i) Assume energy-momentum conservation at each vertex. Integrate with $d^4q/(2\pi)^4$ over an *undetermined* four-momentum (that is, four-momentum not fixed by energy-momentum conservation). If the internal four-momentum is already fixed, no integration is necessary.

ii) For each closed fermion loop take the trace with a factor (-1).

iii) Put together the various pieces so that as we read from the right to the left, we just "follow the arrow."

iv) Multiply each matrix element corresponding to a particular diagram by the permutation factor δ_P. For example, δ_P is $+1$ and -1 for the direct and the exchange term of fermion-fermion scattering, -1 and $+1$ for the direct and the annihilation term of fermion-antifermion scattering. The sign of the matrix element is of physical significance only when interference between two or more diagrams is considered or when we are constructing an effective potential.

IV. Properties of free-particle spinors (*see also* Appendix B, Section IV):

$$(i\gamma\cdot p + m)u^{(s)}(\mathbf{p}) = 0, \qquad \bar{u}^{(s)}(\mathbf{p})(i\gamma\cdot p + m) = 0,$$

$$(-i\gamma\cdot p + m)v^{(s)}(\mathbf{p}) = 0, \qquad \bar{v}^{(s)}(\mathbf{p})(-i\gamma\cdot p + m) = 0,$$

$$(p' - p)_\nu \bar{u}'\sigma_{\nu\mu}u = i(p' + p)_\mu \bar{u}'u + (m' + m)\bar{u}'\gamma_\mu u,$$

$$(p' - p)_\mu \bar{u}' i\gamma_5\gamma_\mu u = (m' + m)\bar{u}'\gamma_5 u,$$

$$\sum_{s=1}^{2} u_\alpha^{(s)}(\mathbf{p})\bar{u}_\beta^{(s)}(\mathbf{p}) = \left(\frac{-i\gamma\cdot p + m}{2m}\right)_{\alpha\beta},$$

$$\sum_{s=1}^{2} v_\alpha^{(s)}(\mathbf{p})\bar{v}_\beta^{(s)}(\mathbf{p}) = -\left(\frac{i\gamma\cdot p + m}{2m}\right)_{\alpha\beta}.$$

If $u^{(w)}(\mathbf{p})$ and $v^{(w)}(\mathbf{p})$ are free-particle spinors satisfying

$$i\gamma_5\gamma\cdot w u^{(w)}(\mathbf{p}) = u^{(w)}\mathbf{p}, \qquad i\gamma_5\gamma\cdot w v^{(w)}(\mathbf{p}) = v^{(w)}(\mathbf{p}),$$

then

$$u_\alpha^{(w)}(\mathbf{p})\bar{u}_\beta^{(w)}(\mathbf{p}) = \left[\left(\frac{-i\gamma\cdot p + m}{2m}\right)\left(\frac{1 + i\gamma_5\gamma\cdot w}{2}\right)\right]_{\alpha\beta},$$

$$v_\alpha^{(w)}(\mathbf{p})\bar{v}_\beta^{(w)}(\mathbf{p}) = \left[-\left(\frac{i\gamma\cdot p + m}{2m}\right)\left(\frac{1 + i\gamma_5\gamma\cdot w}{2}\right)\right]_{\alpha\beta},$$

where

$$w\cdot p = 0, \qquad w^2 = |\mathbf{w}|^2 - w_0^2 = 1, \qquad [i\gamma_5\gamma\cdot w, i\gamma\cdot p] = 0,$$

$$w = \begin{cases} (\hat{\mathbf{n}}, 0) & \text{in the rest frame,} \\ (\hat{\mathbf{n}}, 0) & \text{for transverse polarization,} \\ \left(\dfrac{E\hat{\mathbf{p}}}{m}, \dfrac{i|\mathbf{p}|}{m}\right) & \text{for longitudinal polarization.} \\ \left(\hat{\mathbf{n}} + \dfrac{\mathbf{p}(\mathbf{p}\cdot\hat{\mathbf{n}})}{m(E+m)}, \dfrac{i\mathbf{p}\cdot\hat{\mathbf{n}}}{m}\right) & \text{in general.} \end{cases}$$

APPENDIX E

FEYNMAN INTEGRALS; THE COMPUTATIONS OF THE SELF-ENERGY AND THE ANOMALOUS MAGNETIC MOMENT OF THE ELECTRON

When the energy-momentum of an internal line in a Feynman diagram is not completely fixed, there is a nontrivial integration to be performed in four-dimensional k-space. To make an attack on such a four-dimensional integral, R. P. Feynman invented a set of very clever tricks.

Useful formulas. We first prove

$$\int \frac{d^4k}{(2\pi)^4} \frac{1}{(k^2 + s - i\epsilon)^3} = \frac{i}{32\pi^2 s}, \tag{E–1a}$$

$$\int \frac{d^4k}{(2\pi)^4} \frac{k_\mu}{(k^2 + s - i\epsilon)^3} = 0, \tag{E–1b}$$

from which many other useful formulas can be derived. The second one (E–1b) is obvious, since the integrand is an odd function of k_μ and the integration is over all values of $\mathbf{k}$ and k_0. To prove (E–1a) we start with a simpler one-dimensional integral

$$\int_{-\infty}^{\infty} \frac{dk_0}{k^2 + s - i\epsilon} = \int_{-\infty}^{\infty} \frac{dk_0}{|\mathbf{k}|^2 - k_0^2 + s - i\epsilon}. \tag{E–2}$$

We can handle this integral by the usual contour method. Noting that there are poles at $k_0 = \pm(\sqrt{|\mathbf{k}|^2 + s} - i\delta)$, we get the result

$$\int_{-\infty}^{\infty} \frac{dk_0}{(k^2 + s - i\epsilon)} = \frac{i\pi}{\sqrt{|\mathbf{k}|^2 + s}} \tag{E–3}$$

(whether we close the contour in the upper half-plane or in the lower half-plane). We now differentiate both sides of (E–3) twice with respect to s to obtain

$$\int_{-\infty}^{\infty} \frac{dk_0}{(k^2 + s - i\epsilon)^3} = \frac{3i\pi}{8} \frac{1}{(|\mathbf{k}|^2 + s)^{5/2}}. \tag{E–4}$$

The remaining three-dimensional ($\mathbf{k}$-space) integration is elementary since the integrand is spherically symmetric. Using

$$\int \frac{d^3k}{(|\mathbf{k}|^2 + s)^{5/2}} = 4\pi \int_0^\infty \frac{|\mathbf{k}|^2 \, d|\mathbf{k}|}{(|\mathbf{k}|^2 + s)^{5/2}} = \frac{4\pi}{3s}, \tag{E–5}$$

we readily obtain (E–1a).

We now substitute $k \longrightarrow k - p$ for the integration variable in (E–1a) and (E–1b); we also let $s \longrightarrow s - p^2$. We then get variations of (E–1a) and (E–1b) as follows:

$$\int \frac{d^4k}{(2\pi)^4} \frac{1}{(k^2 - 2k\cdot p + s - i\epsilon)^3} = \frac{i}{32\pi^2(s - p^2)}, \tag{E–6a}$$

$$\int \frac{d^4k}{(2\pi)^4} \frac{k_\mu}{(k^2 - 2k\cdot p + s - i\epsilon)^3} = \frac{ip_\mu}{32\pi^2(s - p^2)}, \tag{E–6b}$$

which are quite useful. Furthermore, differentiating both sides of (E–6a) and (E–6b) with respect to s and p_ν, we obtain more useful formulas:

$$\int \frac{d^4k}{(2\pi)^4} \frac{1}{(k^2 - 2k\cdot p + s - i\epsilon)^4} = \frac{i}{96\pi^2(s - p^2)^2}, \tag{E–7a}$$

$$\int \frac{d^4k}{(2\pi)^4} \frac{k_\mu}{(k^2 - 2k\cdot p + s - i\epsilon)^4} = \frac{ip_\mu}{96\pi^2(s - p^2)^2}, \tag{E–7b}$$

$$\int \frac{d^4k}{(2\pi)^4} \frac{k_\mu k_\nu}{(k^2 - 2k\cdot p + s - i\epsilon)^4} = \frac{i}{96\pi^2(s - p^2)^2}\left[p_\mu p_\nu + \frac{1}{2}(s - p^2)\delta_{\mu\nu}\right]. \tag{E–7c}$$

In practice the denominator of the integrand of a typical four-dimensional integral is made up of different invariant factors multiplied together, for example, $(k^2 - 2p\cdot k)(k^2 + \Lambda^2)^2$. Such a product can always be rewritten so that formulas (E–6a) through (E–7c) are immediately applicable. To see this we make use of what is known as the *parametrization method* (also introduced to covariant perturbation theory by Feynman.) The simplest relation along this line is

$$\frac{1}{ab} = \int_0^1 \frac{dx}{[ax + b(1 - x)]^2}, \tag{E–8}$$

which can easily be proved by considering

$$\frac{1}{ab} = \frac{1}{b - a}\int_a^b \frac{dz}{z^2} \tag{E–9}$$

with $z = ax + b(1 - x)$. Differentiating (E–8) with respect to a we get

$$\frac{1}{a^2b} = 2\int_0^1 \frac{x\,dx}{[ax + b(1 - x)]^3}. \tag{E–10}$$

A slight generalization of (E–8) is

$$\frac{1}{abc} = 2\int_0^1 dx \int_0^{1-x} dy \frac{1}{[a + (b - a)x + (c - a)y]^3}, \tag{E–11}$$

the proof of which is left as an exercise for the reader. From this relation we further obtain

$$\frac{1}{a^2bc} = 6\int_0^1 dx \int_0^{1-x} dy \frac{(1 - x - y)}{[a + (b - a)x + (c - a)y]^4}, \quad \text{etc.} \tag{E–12}$$

Electron self-energy. Let us see how we can apply Feynman's method to compute the electron self-energy. According to (4.385), (4.386), and (4.389) we have

$$\delta m = -ie^2 \int \frac{d^4k}{(2\pi)^4} \gamma_\mu \frac{[-i\gamma\cdot(p - k) + m]}{-2p\cdot k + k^2} \gamma_\mu \left(\frac{1}{k^2}\right)\left(\frac{\Lambda^2}{k^2 + \Lambda^2}\right)\Bigg|_{i\gamma\cdot p = -m}, \tag{E–13}$$

where the prescription $k^2 \to k^2 - i\epsilon$ is implicit. But

$$\begin{aligned}\gamma_\mu[-i\gamma\cdot(p-k)+m]\gamma_\mu &= 2i\gamma\cdot(p-k)+4m\\ &= -2i\gamma\cdot k + 2m,\end{aligned} \tag{E–14}$$

provided that the whole expression is sandwiched by $\bar{u}^{(s')}(\mathbf{p})$ from the left and $u^{(s)}(\mathbf{p})$ from the right. We can, of course, use (E–11) to combine the three factors in the denominator of (E–13), but instead we choose to use

$$\left(\frac{1}{k^2}\right)\left(\frac{\Lambda^2}{k^2+\Lambda^2}\right) = \int_0^{\Lambda^2} dt \frac{1}{(k^2+t)^2} \tag{E–15}$$

and take advantage of (E–10). We then have

$$\begin{aligned}\delta m &= -ie^2 \int \frac{d^4k}{(2\pi)^4} \int_0^{\Lambda^2} dt \frac{(-2i\gamma\cdot k+2m)}{(-2p\cdot k+k^2)(k^2+t)^2}\\ &= -ie^2 \int \frac{d^4k}{(2\pi)^4} \int_0^{\Lambda^2} dt \int_0^1 \frac{2x dx(-2i\gamma\cdot k+2m)}{[(k^2+t)x+(k^2-2p\cdot k)(1-x)]^3}\\ &= -ie^2 \int_0^1 dx \int_0^{\Lambda^2} dt \int \frac{d^4k}{(2\pi)^4} \frac{(4mx-4i\gamma\cdot kx)}{[k^2-2k\cdot p(1-x)+tx]^3},\end{aligned} \tag{E–16}$$

to which (E–6a) and (E–6b) are immediately applicable when we recall that k^2 actually means $k^2 - i\epsilon$. The k-space integration gives

$$\begin{aligned}\delta m &= e^2 \int_0^1 dx \int_0^{\Lambda^2} dt \frac{4mx - 4i\gamma\cdot px(1-x)}{32\pi^2[tx - p^2(1-x)^2]}\bigg|_{i\gamma\cdot p=-m}\\ &= \frac{e^2 m}{8\pi^2}\int_0^1 dx \int_0^{\Lambda^2} dt \frac{(2x-x^2)}{tx+m^2(1-x)^2}.\end{aligned} \tag{E–17}$$

We integrate over t, remembering $\Lambda^2 \gg m^2$, as follows:

$$\begin{aligned}\int_0^{\Lambda^2} \frac{dt}{tx+m^2(1-x^2)} &= \frac{1}{x}\log\left[\frac{\Lambda^2 x + m^2(1-x)^2}{m^2(1-x)^2}\right]\\ &\approx \frac{1}{x}\log\left[\frac{\Lambda^2 x}{m^2(1-x)^2}\right]\\ &= \frac{1}{x}\left\{\log\left(\frac{\Lambda^2}{m^2}\right) + \log\left[\frac{x}{(1-x)^2}\right]\right\}.\end{aligned} \tag{E–18}$$

(When x is extremely close to zero, the approximation we have made breaks down but note that we later integrate over x.) The final integration over x is elementary; we end up with

$$\delta m \simeq \left(\frac{e^2}{4\pi}\right)\frac{3m}{4\pi}\log\left(\frac{\Lambda^2}{m^2}\right) = \frac{3\alpha m}{2\pi}\log\left(\frac{\Lambda}{m}\right) \tag{E–19}$$

up to an additive constant, which, of course, is insignificant when it appears together with a divergent expression.‡ We may mention that this covariant calculation is several times shorter than the original calculation of the same quantity by V. F. Weisskopf who used the old-fashioned perturbation theory.

---

‡In fact, the actual value of the additive constant is characteristic of the particular form of the cut-off function we have used.

Vertex correction; the anomalous magnetic moment of the electron. As a second application of Feynman's method, we consider the four-dimensional integral which we encountered in connection with the vertex correction (cf. Eq. 4. 446). Setting $p^2 = p'^2 = -m^2$, we get

$$\Lambda_\mu(p', p)|_{i\gamma\cdot p'=-m,\, i\gamma\cdot p=-m}$$
$$= ie^2 \int \frac{d^4k}{(2\pi)^4} \frac{\gamma_\nu[-i\gamma\cdot(p'-k)+m]\gamma_\mu[-i\gamma\cdot(p-k)+m]\gamma_\nu}{k^2(k^2-2p'\cdot k)(k^2-2p\cdot k)}. \tag{E–20}$$

This integral is divergent at both large and small values of $|k^2|$. We therefore modify the photon propagator as follows:

$$\frac{1}{k^2} \longrightarrow \left(\frac{1}{k^2+\lambda_{\min}^2}\right)\left(\frac{\Lambda^2}{k^2+\Lambda^2}\right) \approx \frac{\Lambda^2-\lambda_{\min}^2}{(k^2+\lambda_{\min}^2)(k^2+\Lambda^2)}$$
$$= \frac{1}{k^2+\lambda_{\min}^2} - \frac{1}{k^2+\Lambda^2} = \int_{\lambda_{\min}^2}^{\Lambda^2} \frac{dt}{(k^2+t)^2}, \tag{E–21}$$

where $\lambda_{\min}$ is a fictitious "photon mass" assumed to be much smaller than m. We then have

$$\Lambda_\mu(p', p)|_{i\gamma\cdot p'=-m,\, i\gamma\cdot p=-m} = ie^2 \int \frac{d^4k}{(2\pi)^4} \int_{\lambda_{\min}^2}^{\Lambda^2} dt \frac{N_\mu}{D}, \tag{E–22}$$

with

$$N_\mu = \gamma_\nu[-i\gamma\cdot(p'-k)+m]\gamma_\mu[-i\gamma\cdot(p-k)+m]\gamma_\nu, \tag{E–23a}$$
$$D = (k^2+t)^2(k^2-2p\cdot k)(k^2-2p'\cdot k). \tag{E–23b}$$

Using (4.341), we can simplify the numerator N_μ:‡

$$N_\mu = 2\gamma\cdot p\gamma_\mu\gamma\cdot p' - 2(\gamma\cdot p\gamma_\mu\gamma\cdot k + \gamma\cdot k\gamma_\mu\gamma\cdot p')$$
$$- 2im[\gamma_\mu\gamma\cdot(p+p'-2k) + \gamma\cdot(p+p'-2k)\gamma_\mu]$$
$$+ 2\gamma\cdot k\gamma_\mu\gamma\cdot k - 2\gamma_\mu m^2. \tag{E–24}$$

As for the denominator we can combine the various factors in D, using (E–12):

$$\frac{1}{D} = 6\int_0^1 dx \int_0^{1-x} dy \frac{(1-x-y)}{[k^2+t-(2p\cdot k+t)x-(2p'\cdot k+t)y]^4}. \tag{E–25}$$

We propose to perform the indicated integrations in the order k, t, y, and x. The k-space integration can be performed immediately if we take advantage of (E–7):

$$\int \frac{d^4k}{(2\pi)^4} \frac{N_\mu}{[k^2-2k\cdot(px+p'y)+t(1-x-y)]^4}$$
$$= \frac{iM_\mu}{96\pi^2[t(1-x-y)-(px+p'y)^2]^2}, \tag{E–26}$$

‡The reader may wonder why we do not simplify further the third term of (E–24) using $\gamma_\mu(\gamma\cdot b) + (\gamma\cdot b)\gamma_\mu = 2b_\mu$. The reason is that eventually we wish to express $\Lambda_\mu(p', p)$ in terms of $q_\nu\sigma_{\nu\mu}$ and γ_μ, not in terms of $p_\mu + p'_\mu$ and γ_μ.

where

$$\begin{aligned} M_\mu = {} & 2\gamma\cdot p\gamma_\mu\gamma\cdot p' - 2[\gamma\cdot p\gamma_\mu\gamma\cdot(px + p'y) + \gamma\cdot(px + p'y)\gamma_\mu\gamma\cdot p'] \\ & -2im[\gamma_\mu\gamma\cdot(p + p' - 2px - 2p'y) + \gamma\cdot(p + p' - 2px - 2p'y)\gamma_\mu] \\ & +2[\gamma\cdot(px + p'y)\gamma_\mu\gamma\cdot(px + p'y)] \\ & -2[t(1 - x - y) - (px + p'y)^2]\gamma_\mu - 2m^2\gamma_\mu. \end{aligned} \tag{E–27}$$

We can eliminate p and p' in favor of m, q, and q^2, remembering $p' = p + q$, $p\cdot p' = -[m^2 + (q^2/2)]$ and keeping in mind that $-i\gamma\cdot p(-i\gamma\cdot p')$ standing on the right (left) can always be replaced by m, for example,

$$i\gamma\cdot p'|_{\text{right}} \longrightarrow -m + i\gamma\cdot q. \tag{E–28}$$

We then have

$$\begin{aligned} M_\mu = {} & \gamma_\mu\{-2t(1 - x - y) - 4m^2[(x + y)^2 - (1 - x - y)] - 2q^2xy\} \\ & + im\gamma_\mu\gamma\cdot q[-2x + 2(x + y)y] + im\gamma\cdot q\gamma_\mu[2y - 2(x + y)x] \\ & + \gamma\cdot q\gamma_\mu\gamma\cdot q(-2xy + 2x + 2y - 2). \end{aligned} \tag{E–29}$$

We further note that the physically observable effects such as the magnetic moment must be independent of the particular gauge which we use for the external potential. So we may as well assume that the external potential satisfies the Lorentz condition which reads in momentum space as follows:

$$q_\nu\tilde{A}_\nu(q) = 0. \tag{E–30}$$

This means that we can replace $\gamma_\mu\gamma\cdot q$ by $-\gamma\cdot q\gamma_\mu$ (or vice versa) since the difference is just $2q_\mu$ which gives zero when contracted with $\tilde{A}_\mu(q)$. Using

$$\begin{aligned} \gamma_\mu\gamma\cdot q & \rightleftarrows -\gamma\cdot q\gamma_\mu, \\ \gamma\cdot q\gamma_\mu\gamma\cdot q & \longrightarrow -q^2\gamma_\mu, \end{aligned} \tag{E–31}$$

we can write (E–29) as

$$\begin{aligned} M_\mu = {} & \gamma_\mu\{-2t(1 - x - y) - 4m^2[(x + y)^2 - (1 - x - y)] - 2q^2(1 - x - y)\} \\ & + \frac{(\gamma\cdot q\gamma_\mu - \gamma_\mu\gamma\cdot q)}{2i}[-2m(x + y)(1 - x - y)]. \end{aligned} \tag{E–32}$$

Recall that Λ_μ is given by

$$\begin{aligned} & \Lambda_\mu(p', p)|_{i\gamma\cdot p' = -m,\, i\gamma\cdot p = -m} \\ & = -\frac{6e^2}{96\pi^2}\int_0^1 dx \int_0^{1-x} dy \int_{\lambda^2_{\min}}^{\Lambda^2} dt \frac{(1 - x - y)M_\mu}{[t(1 - x - y) + m^2(x + y)^2 + q^2xy]^2}. \end{aligned} \tag{E–33}$$

It is now clear that when the x-, y-, and t-integrations are performed, Λ_μ can be written as‡

$$\Lambda_\mu(p', p)|_{i\gamma\cdot p' = -m,\, i\gamma\cdot p = -m} = \gamma_\mu G_1(q^2) + (1/2m)q_\nu\sigma_{\nu\mu}G_2(q^2). \tag{E–34}$$

---

‡This form of Λ_μ can be shown to follow from Lorentz invariance, parity conservation, and charge conservation (gauge invariance) alone.

As we explained in the main text, the Schwinger correction to the magnetic moment is given by the coefficient of the $q_\nu \sigma_{\nu\mu}$-term at small values of q^2. So let us calculate $G_2(0)$:

$$G_2(0) = \frac{\alpha}{4\pi}(2m)^2 \int_0^1 dx \int_0^{1-x} dy \int_{\lambda^2_{\min}}^{\Lambda^2} dt \frac{(x+y)(1-x-y)^2}{[t(1-x-y)+m^2(x+y)^2]^2}. \quad \text{(E–35)}$$

The t-integration gives

$$\int_{\lambda^2_{\min}}^{\Lambda^2} dt \frac{1}{[t(1-x-y)+m^2(x+y)^2]^2} = \frac{1}{1-x-y}$$
$$\times \left[\frac{1}{\lambda^2_{\min}(1-x-y)+m^2(x+y)^2} - \frac{1}{\Lambda^2(1-x-y)+m^2(x+y)^2}\right]. \quad \text{(E–36)}$$

The integral is completely convergent at the upper limit; we can safely let $\Lambda^2 \longrightarrow \infty$. We may also be tempted to set $\lambda^2_{\min} = 0$, but, because the domain of the x-y integration includes the point $x = y = 0$, it is safer to keep $\lambda^2_{\min}$ at this stage. Thus

$$G_2(0) = \frac{\alpha m^2}{\pi} \int_0^1 dx \int_0^{1-x} dy \frac{(x+y)(1-x-y)}{\lambda^2_{\min}(1-x-y)+m^2(x+y)^2}. \quad \text{(E–37)}$$

Now this integral is seen to be finite even if $\lambda^2_{\min} = 0$, since

$$\int_0^1 dx \int_0^{1-x} dy = \tfrac{1}{2}, \quad \text{(E–38a)}$$

$$\int_0^1 dx \int_0^{1-x} dy \frac{1}{x+y} = -\int_0^1 \log x \, dx = 1. \quad \text{(E–38b)}$$

It is therefore legitimate to set $\lambda^2_{\min} = 0$ after all. This would not be the case if the numerator in (E–37) contained a term independent of x and y; in fact, in integrating some of the γ_μ terms in (E–32) it is essential to keep $\lambda^2_{\min}$ to obtain a result that goes as $\log(m^2/\lambda^2_{\min})$ (which is responsible for the "infrared catastrophe" as $\lambda^2_{\min} \longrightarrow 0$). With $\lambda^2_{\min} = 0$, the integral (E–37) can be trivially performed using (E–38). We get

$$G_2(0) = \alpha/2\pi, \quad \text{(E–39)}$$

which is just Schwinger's result.

We emphasize again that the anomalous magnetic moment which we obtain from (E–35) becomes completely independent of the cut-off Λ and the fictitious photon mass $\lambda_{\min}$ as we let $\Lambda \longrightarrow \infty$ and $\lambda_{\min} \longrightarrow 0$. If we are just interested in computing the anomalous moment, it is actually not necessary to modify the photon propagator according to (E–21). We have deliberately calculated it in this way because (a) the method we have used also enables us to calculate $G_1(q^2)$ in a straightforward way, and (b), using this method, we can easily see how the computed value of the magnetic moment of the electron (muon) would be modified if the cut-off Λ were not too high compared to the electron (muon) mass (cf. Problem 4–17). As we mentioned in the main text, $G_1(0) = L$ (which is also equal to B because of the Ward identity) is logarithmically dependent on both Λ and $\lambda_{\min}$;

the logarithmic dependence on Λ is readily seen to arise from (E–32) and (E–33) since the integrand contains a term that goes like $1/t$ for large values of t. On the other hand, the *difference* $G_1(q^2) - G_1(0)$ [which is the coefficient of the γ_μ part of $\Lambda_\mu^f(p'\,p)$] depends only on λ_{min} (cf. Eq. 4.462). If any strength remains, the reader is encouraged to complete the evaluation of the integral (E–33) by computing the coefficient L and $G_1(q^2) - G_1(0)$ (for small values of q^2).

BIBLIOGRAPHY

Throughout the book no attempt has been made to quote original papers. Some of the most important papers on the quantum theory of radiation and covariant quantum electrodynamics are reprinted in Schwinger (1958).

A. I. AKHIEZER and V. B. BERESTETSKII, *Quantum Electrodynamics* (translated by G. M. Volkoff). Interscience Publishers, New York (1965).

H. A. BETHE and E. E. SALPETER, *Quantum Mechanics of One- and Two-Electron Atoms.* Academic Press, New York (1957).

J. D. BJORKEN and S. D. DRELL, *Relativistic Quantum Mechanics.* McGraw-Hill Book Co., New York (1964).

J. D. BJORKEN and S. D. DRELL, *Relativistic Quantum Fields.* McGraw-Hill Book Co., New York (1965).

J. M. BLATT, *Superconductivity.* Academic Press, New York (1964).

J. M. BLATT and V. F. WEISSKOPF, *Theoretical Nuclear Physics.* John Wiley and Sons, New York (1952).

N. N. BOGOLIUBOV and D. V. SHIRKOV, *Introduction to the Theory of Quantized Fields* (translated by G. M. Volkoff). Interscience Publishers, New York (1959).

E. V. CONDON and G. H. SHORTLEY, *The Theory of Atomic Spectra*, 2nd ed. Cambridge University Press, Cambridge, England (1951).

R. H. DICKE and J. P. WITTKE, *Introduction to Quantum Mechanics.* Addison-Wesley Publishing Co., Reading, Mass. (1960).

R. P. FEYNMAN, *Quantum Electrodynamics.* W. A. Benjamin, New York (1961a).

R. P. FEYNMAN, *The Theory of Fundamental Processes.* W. A. Benjamin, New York (1961b).

S. GASIOROWICZ, *Elementary Particle Physics.* John Wiley and Sons, New York (1966).

H. GOLDSTEIN, *Classical Mechanics.* Addison-Wesley Press, Cambridge, Mass. (1951).

W. HEITLER, *Quantum Theory of Radiation*, 3rd ed. Clarendon Press, Oxford, England (1954).

J. D. JACKSON, *Classical Electrodynamics.* John Wiley and Sons, New York (1962).

J. M. JAUCH and F. ROHRLICH, *The Theory of Photons and Electrons.* Addison-Wesley Publishing Co., Cambridge, Mass. (1955).

G. KÄLLÉN, "Quantenelektrodynamik" in *Handbuch der Physik* (S. Flügge, Ed.), Band V, Teil l. Springer-Verlag, Berlin (1958).

G. KÄLLÉN, *Elementary Particle Physics.* Addison-Wesley Publishing Co., Reading, Mass. (1964).

C. KITTEL, *Elementary Statistical Physics.* John Wiley and Sons, New York (1958).

F. MANDL, *Introduction to Quantum Field Theory.* Interscience Publishers, New York (1959).

E. MERZBACHER, *Quantum Mechanics.* John Wiley and Sons, New York (1961).

A. MESSIAH, *Quantum Mechanics* (translated by J. Potter). North-Holland Publishing Co., Amsterdam, The Netherlands (1962).

P. M. MORSE and H. FESHBACH, *Methods of Theoretical Physics.* McGraw-Hill Book Co., New York (1953).

N. F. MOTT and H. S. W. MASSEY, *The Theory of Atomic Collisions*, 2nd ed. Clarendon Press, Oxford, England (1949).

K. NISHIJIMA, *Fundamental Particles.* W. A. Benjamin, New York (1964).

W. K. H. PANOFSKY and M. PHILLIPS, *Classical Electricity and Magnetism.* Addison-Wesley Publishing Co., Cambridge, Mass. (1955).

M. A. PRESTON, *Physics of the Nucleus.* Addison-Wesley Publishing Co., Reading, Mass. (1962).

P. ROMAN, *Advanced Quantum Theory.* Addison-Wesley Publishing Co., Reading, Mass. (1965).

M. E. ROSE, *Elementary Theory of Angular Momentum.* John Wiley and Sons, New York (1957).

M. E. ROSE, *Relativistic Electron Theory.* John Wiley and Sons, New York (1961).

J. J. SAKURAI, *Invariance Principles and Elementary Particles.* Princeton University Press, Princeton, N. J. (1964).

S. S. SCHWEBER, *Introduction to Relativistic Quantum Field Theory.* Row and Peterson, Evanston, Ill. (1961).

J. SCHWINGER, *Selected Papers on Quantum Electrodynamics.* Dover Publications, New York (1958).

E. SEGRÈ, *Nuclei and Particles.* W. A. Benjamin, New York (1964).

R. F. STREATER and A. S. WIGHTMAN, *PCT, Spin and Statistics, and all that.* W. A. Benjamin, New York (1963).

E. C. TITCHMARSH, *Introduction to the Theory of Fourier Integrals.* Clarendon Press, Oxford, England (1937).

G. WENTZEL, *Quantum Theory of Fields* (translated by J. M. Jauch). Interscience Publishers, New York (1949).

INDEX

INDEX

D

O

P

T